THE PLANTS OF PEHR FORSSKÅL'S 'FLORA AEGYPTIACO-ARABICA'

Pehr Forsskål
Engraving reproduced by Lagus (1877)
Probably an idealised portrait

THE PLANTS OF PEHR FORSSKAL'S 'FLORA AEGYPTIACO-ARABICA'

Collected on the Royal Danish Expedition
to Egypt and the Yemen 1761–63.

F. Nigel Hepper and I. Friis

F. Nigel Hepper

Royal Botanic Gardens, Kew
in association with the
Botanical Museum, Copenhagen

First published 1994

Addresses of Authors:

F.N. Hepper, c/o The Herbarium, Royal Botanic Gardens, Kew,
Richmond, Surrey, TW9 3AE, U.K.

Professor Ib Friis, Botanical Museum and Herbarium, University of Copenhagen,
Gothersgade 130, DK-1123, Copenhagen K, Denmark.

General Editor: J.M. Lock

Cover Design by Media Resources, Royal Botanic Gardens, Kew.

ISBN 0 947643 62 1

Typeset at The Royal Botanic Gardens, Kew by Pam Arnold, Christine Beard,
Dominica Costello, Margaret Newman and Helen Ward

Printed and Bound in Great Britain by Whitstable Litho, Whitstable, Kent

To the memory
of Carsten Niebuhr, Martin Vahl and Carl Christensen,
who contributed so much to our knowledge of
Forsskål's botanical studies and collections.

Contents

Foreword ix

Preface xi

General Introduction 1

The 'Arabian Journey' — a multidisciplinary Royal Danish expedition 1

Pehr Forsskål — a biographical note on the expedition's naturalist 19

Forsskål's botanical library and field methods 25

Forsskål's botanical manuscripts and their publication 29

The Introduction to the 'Flora Aegyptiaco-Arabica.' 35

The layout of the 'Flora Aegyptiaco-Arabica', and its names in modern nomenclature 38

The plants illustrated in the 'Icones rerum naturalium' 42

Forsskål's botanical collections 46

Catalogue of species, excluding algae, fungi and lichens 53

Introduction to Catalogue: its layout and content 53

Gazetteer of Forsskål's collecting localities 57

Dicotyledons 63

Monocotyledons 253

Ferns 289

Gymnosperms 295

Index to herbarium numbers in the 'Herbarium Forsskålii' 299

Index to the IDC microfiche edition of the 'Herbarium Forsskålii' 337

References to literature 371

Taxonomic Index 375

Foreword

Over the centuries, Denmark has been the basis of several important expeditions to remote parts of the globe, expeditions with highly different purposes but in many cases with epoch-making results for various fields of research. In 1619–20 Jens Munk made a heroic effort to find the Northwest Passage to India between Greenland and America. Well over a hundred years later — about the time when the Dane Vitus Bering explored the waters between Asia and America for the Czar — the naval officer Frederic Ludvig Norden brought many-sided knowledge from Africa. His posthumous *Voyage d'Egypte et de Nubie*, 1750–55, became a landmark in several respects; it was recently made topical by Dr. Marie-Louise Buhl's publication of Norden's original drawings (*Les dessins archéologiques et topographiques de l'Egypte ancienne fait par F.L. Norden 1737–1738*, 1993), most properly published at the 250th anniversary of the Royal Danish Academy of Sciences and Letters which had published the original work as one of its first and most important scholarly duties. In the 1840s, a committee under the Academy planned the circumnavigation of the globe, of which *Galathea's* captain Steen Bille gave his report in 1849–51, an expedition not least known for its contributions to botany. And over a quarter of our century, Henning Haslund-Christensen brought himself and his country in close contact with treasures of Mongolian culture.

An outstanding link was omitted in this brief chain of achievements: the German-born Carsten Niebuhr's expedition to Arabia and other countries, 1761–67. Not because it has been forgotten, on the contrary, the sensitive and widely travelled author Thorkild Hansen has seen his book on *Arabia Felix* translated into several languages (1962ff), and undeniably, the expedition attracts existential interest because of the fate of its members of whom only Niebuhr returned after harassing experiences and with weakened eyesight because of his transcriptions of the cuneiform inscriptions at Persepolis, basis for the deciphering by philologists several decades later.

Like the African expedition of Norden, Niebuhr's travel was sponsored by the Danish King; the initiative belonged to the German orientalist J.D. Michaëlis. His idea was favourably received, and an international group of four scholars and an artist and engraver left Copenhagen for Egypt and Arabia (Yemen) in January 1761. And while, from a human viewpoint, the Journey became a tragedy, the scientific results were exceptional, and large amounts of manuscripts, observations, and botanical and zoological material reached Denmark.

Niebuhr worked assiduously to have the results published. His own books about the expedition became classics, translated into other languages: *Beschreibung von Arabien*, 1772, and *Reisebeschreibung nach Arabien und andern umliegenden Ländern*, 1774–78, and, not less important, his publication of Pehr Forsskål's works 1775–76. His many observations of longitude and latitude were postponed until 1801, and his book about Palestine and Syria did not see light until 1837.

The Academy can claim no participation in the planning and later publication of the Arabian Journey. It is true that the first president of the Academy, J.L. Holstein, was also the president of the Government's Department of the Interior, under which the Academy belonged, but the young institution was not in a position to carry out an international project of such magnitude — this obligation fell of necessity to the Foreign Department. However, its member Professor C.G. Kratzenstein was instrumental in preparing the scientific instructions for the expedition. (See Chapter VI in Olaf Pedersen's recent Academy history, *Lovers of Learning*, 1992).

Although the Academy's part in the actual expedition was minimal, many members have since been engaged in investigations based on material from the journey. Among those who have published detailed studies of Forsskål's plants were the Danish members C. Friis Rottböll and M. Vahl. Also associated with such research were the Academy's foreign members Carl von Linné (filius), A.J. Retzius, and Sir Joseph Banks. In this century, the Academy's Swedish member Henrik Schück has written an excellent biography of Forsskål. But many other Danish, Swedish, English and German botanists have examined these collections which are probably the most precious of their kind in Denmark.

This book, which explores and summarizes for the first time all Forsskål's botanical results, is an effort in the international tradition of the Arabian Journey itself, and is published by the Royal Botanic Gardens, Kew, in collaboration with the Botanical Museum, Copenhagen. The Carlsberg Foundation has supported the printing.

Once again, the Academy must face the fact that we have had no direct influence upon the Arabian Journey, but we are satisfied to see one of our members, Professor Ib Friis, as co-author of an important work about the botanical collections of Pehr Forsskål, and to pay tribute to the old and the present efforts to augment our knowledge of the fields in question.

Erik Dal
President of the Royal Danish Academy
of Sciences and Letters

Preface

Plant taxonomists today still need to study the herbaria and botanical publications of Pehr Forsskål, the results of his participation in a scholarly expedition sent to Egypt and the Yemen in 1761–1763 on behalf of the King of Denmark and Norway.

Students of the flora of the Eastern Mediterranean and Egypt have long taken an interest in Forsskål's botanical writing and material. The fact that species described by Forsskal from Egypt and the Yemen occur throughout tropical Africa has caused authors of flora accounts for all tropical African regions also to take an interest in the correct identity of his species and genera. Current work on a critical flora of the entire Arabian Peninsula (Miller et al. 1982; Wickens 1982) requires a thorough reconsideration of Forsskål's work. The same applies to the new floras of Ethiopia (Tewolde & Hedberg 1981; Hedberg & Edwards 1989) and Somalia (Thulin 1993), countries very close to Yemen both in geographical location and in the composition of their vegetation. Connected with the writing of these new floras a number of papers have recently been published, dealing with identification of hitherto unidentified or misinterpreted species proposed by Forsskål (Heine 1968; Brummitt 1978; Friis 1981; Friis 1983b; Wood 1982; Hepper & Wood 1983; Wood, Hillcoat & Brummitt 1983; Gilbert 1989).

However, unfortunate circumstances, particularly Forsskål's tragic death during the expedition and the subsequent difficulties with publication of his results by others, have made both the herbaria and the publications rather complicated to use. The organization of Forsskål's 'Flora Aegyptiaco-Arabica' is confusing, the editing is often inadequate, and there is no index. There are no references from the 'Flora Aegyptiaco-Arabica' to the botanical plates published in the 'Icones rerum naturalium.' In order to help taxonomists in their work with Forsskål's books and herbaria the Danish botanist Carl Christensen (1872–1942) published, in 1922, an index to the 'Flora Aegyptiaco-Arabica', and a revision of the specimens then kept in the 'Herbarium Forsskålii' in the Botanical Museum, Copenhagen. This index and revision greatly facilitated the use of both publications and herbaria, but Christensen's study is now both out of print and out of date. A completely new account is clearly needed.

The present book is intended as a comprehensive guide to the general botanical work of Forsskål, to the names published in the 'Flora Aegyptiaco-Arabica,' and to the Forsskål specimens present in the 'Herbarium Forsskålii' in Copenhagen, and the duplicates which have been distributed to other herbaria, such as the Natural History Museum (BM), the Botanical Museum of Lund (LD), the herbarium of the Linnean Society of London (LINN), the Jussieu herbarium of the Jardin des Plantes, Paris (P-JUSS), the Thunberg herbarium of the Institute of Systematic Botany, Uppsala (UPS-THUNB), the general herbarium of the Komarov Institute of St. Petersburg (LE), the Herbarium of Merseyside County Museums (LIV) and a few other institutions known to hold Forsskål specimens.

Work with this book has been distributed between the two authors as follows: Forsskål's material has been identified by Hepper at the Herbarium, Royal Botanic Gardens, Kew, to whom all specimens currently in the 'Herbarium Forsskålii' at the Botanical Museum, Copenhagen, have been sent on loan. Hepper compared Forsskål's specimens with modern material from the Mediterranean, Egypt, Yemen, and from surrounding countries. This work was carried out in collaboration with specialists in various groups, especially Dr. T. Cope (grasses), P. Edwards (ferns), Dr. I. Hedge & Dr. A. Paterson (Labiatae), C. Jeffrey (Compositae, Cucurbitaceae), and Dr. D. Simpson (Cyperaceae). J.R.I. Wood and Dr. B. Verdcourt read through various versions of the Catalogue and made many helpful suggestions. The help of all is gratefully acknowledged. Unfortunately, it has not been possible to find a specialist to help with the identification of Forsskål's collection of algae, and references to this material must therefore rest entirely on what has been done previously. Hepper has also searched the Natural History

Museum, London, and the Herbarium of the Botanical Museum, Lund, for Forsskål specimens. Typing of the Catalogue and typesetting of the entire volume was done at Kew.

Friis wrote the general introduction. Together with Lene Fugmann, secretary and assistant at the Botanical Museum, he checked the manuscript against the 'Herbarium Forsskålii' after it had been returned to Copenhagen. Friis and Fugmann also organized an extension of the old numbering system for specimens of vascular plants in the entire herbarium, and prepared the two appendices: a numerical list of all numbered specimens at present in the 'Herbarium Forsskålii', and a list of identifications of the IDC microfiche edition of the herbarium. The authors are grateful to Mrs. Anne Fox Maule, former curator at the Botanical Museum, who has kindly helped with translation from Latin, especially of the introductions to the 'Descriptiones animalium,' the 'Flora Aegyptiaco-Arabica,' and quotations from the texts by Vahl; she also arranged the main series of loans of the Forsskål-material to Kew.

The writing and publication of the work has therefore been a joint undertaking by staff members at the Royal Botanic Gardens, Kew, and at the Botanical Museum, Copenhagen. The Carlsberg Foundation has helped with a substantial grant to support the printing of the book.

F. Nigel Hepper
Herbarium, Royal Botanic Gardens, Kew

I. Friis
Botanical Museum, Copenhagen

General Introduction

The 'Arabian Journey' — a multidisciplinary Royal Danish Expedition

In the 17th Century, and well into the 18th, the scholarly study of geography, ethnology, language, archaeology and natural history of non-European countries was largely a by-product of search for new trade routes. By the middle of the 18th Century a few journeys were dedicated to a scholarly quest for knowledge, some of which might —with good luck — turn out to be useful, without that being the main point of the enterprise. The King of Denmark and Norway, Christian VI, had supported such a journey, undertaken by the Danish naval officer Frederik Ludvig Norden (1708–1742) in 1737–38 to the Nile as far as the cataracts in Nubia. Norden's results included one of the first carefully surveyed maps of the Nile, supplemented with observations of ruins of ancient temples, pyramids, obelisks and hieroglyphic inscriptions, as well as much ethnological information about the contemporary population of Egypt (Norden 1750–55). The work has been posthumously published by the Danish Royal Academy of Sciences and Letters, and is amply illustrated with detailed engravings (Pedersen 1992).

First Plans for an 'Arabian Journey'

The next expedition from Denmark to Arabic speaking countries, the 'Arabian Journey' as it became known, went to Egypt and the Yemen, and had the participation of a philologist, a naturalist, and an astronomer. It was from the very beginning a venture of academic interest, and was initiated by a German professor writing and suggesting the project to the Danish-Norwegian King. The ease with which the plans became accepted in both political and academic circles in Copenhagen is an indication of how ripe the times were for such learned projects, most likely helped by the success of Norden's two stately volumes on Egypt, and perhaps also by a certain envy in Copenhagen of the spectacular progress which Swedish science made under the Royal Swedish Academy of Science, created in 1739, and with Linnaeus as one of its most distinguished members and first president.

The two driving forces behind the plans for the 'Arabian Journey' were Professor Johann David Michaëlis (1717–1791) at the University of Göttingen, and Johann Hartwig Ernst Bernstorff (1712–1772), the foreign minister of Frederik V, the new king of Denmark and Norway. Michaëlis was one of the leading authorities on Oriental languages, especially philological questions of the Bible. Bernstorff was probably one of the most able ministers of foreign affairs in the history of absolute monarchy in Scandinavia. Michaëlis suggested in a letter to Bernstorff, dated 20 May 1756 (Michaëlis 1762, introduction; 1794–96), that a person qualified in linguistics and various sciences should be sent to the Yemen, then according to ancient tradition called 'Arabia Felix,' in order to study the linguistic and ethnological background of the Bible. The journey should be paid for by Frederik V, and would reflect honour on his kingdom as an enlightened country and a centre of learning. Michaëlis wrote about the Yemen: "The nature of this country is still rich in gifts which are unknown to us; its history goes back to ancient time; its dialect is as yet quite different from the familiar western Arabic, and because this form of Arabic, which is so far the only one we know well, has until now provided the best means of explaining the Hebrew language, what light can we not reasonably assume will be thrown on the Bible, the most important book of Antiquity, if we learn about the eastern dialect of Arabic as much as we already know about the western?" In the letter Michaëlis recommended two of his suitable students for this task, a Norwegian named Strøm, and the Danish philologist F. C. von Haven. Michaëlis also listed scientific and scholarly questions which it would be desirable to investigate during the journey.

The suggestion was accepted by Frederik V and his government, and the plans were elaborated through correspondence between Bernstorff and the Chancellor A. G. Moltke on one side, and Michaëlis on the other (Michaëlis 1794–96). The most detailed proposal appeared in a letter from Michaëlis to Bernstorff dated 30 August 1756, in which it was still assumed (although with some doubt) that the entire task could be undertaken by one person, who would have to be a scholar of Arabic and Hebrew philology, knowing botany, with a basic knowledge of fossils, and capable of surveying with the purpose of making maps of Arabia. He or his servant should also make drawings of towns, monuments, views and other things of interest.

Eventually in 1759 it was accepted to send three scholars to the Yemen, and an international staff of academics for the expedition was therefore recruited, all previous students of Michaëlis. The naturalist, Pehr Forsskål (1732–1763), was a Swede who had studied under Linnaeus in Uppsala and under Michaëlis in Göttingen. His background will be outlined in the next chapter. The other members of the party were the astronomer, surveyor and mathematician Carsten Niebuhr (1733–1815), Hanoverian born, and the above mentioned Danish philologist Frederik Christian von Haven (1727–1763).

The members of the party were given about a year to prepare for the expedition, especially to improve their knowledge of Arabic, and finally late in 1760 they were ready to leave. On 1 October 1760 Forsskål arrived in Copenhagen and was introduced to the professors of natural history at the University, at the newly established Royal Natural History Cabinet, and at the new Royal Botanic Garden. The Linnaean methods, which were strongly advocated by Forsskål, were not unanimously supported at these scientific institutions; probably one of the reasons why Forsskål was rather, from the outset, critical and suspicious of his new colleagues.

Instructions for the Expedition

The expedition was entirely paid for by the private revenue of the King, who therefore through his advisers gave instructions for the scholarly work of the party. These instructions were drafted by Michaëlis, on the suggestion of some of the Copenhagen scientists and of the King's advisers, chiefly Bernstorff, but also Count Moltke, who himself was a collector of objects of natural history. In Copenhagen, interest in the project was considerable. The zoological and botanical professors took part in making proposals for the work of the naturalist. G. C. Oeder at the Royal Botanic Garden, and P. Ascianus at the Royal Natural History Cabinet proposed that an artist should be employed to draw animals and plants which could not easily be preserved, and that a physician should look after the health of the party and assist the naturalist in collecting objects of natural history (Rasmussen 1990, p. 42). All this was accepted by the government, and Bernstorff prepared to appoint the two assistants.

However, when Forsskål arrived in Copenhagen he immediately submitted an application to the government (Christensen 1918, p. 98–99), asking for (1) an assistant, for which post he suggested a Swede, Johann Peter Falck (1733–1774), another student of Linnaeus', whom he had brought with him to Copenhagen with the intention of having him appointed as his assistant (Christensen 1918, p. 15); (2) to purchase, at the King's expense, a number of books for the journey (these will be discussed later in a chapter on the books used on the expedition); (3) that the King should order the Captain of the naval vessel carrying the expedition to Egypt to be of assistance when Forsskål needed to send chests with specimens to Copenhagen, and when he depended on assistants for inland trips; (4) that the King should allow him to send specimens to Linnaeus on Forsskål's own discretion; (5) that a long list of specified equipment should be put at his disposal for the collecting of plants and animals, including fishing nets to be used from the ship; (6) that he should be allowed to draw money from Danish Envoys and agents along the travel route; (7) that he should be given adequate pocket money, eventually as an advance on the life-pension he had been granted after his return from the expedition; and finally, (8) that an artist should be employed. In Bernstorff's draft to a reply, point (1) and (8) are omitted due to the fact that the government already had decided to employ a physician with the duty to help Forsskål, and an artist. The points (2), (3), (5), and (6) of the application are largely approved of, although an upper limit is stipulated for the amount Forsskål could draw per year. No reply was given to point (4), parcels allowed sent to Linnaeus at Forsskål's discretion, and (7), an allowance of pocket money. Instead, Bernstorff drafted a new paragraph for the Royal instruction, stating that all natural objects collected during the expedition should be sent to Copenhagen, before they were passed on to anybody else. Correspondence in connection with the instructions (Christensen 1918, p. 22) indicates what was probably one of the reasons for this strictness: Loefling, another student of Linnaeus, had made an expedition some years before to South America at the expense of the King of Spain, but natural history material from that expedition had gone to Linnaeus in Uppsala.

These issues developed in a precarious way during the last months before the party was due to depart. Forsskål made a further attempt at having Falck employed as his assistant (Christensen 1918, p. 17–18). Instead another physician, a Dane, Christian Carl Cramer (1732–1764) was appointed. With his degree from the University of Copenhagen, Cramer was the only academically trained member who had neither studied with Michaëlis, nor with Linnaeus. To assist with the drawing of objects of ethnographical or scientific interest the German artist Georg Wilhelm Baurenfeind (1728–1763) was recruited. Finally, the Swedish ex-dragoon Berggren (?–1763) was appointed to assist with the daily needs of all the members of the expedition. Forsskål was very upset with the appointment of Cramer instead of Falck, and complained about the choice in a letter to Bernstorff (Christensen 1918, p. 105–106), in which he expressed doubt about the qualifications of Cramer, and if he could not be allowed to employ Falck, then at least he should to be allowed to test the qualifications of Cramer. The reply (Christensen 1918, p. 107) did not grant this; furthermore, it was underlined that all specimens of natural objects should be sent to Copenhagen. However, Forsskål could propose to whom the duplicates should be distributed.

Meanwhile a significant contribution to the natural history programme was made by C. G. Kratzenstein, professor of physics and medicine at the University, who drafted a long memorandum including suggestions that detailed descriptions should be made of all interesting plants, incorporating notes on the natural size of the trees and herbs, the colour of the flowers, the structure of the fruits, the nature of the roots, etc., not only brief diagnoses for purely taxonomic purposes, as in the works of Linnaeus. At sea, marine plants and animals should be caught in nets, floating algae should be studied, etc. (Christensen 1918, p. 5; Rasmussen 1990, p. 46). Especially Kratzenstein's vision relating to marine biology has later been praised (Wolff in Rasmussen 1990, p. 232). Unfortunately, Kratzenstein had been the teacher of Cramer, and was believed to support him. This caused a conflict between Kratzenstein and Forsskål which the former described his point of view in a letter to Linnaeus (Christensen 1918, p. 168–172) in order to relieve the uncomfortable situation.

Michaëlis followed the preparations carefully and drafted a synthesis of all questions raised by his correspondents, "Fragen an eine Gesellschaft gelehter Männer, die auf Befehl Ihro Majestät des Königs von Dänemark nach Arabien reisen" (Questions to a party of learned men who travel to Arabia on the order of His Majesty, the King of Denmark). It was decided to publish the ultimate version of this draft (Michaëlis 1762) which formed an excellent survey of what was known and not known about Arabian philology, history, and natural history. However, this printed version only appeared when the party had left Arabia, and the book did therefore not as such have much influence on the work of the expedition, although many individual questions had previously reached the expedition by letter. The Royal instruction (based on earlier versions of Michaëlis' draft) must have been given to all, and was certainly entered in full in the journals of some of the expedition's members. The section of the instruction that deal with objects of natural history are part of sect. 15, and sect. 16 to 22 deal with Forsskål's duties.

Sect. 15 is the most important with regard to the natural history collections, and it is clearly influenced by Forsskål's request to send material directly to Linnaeus; it states that: "... all answers to the questions put to the travellers, or questions sent to them on the journey, are to be forwarded to Copenhagen, to the above stated address [that of Count Moltke] and in such a way that the answers to all questions posed by foreign scientists should be sent *sub sigillo volante*, or unsealed, so that copies can be made of them. This applies equally to all drawings, maps ... and furthermore absolutely every kind of natural object collected by the expedition, nothing excepted. The same applies to anything of the above which the travellers themselves may have with them when they return: it has all to be delivered nowhere but at the above stated place. About the future fate of all these documents and objects We [the King] will subsequently, according to the circumstances, make decisions about the keeping or distribution of everything."

Sect. 16. deals with previous works which Forsskål has to refer to: "We expect from Prof. Forsskål that he in his capacity of naturalist will follow the proscriptions given by Linnaeus in his thesis 'Instructiones peregrinatoris' (Instructions to travellers). His main works of reference will be Bochartus' 'Hierozoicon' and Celsius' 'Hierobotanicum' which he will compare with nature, and he will improve their mistakes and fill their gaps while he — in order to do so — seeks information about animals and plants which are mentioned in these works and not yet sufficiently known. The books necessary for his purpose will be purchased on Our account and transferred to him."

Sect. 17 and 18 deal chiefly with the correct identification of vernacular names for natural products: "The attention of the naturalist [Forsskål] will be sharpened and of greater use to others if he on the journey points out the names of all Arabian natural products which are still incompletely understood in Golius' dictionary and which should therefore be further studied in nature. ..." [Sect. 18]: "He should attempt to add Arabic names to all natural products in both Arabic and Latin characters. If the same natural objects have different names in different parts of the country, he should not fail to record this."

Sect. 19 deals with the obligation to collect specimens and seeds of plants: "If possible without additional costs he [Forsskål] should bring with him or send natural products from the mineral and plant kingdoms, and, where the latter is concerned, both as dried specimens and as seeds, and in order to make it possible to use and cultivate the latter he should pay as much attention as possible to the conditions necessary in order to make them germinate. He will not lack opportunity to do so, because he will stay two or three years in Arabia. The prolonged stay he should also utilize to make himself familiar with the different conditions [flowering, fruiting, etc.] of the plants, in which connection he should make careful notes about the seasons for each condition."

Sect. 20 stipulates the way the material is to be sent to Copenhagen: "In order to fulfil these obligations the chairmen and the board of directors of our East Indian and General Trading Company will order their captains to receive and forward trunks with natural products on board the Company's ships. We also order the Captain on Our naval vessel that takes the travellers to Constantinople to take with him Prof. Forsskål and his assistant Cramer each time the crew goes ashore in order to collect natural products. We furthermore order the sailors in calm weather to help with the fishing of corals, shells, marine animals, etc. and ashore to help with the cutting of trees, to carry natural products, etc., for which purpose they are given the necessary equipment such as oyster-nets etc."

Sect. 21 requires coloured drawings to be made of natural objects which are inadequately represented by descriptions alone.

Sect. 22 rounds off the instructions directed specifically to Forsskål: "His attention should also be given to the weather, heat, the tide, especially in the Arabian Sea, the influence of certain traditions on human procreation, and certain illnesses which are unknown to us but often mentioned by oriental writers — in short everything which Linnaeus has recommended a naturalist who attempts to improve on our knowledge."

Sect. 23–26 order Cramer to help Forsskål with the natural history collections and make independent observations on medicine and general health in the countries visited.

Departure of the Expedition and the Stay in Egypt

On 4 January, 1761, the party sailed from Copenhagen on board the Danish naval vessel 'Groenland'. After delays at Elsinore, and severe gales in Danish and Norwegian waters which lasted right into February, a difficult voyage began. More violent storms in February and March made it necessary to make a detour into the North Atlantic not far off the coast of Iceland. The route of the expedition, from the beginning to the end, is shown in Fig. 1.

The ship finally reached Marseille on 14 May, where it docked for several days. Forsskål, who had collected algae and marine animals from the Atlantic by dragging nets from the gun-ports, now made trips inland. One of these trips took him to Montpellier, where he visited the botanical garden and discussed with the botanists Antoine Gouan and François Boissier Sauvages, both strong supporters of Linnaeus. It was the head gardener at Montpellier, Antoine Banal, who gave Forsskål the idea of how he could send material to Linnaeus in spite of the Royal instruction (Christensen 1918, p. 27): Banal asked for seeds from the journey, and Forsskål therefore wrote to Copenhagen in order to obtain permission to distribute seeds by letter directly from the field to a number of European botanical gardens, including those of Montpellier, London, Paris, Göttingen, Copenhagen, and Uppsala. This was granted, and at least the latter three gardens were successful in growing plants from these seeds.

At Marseille the expedition was joined by the philologist, von Haven, who had travelled by stage coach from Copenhagen. von Haven had, since he was first appointed to the expedition, demanded to be the leader and in control of the finances, and a somewhat similar claim had been made by Forsskål. Bernstorff had reacted to this budding conflict by giving all academic members of the party an equal status, except for entrusting Niebuhr with the expedition's cash box and account-keeping (Christensen 1918, p. 30–31).

Schück (1923) has, in a report from the Danish Envoy in Constantinople to Bernstorff, traced how this conflict broke out again after von Haven and Forsskål met once more in Marseille. During a meal at the Captain's table, Forsskål and von Haven discussed the latest development in the succession to the Swedish crown (of two candidates, a Danish and a North German prince, the latter had been chosen). von Haven claimed that the Swedish prime minister, who belonged to the 'Hat' party (of which more in the chapter on Forsskål's early years), had acted in an infamous way when chosing the German. Forsskål, who had no reason to defend the 'Hat' party, found this accusation against his countryman outrageous, and said so in strong words, and the quarrel ended with the use of language which it was 'beyond me [the Danish Envoy] to report, but which referred to the nature of dogs.'

On 28 May Forsskål visited the coastal mountains of Estac west of Marseille, and made a list of the flora which later appeared as the first 'Florula' in Forsskål's printed work the 'Flora Aegyptiaco-Arabica.' The view from Estac towards Marseille may have been that shown in Fig. 2. Legré (1900) has analysed Forsskål's list of the Flora of Estac and found it surprisingly complete and accurate. On 3 June the ship left Marseille and approached Malta on 14 June. The stay on Malta lasted until 20 June; Forsskål collected plants on the island, and compiled a list of 87 species, later printed as 'Florula insulae Maltae.' He was interested in plants specially adapted to high salinity, and studied therefore the flora of the salt-works, an interest he maintained during the entire journey. He also visited the Cathedral of St. John, with its many monuments commemorating Maltese knights, which he found extremely extravagant, and he saw the cave where St. Paul is said to have taken refuge after being shipwrecked at Malta. In his diary, he made a number of ironic remarks on how the cave was being used for various commercial ventures by the Maltese clergy (Forsskål 1950, p. 25–27). He also studied gardens, which were not to his taste because they were too stiff and formal; the best, that of the country seat of the Maltese knights at Boschetto, had straight paths, "adorned with ugly limestone columns covered with grapevine, and with all the paths paved. The garden design reflects the manners of the Orient" (Forsskål 1775b, introduction, p. 7–*8*). Fortunately, he met 'a botanical

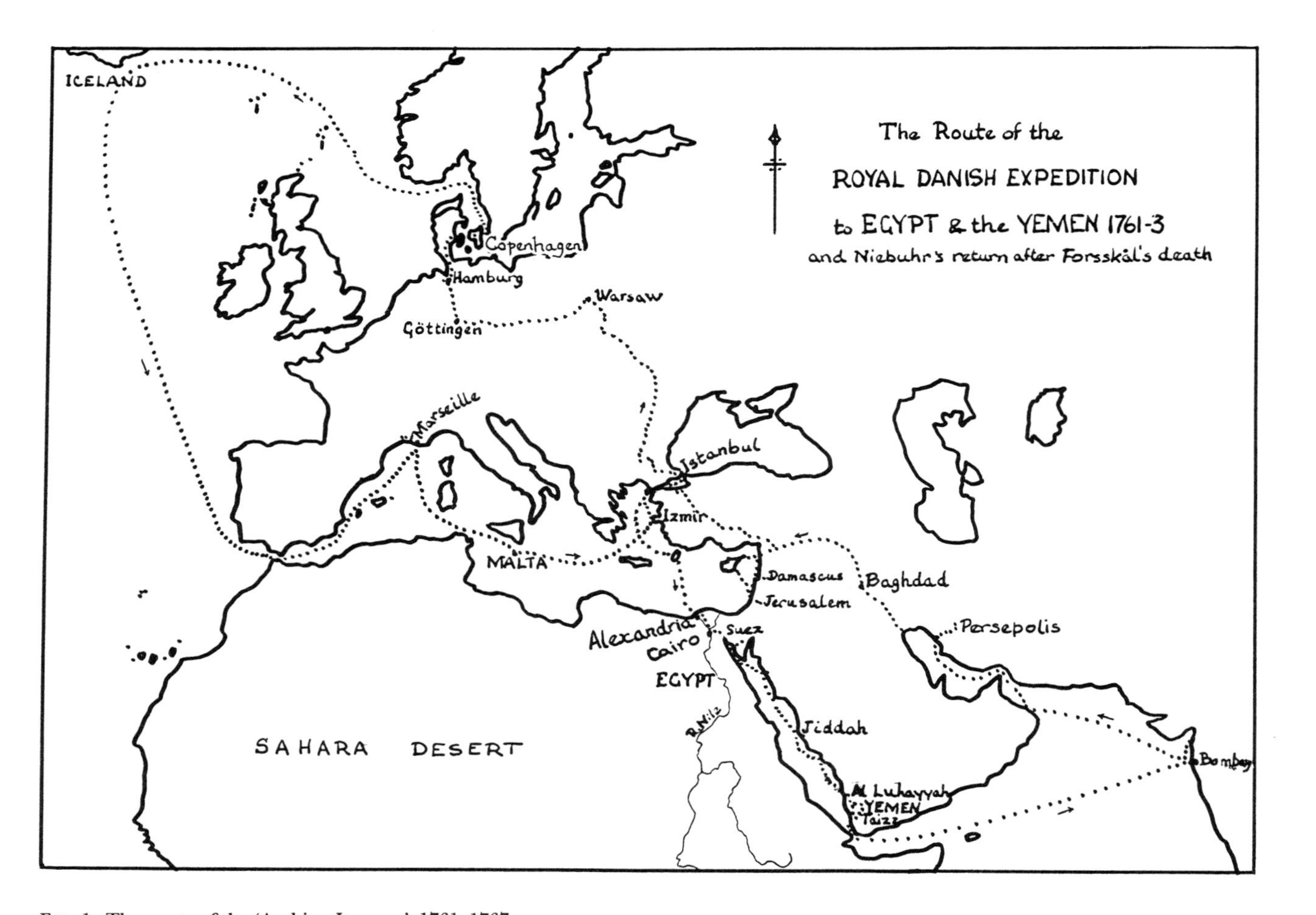

Fig. 1. The route of the 'Arabian Journey', 1761–1767.

colleague and eminent doctor,' a learned Maltese, the name of whom is not known, but who gave him a list of species of fish from the waters around Malta, reproduced in Forsskål's zoological publication (Forsskål 1775a; 1775b, introduction, p. *8*).

From Malta the ship sailed through the Aegean Sea, calling at the Turkish port of Smyrna [Izmir] on 3 July, and for the first time Forsskål was able to botanize in Asia Minor. The ship left Smyrna on 7 July and reached the island of Tenedos [Bozcaada] on 13 July. Here the party left the Danish naval vessel, which was not going to Constantinople [Istanbul] as planned, because pest had broken out in the town. Therefore they embarked 20 July on a voyage with a Turkish vessel which brought them to Constantinople. This ship called briefly at the island of Imros [Gökçeada] on 21 July in order to collect fresh water, and on 24 July Forsskål was able to botanize along the Turkish coast of the Dardanelles (which he called Natolia), while the ship cruised against the wind. Other localities visited on the journey to Constantinople were the small Turkish towns of Borghas [Burgaz], Tjärde [Tekirdag], Märafte [Mürefte], and Eraclissa [ruins of the ancient town of Heracl near Tekirdag, adjacent to Marmaraeglisi].

The ship docked at Constantinople on 30 July. The party stayed at the house of the Danish Envoy, von Gähler, who tried to reconciliate Forsskål and von Haven, persuading them to embrace in public. However, their disagreement was not yet over, and Schück (1923) quotes a statement from von Haven, according to which he would 'make Forsskål regret the humiliation he had brought him [von Haven].'

Forsskål eagerly studied the surroundings of the town, especially some wooded areas nearby, around Belgrad. Other nearby botanical localities visited by him were recorded as Bujuchtari [Büyükdere], Ortacui [Ortaköy] and Baltaliman. Niebuhr has described Forsskål's botanical activities at Constantinople:

"The flora of Constantinople brought the author to a richer field and gave him more leisure for thorough investigations [than had Estac and Malta]. Therefore a richer harvest was gathered than the season would suggest. In order not to tire the reader with an empty enumeration of names, he often gathered additional information and inserted annotations; and Forsskål studied his plants so carefully that he discovered new species among them. The Greek names which are often added are not due to casual information. Some of these names he has received from peasants, others from a Greek who was well versed in pharmacology. In the town he met a community of priests belonging to a distinguished order, and visited their gardens. In the hope of furthering knowledge of the flora in these parts, I have wished to preserve the herbaria which were gathered by the well known M. Höckert from the Swedish legation of the holy order, whom our author has praised as a lover of botany" (Forsskål 1775b, introduction, p. *16*).

The gardens of Constantinople reminded Forsskål of those of Malta, but were, if possible, even less to his taste: "The gardening there was careless, and no better where the gardens were in charge of Europeans, who without reason blamed the deficiency on the climate. The gardens and their ornaments were as we had already seen them on Malta, and the paths were in the most artful manner paved with multicoloured stones from the beach" (Forsskål 1775b, introduction, p. *8*).

On 8 September the party left Constantinople on board another Turkish vessel bound for Rhodes and Egypt. In his diary, Forsskål (1950, p. 39) commented on the cargo: "We travelled together with a merchant with goods destined for the markets of Egypt; the goods were of a kind that would have been very unusual in Europe: young female slaves. They were in a cabin above ours. ... A piece of fabric covered the door, so that they were not to be seen when the door was opened."

On about 15 September some localities at the Dardanelles were re-visited, presenting a seasonal aspect different to that of the first visit in July, and the ship called at Rhodes on 21–22 September. The entire botanical material from the Turkish possessions is listed as the 'Flora Constantinopolitanae ...' in the 'Flora Aegyptiaco-Arabica.'

At Rhodes, Forsskål, Niebuhr and Baurenfeind left the ship and reported in a letter to van Gähler, the Danish Envoy in Constantinople, about a shocking discovery revealed by Cramer. Schück (1923) quotes from their letter: "During the last days we were in

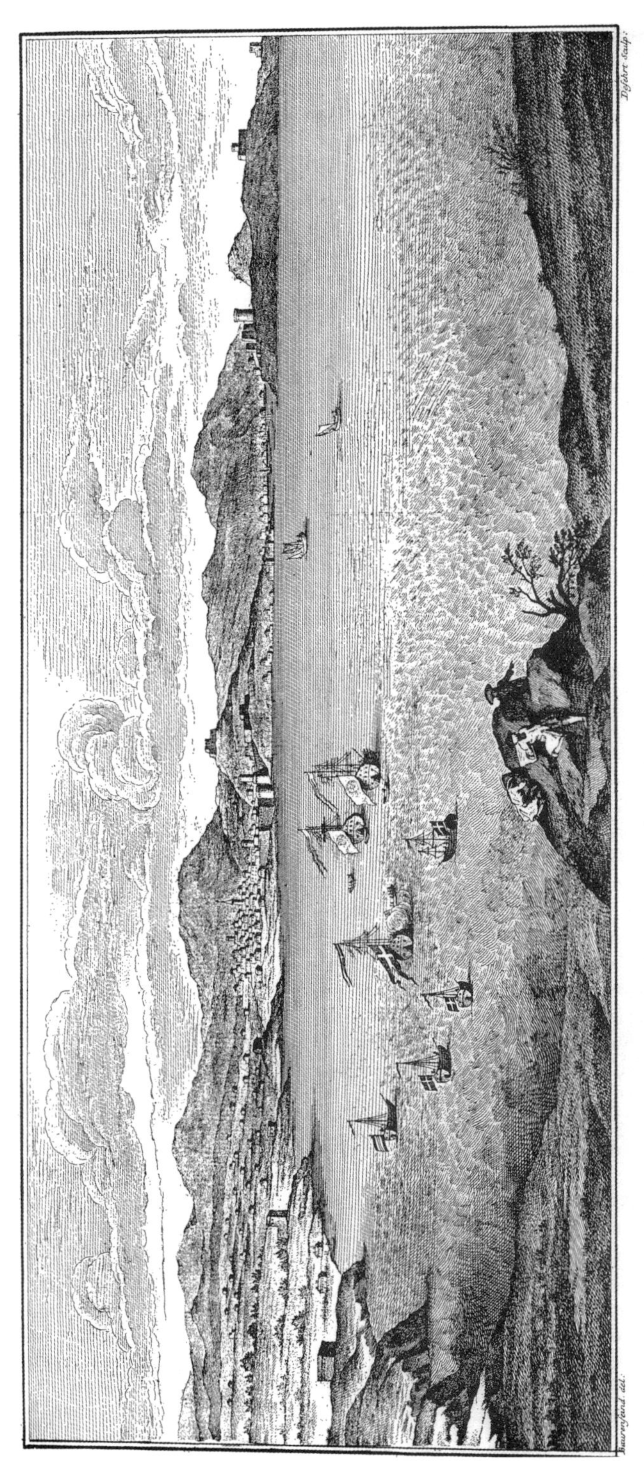

Baurenfeinds tegning af ankomsten til Marseille med „Grønland" for anker mellem de tre danske koffardiskibe, der skal eskorteres til Smyrna. Baurenfeind har øjensynlig tegnet sig selv paa klippen i forgrunden, mens den anden velklædte herre, der ligger bøjet over plantevæksten, næppe kan være andre end professor Forsskål.

FIG. 2. View of Marseille from Estac. 'Groenland' is saluting. In the foreground three men, one drawing the view (Baurenfeind?), and one bending over some natural object (Forsskål?). (From Niebuhr 1774–78, p. 14, Tab. II).

Constantinople, the pharmacist M. Florent delivered to the philologist [von Haven] in the presence of the Doctor [Cramer] two enormous quantities of arsenic, ... which are carried in sealed packets to Egypt. The portions are so large that they would serve three regiments as their last meal. We think your Excellency knows the character of the man, and his wish to control the finances of the expedition. ... We cannot but think that he has had the most vindictive purpose with his purchase. In a country where pest is so frequent, it is very easy to blame sudden deaths on that disease and thus prevent that the bodies of the unfortunate ones are inspected. We do believe that if this is directed against one of us, it is directed against us all, because any survivor would be able to indicate the culprit. We know of no other counsel than deliver this into your hands. ... Any good intention is hardly conceivable with such a provision for the journey. We hope to receive an order in Egypt which will relieve us from this travel companion."

A bill from the pharmacist Florent for drugs sold to 'the party travelling for the Royal Danish court' is indeed preserved in the Danish Public Record office, and has been reproduced by Hansen (1962). The bill shows a list of drugs including a quantity of over 150 g arsenic, a deadly dose for a large group of people, but of course there is nothing revealing the reason for this purchase. The story about von Haven's arsenic has been differently treated by modern authors. Christensen (1918, p. 31–34) tends to play it down; he points out that Forsskål kept writing to von Gähler, complaining about von Haven, while the latter hardly ever mentioned his fellow-travellers in his letters to the Envoy, and that Forsskål's appeals and complaints became nearly pathological in character. Schück (1923) and Hansen (1962), on the other hand, point out that the arsenic had to be seen as a real threat, and that all members of the expedition not only Forsskål, were extremely worried by it.

After leaving Rhodes, the vessel set sail for Alexandria. Forsskål studied marine organisms by fishing from the port-hole of his cabin. This strange activity was noticed by the slaves in the cabin above (Forsskål 1950, p. 43), who dared to look out though their port-hole to watch the activities of the Frank, and the women began to ask him questions in Turkish. "The only sign of friendship I could give them was a lump of white sugar, very rare among the Turks, which they pulled up in a small cloth-bag which had been lowered to my port-hole." This secret understanding was broken when the ship arrived at Alexandria on 26 September, and the party disembarked, while the slaves were taken from the ship only after night had fallen.

At Alexandria, Forsskål (1950, p. 64) found a botanical haven, his "so-called botanical island, or rather a peninsula, which protects the old harbours, and is called Ras-Ettin. The town is built on the isthmus connecting this peninsula with the mainland, and therefore no Arab [beduin] goes there. It is the only area near Alexandria where one can safely botanize alone. The place is well worth a visit. The salinity of the soil attracts herbs which love such ground. The beach, high and low parts of the land, fields, shady and open ground are here found close together." A similar note has been made by Niebuhr: "The same day the illustrious Forsskål had landed, he visited the first gardens and admired the incredible growth and stature of the palm trees. On diligent excursions he explored the surrounding territories, although always with consciousness of the danger, until he reached the safe Ras-Ettin, a place so rich in plants" (Forsskål 1775b, introduction, p. *9*).

On 31 October they left Alexandria on a small ship to Rosette [Rashid], where they stayed until 6 November. On 10 November they arrived in Cairo. The party stayed for almost a year, until the end of August 1762, in Cairo, sometimes working or travelling together, but mostly occupied individually with their tasks. Forsskål usually made his excursions with a guide and riding a donkey, or travelled by boat in the Nile delta. "To ride a donkey is no disgrace here, where it is so widely done" (Forsskål 1950, p. 44). Having already adopted the dress of a Turk in Constantinople, he once again changed his dress to that of a beduin, following the advice of one of his guides, and he even decided to leave his money at home. Yet, that precaution did not spare him unpleasant assaults either (Forsskål 1950, p. 76), including a serious one at the small town of Caid Bey outside Cairo. Niebuhr has recorded one way of getting round the dangers of botanizing alone in the desert:

"When he made his first excursion to the village of Caid Bey, he immediately discovered that all excursions were connected with danger. When he later followed similar routes, he was violently attacked by an Arab, and was scarcely saved by his companions, who had come to his aid. What good counsel had not brought about, that was now effected by dangerous situations which caused him to accept a proposal, that had many times been made to him by people in the town; he should hire a group of Arabs who could bring him plants while he stayed at home! As Forsskål could not see what use the progress of botany could have from the activities of such simple men, he had rejected these proposals, but later he accepted them reluctantly, and finally, when he had put them into effect, he praised them as being beneficial. For the Arabs have, as country-dwellers, from childhood learnt the plants by name, and when offered the opportunity they quickly understood the art of herborization and the gathering of specimens. In this way he bought himself for small expense the needed peace and security, and made a messenger for the world of science of a robber, who, travelling among his own people, brought rare desert plants that would never have been seen by the stranger" (Forsskål 1775b, introduction, p. *27*).

The formal reasons for this prolonged stay in Egypt were a decision to undertake general studies of the country, and especially the study of various Arabic dialects. But perhaps more important was the need for receiving a decision by post from Bernstorff in Copenhagen on the problem which had been raised in the letter from Rhodes: should von Haven continue to travel with the others?

Despite the grim reason for the delay, the year was very fruitful for the expedition, especially for Forsskål's observations, as he was able to study the flora and vegetation of lower Egypt at every season of the year. He decided to return to Alexandria in the spring: "... he travelled again, this time over land, back to Alexandria in order to see the spring flora, but was captured by robbers and thus learnt dearly to fear their impertinence" (Forsskål 1775b, introduction, p. *9*). The account Forsskål has written on the vegetation of lower Egypt in association with the 'Flora Aegyptiaca' is, together with the essay in the introduction of the entire 'Flora Aegyptiaco-Arabica' (Forsskål 1775b, p. XLVI–XLVIII, and introduction, p. *11–15*), considered early steps in the development of ecological and floristic phytogeography. In the essay with the 'Flora Aegyptiaca' he tried to see the flora of lower Egypt as an assembly of species which were especially well suited to grow under the particular climatic conditions of the region, with change between mild winters with sporadic rain and very hot and arid summers.

At Rosette Forsskål also discovered a monstrosity in the floral development of a species of *Corchorus*. This marvel he described (in a note to *Corchorus olitorius*, Forsskål 1775b, p. 101), and pointed out that the flowers instead of petals had green serrate leaves like those of the stem, and he concluded therefore that the petals were transformed leaves and stated, more generally, that 'flos est compendium tantae caulis massae, quantae foliorum habet' (a flower is a contraction of as much stem-mass as it has leaves). This statement has been taken by Ascherson (1884) as the first expression of a general theory of the flower which explains it as a contracted stem with much reduced internodes carrying transformed leaves, a theory which long after Forsskål's death was expanded considerably by the famous German author and 'Naturphilosoph' J. W. von Goethe (1749–1832).

By June the long delayed decision arrived from Copenhagen via the Envoy von Gähler in Constantinople: The party must stay together! Each individual member of the party was admonished by letter to agree with the others, and jointly they should continue towards 'Arabia Felix!'

On 28 August 1762 the party left Cairo. To celebrate their last evening in the town, they hired a group of musicians and dancers; the group was drawn by Baurenfeind (Fig. 3). The party travelled with a caravan of pilgrims heading for Suez, and eventually bound for Mecca by sea via the port of Djidda [Jedda]. Forsskål and Baurenfeind stayed at Suez,

FIG. 3. Musicians and dancers, presumably those hired for the expedition's last evening in Cairo. (From Niebuhr 1774–78, p. 184, Tab. XXVII).

while Niebuhr and von Haven made a trip to Mt. Sinai in order to study rock-inscriptions and to visit the library in the Greek Orthodox monastery of St. Catarina [Monastery of St. Catherine]. Niebuhr has described the situation:

"But just as the party was entering Suez, the painter Baurenfeind fell seriously ill, and asked Forsskål to avert from his plan of exploring Mt. Sinai, and Forsskål gave human consideration priority over his desire for the collection of plants in places which had not before been explored by botanists. Among the plants which I brought back from Sinai, he recognized the Egyptian, others he referred to as Arabian. Without much profit he has himself gone to Ghobeibe in order to collect."

This is the first evidence we have of Forsskål's ideas of floristic elements; this subject, the early foundations of floristic phytogeography, is dealt with in this book in a later chapter on the introduction to the 'Flora Aegyptiaco-Arabica.'

The excursion to Mt. Sinai was very unsuccessful. The expected long and important inscriptions, which should have been studied by von Haven, were not found (although Niebuhr drew a number of smaller inscriptions), the guides had been more than difficult, and access to the monastery, and to its treasure of manuscripts, was denied due to lack of the necessary documents! This abortive Sinai trip further damaged the other members' respect for von Haven, and perhaps it also damaged his own pride, for during the subsequent part of the journey, the conflict seems partly to have subsided. However, in spite of the antagonistic feelings between von Haven and Forsskål, and the lack of respect which the latter had for the work of the former, modern philologists have rehabilitated von Haven's achievements. Rasmussen (1990) points out that the 116 Oriental manuscripts purchased by von Haven for the Royal collections are well chosen, and that these manuscripts have acted as a strong stimulus in the development of the Danish Oriental studies. Seven very important Hebrew manuscripts bought by von Haven in Egypt were used by the English Hebrew scholar Benjamin Kennicot, who in 1776–80 published a now classic collection of variants of Old Testament texts; Keck (in Rasmussen 1990) has concluded that van Haven, through his purchase of these Hebrew manuscripts, has shown competence and extremely good judgement.

Together again, on 8 October the party embarked on a small Egyptian ship crowded with goods and pilgrims sailing to Djidda *en route* to Mecca. Forsskål (1950, p. 101) was surprised to see the number of people who could be packed on board this ship, even in three small boats on tow:

"In the first [and largest of these three boats] were two horses and more than 30 passengers, and even in the two smaller boats were sheep and some passengers, who had to make the journey with the boats when they were used for errands ashore."

The party had hired one of the only two cabins on the ship for themselves, but had to have all their luggage with them. The other cabin was occupied by an important Turkish civil servant. When some small marine animals were being drawn by Baurenfeind, the Turk suggested in a superior way that the party could not find any worthy occupation in their own country, since they travelled to remote parts of the world in search of nothing. Forsskål (1950, p. 100) felt his academic pride hurt, and recorded his reaction in the diary: "We accepted that reputation as a reminder of our privilege (at least with regard to knowledge) over those who, due to ignorance, cannot raise above the most common human errors and weakness." The ship landed at Tor [El Tur] on the Sinai Peninsula, where Forsskål visited a garden belonging to a Greek Orthodox monastery and continued inland to collect plants in the days between 11 and 13 October.

Another stop was made 23–24 October at Janbo [Yanbu al Bahr] on the Arabian coast. Forsskål went ashore, but was disappoined, both with the plants and the town. Finally, on 29 October, the ship and its load of pilgrims arrived at Djidda. Here the party stayed until 13 December, when they were able to transfer to a small coffee-trading boat sailing to Lohaja [Luhayyah], the northernmost port in the Yemen. This boat was small and primitive; Forsskål (1950, p. 115) noted that the planks were tied together with rope and string, it was "a tailor-made boat", without deck or cabins, but heavily loaded, and with cargo everywhere. A voyage on the Red Sea requires that the traveller brings his own bed. ..." After another stop on 21 December at the Arabian port of Ghomfoda [Al Qunfudhah], where Forsskål again went ashore, they finally arrived at Lohaja on 29 December 1762, almost two years after leaving Copenhagen.

Itinerary in the Yemen (Arabia Felix)

Many Yemen place-names are enumerated in the botanical account of the journey (summarized in Forsskål 1775b, p. LXXXVIII–XC). Localities mentioned on labels or otherwise directly associated with the collection of particular plants are indicated in the separate gazetteer later in this book of Forsskål's collecting localities in the Yemen. The following sequence of places and dates are based on the notes in the 'Flora Aegyptiaco-Arabica', on Forsskål's diary, on the smaller of Niebuhr's two maps of the Yemen which has the travel routes of the party indicated (Niebuhr 1774–78; Forsskål 1775b), the latter one made specially for the 'Flora Aegyptiaco-Arabica' and reproduced here as Fig. 9.

From 30 December 1762 to 20 February 1763 the party stayed in the port of Lohaja [Al Luhayyah] (Fig. 4) where they were well received. Forsskål (1950, p. 119) noted:

"We had in the earlier part of the journey been made to suffer by coarse and contemptuous people; now we found polite behaviour, security and the freedom to move around as we wished ... Our luggage passed the customs on the second day, and the Emir, who himself was always present, looked with great attention at our guns, telescopes, and other rare objects."

This Emir of Lohaja is also referred to by Niebuhr, who praised him highly in the introduction to the 'Flora Aegyptiaco-Arabica':

"... [Forsskål] found himself in a completely new situation in Lohaja, and found a gratifying human understanding in the leader of the town, Emir Farhan, who with his word granted security and unrestrained travel, and issued recommendations for the traveller. Never before had that man [the Emir] been made acquainted with botanical journeys or indeed any enterprise which had the study of plants as its purpose; yet he realised from straightforward behaviour of this strange, flower-gathering man that it would be unjust to suspect him of illicit political activity. Therefore I shall remember this Emir with honour and respect, and relate how he received our botanist with pleasure and eased his work to the best of his ability; in short, he was the originator of discoveries in Yemen" (Forsskål 1775b, introduction, p. *30*).

As in Egypt, Forsskål parted from the group and travelled with a guide and a donkey in the northern part of the coastal plains of Yemen, the Tehama. About the surroundings of Lohaja he remarked: "Calcareous rocky sea shore, interspersed with grey clay and sand. Near rocky calcareous hills. Gardens few and poor" (Forsskål 1775b, p. LXXXVIII). In January he made visits to Näman valley ("only provided with stinking stagnant rainwater, with a foul salty taste of sea water which had inundated the low lying parts ..., the fields everywhere consisting of calcareous clay; close to the sea very sandy, otherwise desert-like, bare" (l.c.)), and to the village of Kudmie [Kudmiyah] ("Gardens in a large valley, irrigated by the benefaction of nature ..." (Forsskål 1775b, p. LXXXIX)), to the small town of Mor [Mawr] ("Village with gardens, two hours from Lohaja. 78 plants ..." (l.c.)), and, further inland, to the valley of Wadi Surdud ("A royal place for the flora of Yemen ..." (l.c.)) In February he visited the village of Moglaf in the Surdud valley ("valley 1 mile [7–10 km] broad at the village of Moglaf, visited at the beginning of February." (l.c.)), and crossed the Tehama to the mountains of Jebel Melhan [Milhan] ("An extremely high mountain. The valley of Schäkr adjacent to it. The conspicuous mountains of Hösasch and Ennebijud nearby. (l.c.))

Again and again Forsskål (1950, p. 133) is surprised how peaceful the places are, for example that the flocks of sheep belonging to the villages are unprotected: "A single shepherd and a dog is enough, although the pasture is a desert. This nobody would have dared in Egypt." He does not comment much on the flora of this part of Yemen in his diary, but his impression has been rendered by Niebuhr: "The number of plants collected

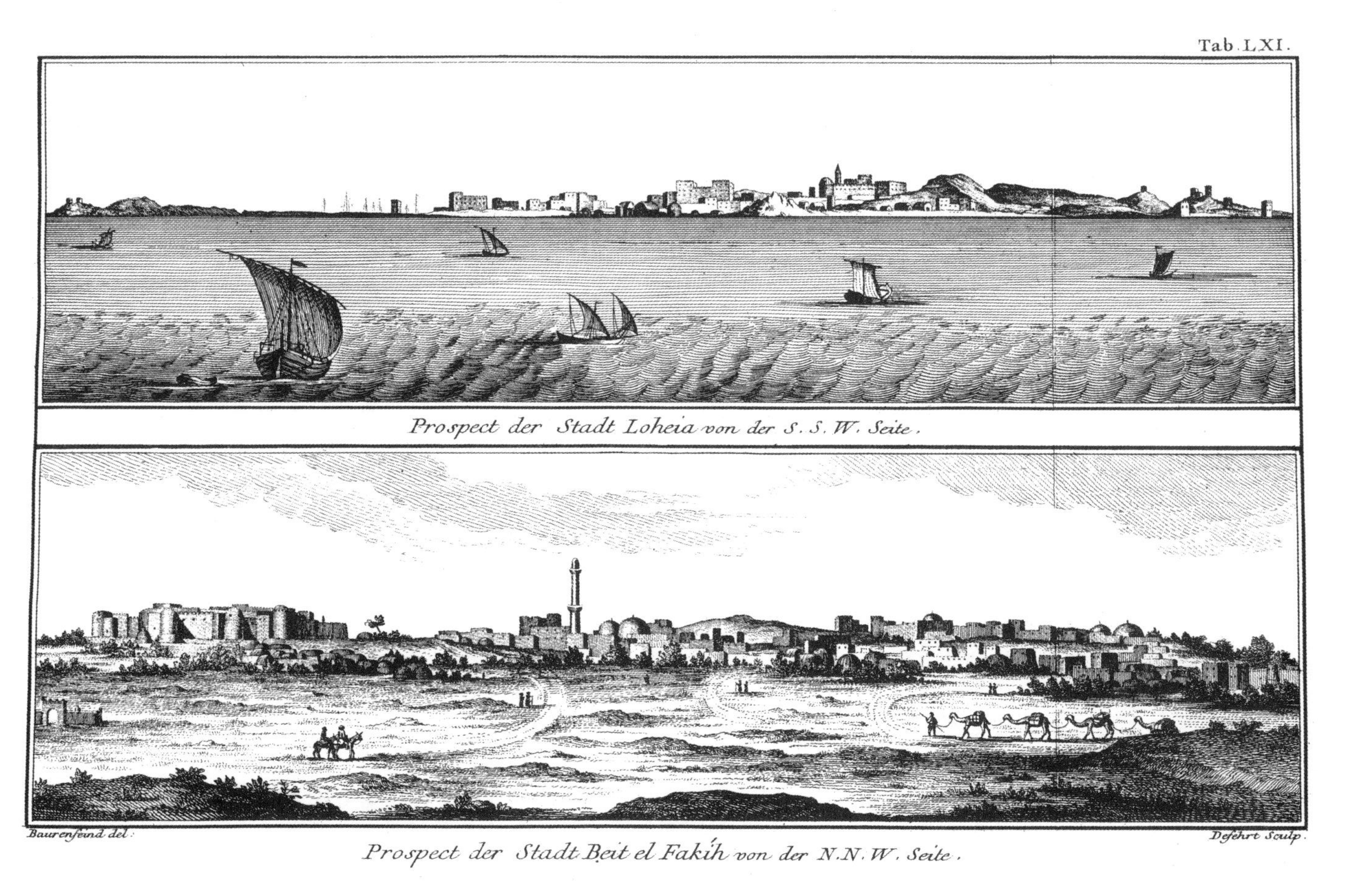

FIG. 4. Prospects of Lohaja and Beit el Fakih. (From Niebuhr 1774–78, p. 312, Tab. LXI).

were lower than expected, but due to the curiosity of the plants he found, and the richness of new species, they were a premonition which encouraged the Botanist to extend his exploration to the valley of Surdud, which, surrounded by mountains, and profiting from cool climate and abundant water, was exuberantly rich in plants" (Forsskål 1775b, introduction, p. *10*).

From 20 to 25 February the party moved base from Lohaja to the town of Beit el Fakih [Bayt al Faqih], further inland and to the south in the Tehama (Fig. 4). "The town with only one very poor garden, establish by a certain Elmas, once a chief of the town" (Forsskål 1775b, p. LXXXIX). On the journey the party made a stop in the small town of Dahhi, and passing the following places mentioned by Forsskål: the village of Okem, Dsjalie, Sabea, Meneyara, and the small town of Ghannemie. The reception in Beit el Fakih was again friendly:

"A sheikh visited us, a learned man according to the tradition of the country, who was curious about everything relating to alchemy. He showed me round the town, during which walk I discovered two new species of *Poa*. In a small garden nothing new was to be discovered, apart from the names of herbs which are here pronounced differently from the way I have heard it pronounced in Egypt" (Forsskål 1950, p. 141).

This sheikh is mentioned no more in the diary, but it is probably the very same sheikh, or fakir, interested in alchemy who is mentioned by Niebuhr in his description of Forsskål's informants in the Yemen:

"The sage (Fakihr) who lived at Beit el Fakih and studied the secrets of alchemy had also a significant knowledge of the herbs of his country, and joined Forsskål when he visited the mountains of Hadie" (Forsskål 1775, introduction, p. *23*).

From 25 February to 2 March the party stayed in Beit el Fakih. Having settled in this town, they split up and made independent excursions, first in the Tehama, and later both independent and joint excursions into the mountains behind the coastal plains.

From 2 to 22 March Forsskål made an excursion to the mountains of Hadie [Al Hadiyah], where he saw, for the first time, coffee being cultivated. "The journey takes 6 hours from Beit el Fakih, and was made in the beginning of March. ... In the valleys the plants are the same as at Surdud ... After Bulgose a flat valley enclosed on all sides in which coffee was grown, and in which was seen various ferns and orchids that had not been collected hitherto. The gardens are on the steep hillsides; coffee in the principal crop" (Forsskål 1775b, p. LXXXIX). The localities visited on this trip included the villages of Hamra, Bulgose, Mokaja, and Örs; Mudje (a stream?); and a number of mountains and peaks in the Hadie region (Djäbbel äsuad, Barah [Jebel Barad], Boka [Bughah], and Mösshöl). These trips took Forsskål as far into the highlands as the village and mountain of Kurma, "... nearly devoid of trees, but with cultivation of herbs and cereals ... (Forsskål 1775b, p. XC). Forsskål's diary contains very little about these journeys, during which he must have been fully occupied with the study of new animals and plants.

From 22 to (?) 26 March Forsskål made an excursion to the town of Djöbla [Jiblah] in the mountains, passing the villages and small towns of Arbaein, Roboa, Machsa, El Uahfad [Wasab] and Maschwara.

From 26 March to 6 April Niebuhr and Forsskål made a joint excursion to the town of Taäs [Taizz], returning to Beit el Fakih, and passing on the way the village and mountain of M'harras, Taäs (1–2 April), village and rest house of Öude and Mt. Chadra [Jebel Khadra] (3 April), "high and with a cold climate. The whole region around Taäs is nearly a homeland for shrubby plants of the genus *Euphorbia* forming thickets in rocky ravines" (Forsskål 1775b, p. XC). The town of Häs [Hays] was visited on 4 April, and on 5 April the valley and temporary stream of Uadi Zebid [Wadi Zabid] was crossed, "with growth of herbaceous plants along the edges" (l.c.) At Öude Forsskål collected and described the first and only balsam tree (*Commiphora gileadensis*) on the entire expedition, the 'Opobalsamum' which he had been specifically asked to identify. Again there is no record of this journey in Forsskål's diary, but Niebuhr recollected:

"The second excursion we [Niebuhr and Forsskål] made together, but we followed different roads, and brought back completely different plants which we had collected along our routes. The culmination of the excursion was the discovery of the tree, from which *Opobalsamum* is produced, with the name of which Forsskål more than once had been deceived in Egypt, because now one, now another tree had been passed as the genuine" (Forsskål 1775b, introduction, p. *10*).

From 6 to 20 April the party stayed at Beit el Fakih. Niebuhr and von Haven were ill. Forsskål must have worked during this time, but did not travel far. On 18 April he wrote in a letter to Linnaeus (Fries 1912, no. 1360) that he had now travelled in the Yemeni 'Alps', "as difficult as the ones in Switzerland, reaching far into the clouds, but with a different climate." A few plants common to the mountains of Yemen and Sweden were found on the journey, "but none of our [Swedish] Alpine plants."

On 20 to 24 April the party, still with some of its members unwell, travelled in two groups from Beit el Fakih to the port of Moccha [Al Mukkha], passing the town of Zebid, the villages of Scherdje and Mauschid, and the towns of Ruäs and Jachtyll. Forsskål (1950, p. 143) was not pleased with the journey: "The greatest disappointment was that I saw hardly any new species of plants, apart from the tree 'Maeru' [now *Maerua*]. After so much herborization in Surdud and Hadie it was difficult to find additions. And had there been more plants, then the drought which has lasted for two years would have disposed of them."

At Moccha, the party was received in quite another manner than they had become used to since their arrival in 'Arabia Felix.' The luggage sent by sea from Lohaja was brutally inspected by the custom officers in presence of the Dola, who was no Emir Farhan. Forsskål (1950, p. 144) details:

"By misfortune it was the chests with the natural history collections which were first brought forward. They were inspected without mercy, and fragile shells were thrown out without any care. An iron rod was pushed through the chest, and the custom officers scowled at such goods which they considered unworthy of being expensively transported in chests. Big shells were broken by pure malice, as if gems had been hidden in them. Finally the chest with *medusae* was brought forward, and being not properly dried they gave off a fetid smell filling the entire room. To complete the disaster, a glass jar with a snake in spirit was found. Everybody was bewildered. The Dola and the people around him claimed that people who had such things should be blamed for illegal sorcery; the goods should be thrown into the sea, and we should not stay one night more in the town. I showed him the bottles with fish to demonstrate that neither the animals nor we were suspicious, and that they were only preserved so that we could show them in our own country as rarities from far away. This was not enough! The barrels with distilled spirits were upset through carelessness, and some alcohol ran out. That smell was even more unendurable among Muslims, who hate liquor. Here was more than necessary to set the whole town up against us. At 11 am, the custom house was closed, and the Dola went angry home. Our personal luggage in our lodgings was thrown in the street, and we were denied access to the house. ... Finally the Dola's fervour calmed down, and he permitted that we again moved into our previous lodgings, and the chests were put into another store house."

Fortunately, Captain Thomas Ringros and Frans Scott, in charge of one of the two British ships from Bombay which called at Moccha once a year, offered the sick members shelter until safer lodgings were found.

From 24 April to 9 June the party stayed at Moccha. von Haven became seriously ill and died on 25 May, the first member of the expedition to succumb to what Niebuhr later called 'the cold,' undoubtedly a fatal form of malaria, a disease unknown to the members of the expedition. During this stay only a trip to the island of Djäbbel Arie was made, and Forsskål found time to write more letters to Linnaeus (Fries 1912, no. 1363), pointing out that he had found the 'Opobalsamum' at Öude and would provide more details, including a specimen, with an open letter sent via Copenhagen. He also continued his comments on the mountains of Yemen, pointing out that he had now established that the plants occurred in distinct zones, but that he did not know the altitude of these zones due to lack of a barometer with which to measure altitudes.

From 9 to 13 June the party travelled from Moccha to the town of Taäs [Taizz] (Fig. 5) in the mountains, passing the towns of Zebid and Dorebat on the way. "Luxurious marvel of *Euphorbia officinalis* mixed with *Euphorbia kert* [*Euphorbia aculeata*, identity uncertain] and *kassar* [not traced]. *Mimosa nilotica* [*Acacia nilotica*] sparingly" (Forsskål 1775b, p. XC). The party stayed at Taäs from 13 to 28 June. Forsskål (1950, p. 149–153) has described the events in Taäs in detail. The location was noteworthy:

"Taäs is considered one on the best strongholds of the Imam, and is indeed so because of the conventions for warfare in this country. It is surrounded with a wall 16 feet thick and with several towers. The wall ascends towards Mt. Saber, and reaches a hill where a citadel called Kähere is located."

But some of the troublesome situations from Moccha were now repeated, and permission to travel on towards Sanaa was not forthcoming. Meanwhile Forsskål wanted to make a trip to Mt. Saber [Jebel Sabar; 3007 m] which he had heard spoken of by the Arabs as the richest area for collecting plants in the country (Fries 1912, no. 1362), but the local Dola did not permit the journey.

Instead, on 18–21 June, Forsskål made an abortive excursion to Mt. Saurek [Sorak], where he found very few plants. Soon after he fell ill, shortly before the departure for Sanaa. On 1 July the party arrived at the town of Äbb [Ibb]. Forsskål was by now too ill to do scientific work. The party proceeded with difficulty to the small town of Mensil, arriving on 3 July. Finally they came to the town of Yerim [Yarim] (Fig. 6) on 5 July, where they stayed, despite the unfriendly attitude of the people, until 13 July. Forsskål died there on 11 July, the second member of the party to perish from the fatal 'cold.'

On 17 July the remaining party of four Europeans arrived at Sanaa, where they were very kindly received by the Imam, the ruler of Yemen. However, in order to arrange for the collections and other scientific results to be sent to Copenhagen as soon as possible the party left Sanaa on 26 July, and travelled at the height of the rainy season along the direct route to Beit el Fakih via Mofhak and Samfur, continuing to Moccha, where they arrived on 5 August.

Niebuhr's Return

When breaking up from Sanaa, the party had decided to leave with the British ships which they knew would probably still be there, and would be able to provide the only passage to India for a long time. Fortunately, the ships were still at Moccha, and the party was able to leave the Yemen on 23 August. Sadly, this rapid departure did not cure the members of the party of their fatal 'cold.' The artist Baurenfeind died on 29 August at sea off Socotra, and the servant Berggren died two days later. Shortly after arrival in Bombay the doctor, Cramer, died, and Niebuhr persevered with the work alone. Being no naturalist, he decided to restrict himself to carrying out geographical studies and to describing and drawing ancient monuments.

After nearly a year of recovery in Bombay, Niebuhr decided to return to Copenhagen overland in order to continue his observations along the Persian Gulf. He travelled through southern Persia, visiting Persepolis en route, and then through Syria, Turkey, and Eastern Europe to Copenhagen, where he arrived on 20 November, 1767, 6 years after the party had set out on the 'Arabian Journey.'

Niebuhr spent ten years in Copenhagen publishing the results of the expedition. His own works consisted of a general description of Arabia (Niebuhr 1772), containing the observations made by the members of the party in reply to questions put to them by Michaëlis and others in the Royal instruction, as well as a more detailed account in two volumes of the journey as far as Mesopotamia (Niebuhr 1774–78). In these works Niebuhr has drawn on his own general observations as well as on notes made by Forsskål. The zoological observations from the journey (Forsskål 1775a) were published by Niebuhr immediately before publication of the botanical. The zoological work has been analyzed by Spärck (1963) and Wolff (in Rasmussen 1990), and has been found to be at least as original and valuable as the botanical account; the work shows that it was especially as a student of marine zoology that Forsskål was a pioneer. The work with the publication of

FIG. 5. View of Taäs, seen from the North, towards Mt. Saber. (From Niebuhr 1774–78, p. 380, Tab. LXVII).

Forsskål's botanical manuscripts (Forsskål 1775b, 1776) will be described below in a separate chapter.

After the monumental effort of publishing five books, making the major part of the observations brought about by the members of the ill-fated 'Arabian Journey' available to the learned world, Niebuhr retired from academic activities and became a civil servant to the King. Having had to make considerable financial sacrifices, especially with regard to the engraving and printing of the plates, he gave up publishing the last volume of his travel account, the one covering the journey from through Syria, Palaestine, Cyprus, Turkey, and Eastern Europe to Copenhagen; this volume was only published in 1837, after Niebuhr's death. He was given a post in the small town of Meldorf in the low marshland of Holstein, where the landscape and proximity of the North Sea resemble the landscape of Hanover where he was born. Here Niebuhr lived for another 35 years, only rarely publishing an occasional paper on subjects relating to Arabia.

Hansen (1962) has in his book 'Arabia Felix' recounted the epic fate of the 'Arabian journey' in literary form. His very readable account of the expedition is based both on the above sources and on archive studies, but the overall view is influenced by the author's talent as a novelist, his personal sympathies, and his experience from participating in archaeological expeditions to the Persian Gulf. In order to underline Forsskål's tragic fate, Hansen has exaggerated the destruction of Forsskål's collections and manuscripts, and overstated the oblivion into which Forsskål's scientific results have fallen. In fact surprisingly much has been preserved. Recently, a large publication has detailed the scholarly achievements of the expedition (Rasmussen 1990), with less emphasis on the human aspects of the journey than in Hansen's poignant account.

Later studies of the flora of the Yemen have been comparatively few until recent years, and for almost seventy years after the 'Arabian Journey' no other European naturalist visited that country. Results of the expedition of M.P.E. Botta were published by Decaisne (1841). Deflers and Schweinfurth published both the results of their own botanical trips to the Yemen nearly contemporarily (Deflers 1889, 1895, 1896; Schweinfurth 1889, 1891, 1894–99, 1912). Blatter (1914–16; 1919–36) included the Yemen in his general account of the Arabian flora, and Schwartz (1939) published the so far most detailed account of the flora of the Yemen, making numerous references to the work of Forsskål. J.R.I. Wood, who has stayed in the Yemen for many of years, has prepared a detailed but so far unpublished account of the flora. Also an account of Yemen ecology, vegetation and flora has been published (in English and Arabic) by A. Al-Hubaishi and K. Müller-Hohenstein (1984).

Pehr Forsskål — a biographical note on the expedition's naturalist

Pehr Forsskål was born in Helsinki, Finland, on 11 January 1732, the third son of a Lutheran vicar of Swedish origin. Finland was then part of the Kingdom of Sweden. Forsskål's family, which included many clergymen, had lived for generations in southern Finland, where its history can be traced back to the later part of the 16th Century (Lagus 1877). Forsskål's mother, Margaretha Kolbeck, died when he was three years old. In 1741 Forsskål's father moved with his family to the parish of Tegelsmora in Uppland, Sweden, near the University town of Uppsala in 1742. Master Forsskål (10 years old!) was registered as a student of philosophy and theology at the University. He was also taught Oriental languages, and a letter in Hebrew, written when he was only 13 years old, was preserved (Lagus 1877).

Although Forsskål seemed at first to have aimed at a career in the church, he also attended the lectures of Linnaeus, and he is known to have taken a considerable interest in the study of plants and animals. There are few sources to our knowledge about Forsskål's early life, but Niebuhr has a short note in the introduction to the 'Flora Aegyptiaco-Arabica,' influenced by the somewhat pompous general style of the introduction:

"... he [Forsskål] found that he had long been predestined and educated for the Orient.

FIG. 6. A house near Sanaa, and view of the town of Yerim where Forsskål died. (From Niebuhr 1774–78, p. 400, Tab. LXVIII).

For this youth, who already as a child had shown his preeminence in intelligence, exceeded under the guidance of his father both the other children of his age and what was expected of him. Hardly had he learnt enough Latin before he embarked on the study of the general works of modern philosophy, and undauntedly he dared to speculate on the ways to the higher truths. Meanwhile he tried his hand with the principles of natural history by describing birds and observing the metamorphosis of insects; in this field he questioned and found general theories. The purpose of his life and his studies had been directed towards theology. But then the well known Professor Aurivillius returned from Paris with a thorough knowledge of Arabic, a language which so few had cared about with exception of Olaus Celsius, who by then was an old man, honoured because of age and great age and merits. Forsskål acquainted himself with the basic principles of this language, which is useful in order to understand the Sacred Language, an auxiliary study of theology. Shortly after, he was attached to the university of Göttingen and climbed towards the eccentric peaks [sic] of philosophy and was even drawn to transcendental dominions before, in due time, he returned to earth, called back by the call of natural history" (Forsskål 1775b, introduction, p. *31–32*).

Forsskål at the University of Göttingen

In 1751 Forsskål applied for the largest scholarship at Uppsala, the 'Stipendium Guthermuthianum,' which included an obligation to travel to a foreign university, and after critical examination he was awarded the scholarship. In October 1753 he went indeed to the University of Göttingen in Germany to study philosophy, theology, and Oriental languages.

The University of Göttingen was new, founded in 1737, but had already gained European reputation for learning at the time when Forsskål became a student. The best known professors were the naturalist Albrecht von Haller, who opposed the Linnaean reform, and J. D. Michaëlis, well known linguist, philosopher, and rational theologian, with whom Forsskål soon became on friendly terms. At that time most teachers at Göttingen belonged to the Wolffian school of philosophy, following the rational teaching of Christian Wolff, who had also greatly influenced many teachers at the University of Uppsala.

Forsskål's thesis for the degree of Doctor of Philosophy at Göttingen aimed at a thorough criticism of some of the fundamental principles of the Wolffian school. It was titled 'Dubia de principiis philosophiae recentioris' (Doubts about the principles of modern philosophy), and was defended in June 1756. Schück (1923), who has studied Forsskål's life in Sweden and Germany in detail, suggests that Forsskål's choice of theme for the thesis was to a large extent due to a strong feature of his character: his wish to scrutinize and if possible oppose established or prevailing opinions. It is generally accepted that Forsskål's thesis did indeed demolish some fundamental principles of the Wolffian philosophy, though it did not propose any new ideas instead. The thesis caused considerable interest at Göttingen and at other German universities, and Forsskål was elected a corresponding member of Göttingen Royal Academy of Science at the early age of 24.

Forsskål and the Freedom of the Press in Sweden

In December 1756 Forsskål was back in Uppsala, and obtained a post as a private tutor for the son of a certain Count Horn, most likely a relative of count Arvid Bernhard Horn (1664–1742), one of Sweden's most influential statesmen and a central figure of the Swedish 'Cap' party, which advocated a flexible and peaceful foreign policy. Horn was strongly opposed by the 'Hat' party which advocated an aggressive foreign policy, especially directed against Russia. It seems that at this point Forsskål had given up the idea of becoming a clergyman, and had begun to study the new 'economic' disciplines, such as agriculture, and natural history, and its applied aspects, under Linnaeus. He also seems to have developed a political inclination and was definitely against the mainly aristocratic 'Hat' party. At this point, Sweden was not an absolute monarchy, but ruled by representatives of its 'Three Estates' organized in two parties, the 'Caps' and the 'Hats.' In connection with his political interests, Forsskål wrote papers in Swedish on philosophical and political subjects. In 1758 he applied for a new teaching post in economy at the Uppsala Faculty of Law, but was not appointed. At least formally, this was due to lack of necessary academic qualifications in the subject, but his radical political views undoubtedly also played a part.

In order to obtain the required formal qualifications in law Forsskål handed in one of his papers, dealing with civil rights, as a thesis at the University of Uppsala, where it was refused for the formal reason that it was written in Swedish, not in Latin, the language for academic discussion. The real reason was that it expressed opinions which would not be tolerated by the government and the 'Hat' party. The main idea of Forsskål's highly political thesis was the existence of an elite minority in all countries, also in Sweden, which oppressed the majority of the population; this oppression could, according to Forsskål, only be counterbalanced by a total freedom of the press.

Already at this point Forsskål had been mentioned by Michaëlis as a suitable member of the expedition which was going to Arabia at the expense of the King of Denmark and Norway. But Forsskål was apparently still indecisive and not sure about his possibilities, for in April 1759 he applied for permission to hand in a new thesis to the University of Uppsala, this time in economy and with the inoffensive title 'De pratis conserendis' (On the sowing of meadows). Now the thesis was accepted, but it was never printed, nor defended. The acceptance and defence of the thesis would have given him the necessary formal qualifications in economy which he had lacked when he first applied for a post in that subject.

Surprisingly enough, in May 1759 Forsskål received the title of Reader in Economy although he still had not yet been granted a degree in this subject. However, he did not settle to lecture on economy for long, for almost simultaneously a serious political dispute broke out, and he decided to leave Sweden — perhaps for good. The reason for this new dispute was that he had handed in a new version of his philosophical-political thesis for defense at the University of Uppsala, this time with the text in both Swedish and Latin, and with the title 'De libertate civilii' (On civil rights). It was again rejected by the Faculty, and this time for obvious political reasons.

Forsskål would now either have to tolerate the University's rejection of his thesis or try to force its acceptance by complaining to the Royal Chancery Council. He chose the latter, but his complaints were refused. After more complaints over matters of formality, the Swedish government resolved in September 1759 to confirm the rejection, and the thesis was returned to Forsskål. At this time negotiations with the Danish government about a post on the expedition to Arabia had gone quite far. An offer from Michaëlis of a post on the 'Arabian journey' had been received as early as January 1759, and perhaps because Forsskål could thus secure an exit, he took the drastic step of publishing the Swedish part of his thesis as a private publication with the title 'Tankar om borgerliga friheten, i anledning af den nu så allmänt omtalade frihetsprincipen hos Fransoserne' (Thoughts on civil liberty, on the occasion of the now so much debated principles of liberty of the French). It appeared in November 1759; it was not easy to have it printed, as no printer dared to publish it without an explicit and written permission from the Censor, but for some surprising reason Forsskål's work was nevertheless accepted by the Censor after deletion of a few of the most controversial statements.

The government immediately decided that the publication should be confiscated, a task which fell to Forsskål's own teacher Linnaeus who was pro-'Hat' and held the post as vice-chancellor of the University of Uppsala that year. Forsskål complained about the confiscation to the government, stating correctly that the book was published with the Censor's permission, and, as the complaint was rejected, complained again, this time directly to the Swedish King, to no avail. On the contrary, in February 1760 a decree was issued in the name of the King, to be made publicly known from the pulpits of all Swedish churches. In this decree it was stated that Forsskål's 'Thoughts on civil liberty ...'

generated wrong ideas and disseminated these ideas among the Swedish people. It was therefore illegal to sell or buy this publication, and those who had a copy should immediately send it to the Chancery.

After this blow, Forsskål prepared to leave Sweden. It might thus seem that he went to Arabia in order to make a retreat, but this is probably not the entire truth. Many people felt sympathy for Forsskål and his views, and it is deeply ironic that he actually won his struggle for freedom of the press in Sweden after he had left the country; in 1766 the censorship was abolished in Sweden —three years after Forsskål had died in Yerim.

Into Service of the Danish King

Schück (1923) has shown that it was only Forsskål's thesis and the opinions he expressed in this, not his person, which was prosecuted in Sweden. He was still very highly thought of by Linnaeus and other important or learned members of Swedish society. There is no doubt that he was genuinely interested in the purpose of the 'Arabian Journey,' and very eager to discover new plants and animals in order to improve knowledge of natural history, the science which he had studied under Linnaeus. It is clear from this description of Forsskål's years at the universities at Göttingen and Uppsala that he was in no doubt about his excellent qualifications, and that he was usually right when he debated a point with others. His loyalty to Linnaeus was unchanged, but he was deeply suspicious of his new colleagues in Copenhagen, as some of them were critical towards Linnaeus, and keen to improve the reputation of Danish science.

This new conflict is reflected in a letter he wrote to Linnaeus from Constantinople (Fries 1912, no. 1352), in which he suggested a secret code should be used in his letters sent to Linnaeus *via* Copenhagen:

"[In future communications] I will refer to the species by number [from ed. 10 of the 'Systema Naturae'] in such a way that the second figure is always first, for example for 1236 I will write 2136. If the second figure is a zero, I will put a dash, for example 1000 will be written -100, so that nobody will discover the code. I will cite the numbers in the new 'Systema Naturae,' for example -83.31 (for 803.13) will mean *Lotus corniculatus*, 958.62 (for 598.26) *Cistus helianthemum*, etc. I will refer to new species, which are not in the 'Systema Naturae', like this: If it is a new species of *Leontodon*, I shall write 187 (for 817) and species 7, as there are only 6 species in the 'Systema Naturae'."

But Forsskål hardly found an opportunity to use this code, and we have no evidence of it being put to practical use during the journey. From Cairo he wrote to Linnaeus (Fries 1912, no. 1358), with continued suspicion of his colleagues in Copenhagen:

"From here I send a large amount of dried plants, and have not put names on them, only their numbers [the code for locality and collection mentioned above], so that they will not be plundered by the wrong hands."

This summary of Forsskål's biography up to his departure for Copenhagen is based on the biographical literature which exists in three Scandinavian languages on the subject: Lagus (1877), Christensen (1918, 1924–26, 1935), Schück (1923), Forsskål (1927), Matinolli (1960), and Wolff (1980).

Forsskål's reputation and the genus Forsskaolea

Finally a note on the naming of the new genus 'Forsskålea' which throws additional light on contemporary opinions of Forsskål. Linnaeus was aware that Forsskål would be pleased to have a new genus discovered on the 'Arabian journey' named after him. This Forsskål had made clear in a letter written from Cairo (Fries 1912, no. 1356), in which he expressed the hope that some easily grown bulbous plant might be given his name. However, on 1 April 1762, while on one of his journeys from Cairo to Alexandria, and just having been released after being kidnapped by a gang of robbers, he found "a species belonging to the *Tetradynamia* ... a species which in its class is the greatest anomaly possible. ... If it pleases you, it should have name after me, who bought the knowledge about it with the experience of having to live by the grace of robbers" (Fries 1912, no. 1357).

Forsskål distributed seeds of this plant to botanical gardens under the name *Thaumastra prostrata* n. gen. The seeds germinated in Copenhagen, where Oeder identified the plant as *Hypecoum pendulum* L., and informed Linnaeus about it in a letter dated 18 April 1764 (Schioedte 1871–72, no. 14). It is thus nearly certain that very early on Linnaeus knew about the true identity of the plant, and was therefore unable to name it after Forsskål. It is unlikely that Linnaeus managed to inform Forsskål about the identity of the *Thaumastra,* because Forsskål, instead of adopting the name *Hypecoum,* changed the name in his notes to *Mnemosilla aegyptiaca,* and it is thus named in the 'Flora Aegyptiaco-Arabica'.

Also Carsten Niebuhr knew about Forsskål's wish to have this presumed new genus associated with his name. In the biography of his father, B. G. Niebuhr (1817) wrote: "Linnaeus showed himself very hostile to his former pupil. Forsskål had told my father that he would like a new genus of plants (in his Flora called *Mnemosilla*) to be given his name. My father wrote about this wish of a great man to Linnaeus, but instead of following this wish, Linnaeus gave Forsskål's name to another of his new genera; the name of the most important species in this new genus referred in a spiteful manner [sic] to the deceased. My father could never forget this trick."

The dispute is reflected at several places in the 'Flora Aegyptiaco-Arabica'. It is referred to in the preface, at the end of p. 21, where the editor suggests that the *Mnemosilla* is the true *Forsskaolea,* and it is also alluded to in comments or footnotes in the relevant places in the text, i.e. at species no. 221 on p. LXV, at species no. 314 on p. LXIX, in a footnote on p. 83, and on p. 122.

One reason for this disagreement is probably that Niebuhr, who was not a botanist, did not realise that Linnaeus was unable to give Forsskål's name to the *Mnemosilla,* as this genus was already named. The other reason is the well known assumption that Linnaeus, when he chose to give personal names for his new genera, liked to refer to the character of the person so honoured by chosing a plant with suitable attributes. The plant which received Forsskål's name was another species from the vicinity of Cairo; in his seed list Forsskål called it *Laniflora adherens,* but he later changed the name to *Caidbeja adherens* (p. 82 in the 'Flora Aegyptiaco-Arabica'). The seeds sent to Uppsala germinated, and Linnaeus realised that the plant represented a new genus. Shortly after, Linnaeus heard the news about Forsskål's death, and published a description and a name for the plant, *Forsskaolea tenacissima,* in a 'Corollarium' to his thesis 'Opobalsamum declaratum' (Linnaeus 1764). The plant is a prostrate herb with insignificant flowers; it belongs to the Urticaceae, and is densely provided with stiff, pointed hairs and also with hairs with hooked tip which make the plant adhere to fur and fabric.

It was not only Niebuhr who took offence of Linnaeus' action. Christensen (1918, p. 78) quotes another author, the German Johann Beckmann, who in 1770 wrote that Linnaeus named this particular plant after Forsskål because not only were the seeds sent to him by Forsskål, but also the name suited Forsskål's rough character, and Linnaeus used words like 'hispida', 'adhaerens' and 'uncinata' in the description. Lagus (1877, p. 49), who seems also to have been offended on behalf of Forsskål, wrote to the Professor of Botany in Helsinki, S. O. Lindberg, and asked his opinion. Lindberg (in Lagus 1877, p. 75) explained in a letter how Linnaeus was unable to fulfil Forsskål's wish because of the priority of *Hypecoum,* and suggested that the epithet 'tenacissima' might possibly have been given because of the tough fibres in the stems, but admitted that there might be a double meaning in the word, as it could also refer to the fact that Forsskål was 'a tough dialectic, who often argued against the advice of Linnaeus.'

Spelling and Use of Forsskål's Name in Botanical Nomenclature

Not only was there some dispute over the choice of plant to commemorate Forsskål; also the spelling of Forsskål's surname (and the genus named after him) has caused confusion and difference of opinion. The list of variant spellings of the epinome *Forsskaolea* Linn. is probably the longest in the 'Index Nominum Genericorum' (Farr, Leussink, and Stafleu 1979), where 13 different variant spellings are recorded.

Furthermore, Christensen (1918, p. 80) points out that in the more than 50 plant species which have been named after Forsskål, the name is often used in a corrupted form.

The spelling has to be settled under the Code (which concerns the use of personal names in the formation of Latin plant names (Art. 73), and the use of personal name should be standardized for the citation of authors of scientific names (Art. 46)), as well as for bibliographical purposes and when referred to in scientific papers. Also when authors referred to Forsskål in publications, a wide range of different spellings have been used (Friis & Thulin 1984). The most frequent are: 'Forskål', 'Forsskåhl', 'Forsskaal', 'Forskal', and 'Forsskål', but a long list of other variants can be compiled, explaining why there is such an inconsistency when the name is used for the formation of Latin plant names. Friis & Thulin (1984) have shown that the spelling 'Forsskål' is to be preferred, and that the abbreviation 'Forssk.' is preferable to indicate Forsskål as author of plant names. This has been followed in Kew's 'Authors of Plant Names' (Brummitt & Powell 1992).

Regarding the epithets formed from Forsskål's name, it is very difficult to determine if emphasis should be put of the spelling used by the original author (first sentence of Art. 73.1 of the Code; Art. 73.3), or whether a standardization should be preferred, based on the assumption that all deviations from the established standard represent 'orthographic errors.' Friis & Thulin (1984) have given priority to the restraint recommended by the Code when 'orthographic errors' are to be corrected. Although they have established a 'recommended form' of Forsskål's name, this does not mean that it is the only correct one, as Forsskål himself used several variant spellings. They therefore recommend that the original authors' deviations from the sequence of consonants used in the recommended (but not necessarily correct) spelling should be accepted. The most frequent deviations are the use of -s- rather than -ss-, and the use of -h- at various places in the name, for example 'forskaolii,' 'forsskaolii,' or 'forskaohlii'. Nearly all these variants have been used by Forsskål himself. The same liberty is not admissible in the vowels to be used in the name, as there is never doubt that the first vowel in the name is -o-, and that the second one is always meant to be -å- (transcribed -ao- according to the Code). The fact that Forsskål, in letters to Danes, sometimes wrote 'Forsskaal' only reflects that -å- was written -aa- in older Danish.

Forsskål's Botanical Library and Field Methods

Before setting off on the 'Arabian Journey,' Forsskål had been lent a considerable number of botanical books from the Danish Royal Library to use on the expedition. We know that he wished to use C. Plumier's 'Plantarum americanarum', fasc. primus -decimus (1755–60), G. E. Rumphius' 'Herbarium Amboinense, cum Auctuario', vol. I–VII (1741–55), H. A. Rheede-tot-Draakestein's 'Hortus Malabaricus', vol. I–XII (1678–1703), J. Burman's 'Thesaurus Zeylanicus' (1737), 'Rariorum africanarum plantarum decas' I–X (1738–39), and J. J. Dillenius' 'Historia muscorum' (1741). Forsskål stated in his correspondence concerning these loans that he did not need to borrow the books by Linnaeus, as he possessed those already (Christensen 1918, p. 99–100). We do not know if he took all the above mentioned books with him on the journey, but it is known that he handed in a number of folio books and other large volumes relating to natural history to the Danish envoy in Constantinople so that they could be returned to Copenhagen. The smaller books on natural history were later sent home by Niebuhr from the Yemen through the Danish East India Company in Bombay.

It is of considerable interest for the understanding of Forsskål's taxonomic concepts that he expected to find American and Asian plants in Egypt and Arabia, especially because he gave much thought to the distribution of plants (see a later chapter in this book on the introduction to the 'Flora Aegyptiaco-Arabica'). He seems often to have accepted an identification if the plant exactly matched a description in the Linnaean works he had with him, no matter where the species was described from. He used the botanical books of Linnaeus as a world flora. In a letter to Linnaeus from the Yemen

(Fries 1912, no. 1360) he wrote, after having seen the balsam tree at Öude:

"Now I know the genus of the 'Opobalsamum'; the tree grows in the Yemen. ... It is not *Pistacia*, not *Lentiscus*, but one of [Patrick] Browne's genera [i.e. an American genus], and it will have its most natural relatives among species already described in that genus. ... Here I have found a lot of American, Indian, and new plants ..."

In a subsequent open letter to Linnaeus sent via Copenhagen (Fries 1912, no. 1363), Forsskål was able to give more specific information, and pointed out that 'Opobalsamum' belonged to the genus *Amyris* P. Browne.

It seems safe to conclude that the 10th edition of Linnaeus' 'Systema Naturae' was the most important work Forsskål used during the expedition to identify his collections, and that this work is the most important bibliographic help in understanding his taxonomy. An example of this is a species of fig now called *Ficus exasperata* Vahl, and widely distributed in tropical Africa as well as in the Yemen; Forsskål identified this species with the Linnaean species *Ficus serrata*, which is only mentioned in the 10th edition of 'Systema Naturae'; neither in 'Species Plantarum' nor in any other of Linnaeus' earlier works. *Ficus serrata* is an American plant, based on a plate by Plumier.

Forsskål's Study of Arabic Plant Names and Other Field Methods

The Royal instruction included orders to collect field data about objects of natural history as well as their vernacular names, as this information would be of importance for the understanding of Oriental manuscripts and of the Bible. It appears from his diary, the published manuscripts, and the herbarium collections, that Forsskål took a keen interest in being as correct as possible, and took great care in collecting information from local informants about uses and vernacular names. These data can therefore, with precaution, be used in the identification of his species, as Schweinfurth (1894–99, 1912), Friis (1981), and Wood (1982) have found.

There is little evidence of Forsskål's field methods in his diary; more can be culled from Niebuhr's notes. His search for people who knew the Greek names of plants around Constantinople and his problems with botanical excursions in Egypt, both around Alexandria and Cairo, have been referred to in the general description of the journey. The journeys in Yemen were completely different, and Niebuhr has sketched how Forsskål, after his disappointments in Egypt, had to change his opinion of the Arabs as informants, fellow travellers, and indeed as human beings:

"It is certainly wonderful to penetrate a country worth seeing for its different plant growth; but it is more praiseworthy to change from a prejudiced idea of a barbaric nation to a friendly relationship with a simple, but well mannered people, who may disagree between themselves, but nevertheless are amicable to strangers, who are uneducated in sciences, but not restricted in intellect, who are poor, but yet hospitable. This tradition has not grown from a void, but is due to a general convention which prevents the violation of strangers. This is how the Arab Yemen is different from the Arab Egypt; here the noble live far from the centres, ... and the life of the common people is even more parochial. The inhabitants of Arabia Felix are satisfied with durable alliances, and do not want unjust conquests. To this can both Forsskål and I bear testimony, saluted as we have been among unknown people; we have always achieved complete protection. He came to places hardly touched before by a European foot, and certainly not by that of a botanist. Here came a man, unknown and apparently eccentric, one for whom not trade but the names and uses of plants was the singular purpose. Encouraged by these principles of the nation, he lived a shepherd's life among his Arabs. Admirably were both the older Arab, the younger one, and the girl equally eager to provide information about the flora of their country. When young people each day gather fodder plants, they involuntarily carefully investigate the botany of their field; and what they have in youth counted as play, they still remember when old. This is the reason why Flora in her idiom has named very few plants that the Arab has not given a name to in his own language" (Forsskål 1775b, introduction, p. *28–29*).

Forsskål's interest in vernacular names has in fact considerably influenced his choice of scientific names for his new genera, many names for which have been formed by a slight modification of the Arab names to make them acceptable in a Latin context. The introduction to the 'Flora Aegyptiaco-Arabica' contains some notes on Forsskål's principles:

"The new names for the genera are a choice subject for critics who are eager not to leave any stone unturned, but rather flood [the literature] with alternative names or with differently worded descriptions; from a discussion of this I will abstain. Such [synonymies] are just lists of words which can be rejected or renewed, if only the core of the matter remains certain. The nomenclature [of the 'Flora Aegyptiaco-Arabica'] has a double origin. [First discussion of the generic names based on place names, which will be given below, then:] If the names of natural objects in Arabic agree with our pronunciation and flow freely in the form outlined by the system, then they should be adapted [as scientific names] and not lack protection from my part, as this language when priority of age is concerned far exceeds Greek and Latin, from which we often adopt names for species which neither a Greek nor a Roman can possibly have known. Nor are the Arab names meaningless. In Yemen, as in Europe, each province often uses 'nomina trivialia' which is formed on the basis of common names recognized in the entire region. These common names are in no way of recent date but have come about through an long development, and can be found both in the works of learned people and in the oral tradition" (Forsskål 1775b, introduction, p. *22*).

How Forsskål has put these ideas into practice can be seen from the following list of generic names for plants, the scientific names of which have been based on the commonly used Arabic name:

Adenia (from 'aden'), a genus of succulent shrubs in the Apocynaceae. *Aerva* (from 'ärua'), a genus of herbs in the Amaranthaceae. *Antura* (from 'antur'), now a synonym of *Carissa* (Apocynaceae). *Arnebia* (from 'sagaret el arneb'), a genus of herbs in the Boraginaceae. *Cadaba* (from 'kadhab'), a genus of shrubs and small trees in the Capparaceae. *Cadia* (from 'kadi'), a genus of shrubs and small trees in the Leguminosae. *Catha* (from 'kat'), a small genus of trees in the Celastraceae. *Caucanthus* (from 'kauka'), a genus of shrubs and lianas in the Malpighiaceae. *Cebatha* (from 'kebath'), now a synonym of *Cocculus* (Menispermaceae). *Ceruana* (from 'käruan'), a small genus of herbs in the Compositae. *Chadara* (from 'chadar'), now a synonym of *Grewia* (Tiliaceae). *Culhamia* (from 'kulhåm'), now a synonym of *Sterculia* (Sterculiaceae). *Digera* (from 'didjar'), a small genus of herbs in the Amaranthaceae. *Elcaja* (from 'djous elkai'), now a synonym of *Trichilia* (Meliaceae). *Geruma* (from 'djerrum'), a small genus of herbs in the (Aizoaceae). *Keura* (from 'keura'), now a synonym of *Pandanus* (Pandanaceae). *Kosaria* (from 'kosar'), now a synonym of *Dorstenia* (Moraceae). *Leaeba* (from 'lbach el djebbel'), now a synonym of *Cocculus* (Menispermaceae). *Maerua* (from 'meru'), a genus of shrubs in the Capparaceae. *Maesa* (from 'maas'), a genus of small trees and shrubs in the Myrsinaceae. *Oncoba* (from 'onkob'), a genus of small trees in the Flacourtiaceae. *Rocama* (from 'rokama'), now a synonym of *Trianthema* (Aïzoaceae). *Rokejeka* (from 'rokäjeka'), now a synonym of *Gypsophila* (Caryophyllaceae). *Saelanthus* (from 'säla'), now a synonym of *Cissus* (Vitaceae). *Sceura* (from 'schura'), now a synonym of *Avicennia* (Avicenniaceae). *Sehima* (from 'sähim'), a small genus of Gramineae. *Simbuleta* (from 'symbulet ennesem'), now a synonym of *Anarrhinum* Desf. *Sodada* (from 'sodad'), now a synonym of *Capparis* (Capparaceae). *Themeda* (from 'thämed'), a small genus of Gramineae. *Turia* (from 'turia'), now a synonym of *Luffa* (Cucurbitaceae). *Zilla* (from 'zillä'), a small genus of subshrubs and herbs in the Cruciferae.

In comparison with this, only very few of Forsskål's new genera are based on place names. This is somewhat surprising, because of the warm recommendation such names are given in the introduction to the 'Flora Aegyptiaco-Arabica': "If a new plant has been associated with the place where it has first been found, in commemoration of that place, it is a duty out of gratitude to that soil which with nourishment and care has nursed the kin of the plant not only that year, but through centuries. Therefore I have retained *Kahiria, Caidbeja, Melhania* and *Eraclissa* as names" (Forsskål 1775b, introduction, p. *21*).

The following is a list of such generic names:
Caidbeja, now a synonym of *Forsskaolea* (Urticaceae), from 'Caid Bey', a place near Cairo. *Eraclissa*, now a synonym of *Andrachne* (Euphorbiaceae), from 'Eraclissa', a place near Constantinople. *Kahiria*, a synonym of *Ethulia* (Compositae), from 'Kahira', Cairo. *Melhania* (Sterculiaceae), from 'Djebbel Melhan', a mountain in the Yemen.

Very few names of new genera have been formed from words of technical botanical origin, signifying morphological features of the plant, for example:
Alternanthera, a genus of herbs in the Amaranthaceae. *Binectaria*, now a synonym of *Mimusops* (Sapotaceae). *Gymnocarpos*, a small genus of herbs in the Caryophyllaceae. *Hyperanthera*, now a synonym of *Moringa* (Moringaceae). *Papularia*, now a synonym of *Trianthema* (Aïzoaceae). *Pentaglossum*, now a synonym of *Lythrum* (Lythraceae). *Polycephalos*, now a synonym of *Sphaeranthus* (Compositae). *Pteranthus*, a small genus of herbs in the Caryophyllaceae. *Siliquaria*, now a synonym of *Cleome* (Capparaceae).

As it appears from Forsskål's correspondence, diary (Forsskål 1950), and printed botanical works, his method was to study the plant very carefully on the spot, gather information from local informants, and attempt an identification by means of Linnaeus' works before entering the information in his manuscript notes on slips of paper and preserving a specimen. If the species appeared to be new, or if the Linnaean diagnosis of the plant was incomplete, he would prepare a more detailed description. He went even as far as to correcting Linnaeus by letter (Fries 1912, no. 1352): 'In your 'Genera Plantarum' Hieracium needs an extended diagnosis, as every word in it also fits Sonchus'.

In a number of cases Forsskål was unable to identify his plants in spite of the fact that they were already described by Linnaeus, because the Linnaean diagnosis was incomplete. An obvious example of this is Forsskål's genus *Binectaria* (identical with the Linnaean genus *Mimusops*), a case which has been discussed by Friis (1981). Forsskål pointed out in his notes that the flower of this genus was characterised by its two 'nectaries' (actually whorls of staminodes), but these were not recorded by Linnaeus, not being directly associated with stamens or pistil. Forsskål's description of the very complicated flower of *Mimusops* is very well made, and contains more detail than is shown by the specimens which have come down to us. Also his descriptions of the minute flowers of his new genera in the Chenopodiaceae are of very high quality (Christensen 1918, p. 48), and show his abilities as a very competent field observer.

Forsskål's herbarium specimens are often rather fragmentary, and not good when compared with the competent and detailed descriptions. If the specimens are labelled at all, the labels are small and insignificant, usually just a small scrap of paper torn from a sheet, and the information on them very scanty. However, when compared with other contemporary herbaria, the material in Forsskål's herbaria does not stand out as particularly insufficient.

Forsskål's small labels are basically of two different kinds, which may reflect the way specimens were sent home. The Egyptian plants were sent directly to Copenhagen, and would therefore become accessible to others before Forsskål's return. These Egyptian specimens in the 'Herbarium Forsskålii' are sometimes provided with the original label, which is tiny, and on which is written only a capital letter and a number, for example C25, which means that the collection was made in or around Cairo, and that the collection number was no. 25. The abbreviations on the Egyptian labels are: A (Alexandria), C (Cairo), and R (Rosette). Similar abbreviations can be seen in the so called 'Florulae', the section numbered with Roman pagination in the printed flora. These are further discussed in the chapter dealing with the publication of the 'Flora Aegyptiaco-Arabica,' and are listed in the gazetteer.

The Yemen plants, which Forsskål presumably expected to carry home himself, are provided with more information on the preserved labels, sometimes the locality is written in full, and a vernacular name is indicated. An example of a label from the Yemen collections in the 'Herbarium Forsskålii' is reproduced here on Fig. 8, and another reproduced by Christensen (1924–26, p. 152, Fig. 41).

A number of plants, especially succulents, may not have been collected at all, or only the parts which could be preserved without difficulty were collected. Several succulent plants, especially in the Aizoaceae, Asclepiadaceae, Chenopodiaceae, Vitaceae, and Zygophyllaceae, were drawn by Baurenfeind, the artist, under Forsskål's guidance, as stated in the introduction to the 'Icones rerum naturalium' (Forsskål 1776). See further in the chapter on the 'Icones ...' below.

Forsskål's Botanical Manuscripts and their Publication

At his death at Yerim in Yemen, Forsskål left seven parcels with notes and manuscripts (Christensen 1918, p. 59). In these parcels were a diary (Forsskål 1950) and at least as many scraps of paper as Forsskål had collected species. These notes were saved by Niebuhr who sent them home together with the herbarium from India. It appears from an inventory of the seven parcels, that two of these contained herbarium material and the manuscripts which were later to be published as the Flora Aegyptiaco-Arabica: "One parcel in small quarto, entitled 'Plantae descriptae'; one parcel in small quarto entitled 'Plantae piccatae, semina, Caidebegensis aliquae herbationes, Florae.'" (Schück 1923, p. 476–477; 'plantae piccatae' would seem to be an error for 'plantae siccatae' (dried plants)).

The Unknown Editor of the 'Flora Aegyptiaco-Arabica.'

After his return to Copenhagen in 1767, Niebuhr was preoccupied with the publication of his own works, and was only free to attend to the publication of Forsskål's papers in the mid-1770's. It would appear that Forsskål had, on the journey, compiled the flora-lists for Egypt and the Yemen, as well as for a number of localities visited in the Mediterranean, and written a detailed essay on the flora of lower Egypt, and a somewhat shorter one on the flora of the Yemen (Christensen 1918, p. 66). The remaining manuscripts consisted of unorganized scraps of paper with descriptions of animals and plants. Niebuhr has described them:

"The illustrious author used to transfer his notes to separate and very small slips of paper according to a method which he found convenient because he would, when the number of species increased, be able to rearrange them in their new and correct place like sheets in a herbarium, while one volume with no possibility for rearrangement is inflexible and easily becomes confused with addition of new observations and experience. ... But how easily could not a small page disappear without the author having possibility to reconstruct it from memory, and even more incapable are we who are ignorant of the loss! ... A heavy burden has accumulated for the scholar who has to edit this, heavy not only due to its volume, as the manuscripts include more than 1800 scraps of paper, and some have almost certainly disappeared for the original owner. That more has disappeared after his death is no less certain. ... [Back in Copenhagen from the journey] I left these manuscripts in the hand of a man very knowledgeable in natural history and did not give them more thought for a long time. But as the years passed by, I felt sad that Forsskål was forgotten and was afraid that his papers should vanish altogether. ... I asked the advice of learned friends on how best to publish these writings. The friends unanimously recommended that the words of the author would be more valuable than any revision, and that the public would want to hear precisely what Forsskål had discovered. ..." (Forsskål 1775a, introduction).

As Niebuhr was not a naturalist, and not very well versed in Latin, he had to find a collaborator to organize the manuscripts and notes, and to edit and translate into Latin the text of the 24 pages long introduction, which is discussed in more detail in a following

FLORA
ÆGYPTIACO-ARABICA.
SIVE
DESCRIPTIONES
PLANTARUM,
QUAS
PER
ÆGYPTUM INFERIOREM
ET
ARABIAM FELICEM
DETEXIT, ILLUSTRAVIT
PETRUS FORSKÅL.
PROF. HAUN.

F. Möller

POST MORTEM AUCTORIS
EDIDIT
CARSTEN NIEBUHR.

ACCEDIT
TABULA ARABIÆ FELICIS GEOGRAPHICO-BOTANICA.

HAUNIÆ, 1775.
EX OFFICINA MÖLLERI, AULÆ TYPOGRAPHI.

FIG. 7. Title page of 'Flora Aegyptiaco-Arabica.' (Forsskål 1775a).

chapter. The flora was published in 1775; the most likely date of publications is discussed below. Niebuhr appeared as editor on the title page, and is said to have paid for the publication out of his own pocket (Niebuhr 1817). He seems to have been unhappy with the book, and felt deceived by his naturalist collaborator.

In the introduction to volume one of his 'Symbolae botanicae' Vahl (1790–94) criticized the person who had been putting Forsskål's notes into order, but does not mention any names:

"Everyone with experience in these matters [natural history], however, is greatly indebted to the illustrious Niebuhr for publishing the observations by his friend [Forsskål], which otherwise would have been lost together with the discoverer himself. Even if several particulars in this posthumous work, lacking the ultimate correction from its author's hand, could be criticized for inaccuracy, it is still worthwhile to pay considerable attention to most of the observations in his Fauna and his Flora as well, but it is desirable that the specimens themselves be compared with the appropriate descriptions. In this, inevitable errors due to travelling, when notes have to be made in a hurry and not always with extreme exactitude, can be corrected, and material referred to wrong places [in the system] can be correctly determined before the observations are published. This work is most easily accomplished here in our city [Copenhagen] where his collections of animals and plants are still kept. In this way errors may be avoided, both these that have crept in through the mistakes of the man to whom the task of arranging Forsskål's notes was committed, those that Forsskål himself could not avoid due to the difficulties of travelling, and finally those that have been and in the future may be, caused by Forsskål's books."

It has been suggested by Warming (1880, p. 73) that it was the Danish botanist J. Zoëga, a pupil of Linnaeus, who edited Forsskål's flora, but Zoëga was a competent botanist and a friend of Vahl's. It is therefore unlikely that Vahl would have written as quoted, had the editor been Zoëga. Other evidence against Warming's theory is that we know from a statement in Zoëga's obituary (Anonymous 1789) that he wrote the explanatory text to the plates and compiled the synonymy of the 'Icones rerum naturalium' — a smaller task, but done with much greater care than the huge task of editing the 'Flora Aegyptiaco-Arabica'; on the other hand the obituary does not mention anything about his editing the Flora, which would have been reasonable to mention, had it really been done by Zoëga. It is also worth mentioning that Zoëga was employed at the Royal Botanical Garden, which received Forsskål's seeds, not at the Royal Natural History Cabinet, which received Forsskål's dried plants and other collections. This fits with a remark in the 'Icones,' in which it is stated that the synonymy is based on observations made 'at the Botanical Garden.'

Christensen (1918, p. 67–68) has compiled a considerable amount of evidence in favour of a different editor: Most important is a statement by G. B. Niebuhr, Carsten Niebuhr's son (Niebuhr 1817) who wrote in a biography of his father: "... the papers could not be published as they were, nor could my father undertake to edit them, as he was unfamiliar with natural history, and not very competent in Latin. He therefore employed an erudite Swede, who asked a considerable fee for his service. This Swede was an eccentric man, who amongst other things pressed my father to allow the introduction to appear in his [Niebuhr's] name, a compliance my father has afterwards much regretted."

The identity of this Swede is not known; Schück (1923) suggested that Dr. Christian Wåhlin, at one time reader of Oriental languages, and from 1775 Professor of Medicine at the University of Lund, not far from Copenhagen, would have been qualified for the job. But Uggla, in the introduction to the publication containing Forsskål's diary from the journey (Forsskål 1950), has speculated on its being Daniel Rolander, a theory supported by Fox Maule (1979). Daniel Rolander was an ill-fated pupil of Linnaeus who went to Surinam and fell out with his former teacher on his return in 1765 to Sweden, after which he lived a miserable existence in Copenhagen and in Lund until his death. He seems to have been permanently short of money, and sold his herbarium and manuscripts from

Surinam to Rottböll, who published the manuscripts in 1776 (Christensen 1924–26, p. 151). There is no proof of this speculation, but the younger Niebuhr's comments on the unknown collaborator would seem to fit the known facts of Rolander better than those of Wåhlin.

There is, however, no doubt that Niebuhr played a part in the practical work with the publication, as he would be the only person who could supply some of the information about the journey both in the main text of the introduction and in the footnotes stated to be by the editor, for example on pp. LXXXII–LXXXIII, p. LXXXVIII, and pp. LXXXIX–XC. He also drew the map, and must have made a considerable contribution to that part of the introduction which required personal recollection of Forsskål during the journey. He must also have prompted the note about *Forsskaolea* mentioned previously.

The flora was, as stated in the introduction to the 'Descriptiones animalium,' and mentioned in the above quotation from Vahl (1790–94), published largely as written by Forsskål, without comparing the manuscripts with the herbarium, and without checking whether any of the new species had in the meantime been described by others. This is confirmed by the introduction to the 'Flora Aegyptiaco-Arabica':

"Written in the field, it is now handed over to the guardianship of the erudite community. The accurately fixed Polar Star around which this process [the future work with Forsskål's book and taxa] revolves is the preserved herbarium. Very many plants are today not new, but long ago described or illustrated, all with the illustrious Forsskål as their first originator, because he gave them access to the botanical gardens by sending seeds of the rare ones" (Forsskål 1775b, introduction, p. *32*).

The situation is, as mentioned above, completely different with Zoëga's edition of 'Icones rerum naturalium', where synonymy is carefully indicated, and it is pointed out that 17 species described by Forsskål should correctly have other names. These 17 species are enumerated in a following chapter on the 'Icones ...'

Perhaps the worst fault which must be blamed on the editor, is that the flora is very inconsistently edited, and that there is no attempt at compiling an index, which makes the book almost impossible to use, as many species are treated in several different, but often partly connected parts. The two major parts, the systematic lists and the descriptions, could well represent the two manuscript parcels mentioned above and entitled '... Descriptiones' and '... Florae'.

'Florulae' and 'Florae'

The first part of the book has pagination with Roman numerals, running from page I to page CXXVI. This part contains the 'Florae' and the 'Florulae', which are gathered in sections entitled (1) 'Florula littoris gallie prope Massiliam; Florula insulae Maltae.' (2) 'Flora Constantinopolitana, littoris ad Dardanellos et insularum Tenedos, Imros, Rhodi.' (3) 'Flora Aegyptiaca: Sive Catalogus plantarum systematicus Aegypti inferioris: Alexandriae, Rosettae, Kahirae, Sues.' (4) 'Flores Arabico-Yemen. sive Catalogus Plantarum Arabiae Felicis systematicus.' These 'Florae' or 'Florulae' are numbered floristic lists, each organized according to the Linnaean sexual system, and with a great deal of additional information, such as the essays on the floras of Egypt and the Yemen, information on uses of the plants, etc.

The additional information is scanty for the lists of plants from Estac, Malta, Constantinople and the Aegean Islands. The Egyptian lists contain the description of the flora of Egypt from an ecological and geographical point of view, and a list of plants classified according to uses, as well as a full list of all species observed. The lists from Yemen are similar to those of Egypt; there is a short and more fragmentary account of the flora from a ecological and geographical point of view, a list of plants classified according to use, and a complete floristic list.

The collecting locality is one detail relating to the plants mentioned in the 'Florulae' and 'Florae' which is often very clearly indicated. This locality is usually indicated by a system of abbreviations. None have been proposed for the floras of Estac and Malta, but for the Turkish collections the following locality abbreviations are used: Td: Tenedos; Imr: Imros; Brg: Borghas; Ecl: Eraclissa; Dd: Dardanelles; Cph: Constantinople; Bg: Belgrad; Bj: Bujuchtari; Sm: Smyrna; Rh: Rhodes. The modern equivalents and precise positions of these places will appear in the gazetteer later in this book.

The Egyptian abbreviations are simple: A: Alexandria; C: Cairo; and R: Rosette. The capital letters are provided with a qualifying non-capital letter: d: desert; h: ('horti') gardens; and s: spontaneous. The list of locality abbreviations in the 'Flora Arabico-Yemen' is not long: Blg: Bulgose or Hadie; Btf: Beit el Fakih; Lhj: Lohaja; Mlh: Mt. Melhan; Srd: Wadi Surdud; Uhf: Uahfad. Again, the precise positions of these place names will appear from the gazetteer later in this book. However, there is also a slightly different, ecologically orientated system for indication of localities, where P means that the species occurred in the clay plains (the Tehama), and M that it occurred in the mountains. Again these capital letters can be combined with non-capital letters to qualify the habitat: Mi: Lower ('inferiora') zone of mountains; Mm: Middle ('media') zone of mountains; Ms: Upper ('superiora') zone of mountains; Ma: Mountains, in water; Mma: Middle zone of mountains, in water; Mia: Lower zone of mountains, in water; Msa: Upper zone of mountains, in water; Pm: Maritime part of clay plains; Pu: Inundate parts of clay plains; Ps: Dry parts of clay plains; Ph: Humid parts of clay plains, near the mountains; Pa: Sandy clay plains; Pr: Clay plains near rivers. C means 'cultivated,' and can be combined with both M and P: C.P.: Cultivated in the clay plains; C.M.: Cultivated in the mountains; C.M.M.: Cultivated in the middle zone of the mountains.

The localities in Yemen can also be found on the map of Forsskål's journeys in that country (Fig. 9) especially prepared by Niebuhr for the book, as he described in the introduction:

"I have, out of fear that future travellers should be misdirected in a country which has not previously been represented by any hand, attempted to order roads and areas of plant growth in a 'Tabula Geographico-Botanica,' designed in a form showing in particular places on our journey, especially those places which the illustrious Forsskål has made famous. Meanwhile the geographical excursions come back to my memory, excursions which offered me bouquets of plants, which I noted with pleasure that I had collected for the benefit of my travelling companion" (Forsskål 1775b, introduction, p. *30*).

'Descriptiones'

The second part of the 'Flora Aegyptiaco-Arabica', the 'Descriptiones plantarum florae Aegyptiaco-Arabica', has pagination with Arabic numerals which runs from 1 to 219. Here all the plants are arranged in one sequence, without much consideration for geographical origin, and basically according to the Linnaean system, but with a curious system of 8 'Centuriae', or groups of hundred species, superimposed on the Linnaean sequence:

In the 'Centuria I' to 'Centuria VI' of the 'Descriptiones Plantarum' the plants are arranged in a sequence according to the Linnaean system and contain detailed descriptions of higher plants, mainly from Egypt and the Yemen, but also from a few of the other localities visited during the journey. In 'Centuria VII' there are new cryptogams from the whole expedition, including algae from the Sound at Elsinore, from the very beginning of the voyage to the Mediterranean, as well as an 'Appendix Plantarum Arabiae Felicis indeterminat', a list of 55 species which Forsskål had been unable to place in a genus according to the Linnaean principles. The plants in this list are either 'sub-obscurae' because of imperfectly known stamens and pistil, or 'obscurae' because these organs were impossible to study, due the plants being sterile when observed. 'Centuria VIII,' called 'Descriptionum Plantarum supplementum,' would appear to be a general supplement to 'Centuriae I–VI', with a selection of plants from all stages of the journey.

Christensen (1918, p. 74) thought it likely that the editor attempted to finish each 'Centuria' at the same point as a Linnaean class, so that the following 'Centuria' could begin with the first species in a new class. This attempt has not always been successful. 'Centuria VII' contains only 61 named species and probably for this reason it has been supplemented with the short descriptions of unnamed, sterile plants mostly referred to with their vernacular names only. More often, however, there would appear to have been too many species in the Linnaean classes to fit neatly with the division into Centuriae, and Christensen (1918, p. 74) thinks that the supernumerary species have therefore been assembled in 'Centuria VIII', or perhaps even inserted with short descriptions in the 'Florulae'. This view is supported by a statement in the introduction of the work:

"In order not to enforce an undigested mass of information on the readers I have found it advisable to organize the descriptions so that they were more clear. In this way the *Centuriae* came about: The integrity of the classes and the homogeneous members of the genera have been preserved as far as possible. In this connection, when it is very useful to gather all species under their genus, it seemed less necessary to separate the Arabian flora from the Egyptian one. The previous lists of the richness of species [for each region] have already given the occasion unambiguously to enumerate all the species which the author has found" (Forsskål 1775b, introduction, p. *21*).

In the 'Descriptiones' the locality is usually stated in full, and there may be a short statement of the habitat or other notes about the plant.

Date of Publication of the 'Flora Aegyptiaco-Arabica'

The date of publication of Forsskål's 'Flora Aegyptiaco-Arabica' has been indicated by Stafleu & Cowan (1976) as 1 October 1775, but there is no source for that date, and the problem has not been dealt with by Christensen. However, it is possible to settle the date of publication with a reasonable degree of accuracy by means of Danish review journals, as described by Friis (1983a). The book was reviewed in two Copenhagen review journals: 'Efterretninger om laerde Sager' (News about erudite matters), and 'Nye kritiske Journal' (New critical journal). The former carried only a review in the issue dated 11 April 1776, the year after that indicated on the title page of the 'Flora Aegyptiaco-Arabica.' The latter journal, which unfortunately was issued in undated instalments, published a review in instalment no. 26 of the year 1775. The two journals were more or less parallel publications, published by competing Copenhagen printing houses, and both appearing with 52 issues a year. It seems therefore safe to assume that they both appeared weekly, as the 'News ...' definitely did, and it is therefore possible to infer the date of the earlier, undated review, that of the 'Critical journal'.

In issue no. 11 of the 'Critical Journal' there is an advertisement for the new issue of the prestigious 'Flora Danica', which the editor of 'Flora Danica' has signed and dated 8 March 1775. If we assume that the advertisement was published in the 'Critical Journal' shortly after the draft was signed and dated, then it fits with the calculation that the 11th issue should appear in the second or third week of March. A review of Forsskål's 'Descriptiones animalium...' appeared in issue no. 22, which would therefore have apparently been published in the last week of May or the first week of June; the review concluded with the note: "The publisher has announced that Forsskål's 'Flora Orientalis' ['Flora Aegyptiaco-Arabica'] will appear very soon, and that the work on the 'Icones ...' is making good progress." Finally, a review of the 'Flora Aegyptiaco-Arabica' itself appeared in issue no. 26, which must have come out in the last week of June or in the first week of July.

There is no reason to believe that the book was not available to the Copenhagen public by the time it was reviewed in a commercial review journal, and June 1775 is therefore the date of publication of the 'Flora Aegyptiaco-Arabica' according to the International Code of Botanical Nomenclature.

The Introduction to the 'Flora Aegyptiaco-Arabica'

The introduction to the 'Flora Aegyptiaco-Arabica' has not previously been much referred to as a source of information about the 'Arabian Journey,' but is nevertheless useful for this work as it provides details not available elsewhere. The organization of the introduction is somewhat confused; it would seem to represent a compilation of notes by Forsskål, recollections and comments by Niebuhr, and notes on systematic botany, presumably by the unknown editor. The parts of the text not previously in Latin have been translated into that language; unlike Forsskål's prose it is quite verbose and provided with embellishments, and is sometimes not altogether clear. In spite of that, it seems useful here to present a free rendering of the introduction, especially the parts which have not been covered by other chapters in this book.

First the introduction points out that the 'Descriptiones animalium ...' (Forsskål 1775a) was published before the 'Flora Aegyptiaco-Arabica,' a point which we have just seen confirmed by contemporary review journals, and that there were considerable problems with the editing of the complicated manuscripts of the two works. Then follows a fairly detailed account of the expedition up to Forsskål's death in Yerim. Some of that travel account, which must have been drafted by Niebuhr, has been quoted in the first chapter.

A Phytogeographical Essay

A treatise on the difference between the localities and habitats of plants (Forsskål 1775b, introduction, p. *11–15*) is interesting. It is likely that this essay rests on some manuscript by Forsskål, although no such manuscript is mentioned anywhere. It seems less likely that these comparatively complicated ideas should represent Niebuhr's recollections of discussions with Forsskål on the subject, as has been suggested by Christensen (1918, p. 52). However, it is natural that Niebuhr, being an astronomer and geographer, has taken an interest in the subject. Christensen (1918, p. 52) quotes, as evidence of Forsskål's general interest in habitats and distributions, an unpublished letter to Linnaeus sent by Forsskål from Constantinople in which he argues that more attention should be given to the habitats and geographical distribution of plants in dissertations and floras.

The essay in the introduction shows basic, but not yet clearly developed ideas of relevance to floristic and ecological biogeography:

"In these districts, in these varied plains, on these mountain slopes are plants gathered which penetrate into both floras. A geographical study alone will demonstrate that they do not have the same origin. Another and higher purpose now makes itself noticed, that is to explore the habitats of the plants ['Situs plantarum'] and to distinguish between the changing provinces of the habitats. Botanists normally attribute natural localities ['locus natalis'] to their species, but often these localities are much more extensive than the small areas which the plants cover with their stems and roots. When considerations of this kind are reliable, they can by the help of comparisons form the Parallel plant geography ['geographia vegetabilium parallela' (?)], which within the individual areas is not everywhere the same, in more remote areas is more manifest, and finally, in clearly separated countries is quite different. When the plants are known, it is therefore possible to draw conclusions about the latitude of the country, about the variation in altitude, further about the zonation of the vegetation, from their base to the highest peaks. Such observations are equally profitable whether they relate to gardening or to botany. From these rules deductions can be made about place of origin and colonization, and about the history of the plant migrations, how they have moved over the world, as foreigners, migrants, harmful, or as cultivated, harmless, splendid or home-born [not straying from home].

"I have wished to stress this to those who do not, with the same conviction as Forsskål feel the necessity to add to the description of each species information of its habitat, which could be called the dialect with which it speaks to the botanist. These words, which are not unimportant, but of great significance, I have preserved in connection with the actual localities. In order to give the reader a clear impression of the composition of the flora, from where it has originated, and on which principles it rests in this overseas part of the world, I shall add a general physical-geographical sketch. [Such sketches are in fact present both for the 'Flora Aegyptiaca' and for the 'Flora Arabico-Yemen].

"[People in] the temperate zones and the cold regions have rarely recognized how much water means for the flora. [People in] the southern countries naturally pay much more attention to this, burnt as they are by the blazing sun. Gardens and fields owe to water their life and the salvation of the people who live by them. Economic considerations are largely attached to rivers, wells, and springs; how they are controlled, fortified and used by hydraulic methods. This kind of irrigation is available to the ingenuity of mankind when the rain fails or is delayed, the rain which can revive the country by invigorate drying plants or wake up those lying dormant in the ground. All this the illustrious Forsskål noted in Egypt, and even more in the Tehama in Arabia, which he visited at a time when it was desiccated from a drought which had lasted a whole year, while the water-rich regions in the adjacent valleys and in the mountains flowered luxuriantly.

"The Byzantine territory demonstrates a mixture of two floras, the Northern and the Southern, as the first is decreasing, while the latter is increasing, even dominating. From a preliminary calculation it appears, though, that the northern flora here is not poorer than the Egyptian or the Arabian, but who would look for it in the African or the Asian world? There these plants are few and insignificant, and seem to deny alliance with the form which grows in their natural home ground. But nor does the North want to call them its own, because they due to tenderness to cold must be called adventive, as they stick to the border of the region and seem to bargain with the climate.

"It is not possible to mention Egypt without mentioning the Nile; it produces a ground rich in moisture and clay, fertilised with lime from the subsoil. The country formed by these factors offers, due to the yearly sedimentation, possibility for the making of garden after garden. The plain is of vast extent, quite uniform and nearly level, without the rises which the [wild] flora would often occupy as its ground. It is divided into cultivated areas and deserts, and is subject to overwhelming metamorphoses, brought about by a change between inundation and unbearable drought. On these conditions the plants live and die; a study of great interest for the plant physiology. However, it is completely outside the normal ways of nature when *Arundo* is able to grow prolificly in the hot springs at Ghobeibe.

"The Arabian flora produces plants of variable origin, every group adapted to a habitat in its own special area. Tehama offers habitats on the beach and on the plain if the rain returns there regularly year after year. The area between the mountains is favoured by shade and moisture. And finally Yemen can be pleased by having mountain plants which have climbed into the heights by themselves or by human help. Here terrace-farming is practised, induced by the same natural conditions which in the remote past caused Palaestine, and now the people of China, to introduce the same methods. Egypt can boast of highly cultivated plants; Arabia has mainly wild and more simple plants, among which, however, we can note some very productive and nearly noble ones, which with a little more culture could be made very valuable for the economy. Together, the two countries house the extremes in the natural range of plants; Egypt condemns its sand plants to thirst and exile, the Arabian mountains protect their alpine plants on narrow strips of soil and nearly in the embrace of the clouds.

"The available lists of plants from Egypt and Arabia have been made with diligent care, but I have not dared to establish which plants are definitely indigenous in one or the other country; this is only possible after a thorough comparison with the normal flora of the surrounding countries, such as Abyssinia, Persia, and Asia north of Arabia. Egypt has received nearly all its garden plants as gifts, and has hardly any enumeration of its agricultural plants, let alone their origin. Nor does Yemen lack a flora of guests. The Banyans, who from the most ancient time have travelled in these parts, have introduced very rare plants, nearly all Indian, at first for their own use, later for the general improvement of the society."

Taxonomic Considerations

The introduction then continues with a more cursory description of how and to which extent studies were carried out in various parts of the journey, unavoidably with a certain overlap with the travel description in the beginning.

The question of the scientific reliability of Forsskål genera and species and other taxonomic questions are raised; from the comparatively technical botanical language of the following, one may assume that it has been written by the editor (Forsskål 1775b, introduction, p. *17–20*):

"The question about the validity of the species I will leave to the further inquiries of the botanists. It is not my task to topple genera, let alone classes, or in any way make them unstable. Who will claim that the illustrious Forsskål, roaming along roads and through forests, has been able to name all plants correctly. However, he did not knowingly perpetuate his mistakes, and often he blamed himself an error in subsequent corrections. ... The illustrious author is not to be blamed that through this many and clearly marked varieties have been established; it is, on the contrary, rather to be counted in his favour that he with his natural talent and experience has been able to produce such accurate descriptions. There is a not insignificant difference between zoological and botanical descriptions. Animals show their species characters at the first glance, they have them so to speak in their skin. Although insects during their triple metamorphosis have completely different forms, they do nevertheless make it possible at each stage to prepare accurate descriptions which can be identified with the correct animal. A different law rules over the science of plants. Before flowering the stem and leaf gives no message about species characters. The flower is also tardy in opening. And what work it is to read in the fine disc of the flower! The description of the ripe fruits is determined by fixed numbers. When the life span is out, the plant dies, and only after many months it is resurrected from the root or from seeds. Therefore it cannot be avoided that the botanist, when he is not able to follow this life-span to the end, must of necessity collect many indeterminable [specimens] which cannot be fitted in the system. ...

"For a naturalist it is a glorious achievement to describe new genera, and he prefers light and order in the great crowd of species. But it will fall to the learned botanical world to judge if all the genera which the great Forsskål has proposed may stand a critical examination. But what would it matter if some new species had to be reduced to the lower status of variety, or even to serve under another, already known flag, or even if some genera should be deemed invalid; on the contrary, some species, which are here placed under an older genus, may with justification reach the higher status of genera. The status of this work is so that it will stand smaller changes without succumbing altogether."

Then the introduction continues with a description of how the manuscript has been edited (Forsskål 1775b, introduction, p. *21–22*), that the number of species in the 'Centuriae' were made to fit the Linnaean classes, and how the new names for genera were coined. Parts of this has been quoted in the relevant chapters of this book. There are also discussions of Forsskål's filing methods (p. *22–25*), especially his work with the collection of vernacular names and information about uses. Sections of this, presumably with considerable input from Niebuhr, have been quoted elsewhere.

Medicinal Uses of Plants; Health in Yemen

A section on medicinal plants compares the practice of Egypt with that of Yemen and describes the general health of the country, surprisingly enough without any mentioning of the 'cold' that was so fatal for the expedition (Forsskål 1775b, introduction, p. *25–26*):

"The information by Alpinus about the use of plants in [Egyptian] medicine is not to be undervalued; and other travellers have gathered more; Forsskål has added his 'Materia Medica Kahirina' to the earlier information. Arabia Felix offers a very different and more original kind of medicine, based on wild plants. People do not know about the wisdom of the science of medicine, and are satisfied with household remedies. But that does not mean that the countrified pharmacology [of Yemen] is quite arbitrary and its *formulae* altogether simple; on the contrary, these are often courageous and less suited for a weak stomach. This experience is not concealed as a secret, no matter how small it might be, but is given to the inquisitive traveller without suspicion. There are also some doctors who cultivate the art of chemistry and, less fortunately, also alchemy; but the people, who have not spoilt their health by a way of life like ours, are happily ignorant of many of our diseases. Wounds of all kinds are a serious nuisance. The legs of the inhabitants of Mora [?] are deformed by abscesses. Piles and abscesses are common afflictions. It is common to hear complaints about an affliction of the stomach due to badly baked barley and poor digestion; to these should be added inactive life and superfluous polygamy which ultimately leads to frustration. Inhabitants of closed harems often complain bitterly over diseases introduced from free [?] harems. The inhabitants of the mountains, who breathe a fresh and cool air, are pleased by the privilege of obvious good health. Throughout the Orient vegetarian food is preferred to meat. That this way of life with a perpetual vegetarian diet should be harmful for the body, as has been claimed by a school of medicine, is utterly refuted by the example of the Banyan people, because they reject all food of animal origin, except honey and milk, and yet they are healthy and fit."

A Botanical Garden in Alexandria; Conclusion

The plans for a botanical garden at Alexandria (Forsskål 1775b, introduction, p. *27–28*) are outlined in some detail. Such a garden could be made safe, would be near a harbour used by Europeans, and would allow the botanists to study the plants cultivated there at all seasons. It would be an excellent centre of exchange for useful plants, likely in the long run to benefit both Oriental and European nations. Plants could through such a garden be exchanged between Yemen, India, and Africa. But unfortunately such a botanic garden will most likely only remain an idea or a vision.

The remaining sections (Forsskål 1775b, introduction, p. *28–32*) contain a eulogy on Forsskål, an acclaim of the help accorded the expedition by Emir Farhan of Lohaja (quoted in the first chapter of this book), and a celebration of the deceased King Frederik V and his ministers Bernstorff and Moltke. Finally, after some further information on editorial policy, the introduction concludes:

"We have seen the Flora of Arabia rooting, and sprouting branches and leaves. Such activity has the few months [which were spent in Yemen] caused. May more joyful suns encourage it to open its flowers. As long as it lives, as long as it grows, as long as it produces ripe fruits, it will bear testimony to its first discoverer."

The layout of the 'Flora Aegyptiaco-Arabica', and its names in modern nomenclature

The botanical works of Forsskål appear as publications which largely follow the Linnaean principles. The text does not have the same concise format as the large Linnaean review publications such as the 'Species Plantarum' and 'Systema Naturae,' in which a very concentrated format is used. On the contrary, the detailed lay-out of the treatment of each species in Forsskål's 'Descriptiones' resembles some of the formats used by Linnaeus in his 'Dissertationes.'

Complete and Incomplete Protologues

The descriptive matter in the 'Descriptiones' is normally structured according to a general formula, as described in the following: New genera have, if their protologue is complete, as is for example the case with the genus *Adenia* on p. 77, a centred generic name, the same name provided with a number, and flushing to the left, followed by a generic diagnosis in italics, with the key characters in the nominative case; these short diagnoses are parallel to those given by Linnaeus in for example 'Systema Naturae.' This is followed by a generic description in Roman characters, headed with the words 'Charact. Genericus', and clearly parallel to the descriptions of genera in for example Linnaeus' 'Genera Plantarum.' A new line with the heading 'Species' contains, if the entry is complete, a binary name ('nomina trivialia' in the Linnaean sense), followed by a more detailed specific diagnosis (a phrase name, 'nomina specifica legitima') in the ablative case, and often a specific description, headed 'Descr', to which information on habitat, vernacular name, etc., has most often been added.

This formula may be truncated in various ways, as almost all the above mentioned elements can have been left out, presumably due to lack of notes by Forsskål, or perhaps due to carelessness of the editor. It is, however, nearly always possible to recognize which are the elements included, and which are left out. Often the first entry in a new genus consists of a numbered generic name, followed by a generic diagnosis and a specific description, but without a binary specific name, while the following species entries mostly consist of binary species names, specific diagnoses, and specific descriptions.

Very frequently the lacking first binary specific name at the description in the 'Descriptiones' can be found at the corresponding entry in one of the 'Florae' or 'Florulae'. Names of this type are all indicated in Christensen's 'Key to Flora Aegyptiaco-Arabica' (Christensen 1922, p. 8–9), but a few examples may appropriately be described here:

On p. 63 of the 'Descriptiones' the generic name *Catha* appears without an epithet at the first description, which would appear to be the description of a species, while the name *Catha spinosa* occurs at the second species description. There is no separate diagnosis or description of the genus *Catha*. On p. CVII in the 'Flora' of the Yemen two species names occur, *Catha edulis* and *Catha spinosa*. From the vernacular name of the former and the locality of the latter it is clear which name belongs to which description. However, due to the lacking generic description, the epithet 'edulis' only dates from Vahl's 'Symbolae' (Vahl 1790–94), and the generic name *Catha* is dated from Scopoli and to be cited as *Catha* Forssk. ex Scop. (1777).

A validating generic description is also lacking from the protologue of the new genus *Suaeda* on pp. 69–71, where 6 named and one unnamed species are described, while no generic description is provided. For this reason the genus *Suaeda* is dated from Gmelin, and to be cited as *Suaeda* Forssk. ex J. F. Gmel. (1776). A discussion of this case has been given by Pedrol (1992). The same applies to the new genus *Saelanthus* on pp. 33–35, under which 5 named species are described, but no generic description is provided.

The new genus *Maesa* on p. 66 of the 'Descriptiones', is a monospecific genus, but there are descriptions of both the genus and of the species. There is no specific name on p. 66, but the specific name *Maesa lanceolata* is provided on p. CVI in the 'Flora' of Yemen. Hence both genus and species date from Forsskål's book and are to be cited as *Maesa* Forssk. (1775) and *Maesa lanceolata* Forssk. (1775).

The new genus *Arnebia* on p. 62 is apparently provided with one described species, *Arnebia tinctoria*, and one undescribed species, *Arnebia tetrastigma*, but closer study of the text shows that the word 'tetrastigma' is not an epithet, but is part of the generic diagnosis, and only *Arnebia tinctoria* is intended as a species name. This is also the way it has been interpreted by Farr, Leussink & Stafleu (1979). In the 'Florae' on p. LXII only one species of *Arnebia* is indicated, but it has been provided with the wrong word as epithet, and has become 'Arnebia tetrastigma.'

The case of *Orygia*, on p. 103, is yet again different; the new generic name is provided with a very short generic diagnosis and two named species in the 'Descriptiones', while a third species, *Orygia villosa*, is described briefly in the 'Florae', on p. CXIV.

There is a rather consistently used system of references from the 'Florae' and 'Florulae' to the 'Descriptiones': One asterisk at a name in the lists of the 'Florae' and 'Florulae' means that there is a specific description in the 'Descriptiones'; two asterisks mean that there is a generic description in the same place. Thanks to this, it is usually possible to relate epithets in the 'Florae' and 'Florulae' with descriptions in the 'Descriptiones'. However, not even this system with asterisks has unfailingly been applied in every single case where it would seem appropriate.

These 'asterisked' names, which have no specific epithet at the description, but only elsewhere in the book, were overlooked by a number of 18th Century botanists, for example Lamarck, J. F. Gmelin, and Roemer & Schultes, but most are now well established (Friis 1981 & 1983b).

A few new genera are not provided with any named species at all, for example *Culhamia* on p. 96, and *Simbuleta* on p. 115. They have only been provided with species names by later authors.

Forsskål's new genera can be classified into the following categories:

1) Genus separately designated with diagnosis and description, or with diagnosis alone, either monotypic or with several named and described species.
2) Genus without separate designation, but only with one named species.
3) Genus without separate designation, but with more named and described species, either in the 'Descriptiones' or in the 'Florae' or 'Florulae'.

Category 1) and 2) are valid under the International Code of Botanical Nomenclature, while 3) is not. Forsskål's new genera have been analyzed by Friis (1984).

Distinction of Various Categories of Names

In the 'Florae' and the 'Florulae' it is not always possible to tell the phrase names (*nomina specifica legitima*) apart from the binomial names (*nomina trivialia*) just by the number of words they contain. However, C. Jeffrey of the Royal Botanic Gardens, Kew, has carefully analyzed the names, and found that such a distinction is possible by a more complicated set of principles, based on the typography (Jeffrey 1985; we are grateful to Mr. Jeffrey for permission to repeat the principles here):

For pp. I–XIV non-epithets are not in italics.

For pp. XVIII–XXXVI non-epithets are not in italics, or if in italics they are in the ablative case.

For pp. LI–LVIII the epithets are in Roman letters, or if in italics then they consist of one word only.

For pp. LIX–LXXVIII non-epithets are not in italics, or if in italics, then in the ablative case or vernacular in form. An exception is the list of Cucurbitaceae on pp. LXXV–LXXVI where the epithets are in Roman letters.

For pp. XCI–CXXXI non-epithets are not in italics (not in Roman in Sect. VII), or if in italics (Roman in Sect. VII), then in the ablative case or vernacular in form.

For pp. 1–219 non-epithets are not in small upper case letters.

In all cases 'dubia' and a few other similar words implying doubt, and not intended as epithets, are to be excluded as such.

On this basis, the following distinct categories of taxa can be found in the 'Florae' and 'Florulae':

1) Binominally named species (some of which have epithets consisting of more than one word, to be hyphenated according to the Code).
2) Un-named species without any adjunct to the generic name.
3) Un-named species with an epithet-like vernacular.

4) Un-named species with a polynomial ablative descriptive phrase.
5) Un-named species with 'dubia' or other adjunct signifying doubt.
6) Un-named subordinate taxa (conventionally treated as varieties), with or without a vernacular-adjunct and/or an ablative polynomial adjunct.
7) Trinominally-named subordinate taxa.

There is a last category, consisting of named species with abbreviated epithets, but these refer to already established Linnaean species or species described elsewhere in the 'Flora Aegyptiaco-Arabica'; these have to be spelt out in full when referred to in modern publications.

There is sometimes difficulty concerning the distinction between category 1) and 3). Forsskål, as a good Linnaean, would presumably not have used what were obviously intended as provisional distinguishing vernacular names as epithets — this is made clear by comparison between on one hand the 'Florae' and 'Florulae', and on the other the 'Descriptiones', for in the latter are examples of replacement of the provisional vernaculars by proper epithets. A list of these has been provided by Heine (1968, p. 184).

Apparently Forsskål's editor, when publishing the 'Flora Aegyptiaco-Arabica', also adopted some of his vernacular names as epithets in the 'Descriptiones,' so that in this part of the work, unlike what is mostly the case in the 'Florulae', vernaculars are sometimes used as intended epithets; a selection of these has also been listed by Heine (1968, p. 184).

Validity of Forsskål's Scientific Plant Names

We are, with the 'Flora Aegyptiaco-Arabica,' presented with a nomenclatural *fait accompli*, and must accept the vernacular epithets in the 'Descriptiones' as intended names which were accepted as such by Forsskål. The quotations from the introduction demonstrate his approval of names of Arabic origin. Two specific names with vernacular epithets from the 'Descriptiones' have recently been taken up by Hepper and Wood (1983), viz. *Celtis toka* (Forssk.) Hepper & Wood, and *Debregeasia seneb* (Forssk.) Hepper & Wood. Examples of vernacular epithets in the 'Florae' and 'Florulae' are rare, for example *Cassia aschrek* (name on p. CXI, no. 265, description in 'Centuria III', no. 62).

Heine (1968, p. 184) has also suggested that names in the lists of 'useful plants' in the 'Florulae', on p. LI–LVIII, and on p. XCI–C, which are clearly be regarded as non-accepted by Forsskål, and thus not valid according to the Code, Art. 34.1. A troublesome case, that of *Elcaja roka*, is informative to discuss in this connection: The name of the new genus *Elcaja* is proposed on p. 127 in the 'Descriptiones', and occurs in the 'Florulae' on p. C, and again in the numbered flora lists on p. CXVI, no. 409, in both places without a specific epithet. In the list of useful plants this name occurs on p. XCV with the adjunct 'roka'. This has been accepted by Chiovenda (1923) and later authors as validly published, but rejected by F. White and B. Styles in 'Flora Zambesiaca', vol. 2,1: 299 (1963), in which the authors adopted *Trichilia emetica* Vahl. It seems reasonable to treat this as a case which is parallel with those discussed by Heine, and therefore to treat the adjunct 'roka' from the list of useful plants as not accepted by Forsskål.

In a recent paper Burdet & Perret (1983) have suggested that the 18th Century authors Asso, Grimm, and Forsskål have employed 'non-Linnaean' nomenclature in their work, and are therefore to be rejected under the very restricted wording of the present Art. 23.6(c) of the present Code. However, the main thrust of Burdet & Perret is a criticism of the wording of Art. 23.6(c), rather than an attempt to rule out all Forsskål's Latin plant names, and Burdet & Perret's paper has therefore been discussed in these terms by Greuter (1984), Friis et al. (1984), Friis & Jeffrey (1986), and Friis (1992). It can at present be concluded that all authors aim at a wording of Art. 23 which will ensure retention of Forsskål's well known Latin plant names, including those which have a description in the 'Descriptiones' and the complete species name in the 'Florulae'.

The only remaining point of disagreement seems to be over the status of names given only a very short description in the 'Florae' and 'Florulae'; these names are almost impossible to identify with modern species, unless backed by a specimen collected by Forsskål. It is contended by some that most of these have been neglected for centuries, that they are frequently without any specimen to help in their identification, and that no useful purpose is therefore served by trying to adopt them now. The authors of this book agree that Forsskål's 'Flora Aegyptiaco-Arabica' is an important source of early and widely known names for plants from Egypt and the Yemen, and that it should remain so; on the other hand one should not revive the remaining Forsskål names with short and very inadequate descriptions ('nomina seminuda'), and with no herbarium material to back them. A resolution was passed at the International Botanical Congress in Tokyo 1993 to discourage botanists to take up long unused names which will for purely nomenclatural reasons cause changes of well established names in current use. This resolution is incorporated in the foreword to the Tokyo Code and will apply to most, if not all the names with very short protologue in the Flora Aegyptiaco-Arabica which have not yet been taken up in modern floristic or taxonomic literature.

According to another decision made at the International Congress in Tokyo all names in the Flora Aegyptiaco-Arabica which agree with the requirements outlined in the new Art. 23 of the Tokyo Code are validly published. Invalid species names in the Flora Aegyptiaco-Arabica are only the following: (1) phrase names, even short ones, consisting of a generic name followed by descriptive substantives and adjectives in the ablative; (2) generic names followed by several parallel substantives or adjectives in the nominative; (3) generic names followed by a vernacular name not clearly intended as an epithet by use in either the 'Florae', the 'Descriptiones' or both; and (4) generic names followed by the word 'dubia'.

The Plants Illustrated in the 'Icones Rerum Naturalium'

A number of species described in the 'Flora Aegyptiaco-Arabica' are illustrated on plates in 'Icones rerum naturalium, quas in itinere orientali depingi curavit Petrus Forskal', published in 1776 with Carsten Niebuhr indicated as editor on the title page. Plants are shown on the first 20 plates, all based on Baurenfeind's drawings from the field; the remaining plates show marine invertebrates, fishes, and one bird. As mentioned above, it is almost certain that the botanical text, which consists of short notes to each plate, was written by J. Zoëga. It appears from the Latin introduction that there were either no explanations at all or only rather incomplete ones to the figures amongst Forsskål's notes. The plates have therefore been identified by comparison of the illustrations with the descriptions in 'Flora Aegyptiaco-Arabica'. The plates are reproduced in this book as Figs. 10–29 under the currently recognised plant names.

The legends to the botanical plates are reproduced here, as they are often overlooked by botanists. The references to Linnaeus' 'Systema Naturae' refer to the 'Regnum vegetabile' part of the 12th edition (Linnaeus 1767a):

/Pag. 5./

Explicatio tabularum. [Explanation of the plates].

Name in Flora Aegyptiaco-Arabia	Position	Identification by Editor of 'Icones'	Currently accepted name
Tab. I.			
Salicornia perfoliata.	Pag. 3. n. 4.		**Halopeplis perfoliata** (Forssk.) Bunge

Name in Flora Aegyptiaco-Arabia	Position	Identification by Editor of 'Icones'	Currently accepted name
Tab. II. *Saelanthus quadragonus.*	Pag. 33. n. 11.	*Cissus quadrangularis* Linn., S[yst]. N[at]., Ed. XII, II, pag. 124. n. 6	**Cissus quadrangularis** L.
Tab. III. *Saelanthus digitatus.*	Pag. 35. n. 13.		**Cissus digitatus** (Forssk.) Lam.
Tab. IV. *Saelanthus rotundifolius.*	Pag. 35. n. 14.	*Cissus cordifolia* L., S[yst.] N[at]. II, pag. 124. n. 2.	**Cissus rotundifolius** (Forssk.) Vahl,
Tab. VA. *Hyoscyamus Datora.*	Pag. 45. n. 47.	*Hyoscyamus muticus* L., S[yst.] N[at.], II, pag. 170. n. 6.	**Hyoscyamus muticus** L.
Tab. VB. *Parnassia polynectaria.*	Pag. 207. n. 34.		**Swertia polynectaria** (Forssk.) Gilg
Tab. VI. *Stapelia quadrangula.*	Pag. 52. n. 76.		**Caralluma quadrangula** (Forssk.) N. E. Br.
Tab. VII. *Stapelia subulata.*	Pag. CVIII. n. 193.		**Caralluma subulata** (Forssk.) Decn.
Tab. VIIIA. *Salsola articulata.*	Pag. 55. n. 87.		**Anabasis articulata** (Forssk.) Moq.
Tab. VIIIB. *Salsola inermis.*	Pag. 57. n. 89.		**Salsola inermis** Forssk.
Tab. VIIIC. *Salsola imbricata.*	Pag. 57. n. 90.		**Salsola imbricata** Forssk.
Tab. IV. *Sueda fruticosa?*	Pag. 70. n. 19.	*Salsola altissima* L., S[yst]. N[at]., II: pag. 196. n. 10.	**Suaeda fruticosa** Forssk. is an accepted name; the identity of the species shown on the illustration is not absolutely certain
/Pag. 6/			
Tab. X. *Gymnocarpos decandrum.*	Pag. 65. n. 8.		**Gymnocarpos decandrum** Forssk.

Name in Flora Aegyptiaco-Arabia	Position	Identification by Editor of 'Icones'	Currently accepted name
Tab. XI.			
Zygophyllum desertorum.	Pag. 87. n. 66.	*Zygophyllum coccineum* L.,. S[yst] N[at]., II: pag. 295. n. 2.	**Zygophyllum coccineum** L.
Tab. XIIA.			
Zygophyllum proliferum.	Pag. 87. n. 65.	*Zygophyllum album* L., S[yst]. N[at]., II: pag. 295. n. 3.	**Zygophyllum album** L.
Tab. XIIB.			
Zygophyllum portulacoides.	Pag. 88. n. 67.	*Zygophyllum simplex* L., S[yst]. N[at]., II: pag. 295. n. 9.	**Zygophyllum simplex** L.
Tab. XIII.			
Euphorbia retusa.	Pag. 93. n. 84.	An *Euphorbia serrata* L., S[yst]. N[at]., II: pag. 333. n. 47.	**Euphorbia retusa** Forssk.
Tab. XIV.			
Glinus crystallinus.	Pag. 95. n. 97 [sic, should be 98].	*Aizoon canariense* L., S[yst]. N[at]. II: pag. 347. n. 1.	**Aizoon canariense** L.
Tab. XV.			
Annona glabra.	Pag. 102. n. 16.	An *Annona reticulata* L., S[yst]. N[at]., II: pag. 374. n. 3.	**Annona squamosa** L.
Tab. XVIA.			
Lunaria scabra.	Pag. 117. n. 60.	*Cheiranthus farsetia* L., S[yst]. N[at]., II: pag. 442. n. 17.	**Farsetia aegyptia** Turra
Tab. XVIB.			
Siliquaria glandulosa.	Pag. 78. n. 46.	*Cleome arabica* L., S[yst]. N[at]., II: pag. 448. n. 11.	**Cleome amblyocarpa** Barr. & Urb.]
Tab. XVIIA.			
Zilla Myagroides.	Pag. 121. n. 74.	*Bunias spinosa* L., S[yst]. N[at]., II: pag. 446. n. 5.	**Zilla spinosa** (L.) Prantl
Tab. XVIIB.			
Eraclissa hexagyna.	Pag. 208. n. 35.	*Andrachne Telephioides.* L., S[yst]. N[at]., II: pag. 641. n. 1.	**Andrachne telephioides** L.
/Pag. 7/			
Tab. XVIIIA.			
Zilla.	Pag. 121. n. 75.		**Zilla spinosa** (L.) Parl.
Tab. XVIIIB.			
Suaeda vermiculata.	Pag. 70. n. 18.		**Suaeda vermiculata** Forssk., but identity of plate not absolutely certain

Name in Flora Aegyptiaco-Arabia	Position	Identification by Editor of 'Icones'	Currently accepted name
/Pag. 7/ cont.			
Tab. XIX.			
Senecio hadiensis.	Pag. 149. n. 79.		**Senecio hadiensis** Forssk.
Tab. XX.			
Kosaria.	Pag. 164. n. 34.	Ad genus *Dorsteniae* L. referenda videtur. ['To be referred to the genus *Dorstenia* L.'	**Dorstenia foetida** (Forssk.) Schweinf.

/Pag. 15/

[On p. 15 of the 'Icones ...' is a list of the editor's Identifications of Forsskål's species, as named in the 'Flora Aegyptiaco-Arabica', with the names of the species in Linnaeus' 'Systema Naturae', 12th edition].

Appendix loco Synonyma aliquot Plantarum, ex observatis in horto botanico, heic subjungimus.
[Appendix of synonyms, which are based on observations in the botanical garden.]

[left column:] Nomina Forsk[å]liana in flora Aegyptiaco-Arabica.

[right column:] Nomina Linnaeana in Systematis naturae Editione duodecima.

Name in Flora Aegyptiaco-Arabia	Pag. No.	Identification by Editor of 'Icones'	Currently accepted name
Verbena capitata	Pag. 10.	*Verbena nodiflora.*	**Phyla nodiflora** (L.) Greene
Salvia bifida.	Pag. 202.	*Salvia Forskålii*	**Salvia forskahlei** L.
Pentaglossum linifolium.	Pag. 11.	*Lythrum thymifolia.*	**Lythrum hyssopifolium** L.
Phalaris muricata.	Pag. 202.	*Cenchrus racemosus.*	**Tragus racemosus** (L.) All.
Crucianella herbacea.	Pag. 30.	*Crucianella aegyptiaca.*	**Crucianella herbacea** Forssk.
Plantago decumbens.	Pag. 30.	*Plantago cretica.*	**Plantago ovata** Forssk. var. **decumbens** (Forssk.) Zohary
Scoparia ternata.	Pag. 31.	*Scoparia dulcis.*	**Scoparia dulcis** L.
Pteranthus.	Pag. 36.	*Camphorosma Pteranthus.*	**Pteranthus dichotomus** Forssk.
Pharnaceum occultum.	Pag. 58.	*Mollugo verticillata.*	**Gisekia pharnaceoides** L.
Papularia crystallina.	Pag. 69.	*Trianthema monogyna.*	**Trianthema triquetra** Rottl. ex Willd.

Name in Flora Aegyptiaco-Arabia	Pag. No.	Identification by Editor of 'Icones'	Currently accepted name
/Pag. 15/ cont.			
Racoma.	Pag. 71.	*Trianthema pentandra.*	**Zaleya pentandra** (L.) Jeffrey.
Portulaca linifolia.	Pag. 92.	*Portulaca quadrifida.*	**Portulaca quadrifida** L.
Corchorus aestuans.	Pag. 101.	*Corchorus trilocularis.*	**Corchorus trilocularis** L.
Mnemosilla aegyptiaca.	Pag. 122.	*Hypecoum erectum.*	**Hypecoum aegyptiacum** (Forssk.) Aschers. & Schweinf.
Lotus rosea.	Pag. 140.	*Lotus arabica.*	**Lotus arabicus** L.
Erigeron serratum.	Pag. 148.	*Erigeron aegyptiacum.*	**Conyza aegyptiaca** (L.) Ait.
Polycephalos.	Pag. 154.	*Sphaeranthus indicus.*	**Sphaeranthus suaveolens** (Forssk.) DC.

Forsskål's Botanical Collections

The seeds sent by Forsskål to various botanical gardens in Europe have been mentioned above. They represent the earliest botanical material from the journey to reach Europe. It is not known if culture of any seeds sent to the botanical gardens apart from Copenhagen, Uppsala and Göttingen was attempted, but seeds sent to these three places were grown with reasonable success.

Successfully Grown Seeds

Seeds sent to Copenhagen were grown by Oeder in the Royal Botanic Garden, a predecessor of the present University Botanic Garden. They are recorded in a manuscript with 230 numbers, entitled 'Aegyptiaca Forsskålii', now kept in the archives of the Botanical Library of the University of Copenhagen. Unfortunately, Oeder himself did not publish any of the species, and it is not known if specimens were prepared from them. However, they are probably, at least in part, the basis of the list of identifications by Zoëga on p. 15 in the 'Icones'.

The seeds sent to Linnaeus in Uppsala were perhaps put to even better use. One new genus (*Forsskaolea* Linn., with the new species *Forsskaolea tenacissima* Linn.) was described from these seeds after the news of Forsskål's death had reached Linnaeus (Linnaeus 1764), as a tribute to the collector. About 20 additional species from Forsskål's seeds were described in 'Mantissa Plantarum' (Linnaeus 1767b) and elsewhere.

Herbarium

Sending herbarium material directly to Linnaeus was not permitted under the Royal instruction, and the only herbarium material which reached Linnaeus was therefore the branch of the balsam tree (*Commiphora gileadensis*), which Forsskål believed was the *Opobalsamum* of the Bible. Linnaeus had officially put the question of identity of the *Opobalsamum* to the expedition, and it was therefore acceptable that material of this species was sent to him through Copenhagen. The result of Linnaeus' investigation of Forsskål's specimen and communication was the dissertation 'Opobalsamum declaratum' (Linnaeus 1764). The specimen does not appear to be in the Linnaean Herbarium (LINN) today.

All other herbarium material, including both vascular plants and algae, was sent to Copenhagen. From Egypt Forsskål sent three parcels through Leghorn, Italy. They arrived in good condition (Christensen 1918, p. 58). A parcel sent from Suez was apparently lost, but it is not known if it contained botanical material. From the Yemen Forsskål only managed to send seeds and the branch of *Commiphora* for Linnaeus. All other material was still in the Yemen when he died. Niebuhr sent in July 1764 Forsskål's herbarium with a British ship to the Danish colony of Tranquebar in India, and from there it was carried to Copenhagen by a ship belonging to the Danish East India Company (Christensen 1918, p. 59–60). It probably arrived in Copenhagen some time during 1765, but the date has not been traced.

At their arrival in Copenhagen, the boxes with Forsskål's collections were handed over to Professor P. Ascianus of the Royal Natural History Cabinet, but he did not pay much attention to them, and they were not opened before Niebuhr's return in the autumn of 1767. In 1768, J. Zoëga, then employed at the Royal Botanic Gardens, wrote to Linnaeus that he was uncertain what the official decision about Forsskål's herbarium would be, but in a later letter he was able to inform Linnaeus that he had now gained access to the material and gone through both the herbarium and Forsskål's manuscripts (Schioedte 1870–71, no. 104, 105 & 109).

Early Revisions of the Herbarium

Zoëga did not manage to publish any independent work based on his studies of the dried collections (his work on the 'Icones' is mentioned above), but he sent a description of *Justicia forsskalii* to Linnaeus, who published it in 'Mantissa plantarum altera' (Linnaeus 1771). Wood, Hillcoat & Brummitt (1983) have discussed the rather confusing details of this particular point. In 1770 the Royal Botanic Garden was handed over to the University, and both members of the scientific staff, Zoëga and Oeder, were dismissed as they were not wanted by the university professors (Christensen 1918, p. 62). At the University, the Forsskål herbarium came under the care Professor of Medicine with responsibility for botany, C. Friis Rottböll, who published a number of new species of Cyperaceae based on the Forsskål collections (Rottböll 1772 & 1773).

However, a really thorough and systematic work on the Forsskål collections started only when the botanist Martin Vahl became employed as a lecturer at the Botanical Garden in 1779 and gained access to the collections, unfortunately only after the Flora had been published. His work with the Forsskål Herbarium continued until 1783, when he started on a two years tour of European botanical institutions (Christensen 1924–26, p. 157–159). On this journey Vahl had the opportunity to study North African plants in a number of important herbaria, and was thus able to improve his revision of the Forsskål Herbarium.

The results of these studies were published in the 'Symbolae Botanicae' (Vahl 1790–94), in which about 300 of Forsskål's species are dealt with. Also in his 'Enumeratio Plantarum' (Vahl 1804–05) some Forsskål species, especially some belonging to the Commelinaceae, are revised, and in the manuscripts to the remaining part of the 'Enumeratio ...', which are kept as a valuable key to Vahl's herbarium at the Botanical Museum, there are many more references to Forsskål species.

Vahl is also responsible for the mounting of the Forsskål Herbarium on the comparatively small sheets on which the plants are still mounted today; this size of herbarium paper is also known from the Linnaean herbaria and from Vahl's own herbarium. Most of the identifications written on the back of these sheets are by Vahl, a few are by Rottböll and others. Newer identifications are mostly on determination slips on the front of the sheets. As an aid to identification of the handwriting on the sheets one can confer with the samples reproduced in this book in Fig. 8. Other reproductions of relevant examples of handwriting can be found in Christensen (1924–26); Vahl's is found on p. 169, Fig. 46, and Rottböll's on p. 151, Fig. 40.

Vahl has been criticized by posterity for having in some cases given new and, in his opinion, more appropriate names to Forsskål's species. These names are, even when more appropriate than Forsskål's own names, according to the modern International Code of Botanical Nomenclature superfluous and illegitimate. One can hardly reproach Vahl for not following rules which were laid down about a hundred years later, but perhaps it should be pointed out that he has changed more names than he apparently intended, at least according to his introduction to 'Symbolae botanicae':

"I have retained Forsskål's specific names if they were not previously used for another species of the same genus, for as a whole it is preferable to retain names even if they are less apt, instead of always renewing them."

The collections were, when studied by Vahl, in quite good condition. Vahl states in the introduction to 'Symbolae botanicae' (Vahl 1790–94):

"... I thought it worthwhile to publish the observations that I made several years ago when studying the riches of Forsskål's plants. Certainly not undamaged by the long transport from the Indies to this city [Copenhagen], they were nevertheless, with the exception of a few of which nothing but the name was left [sic], in such a condition that I am able to determine most of them because Forsskål collected several specimens of every species, and added names to each one of them."

Distribution of Duplicates

Vahl seems to have felt that many species were represented by ample duplicates, and when he became Professor of Botany and keeper of the Botanical Museum he started distributing these duplicates to his Danish and foreign correspondents. Those in Denmark who received Forsskål duplicates were in particular H.F.C. Schumacher (1757–1830), N. Hofman-Bang (1776–1855), and J.W. Hornemann (1770–1841). Later, some specimens from Hofman-Bang's herbarium passed into the herbarium of F.M. Liebmann (1813–56). These specimens have now mostly returned to the Botanical Museum of the University of Copenhagen (C), and are incorporated in the Forsskål Herbarium, with a stamp indicating from which herbaria they have been incorporated ('Herb. Schumacher', 'Herb. Hornemann', 'Herb. Hofman-Bang', etc.) This stamp is mostly on the back, more rarely on the front of the sheet, and does not therefore usually appear in photographs of the sheets.

Those outside Denmark who received Forsskål duplicates from Vahl were particularly A.L. de Jussieu (1748–1836) in Paris, Joseph Banks (1743–1820) in London, and A.J. Retzius (1742–1821) in Lund (Gertz 1945). According to Stafleu & Cowan (1976) there are now duplicates of Forsskål's specimens at B, BM, DS, P-JU, LD (Retzius herbarium), LINN, LIV, S, SBT, and UPS (Thunberg Herb.). Wood, Hillcoat & Brummitt (1983) discussed a specimen thought to be collected by Forsskål, now at Kew, received in the Forster Herbarium from Liverpool Museum 1881, and now referred to *Dicliptera foetida* (Forssk.) Blatter.

Revisions in the 19th Century; Numbering of the Sheets

After Vahl's premature death in 1804 the curation of the Forsskål Herbarium was taken over by his successor as Professor of Botany and keeper of the Botanical Museum, J.W. Hornemann. Unfortunately Hornemann did not continue Vahl's work on foreign plants, including the Forsskål Herbarium, but concentrated his studies on the Scandinavian flora and the flora of Greenland. Nevertheless, in his spell of duty, which lasted until 1841, the herbarium was frequently studied by foreign botanists, for example G. Bentham (1800–1884), who studied the Labiatae in 1830, and large parts of material was sent on

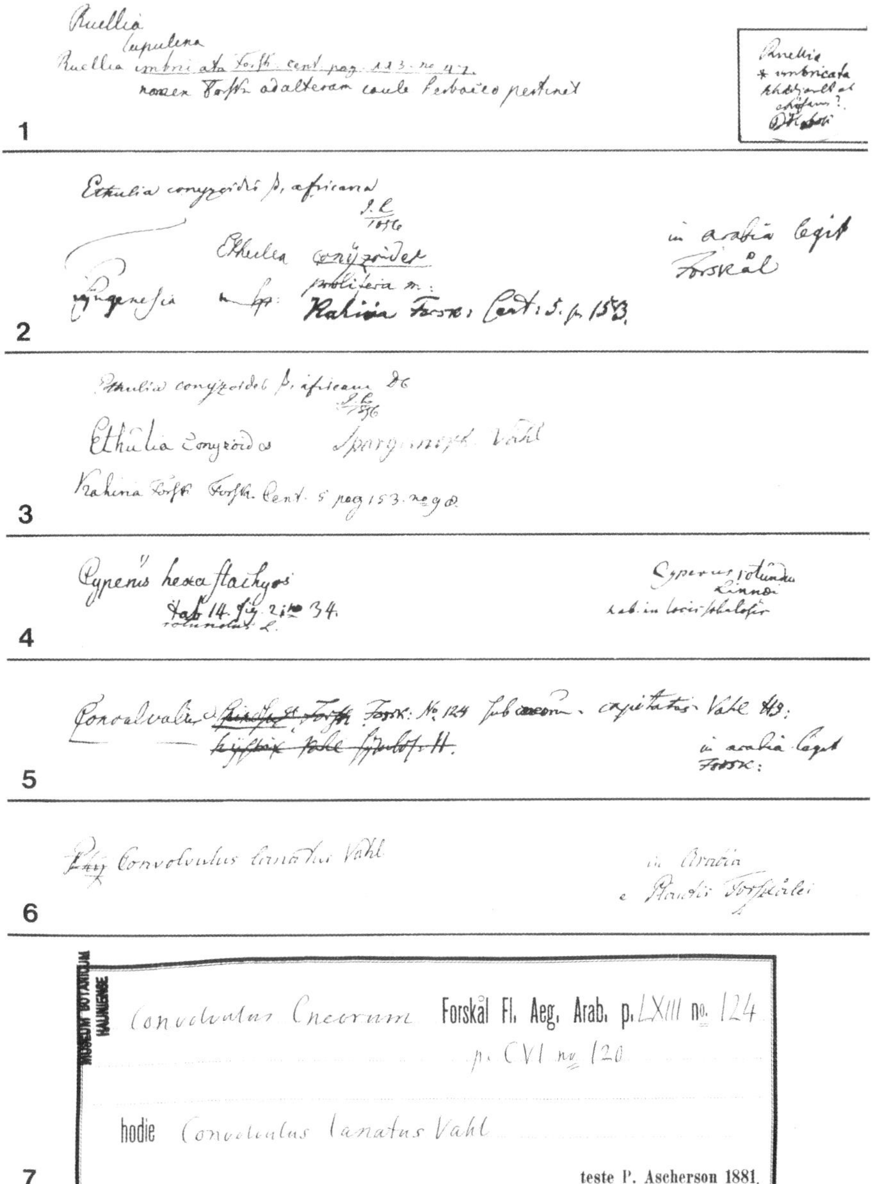

1

2

3

4

5

6

MUSEUM BOTANICUM HAUNIENSE

Forskål Fl. Aeg. Arab. p. LXIII no. 124

p. CVI no 120

hodie Convolvulus lanatus Vahl

teste P. Ascherson 1881.

7

FIG. 8. Original label written by Forsskål during the expedition, and a selection of later handwritings on the sheets in 'Herbarium Forsskålii'. (1) "Ruellia luppulina/Ruellia imbricata ..." by Martin Vahl. To the right a field label by Forsskål with the text: "Ruellia * imbricata ... Bölghose." From sheet no. 368 at Copenhagen. (2) "Ethuilia conyzoides ..." by Johan Lange. The remaining text is written by H.C.F. Schumacher. From sheet no. 1386. (3) First line, "Ethulia conyzoides ...", by Johan Lange. Second line: "Ethulia conyzoides", by Christen Friis Rottböll or Martin Vahl, "Sparganorph. Vahl" by J.W. Hornemann. Third line, "Kahira ...", by Martin Vahl. From sheet no. 1384. (4) "Cyperus hexastachyos ..." by Christen Friis Rottbøll. Below: "rotundus L." by Johan Lange. To the right: "Cyperus rotundus Linnaei ..." probably by H.C.F. Schumacher. From sheet no. 1200. (5) The entire text by H.C.F. Schumacher. From sheet no. 1137. (6) The entire text by J.W. Hornemann. From sheet no. 1135. (7) One of the large labels on which P. Ascherson has annotated the Egyptian collections. From sheet no. 456.

loan outside Denmark. Some of these loans were apparently lost (Christensen 1918, p. 84; 1922, p. 4). The ferns and fern allies (Pteridophytes) were in 1819 sent to G.E. Kaulfuss (1786–1830) in Halle, who included most of Forsskål's taxa in his 'Enumeratio Filicum' (Kaulfuss 1824). According to letters from Kaulfuss to Hornemann in the archives of the Botanical Library, Copenhagen, it seems that the specimens were returned in 1826, but that the parcel did not arrive in Copenhagen. This agrees with the fact that the 'Herbarium Forsskålii' is now rather poor in ferns.

After becoming Professor of Botany and keeper of the Botanical Museum in 1841 J.F. Schouw (1789–1852) reorganized the entire collections of the Botanical Museum (Lange 1875, p. 51; Christensen 1924–26), and many historical herbaria were incorporated in the general herbarium of flowering plants. According to J.M.C. Lange (director of the Copenhagen Botanical Garden; 1818–1898): "Before the merging of the older individual herbaria into a general herbarium, these were distributed on the following separate, larger and smaller collections: (a) a small collection of Forsskål's plants from Egypt and Arabia; (b) Rottböll's and (c) Viborg's herbaria ...; (d) Schousboe's Herbarium from Spain and NW Africa ...; (e) Martin Vahl's large herbarium ...; (f) Schumacher's herbarium ... (g) Isert's herbarium from Guinea; and (h) Hornemann's herbarium ..." (Lange 1875). However, this merge was not carried through with one set of specimens from Forsskål's collections. As pointed out by Christensen (1918, p. 84), this small herbarium continued to be kept separate from the general herbarium as a 'Herbarium Forsskålii', which included both vascular plants and algae.

During the periods of duty of the subsequent curators the 'Herbarium Forsskålii' was neglected or almost forgotten, but in 1880 Warming (1841–1924) again took note of it and sent it on loan to P. Ascherson (1834–1913) in Berlin. Asherson's very large determination slips can be seen on many of the Egyptian sheets in the 'Herbarium Forsskålii.' At some time during the last century, perhaps during Schouw's work with the collections, or perhaps in connection with the new interest for the 'Herbarium Forsskålii' under Warming, the old sheets were stamped with a small, square stamp marked 'Herb. Forskåli No.' and numbered from 1 to c. 880. As pointed out by Christensen (1918, p. 85), a small number of these specimens can be demonstrated not to have been collected by Forsskål.

The German botanist G. Schweinfurth (1836–1925) became interested in the Forsskål collections due to his own travels in Eritrea and the Sudan. In 1889 he equipped an expedition and travelled for 80 days in the Yemen. During that time he collected 930 species, which were unfortunately only partly published (Schweinfurth 1889, 1891, 1894–1899 & 1912), but Schweinfurth's identifications of Forsskål names became widely known through the many duplicates from the expedition which he distributed under the heading 'In memoriam Divi Forskalii'. Some of Ascherson & Schweinfurth's results from the previously mentioned loan of the herbarium to Berlin were included in their two joint works on the flora of Egypt and Ethiopia (Schweinfurth & Ascherson 1867; Ascherson & Schweinfurth 1887–89).

Modern Development

In 1915 Christensen, then keeper of the Botanical Museum of Copenhagen, started a registration and rearrangement of the 'Herbarium Forsskålii'. During this work, he placed the small sheets in the brown annotated covers in which they are now kept. Christensen's efforts resulted in a book about Forsskål (Christensen 1918), and in an index and a revision of the herbarium (Christensen 1922, p. 40–54 (index) & p. 11–37 (revision)), both of which have been frequently cited here. Christensen also started a search in the general herbarium for additional Forsskål-material, including the duplicates which had been in the herbaria of Schumacher, Hornemann, Hofman-Bang, and others. This work has continued, and even today hitherto unnoticed specimens collected by Forsskål occasionally come to light.

Forsskål's algae have been studied and identified both early in the 19th Century by "Mertens [and] Agardh", and later, at "1881 by G. Zeller" (Christensen 1922, p 30). Some of the algae were stamped and inserted in the numbered series from 1 to 880, others were not, and may have been away on loan while the numbering was done. Identifications of Forsskål's algae appear only to have been published by Christensen in his enumeration of Forsskål's names from the 'Flora Aegyptiaco-Arabica.' Forsskål's names for algae occur in the 'Florula Estaciensis', the 'Flora Constantinopolitana', the 'Flora Aegyptiaca', the 'Flora Arabiae Felicis', and in the 'Centuria VII, species no. 21–60' (Christensen 1922, p. 31–32, 36–37). Forsskål had referred his algae to the genera *Conferva*, *Fucus*, *Spongia* [not now considered a plant genus, but included in 'Centuria VII'], and *Ulva*. Many specimens of algae have been preserved in the 'Herbarium Forsskålii', but they have not been studied since Christensen's accounts, except for some notes by Börgesen (1932). Some of Forsskål's names for algae may have priority over currently accepted names. A number of names for lichens (all referred to the genus *Lichen*) and one species of fungus (*Peziza punctata*) are listed by Christensen (1922, p. 32–37); no material of lichens or fungi has been preserved.

Shortly after the works of Christensen, E. Chiovenda (1871–1941) published a revision of some Forsskål species (Chiovenda 1923). The work has since continued in connection with revisions of Arabian and African genera and species.

'Herbarium Forsskålii' Now

The 'Herbarium Forsskålii' now contains about 1750 sheets, including presumed duplicates of the same collections. However, the number of preserved Forsskål sheets is considerably higher, well over 2000, if the so far incompletely recorded sheets of algae are included. A numerical list of all the sheets recorded so far is included at the end of this volume. The specimens of vascular plants which were not previously numbered in the original sequence of 1–880 have now been given modern numbers in continuation of that sequence; this does not apply to the previously unnumbered algae. The number of collections (counting series of duplicates only once) is difficult to estimate; Fox Maule (1974) has as an estimate simply listed the number of plants enumerated from each part of the journey in 'Flora Aegyptiaco-Arabica:' 256 species from the south of France, 87 from Malta, 481 from Turkey, 576 from Egypt, and 693 from the Yemen; a total of 2093, not taking species observed repeatedly in several parts of the journey into account.

The specimens which were in the 'Herbarium Forsskålii' in the 1960s have been photographed and reproduced in a microfiche edition by the Inter Documentation Company, Zug, Switzerland, as IDC 2200. Some of the specimens in this microfiche edition were not collected by Forsskål, and are now removed from the 'Herbarium Forsskålii.' A list of identifications of the sheets in this microfiche edition is given at the end of this volume; specimens not collected by Forsskål are indicated in that list.

A set of colour photographs (Cibachrome) of the 'Herbarium Forsskålii' was made in connection with the preparations for this book, and is now in the herbarium of the Royal Botanic Gardens, Kew.

Catalogue of Species, Excluding Algae, Fungi and Lichens

Introduction to Catalogue: its Layout and Content

The following catalogue is arranged in the major categories of Dicotyledons, Monocotyledons, Gymnosperms and Ferns and Fern Allies. Algae, lichens and fungi are omitted from the Catalogue, but they appear in the indexes to the numbering system of the 'Herbarium Forsskålii' and the IDC microfiche edition of that herbarium. Within the major categories included, the arrangement is alphabetically by families and by genera and species within the families.

Specific Names and their Application

Names occurring in the 'Descriptiones' of the 'Flora Aegyptiaco-Arabica' are always included in the Catalogue, either as the accepted name for a species or as a synonym. Names only listed in the 'Florae' and 'Florulae' are only included if associated with a description there or elsewhere in Forsskål's book, or if supported by one or more specimens collected by Forsskål. The majority of the names which only occur in the 'Florae' and 'Florulae' are completely without any descriptive matter, and are considered *nomina nuda*. Some of these names are only rendered in an abbreviated form for the expansion of which there is no authorisation in the International Code of Botanical Nomenclature. A few names in the 'Florae' and 'Florulae' are provided with one or two descriptive words only. These *nomina seminuda*, as well as the proper *nomina nuda* only listed in the accounts of useful plants have not been given consideration.

A reference to the place of publication of each species is always provided. References to Forsskål's own works include references to the 'Flora Aegyptiaco-Arabica' and the 'Icones ...' References to the 'Florae' and 'Florulae' are in Roman, and to the 'Descriptiones' in Arabic numerals. At least one other reference to literature is given in order to ensure continuity with former publications, especially with Christensen's index to the 'Flora Aegyptiaco-Arabica' and revision of the 'Herbarium Forsskålii.' Frequently reference is also made to relevant literature published shortly after Forsskål's work, for example Gmelin's edition of 'Systema Naturae' (Gmelin 1788–93), 'Description de l'Egypte' (Delile 1813), 'Plante de l'Arabie Heureuse ...' (Decaisne 1841), and 'Voyage aux Yemen' (Deflers 1889). Reference to various editions of the works of Linnaeus and Linn. filius, especially 'Species Plantarum' and 'Systema Naturae' have of necessity been included (Linnaeus 1753, 1758–59, 1762–63, 1764, 1767a, 1767b & 1771; Linné filius 1781). A reference to a modern regional flora account is also included. Thus French, Maltese and some Turkish species bear a reference to the Flora Europaea. Turkish species generally have a reference to 'Flora of Turkey' (Davis 1965–88). (Some references are also made to Meikle's Flora of Cyprus (Meikle 1977, 1985). Egyptian species have references to 'Students' Flora of Egypt' (Täckholm 1974), 'Manual Flora of Egypt' (Muschler 1912), the incomplete 'Flora of Egypt' (Täckholm & Drar 1941–69), and/or 'Flora Palaestina' (Zohary 1966–72; Feinbrun-Dothan 1978, 1986). Yemeni plants have reference to 'Flora of Aden' (Blatter 1914–16), 'Flora Arabica' (Blatter 1919–36), and/or 'Flora des tropischen Arabien' (Schwartz 1939). These references are made by means of much abbreviated titles. References to other, less frequently quoted publications are given where relevant for current discussions of taxonomy or nomenclature; these references are given in moderately abbreviated form and not included in the list of References.

Synonyms are generally included only when appertaining to the nomenclatural history of Forsskål's new genera and species. Other relevant synonyms may be given after the citation of specimens.

Vernacular Names

Vernacular names have been extracted from the 'Flora Aegyptiaco-Arabica' and from Forsskål's field labels. They are transcribed from Greek or Arabic.

Localities

Localities have been extracted from the 'Flora Aegyptiaco-Arabica' and from Forsskål's field labels. The Turkish place names are transcribed in accordance with the conventions used in the 'Flora of Turkey.' The spelling of the Egyptian place names follows recent practice. The rendering of the Yemeni place names follows those of the Gazetteer in this volume.

Material and Forsskål's Original Labels

To facilitate citation of the Forsskål-material in future taxonomic works, such information as is known has been set out in the customary manner with collector, number, locality, date, herbarium location (using the standard abbreviations of the Index Herbariorum), and the type status where appropriate. When a species is known from the 'Flora Aegyptiaco-Arabica' to have been collected at several localities, it has been attempted to distinguish between the specimens from these localities, though most Forsskål sheets are now without indications of locality. The collecting localities can only be identified with absolute certainty when the specimen is still provided with one of Forsskål's original field labels (strongly abbreviated on collections from Egypt, and more detailed on collections from Arabia).

Notes from Forsskål's field labels are quoted where relevant. One asterisk on the field label seems often to identify species which Forsskål considered new. Two asterisks indicated that he considered the specimen to represent a new genus.

Indication is given of the private herbaria where Forsskål collections have previously been deposited, for example the herbaria of Vahl, Schumacher, Hornemann, Hofman-Bang, Liebmann, Banks and Retzius.

Specimens once erroneously included in the 'Herbarium Forsskålii' and cited or photographed in the IDC microfiche edition as such have also been included, but are clearly indicated in the Catalogue.

Numbering

The specimens which have always been in the 'Herbarium Forsskålii' were numbered at some time during the 19th Century. The system involved individual numbers given to all sheets irrespectively of whether they may have been duplicates of the same collection or not. This numbering system has now been extended to cover all specimens of vascular plants presently in the 'Herbarium Forsskålii' in Copenhagen; due to the difficulties of the task, no attempt has been made to indicate possible duplicate numbers. It would be convenient if future authors constantly were to cite the numbers here given to the Forsskål collections. A numbering of the previously unnumbered algae, and of specimens in herbaria other than the 'Herbarium Forsskålii' has not been attempted.

Types

The current terminology of types laid down in the International Code of Botanical Nomenclature includes many categories of type specimens; holotypes, syntypes, isotypes, etc. Mostly it is not logical to apply this terminology to 18th Century material, and there is the added complication that Forsskål prepared his descriptions either from living plants in the field or at least from the entire set of specimens which were subsequently made into duplicates. We have therefore decided to indicate all potential type specimens for Forsskål's names with the designation 'type of...' followed by the relevant scientific name.

It is often equally difficult to designate a holotype or a syntype in cases where later authors, mainly Vahl, have used Forsskål-material as a basis for their descriptions of new taxa. Although it has been attempted here to indicate cases where material is taxonomically heterogeneous, the selection of lectotypes is generally best left to a future monographer. A few already published lectotypes have been indicated.

Forsskål-Material in Microfiche Editions of Historical Herbaria

The main microfiche edition of Forsskål-material is the IDC 2200 which represents the 'Herbarium Forsskålii' at Copenhagen as it was in the late 1960s. One Forsskål specimen has been included in the microfiche edition of the Vahl herbarium of the Botanical Museum, Copenhagen (IDC 2201). Forsskål-material may also be traced in the microfiche editions of the Thunberg herbarium in Uppsala (IDC 1036), and in the Jussieu herbarium in Paris (IDC 6206).

GAZETTEER TO FORSSKÅL'S GEOGRAPHICAL PLACE-NAMES

See the General Introduction for dates of Forsskål's itinerary.

Forsskål's place names	**Modern equivalents and coordinates** (based on the Times Atlas of the World, 1992)	
	EGYPT	
Alexandria Ah. Alexandriae hortenses As. Alexandriae montaneae Caid Bey (east of Cairo)	Alexandria (El Iskandariya)	13°12'N 29°54'E
Cairum, Cairus Cd. Cairi loca deserta Ch. Cairi plantae hortenses Cs. Cairi spontaneae	Cairo (El Qâhira)	30°03'N 31°15'E
Ghobeiba	? (40 miles from Suez)	
Giza, Gizeh, Guise (pyramids)	El Giza (see Cairo)	
Horeb, Mt (used in the sense of Mt. Sinai)		
Kahirum	Cairo (El Qâhira)	30°03'N 31°15'E
Liblab	Wadi al Lablabah	30°03'N 31°18'E
Ras-ettin, Ras Atir	Ras etin (Alexandria Peninsula)	31°12'N 29°53'E
Rosette Rh. Rosettae hortenses Rs. Rosettae spontaneae	Rashid	31°30'N 30°20'E
Santa Catarina	St. Catherine's Monastery, Sinai (Jabal Katrinah)	28°31'N 33°57'E
Sues	Suez	29°58'N 32°33'E
Tor	El Tûr (Sinai)	28°14'N 33°37'E
	FRANCE	
Estac	Estac Mts (Chaîne de l'Estaque)	43°22'N 5°08'E
Massilia	Marseille	43°18'N 5°24'E
Monspele	Montpellier	43°36'N 3°53'E

GREECE

Rhodum	Rhodes (Ródhos)	31°10'N 28°00'E

MALTA

Melita	Malta	35°54'N 14°31'E

SAUDI ARABIA

Djidda	Jiddah (Jedda)	21°29'N 39°12'E
Janbo	Yanbu al Bahr	24°07'N 38°04'E
Ghomfoda, Gomfoda, Gumfoda	Al Qunfudhah	19°09'N 41°07'E

TURKEY

Baltaliman	Baltaliman (N of Rumelihisari)	41°05'N 29°03'E
Belgrad (Bg.)	Belgrad (see Istanbul)	
Borghas	Burgaz (NE of Dardanelles)	40°25'N 26°30'E
Buiuchtari (Bj.)	Büyükdere (S of Sariyer)	41°10'N 29°03'E
Constantinopolis (Cph.)	Istanbul	41°01'N 28°58'E
Dardanelli (Dd.)	Dardanelles	40°15'N 26°25'E
Eraclissa (Ecl.)	Tekirdag (ruins of Heraclea adjacent to Marmaraereğlisa)	40°59'N 27°31'E
Imros (Imr.)	Gökçeada Island (Imroz)	40°10'N 25°50'E
Maerafte	Mürefte	40°40'N 27°14'E
Marmara, Mare	Sea of Marmara	40°40'N 28°15'E
Natolia et regione Tenedos littus	Coastal regions of Turkey (Anatolia) opposite Bozcaada around Truva (Troy) etc.	
Ortacui	Ortaköy (part of Greater Istanbul)	
Smirnam (Sm.)	Izmir (Smyrna)	38°25'N 27°09'E
Tenedos (Td.)	Bozcaada (Greek Tenedos)	39°49'N 26°03'E
Therapiae	Tarabya (between Yeniköy and Büyükdere)	41°08'N 29°03'E
Tiaerde	Tekirdağ	40°59'N 27°31'E

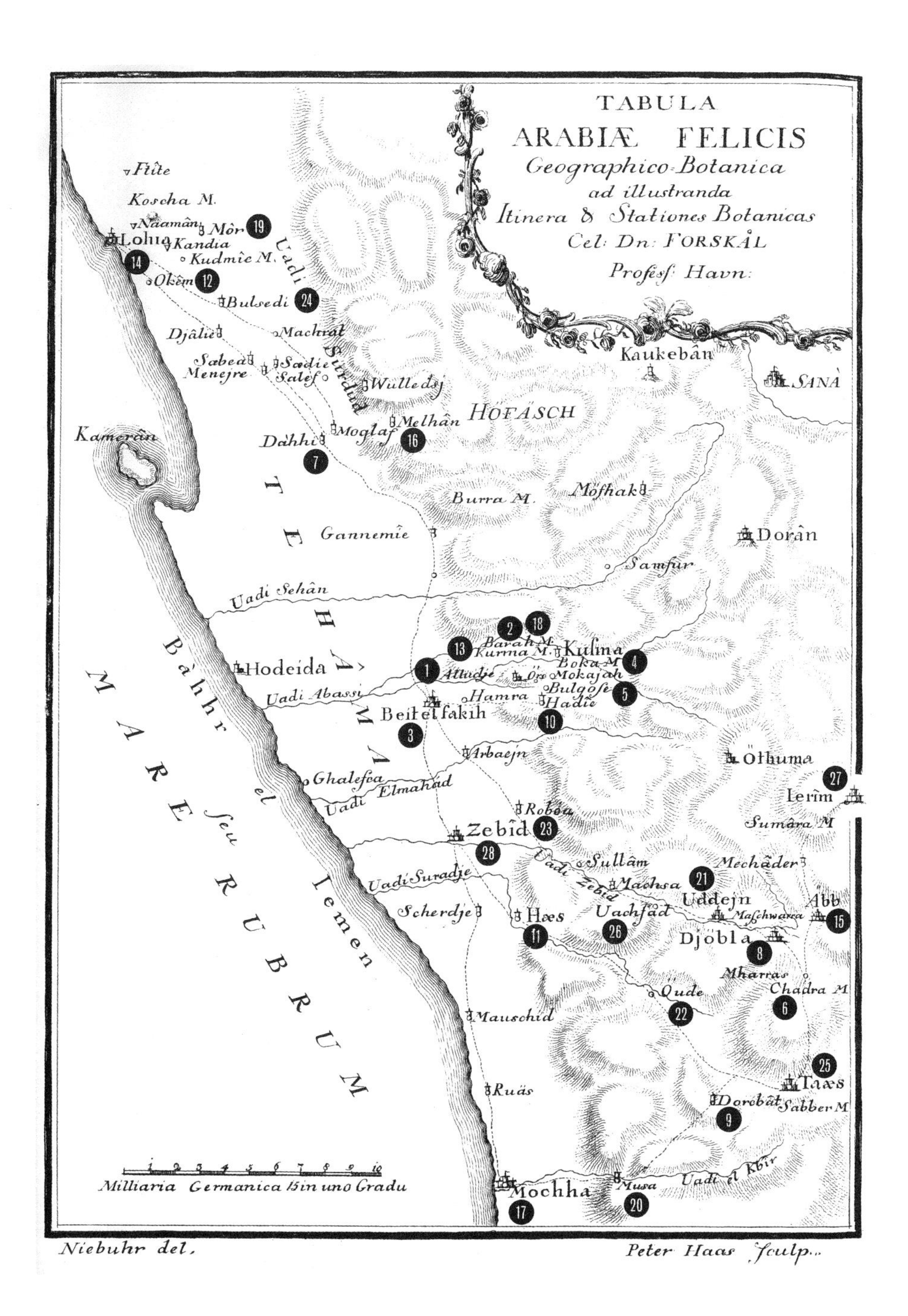

FIG. 9. Map of the Yemen, showing Forsskål's journeys and collecting localities (Forsskål 1775a, frontispiece). The numbers refer to the numbers in the gazetteer.

YEMEN

See reproduction of Niebuhr's maps with Nos. against place-names

Forsskål's place names (with abbreviations)	**J.R.I. Wood's identification and co-ordinates**	
Abb see Ib		
Alludje, Aludje, Aludja (1)	'Alujah	14°35'N 43°33'E
Barah (2)	Barad, Jabal	14°34'N 43°39'E
Beit el Fakih (=Btf.) (3)	Bayt al Faqih	14°31'N 43°19'E
Boka (4)	Bughah	14°34'N 43°37'E
Bolghose, Bulghose (=Blg.) (5)	Bolghose	14°33'N 43°36'E
Chadra (6)	Khadra, Jabal	13°51'N 44°11'E
Dahhi (7)	Dahi, Ad	15°13'N 43°04'E
Djöbla (8)	Jiblah	13°55'N 44°08'E
Dorebat (9)	Modern equivalent not known, near Taizz	
Ersch	? Not known (near 'Alujah), probably the same as 'Örs on Neibuhr's map	
Fatha	? Not known	
Ghorab	? Ghurab Not known (probably a pass)	
Habak	Between ?Mukhajah and Bughah	
Hadie (10)	Al Hadiyah	14°31'N 43°34'E
Haes (11)	Hays	13°55'N 43°29'E
Ib, Äbb (15)	Ibb	13°58'N 44°11'E
Kudmie (12)	Kudmiyah	15°42'N 42°54'E
Kurma, Kusma (13)	Kusmah	14°34'N 43°38'E
Lohaja (=Lhj.) (14)	Al Luhayyah	15°42'N 43°19'E
Melhan (=Mlh.) (16)	Milhan	15°22'N 43°21'E
Meneyre	Munayrah	
Mharras	Modern equivalent not known, near Jabal Khadra ('Chadra')	
Moccha, Mochha (17)	Al Mukha (Mocha)	13°19'N 43°15'E
Mokhaja, Mokajah, Moxhaja, Mt. Barak (18)	? Mukhajah, Jabal Barad near Bolghose	
Mor, Mour (19)	Mawr, Wadi	outflow at 15°41'N 42°43'E

Mochham	? Mukham ? Mukha?	
Musa (20)	between Mocha and Dorebat	
Oddein (21)	Al Udayn	13°58'N 44°00'E
Oudae and Homanae (? 22)	Uday near Taizz	
Öude (22)	Between Taizz and Hays	
Roboa (23)	Suq Ar Rubu' east of Zebid (Zabid)	14°17'N 43°43'E
Saurek	Jabal Sorak near Taizz	
Sordud, Uadi (=Srd.) (24)	Surdud, Wadi (Summarra Pass)	15°10'N 42°52'E
Taaes (25)	Taizz	13°34'N 44°01'E
Uachfad, Uahfad (=Uhf.) (26)	Wasab (as Safal)	14°16'N 43°35'E
Yerim (where Forsskål died) (27)	Yarim	14°18'N 44°23'E
Zebid (28)	Zabid	14°10'N 43°18'E

DICOTYLEDONS

ACANTHACEAE

Acanthus arboreus *Forssk.* 1775: 115 (CXV No. 392; Cent. IV No. 53); Vahl 1790: 47; Christensen 1922: 22; Schwartz 1939: 254.
VERNACULAR NAME. *Senaf* (Arabic).
LOCALITY. Yemen: Al Hadiyah & Kusma ("Hadîe & Kurmae frequens."), Mar. 1763.
MATERIAL. *Forsskål* 379 (C — 1 sheet, type of *A. arboreus*, microf 1: I. 5, 6); 904 (C — 1 sheet ex herb. Vahl, type of *A. arboreus*, microf. 1: I. 7, 8).
Forsskål s.n. (BM — 1 sheet ex herb. Banks, type of *A. arboreus*).

Acanthus hirsutus *Boiss.*, Diagn. 1 (4): 86 (1844); Fl. Turk. 7: 25 (1982).
Acanthus [*sp. without epithet*] Forssk. 1775: XXVIII No. 278.
VERNACULAR NAME. *Zillatros* (Greek).
LOCALITY. Turkey: Dardanelles ("Dardanelli"), July 1761.
MATERIAL. *Forsskål* 906 (C — 1 sheet ex herb. Vahl, with field label "Da. 27", microf. 1: II. 5, 6).

Anisotes trisulcus (*Forssk.*) *Nees* in DC., Prodr. 11: 424 (1847); Deflers 1889: 182; Schwartz 1939: 259.
Dianthera trisulca Forssk. 1775: 7 (CIII No. 28; Cent. I No. 20); Christensen 1922: 11.
Justicia trisulca (Forssk.) Vahl 1791: 10; 1804: 161.
VERNACULAR NAME. *Maddh* (Arabic).
LOCALITY. Yemen: Wadi Sordud ("In *Surdûd* frequens").
MATERIAL. *Forsskål* 1222 (C — 1 sheet ex herb. Vahl, type of *D. trisulca* microf. 38. III. 7, 8); 1223 (C — 1 sheet ex herb. Schumacher, annotated by Schumacher "Nerium obesum Forssk.", microf. 71: I. 1, 2).

Asystasia gangetica (*L.*) *T. Anders.* in Thwaites, Enum. Pl. Zeyl. 235 (1860); Schwartz 1939: 255.
Ruellia intrusa Forssk. 1775: 113 (CXV No. 382; Cent. IV No. 48); Vahl 1790: 45; Christensen 1922: 22.
Asystasia intrusa (Forssk.) Blume, Bijdr. Fl. Ned. Ind.: 796 (1826).
VERNACULAR NAME. *Soudvud* (Arabic).
LOCALITY. Yemen: Bolghose ('Bolghose' on field label), Mar. 1763.
MATERIAL. *Forsskål* 370 (C — 1 sheet, type of *R. intrusa*, microf. 38: I. 7,8); *Forsskål* 371 (C — 1 sheet with field label "Ruella * intrusa. Bolghose", type of *R. intrusa*, microf. 88: III. 1, 2).
Forsskål s.n. (BM — 1 sheet ex herb. Banks, with field label "Soudvud, Bolghose", type of *R. intrusa*).
Forsskål s.n. (LD — 1 sheet ex herb. Retzius, type of *R. intrusa*).

Asystasia guttata (*Forssk.*) *Brummitt* in Kew Bull. 32: 452 (1978).
Ruellia guttata Forssk. 1775: 114 (CXV No. 385; Cent. IV No. 50); Vahl 1791: 72; Deflers 1889: 181; Schwartz 1939: 250.
VERNACULAR NAME. *Kasr, Ghobeire, Ghobîre* (Arabic).
LOCALITY. Yemen: Jabal Khadra ("in monte Chadra"), Mar. 1763.
MATERIAL. *Forsskål* 369 (C — 1 sheet, holotype of *R. guttata*, microf. 88: II. 3, 4).
Forsskål s.n. (LD — 1 sheet ex herb. Retzius, type of *R. guttata*).

Barleria bispinosa (*Forssk.*) *Vahl* 1790: 46.
Justicia bispinosa Forssk. 1775: 6 (CII No. 14; Cent. I No. 16); Christensen 1922: 10.
VERNACULAR NAME. *Schechadd, Schechadh, Kulibe* (Arabic)
LOCALITY. Yemen: Al Hadiyah ("Montium regio inferio"), 1763.
MATERIAL. *Forsskål* 374 (C — 1 sheet, lectotype, microf. 59: I. 5, 6); 1379 (C — 1 sheet, microf. 60: I. 7, 8); 372 (C — 1 sheet, microf. 59: II. 3, 4); 1378 (C — 1 sheet ex herb. Schumacher, microf. 59: II. 5, 6); 1380 (C — 1 sheet, microf. 60: II. 1, 2).
Barleria spinicyma Nees in DC., Prodr. 11: 240 (1847); Deflers 1889: 182; Schwartz 1939: 251.
NOTE. The locality "Hadiâ" is on Forsskål's own label but not in his book.

Barleria lanceata (*Forssk.*) *C. Christensen* 1922: 10; Schwartz 1939: 251; Wood, Hillcoat & Brummitt in Kew Bull. 38: 434 (1983).
Justicia lanceata Forssk. 1775: 6 (CII No. 12; Cent. I No. 18); Christensen 1922: 10.
VERNACULAR NAMES. *Sokaejt* (Arabic).
LOCALITY. Yemen: Wadi Sordud ("Surdûd"), 1763.
MATERIAL. *Forsskål* 377 (C — 1 sheet with field label "Justicia stipulis folioceis spinosis. Vadi Sebid.", lectotype of *J. lanceata*, microf. 59: III. 5, 6); 1832 (C — 1 sheet, type of *J. lanceata*, microf. 59: III. 7, 8).
Forsskål s.n. (BM — 1 sheet ex herb. Banks, type of *J. lanceata*).

Barleria noctiflora sensu Vahl 1790: 46; Schwartz 1939: 251, non L.f.
B. acanthoides Vahl 1790: 147; Deflers 1889: 182.

Barleria prioritis *L.* ssp. **appressa** (*Forssk.*) *Brummitt & J.R.I. Wood* in Kew Bull. 38: 437 (1983).
Justicia appressa Forssk. 1775: 6 (CII No. 15; Cent. I No. 17); Christensen 1922: 10.
Barleria appressa (Forssk.) Deflers 1894: 46.
VERNACULAR NAME. *Schechadh* (Arabic).
LOCALITY. Yemen: Hays ("In montibus ad Haes frequens"), 1763.
MATERIAL. *Forsskål* 378 (C — 1 sheet, lectotype of *J. appressa*, microf. 59: I. 1, 2); 375 (C —1 sheet, type of *J. appressa*, microf. 59: I. 7, 8); 1376 (C —1 sheet, type of *J. appressa*, microf. 59: I. 3, 4); 1377 (C — 1 sheet, type of *J. appressa*, microf. 59: II. 7, 8).
Forsskål s.n. (BM — 1 sheet ex herb. Banks, type of *J. appressa*).

Barleria proxima *Lindau* in Ann. Ist. Bot. Roma 6: 72 (1895).
LOCALITY. Yemen: unspecified, 1763.
MATERIAL. *Forsskål* 1831 (C — 1 sheet ex herb. Schumacher, microf. 116: II. 3, 4).
NOTE. This sheet was kept among the indeterminates at the end of the herbarium. It has been named by K. Vollesen.

Barleria trispinosa (*Forssk.*) *Vahl* 1790: 46; Deflers 1889: 182; Schwartz 1939: 251.
Justicia trispinosa Forssk. 1775: 6 (CII No. 13; Cent. I No. 15); Christensen 1922: 10.
VERNACULAR NAMES. *Schechar, Uzal, Kullibae, Vusar* (Arabic).
LOCALITY. Yemen: Al Hadiyah & Milhan ("Hadie & in sylvis montis Melhân"), 1763.
MATERIAL. *Forsskål* 376 (C — 1 sheet, lectotype of *J. trispinosa*, microf. 59: II. 1, 2).

Blepharis ciliaris (*L.*) *B.L. Burtt* in Notes Roy. Bot. Gard. Edinb. 22: 94 (1956).
B. edulis (Forssk.) Pers., Syn. 2: 180 (1806); Schwartz 1939: 253.
Acanthus edulis Forssk. 1775: 114 (CXV No. 391; Cent. IV No. 52); Vahl 1790: 48; Deflers 1889: 181; Christensen 1922: 22.
VERNACULAR NAME. *Zogaf, Sogaf* (Arabic).

LOCALITY. Yemen: Al Luhayyah ("Lohajae"), Jan. 1763.
MATERIAL. *Forsskål* 905 (C — 1 sheet ex herb. Vahl, type of *A. edulis*, microf. 1: II. 1, 2).

Blepharis maderaspatensis (*L.*) *Hayne ex Roth*, Nov. Sp. Pl. Ind. Or.: 320 (1821); Deflers 1896: 327.
Acanthus maderasp.? — Forssk. 1775: CXV No. 390.
VERNACULAR NAME. *Saebak* (Arabic).
LOCALITY. Yemen: Wadi Surdud ("Wadi Surdûd"), Mar. 1763.
MATERIAL. *Forsskål* 907 (C — 1 sheet ex herb. Vahl, microf. 1: II. 3, 4).

Crossandra johanninae *Fiori* in Bull. Soc. Bot. Ital. 1911: 61 (1911); Vollesen in Kew Bull. 45: 122 (1990).
C. infundibuliformis sensu Deflers 1889: 182; Schwartz 1939: 254.
LOCALITY. Yemen, without locality.
MATERIAL. *Forsskal* 1637 (C — 1 sheet ex herb. Schumacher, microf. 130: III. 1, 2).

Dicliptera foetida (*Forssk.*) *Blatter* 1921: 361; Wood, Hillcoat & Brummitt in Kew Bull. 38: 453 (1983).
Justicia foetida Forssk. 1775: 5 (CII No. 19; Cent. I No. 12) Christensen 1922: 10.
J. bivalvis sensu Vahl 1790: 2 & 1804: 149, non L.
Dicliptera bivalvis sensu Nees in DC., Prodr. 11: 433 (1847); Schwartz 1939: 255; non *J. bivalvis* L.
VERNACULAR NAME. *Tuna* (Arabic).
LOCALITY. Yemen: Al Hadiyah ("In montis Hadiensi")
MATERIAL. *Forsskal* 382 (C — 1 sheet, lectotype of *J. foetida*, microf. 59: III. 3, 4).
Forsskal s.n. (K — 1 sheet ex herb Forster and LIV, isolectotype of *J. foetida*).
Forsskal s.n. (BM — 1 sheet ex herb. Banks, isolectotype of *J. foetida*).

Dicliptera verticillata (*Forssk.*) *C. Christensen* 1922: 11; Wood, Hillcoat & Brummitt in Kew Bull. 38: 450 (1983).
Dianthera verticllata Forssk. 1775: 9 (CIII No. 27; Cent. I No. 22).
Justicia cuspidata Vahl 1791: 9, 16 & 1804: 149.
LOCALITY. Yemen: Wasab ("Uahfad"), Mar. 1763.
MATERIAL. *Forsskål* 393 (C — 1 sheet with field label "Dianthera * verticillata. Uahfat", lectotype of *D. verticillata* & *J. cuspidata*, microf. 39: I. 3, 4).

Dianthera micranthes Nees in Wall., Pl. As. Rar. 3: 112 (1832); Schwartz 1939: 255.
LOCALITY. Yemen: Wasab ("Uahfad"), Mar. 1763.
MATERIAL. *Forsskål* 392 (C — 1 sheet, lectotype of *D. micanthes*, microf. 39: I. 1, 2).
Justicia sexangularis sensu Forssk. 1775: 5 (CII No. 17; Cent. I No. 13), non L.; Christensen 1922: 10.
J. chinensis sensu Vahl 1790: 4 & 1804: 110, non L.
Dicliptera chinensis sensu Christensen 1922: 10, non (L.) Juss.; Schwartz 1939: 255.

D. sexangularis sensu Blatter 1931: 361, non (L.) Juss.
VERNACULAR NAME. *Sövudvud* (Arabic).
LOCALITY. Yemen: Al Hadiyah ("Hadîe"), 1763.
MATERIAL. *Forsskål* no specimen found.

Ecbolium gymnostachyum (*Nees*) *Milne-Redhead* in Kew Bull. 1941: 175 (1941); Wood, Hillcoat & Brummitt in Kew Bull. 38: 445 (1983).
LOCALITY. Yemen: Bolghose, 1763.

MATERIAL. *Forsskål* 380 (C — 1 sheet, microf. 60: II. 3, 4); 386 (C — 1 sheet, with field label "Justicia * viridis. Tana Bolghose", microf. 60: II. 5, 6).

Ecbolium viride (*Forssk.*) *Alston* in Trimen, Handb. Fl. Ceylon 6: 229 (1931); Wood, Hillcoat & Brummitt in Kew Bull. 38: 446 (1983).
Justicia viridis Forssk. 1775: 5 (CII No. 16; Cent. I No. 14); Christensen 1922: 10.
VERNACULAR NAME. *Kossaejf, Chasser* (Arabic)
LOCALITY. Yemen: Wadi Sordud ("Uadi Surdûd"), Feb. 1763.
MATERIAL. *Forsskål* s.n. (BM — 1 sheet, lectotype of *J. viridis*).
Forsskål s.n. (LD — 1 sheet ex herb. Retzius).
NOTE. Wood, Hillcoat & Brummitt (l.c.) point out that the two specimens No. 380 and 386 annotated "*Justicia viridis*" in Vahl's handwriting do not match Forsskål's description and are referable to *Ecbolium gymnostachyum* (q.v.). One of these, No. 386, is even provided with a field label with Forsskål's new name "*Justicia * viridis*". *J. viride* has long been considered to be a synonym of *Justicia ecbolium*, an Indian species.

Hypoestes forskalei (*Vahl*) *R. Br.* in Salt, Voy. Abyss., App.: 63 (1814); Roem. & Schult., Syst. Veg. 1: 140 (1817); Deflers 1889: 184; Schwartz 1939: 256.
Justicia paniculata Forssk. 1775: 4 (CII No. 20; Cent. I No. 4); Christensen 1922: 10; non Burm. f. (1768).
J. forskalei Vahl 1790: 2 & 1804: 109.
Hypoestes paniculata (Forssk.) Schweinf. 1912: 144.
VERNACULAR NAME. *Vusar* (Arabic).
LOCALITY. Yemen: Jabal Milhan ("Prope montem Melhân frequens"), 1763.
MATERIAL. *Forsskål* 387 (C — 1 sheet, type of *J. paniculata* & *J. forskalei*, microf. 60: I. 1, 2).
Forsskål s.n. (BM — 1 sheet ex herb. Banks, type of *J. paniculata*).
Forsskål s.n. (LD — ½ sheet ex herb. Retzius, type of *J. paniculata*).

Hypoestes triflora (*Forssk.*) *Roem. & Schult.*, Syst. Veg. 1: 141 (1817); Schwartz 1939: 255.
Justicia triflora Forssk. 1775: 4 (CII No. 18; Cent. I No. 10); Vahl 1790: 3 & 1804: 11; Christensen 1922: 10.
VERNACULAR NAME. *Chodîe* (Arabic).
LOCALITY. Yemen: Al Hadiyah ("In montibus Hadiensibus"), 1763.
MATERIAL. *Forsskål* 1381 (C — 1 sheet ex herb. Vahl, type of *J. triflora*, microf. 60. I: 3, 4); 1382 (C — 1 sheet type of *J. triflora*, microf. 60: I. 5, 6).
Forsskål s.n. (BM — 1 sheet ex herb. Banks, type of *J. triflora*).
Forsskål s.n. (LD — 1 sheet ex herb. Retzius, type of *J. triflora*).

Isoglossa punctata (*Vahl*) *Brummitt & J.R.I. Wood* in Kew Bull. 38: 448 (1983).
Dianthera punctata Vahl 1790: 4.
Justicia punctata (Vahl) Vahl 1791: 15; Schwartz 1939: 258.
Dianthera β) americana fl. albo Forssk. 1775: 9 (CII No. 25; Cent. I No. 25); Christensen 1922: 11.
LOCALITY. Yemen: Al Hadiyah ("Hadîe"), 1763.
MATERIAL. *Forsskål* 389 (C — 1 sheet, lectotype of *D. punctata*, microf. 38: II. 1, 2).
Forsskål s.n. (LD — 1 sheet ex herb. Retzius, type of *D. punctata*).

Justicia caerulea *Forssk.* 1775: 5 (CII No. 21; Cent. I No. 11); Christensen 1922: 10; Schwartz 1939: 258.
VERNACULAR NAME. *Vusar* (Arabic).
LOCALITY. Yemen: Milhan ("ad montem Melhân"), 1763.

MATERIAL. *Forsskål* 383 (C — 1 sheet, type of *J. caerulea*, microf. 138: II. 4, 5).
NOTE. This type specimen was not found by Christensen.

Justicia dubia *Forssk.* 1775: CII No. 23; Christensen 1922: 37.
LOCALITY. Yemen: Milhan (Mlh. = Mons Melhân).
MATERIAL. *Forsskål* no specimen found.
NOTE. It is uncertain whether "dubia" was intended as the formal epithet of a new species which is characterized by the three phrase description: "Spicis terminalibus, bracteis linearibus; foliorum cilis nullis". Christensen added "an Dianthera violacea Vahl?" Its identity remains obscure,and the epithet is invalid under Art. 23 of the Code.

Justicia flava (*Vahl*) *Vahl* 1791: 15 & 1804: 139; Schwartz 1939: 257; Wood, Hillcoat & Brummitt in Kew Bull. 38: 443 (1983).
Adhatoda flava (Vahl) Nees in DC., Prodr. 11: 401 (1847); Deflers 1889: 183.
Dianthera flava Vahl 1790: 5.
[*D. α*) *americana, flava* Forssk. 1775: 9 (CIII No. 24; Cent. I No. 24); Christensen 1922: 11].
LOCALITY. Yemen: Al Hadiyah ("Hadîe", "Mokhaja" on field label), Mar. 1763.
MATERIAL. *Forsskål* 394 (C — 1 sheet with field label, lectotype, microf. 38: I. 3, 4); 395 (C — 1 sheet, type of *D. flava*, microf. 38: I. 5, 6).
Forsskål s.n. (BM — 1 sheet ex herb. Banks, type of *D. flava*).
Forsskål s.n. (LD — 1 sheet ex herb. Retzius, type of *D. flava*).

Dianthera sulcata Vahl 1790: 4; Christensen 1922: 38.
Justicia sulcata (Vahl) Vahl 1791: 13.
Adhatoda sulcata (Vahl) Nees in DC., Prodr. 11: 401 (1847).
LOCALITY. Yemen: Milhan ("Melhan" on label), 1763.
MATERIAL. *Forsskål* 388 (C — 1 sheet, type of *D. sulcata*, microf. 116: II. 5, 6).

Justicia odora (*Forssk.*) *Lam.*, Encycl. Meth. Bot. 1: 629 (1785); Vahl 1791: 11 & 1804: 127; Schwartz 1939: 258.
Dianthera odora Forssk. 1775: 8 (CIII No. 29; Cent. I No. 21); Christensen 1922: 11.
Adhatoda odora (Forssk.) Nees in DC., Prodr. 11: 399; Deflers 1889: 183.
VERNACULAR NAME. *Kejsemân* (Arabic).
LOCALITY. Yemen: Wadi Sordud ("*Surdûd* in sylvis frequens", "Hadie" on field label), Feb. 1763.
MATERIAL. *Forsskål* 384 (C — 1 sheet with field label, type of *D. odora*, microf. 38: II. 7, 8); 1220 (C — 1 sheet, type of *D. odora*, microf. 38: III. 1, 2).
Forsskål s.n. (LD — 2 sheets ex herb. Retzius, types of *D. odora*).

Justicia resupinata *Forssk.* 1775: CII No. 22; Christensen 1922: 37.
VERNACULAR NAME. *Uufar* (Arabic).
LOCALITY. Yemen: Milhan ("Mlh" = Mons Melhân).
MATERIAL. *Forsskål* no specimen found.
NOTE. Unresolved species; name validly published with the description "Corolla alba, parva, resupinata". This name should not be adopted if it upsets widely accepted nomenclature.

Lepidagathis aristata (*Vahl*) *Nees* in Wallich, Pl. As. Rar. 3: 95 (1832).
Ruellia aristata Vahl 1791: 73.
LOCALITY. Yemen, 1763.
MATERIAL. *Forsskål* 372 (C — 1 sheet, type of *R. aristata*, microf. 116: III. 3,4).

Megalochlamys violacea (*Vahl*) *Vollesen* in Kew Bull. 44: 608 (1989).
Ecbolium violaceum (Vahl) Hillcoat & J.R.I. Wood in Kew Bull. 38: 446 (1983).
Dianthera violacea Vahl 1790: 6.
Justicia violacea (Vahl) Vahl 1791: 15; Deflers 1889: 182; Schwartz 1939: 259.
Monechma violaceum (Vahl) Nees in DC., Prodr. 11: 411 (1847).
Justicia 'dubia' Forssk. 1775: CII No. 23 with short description.
LOCALITY. Yemen: Jabal Milhan ("Mlh.").
MATERIAL. *Forsskål* 390 (C — 1 sheet, lectotype of *D. violaceus*, microf. 59: III. 1, 2), 1383 (C — 1 sheet, isolectotype of *D. violaceus*, microf. 38: II. 5, 6).
Forsskål s.n. (BM — 1 sheet ex herb. Banks, type of *D. violacea*).

Monechma debile (*Forssk.*) *Nees* in DC., Prodr. 11: 411 (1847) as 'debilis'; Wood, Hillcoat & Brummitt in Kew Bull. 38: 444 (1983).
Dianthera debilis Forssk. 1775: 9 (CIII No. 30; Cent. I No. 23); Christensen 1922: 11.
Justicia debilis (Forssk.) Vahl 1791: 15; Deflers 1889: 82; Schwartz 1939: 259.
Gendarussa debilis (Forssk.) Nees in Linnaea 16: 302 (1842).
LOCALITY. Yemen: Taizz ("In montibus humilioribus prope Taaes"), 1763.
MATERIAL. *Forsskål* 391 (C — 1 sheet, Wood's lectotype of *D. debilis*, microf. 38: II. 3, 4).
NOTE. The other sheet, No. 1383, hitherto named *Dianthera debilis* is referable to *Ecbolium violaceum* (q.v.).

Peristrophe paniculata (*Forssk.*) *Brummitt* in Kew Bull. 38: 451 (1983).
Dianthera paniculata Forssk. 1775: 7 (CIII No. 26; Cent. I No. 19); Christensen 1922: 10.
Dianthera bicalyculata Retz. in Acta Holm. 1775: 279 (1776); Vahl 1790: 6.
Justicia bicalyculata (Retz.) Vahl 1804: 113.
Peristrophe bicalyculata (Retz.) Nees in DC. Prodr. 11: 496 (1847); Deflers 1889: 183; Schwartz 1939: 255.
VERNACULAR NAME. *Medhaefaa, Toaejm* (Arabic).
LOCALITY. Yemen: Wadi Sordud ("Surdûd"), Jan. 1763.
MATERIAL. *Forsskål* 385 (C — 1 sheet, lectotype of *D. paniculata*, microf. 38: III. 3, 4); 1221 (C — 1 sheet, type of *D. paniculata*, microf. 38: III. 5, 6).

Phaulopsis imbricata (*Forssk.*) *Sweet*, Hort. Brit. ed. 1, 327 (1827); Schwartz 1939: 249.
Ruellia imbricata Forssk. 1775: 113 (CXV No. 381; Cent. IV No. 47); Christensen 1922: 22.
LOCALITY. Yemen: Mukhajah, Bolghose, Mar. 1763.
MATERIAL. *Forsskål* 367 (C — 1 sheet with field label "Ignota 2-dynam. Mokhaja", type of *R. imbricata*, microf. 88: II. 7, 8); 368 (C — 1 sheet with field label 'Bolghose', type of *R. imbricata*, microf. 88: II. 5, 6).
Forsskål s.n. (LD — 1 sheet ex herb. Retzius, type of *R. imbricata*).

Ruellia alba *Forssk.* 1775: CXV No. 386; Christensen 1932: 37.
LOCALITY. Yemen: Alujah ("Aludje"), 1763.
MATERIAL. *Forsskål* no type specimen found.
NOTE. The name is validated by the description "Flore albo, parvo; foliis acutis", but this remains imperfectly known. This name should not be adopted if it upsets widely accepted nomenclature.

Ruellia grandiflora (*Forssk.*) *Blatter* 1919–36: 353.
Camellia grandiflora Forssk. 1775: 126 (CXVII No. 427; Cent. IV No. 99); Christensen 1922: 23.
Ruellia longiflora Vahl 1790: 45, t. XV a new name for the last, *Camellia* being an error for *Ruellia*; Schwartz 1939: 250.

LOCALITY. Yemen: Taizz ("Ad Taaes"), Apr. 1763.
MATERIAL. *Forsskål* 1041 (C — 1 sheet ex herb. Liebman, type of *C. grandiflora*, microf. 20: I. 3, 4); 1042 (C — 1 sheet ex herb. Vahl, type of *C. grandiflora*, microf. 20: I. 5, 6); 1043 (C — 1 sheet ex herb. Schumacher, type of *R. grandiflora*, microf. 20: I. 7, 8). *Forsskål* s.n. (BM — 1 sheet ex herb. Banks, type of *R. longiflora*).
NOTE. The sheet No. 1041 is the plant illustrated in Vahl 1790: t. XV.

Ruellia hispida *Forssk.* 1775: CXV No. 387; Christensen 1922: 37.
VERNACULAR NAME. *Chommab* (Arabic).
LOCALITY. Yemen: Wasab (Uhf. = Uahfad).
MATERIAL. *Forsskål* no specimen found.
NOTE. The name is validated by the description "Caule hispido perpendiculari", but this taxon remains imperfectly known. This name should not be adopted if it upsets widely accepted nomenclature.

Ruellia patula *Jacq.*, Misc. Austr. 2: 358 (1781); Wood, Hillcoat & Brummitt in Kew Bull. 38: 431 (1983).
R. pallida Vahl 1791: 72; Christensen 1922: 22.
R. strepens sensu Forssk. 1775: 114 (CXV No. 383; Cent. IV No. 49), non L.
VERNACULAR NAME. *Kossejf, Kosseif, Mtaktka, Ghobbar* (Arabic).
LOCALITY. Yemen: Wadi Surdud ("Surdûd"), Feb. 1763.
MATERIAL. *Forsskål* 1638 (C — 1 sheet, ex herb. Vahl, lectotype of *R. pallida*, microf. 88: III. 3, 4).

AIZOACEAE

Aizoon canariense *L.* 1753: 488; Tackholm 1974: 73, pl. 13D. Fig. 10.
Glinus chrystallinus Forssk. 1775: 95 (LXVII No. 262; Cent. III No. 98); & Icones t. 14 (1776); Christensen 1922: 20.
VERNACULAR NAME. *Kusjet el bellâd* (Arabic).
LOCALITY. Egypt: Cairo ("Birket el-hadj"), Aug. 1762.
MATERIAL. *Forsskål* 544 (C — 1 sheet, holotype of *G. chrystallinus*, microf. 51: I. 1, 2).

Corbichonia decumbens (*Forssk.*) *Exell* in Journ. Bot. 73: 80 (1935); Jeffrey in F.T.E.A. Aizoaceae: 9 (1961).
Orygia decumbens Forssk. 1775: 103 (CXIV No. 344; Cent. IV No. 20); Christensen 1922: 21; Schweinfurth 1896: 169; Schwartz 1939: 49.
Talinum decumbens (Forssk.) Willd., Sp. Pl. 2: 864 (1800).
LOCALITY. Yemen: Musa ("Musae"), 1763.
MATERIAL. *Forsskål* 541 (C — 1 sheet, type of *O. decumbens*, microf. 75: I. 5, 6). *Forsskål* s.n. (BM — 1 sheet ex herb. Banks, type of *O. decumbens*, microf. nil).
NOTE. The C sheet was written up as *O. portulacifolia* (*Talinum portulacifolium* q.v.) but there is no doubt as to its identity.

Geruma alba *Forssk.* 1775: 62 (CVII No. 161; Cent. III No. 1); Benth. & Hook., Gen. Pl. 1: 387 (1862).
VERNACULAR NAME. *Djerrum* (Arabic).
LOCALITY. Yemen: Al Hadiyah ("Hadîe"), 1763.
MATERIAL. *Forsskål* no type specimen found.
NOTE. In the absence of a type specimen this new genus and species remains unknown. Airy Shaw in Willis, Dict. Fl. Pl. considers it to be possibly a member of the Aizoaceae. This name should not be adopted if it upsets widely accepted nomenclature.

Tab. XIV.

M. Haas. sc.

Glinus crystallinus.

FIG. 10. **Aizoon canariense** L. — Icones, Tab. XIV *Glinus chrystallinus*.

Gisekia pharnaceoides *L.* 1771: 554 & 562; Schwartz 1939: 48; F.T.E.A. Aizoac.: 5, fig. 1 (1961).
Pharmaceum occultum Forssk. 1775: 58 (CIX No. 218; Cent. II No. 95); Christensen 1922: 17.
LOCALITY. Yemen: Al Luhhaya ("Lohajae"), Jan. 1763.
MATERIAL. *Forsskål* 573 (C — 1 sheet, type of *P. occultum*, microf. 79: I. 3, 4).

Glinus lotoides *L.* 1753: 463; Forssk. 1775: 96 (LXVII No. 261; CXIII No. 320; Cent. IV No. 99); Christensen 1922: 20; Schwartz 1938: 49; F.T.E.A.: Aizoac.: 15 (1961); Tackholm 1974: 70, pl. 12 B.
Mollugo glinus Rich. — Deflers 1889: 141.
VERNACULAR NAME. *Ghobbaejre* (Egyptian Arabic), *Haschfe* (Yemen Arabic).
LOCALITY. Egypt: Cairo ("Cairi loca deserta").
MATERIAL. *Forsskål* 543 (C — 1 sheet, microf. 51: I. 3,4);
LOCALITY. Yemen: Wadi Mawr ("Mour" on field label "Matarea ... Melhan", in book), Feb. 1763.
MATERIAL. *Forsskål* 534 (C — 1 sheet with field label "Mo. 27", microf. 51: I. 5, 6).
Forsskål s.n. (BM — 1 sheet ex herb. Banks).

Glinus setiflorus *Forssk.* 1775: 95 (CXIII No. 319; Cent. III No. 97); Vahl 1794: 64; Christensen 1922: 20; F.T.E.A. Aizoac.: 16 (1961).
G. lotoides L. var. *setiflorus* (Forssk.) Fenzl in Ann. Wien. Mus. 1: 358 (1836).
Mollugo setiflora (Forssk.) Chiov. in Bull. Soc. Bot. Ital. 1923: 114 (1923).
LOCALITY. Yemen: Dahhi, between Al Luhayyah & Bayt al Faqih ("Dahhi" or "Dáhi"), Feb. 1763.
MATERIAL. *Forsskål* 535 (C — 1 sheet, type of *G. setiflorus*, microf. 51: II. 1, 2); 540 (C — 1 sheet, type of *G. setiflorus*, microf. 51: I. 7, 8); 542 (C —1 sheet, type of *G. setiflorus*, microf. 51: II. 4, 5); 543 (C — 1 sheet, type of *G. setiflorus*, microf. 51: II. 6, 7); 1300 (C — 1 sheet ex herb. Schumacher, type of *G. setiflorus*, microf. 51: II. 8); 1301 (C — 1 sheet ex herb. Horneman, type of *G. setiflorus*, microf. 51: II. 3).
Forsskål s.n. (BM — 1 sheet ex herb. Banks).

Limeum humile *Forssk.* 1775: 79 (CX No. 249; Cent. III No. 47); Christensen 1922: 18.
Andrachne telephioides sensu Vahl 1791: 99; non L.
LOCALITY. Yemen: Al Luhayyah ("Lohajae"), Jan. 1763.
MATERIAL. *Forsskål* 1408 (C — 1 sheet with field label "Lhaja", type of *L. humile*, microf. 62: III. 3, 4); 1409 (C — 1 sheet, type of *L. humile*, microf. 62: III. 5, 6); 1410 (C — 1 sheet with field label "Lahaja ad (?) Naeemau", type of *L. humile*, microf. 62: III. 7, 8).

Mesembryanthemum forsskålii *Hochst. ex Boiss.*, Fl. Or. 2: 765 (1872); Fl. Palaest. 1: 77 pl. 94 (1966).
M. forskalei Hochst. in Schimper, Pl. Arab. Exs. ed. 2, (1832), nomen; Tackholm 1974: 73 pl. 13C.
Opophytum forskahlei (Hochst. ex Boiss.) N.E. Br. in Gard. Chron. ser. 3, 84: 253 (1928).
M. geniculiflorum sensu Forssk. 1775: 98 (LXVII No. 273; Cent. IV No. 1); Christensen 1922: 20.
VERNACULAR NAME. *Ghasûl* (Arabic).
LOCALITY. Egypt: Cairo ("Káhirae ad pyramides"), 1761–62.
MATERIAL. *Forsskål* no specimen found.

Mesembryanthemum nodiflorum *L.* 1753: 480; Forssk. 1775: 98 (XIII No. 38; LXVII No. 272; Cent. IV No. 2); Christensen 1922: 20; Fl. Europ. 1: 113 (1964); Tackholm 1974: 73, pl. 13B.
VERNACULAR NAME. *Ghasûl*, *Schaechacha* (Arabic).

LOCALITY. Malta ("Malta ad salinas" on field label), June 1761.
MATERIAL. *Forsskål* 631 (C — 1 sheet, microf. 68: I. 3,4);
LOCALITY. Egypt: Cairo, 1762.
MATERIAL. *Forsskål* 629 (C — 1 sheet, microf. 67: III. 7, 8); 630 (C — 1 sheet with field label "Ca. 194 Ghasul", microf. 68: I. 1–2); 632 (C — 1 sheet, microf. 68: I. 5, 6); 1467 (C — 1 sheet ex herb. Hofman Bang, microf. 67: III. 5, 6); 1468 (C —1 sheet, microf. 68: I. 7, 8).
Forsskål s.n. (BM — 1 sheet ex herb. Banks).
Forsskål s.n. (LD — 1 sheet ex herb. Retzius).

Mesembryanthemum sp. indet.
LOCALITY. ? Yemen
MATERIAL. *Forsskål* s.n. (LD — 1 sheet ex herb. Retzius).
NOTE. Labelled "Mesembryanthemum glaucum".

Trianthema triquetra *Rottl. ex Willd.* in Ges. Naturf. Fr. Berlin, Schr. 4: 181 (1803); F.T.E.A. Aizoac.: 25 (1961).
Papularia crystallina Forssk. 1775: 69 (CVIII No. 199; Cent. III No. 14).
Trianthema crystallina (Forssk.) Vahl 1790: 32; Deflers 1889: 140; Schweinfurth 1896: 169; Schwartz 1939: 51.
LOCALITY. Yemen: "ad Dáhi" (see Note).
MATERIAL. *Forsskål* 144 (C — 1 sheet, ? type of *P. crystallina,* microf. 77: I. 1, 2); 536 bis (C — 1 sheet, ? type of *P. crystallina,* microf. 77: I. 5, 6); 539 (C — 1 sheet, ? type of *P. crystallina,* microf. 77: I. 3, 4)
NOTE. Although the Nos. 144 and 539 are in the Forsskål collection there seems to be some doubt as to whether they are his or Rottler's from India, as indicated by Jeffrey in F.T.E.A. Aizoac.: 25 (1961). In that case they are types of *T. triquetra* and not of *P. crystallina.*

Zaleya pentandra (*L.*) *C. Jeffrey* in Kew Bull. 14: 238 (1960); F.T.E.A. Aizoac.: 28 (1961); Tackholm 1974: 75.
Trianthema pentandrum L. 1767: 70; Schwartz 1939: 50.
MATERIAL. Yemen: cult. Uppsala, grown from seed sent by Forsskål to Linnaeus (LINN 572: 273, microf. 284: II. 6, 7).

Rocama prostrata Forssk. 1775: 71 (CVIII No. 200; Christensen 1922: 18 = *Rocama* [*sp. without epithet*] Forssk. 1775: 71 (Cent. III No. 22).
VERNACULAR NAME. *Rókama* (Arabic).
LOCALITY. Yemen: Al Luhayyah ('Lohajae', 'Surdûd, 'Môr'), Jan., Feb. 1763.
MATERIAL. *Forsskål* 538 (C — 1 sheet with field label "Rocama * prostrata Lokeja, Mour", type of *R. prostrata,* microf. 87: III. 3, 4); 1838 (C — 1 sheet ex herb. Schumacher, microf. nil).
NOTE. This is an example of a new genus and species described without an epithet in the 'Centuriae', while an epithet is provided in the 'Florae'.

AMARANTHACEAE

Achyranthes aspera *L.* var. **sicula** *L.* 1753: 204; Deflers 1889: 194; Schweinfurth 1896: 165; Schwartz 1939: 45.
A. aspera ? — Forssk. 1775: 48 (CVII No. 162; Cent. II No. 63); Christensen 1922: 16.
VERNACULAR NAMES. *Uokkes, Uokes, Höllem, Mahôt, Hamsched* (Arabic).
LOCALITY. Yemen: Wadi Mawr ("*Môr*"), Jan. 1763.
MATERIAL. *Forsskål* 910 (C — 1 sheet ex herb. ? Hornemann, microf. 1: III. 3,4); 911 (C — 1 sheet with 2 specimens each with field label, microf. 116: III. 5).
Forsskål s.n. (BM — 1 sheet ex herb. Banks).

Achyranthes capitata *Forssk.* 1775: 48 (CVII No. 166; Cent. II No. 61); Christensen 1922: 16.
VERNACULAR NAME. *Suaed* (Arabic).
LOCALITY. Yemen: Surdud ("Surdûd"), 1763.
MATERIAL. *Forsskål* no type specimen found.
NOTE. This is a validly described species which remains unknown. This name should not be adopted if it upsets widely accepted nomenclature.

Aerva javanica (*Burm. f.*) *Juss.*, Ann. Mus. Paris 2: 131 (1803); Deflers 1889: 194; Schweinfurth 1896: 165; Schwartz 1939: 43; Tackholm 1974: 134 pl. 34A.
A. tomentosa Forssk. 1775: CXXII No. 584; Christensen 1922: 29. = *Aerva* [*sp. without epithet*] Forssk. 1775: 170 (LXXVII No. 538, Cent. VI No. 66).
VERNACULAR NAMES. *Aerua, Râ, Sadjaret ennaghi, Sadjaret ennadje* (Arabic)
LOCALITY. Egypt: Cairo ("In desertis Káhirinis rarior"), 1762.
MATERIAL. *Forsskål* 193 (C — 1 sheet, type of *A. tomentosa*, microf. 3: II. 7, 8); 194 (C —1 sheet, type of *A. tomentosa*, microf. 3: III. 1, 2); 195 (C — 1 sheet, type of *A. tomentosa*, microf. 3: I. 7, 8); 196 (C — 1 sheet, type of *A. tomentosa*, microf. 3: II. 1, 2); 197 (C — 1 sheet, type of *A. tomentosa*, microf. 3: I. 5, 6); 918 (C — 1 sheet ex herb., type of *A. tomentosa*, microf. 3: III. 3, 4).
LOCALITY. Egypt: Suez, Oct. 1762.
MATERIAL. *Forsskål* 916 (C — 1 sheet ex herb. Hofman Bang, type of *A. tomentosa*, microf. 2: I. 7, 8).
LOCALITY. Yemen: Al Luhayyah, Wadi Mawr, Surdud ("Lhj, Môr, Srd", "Yemen in solo arenoso et calcarea frequentissima"), 1763.
MATERIAL. *Forsskål* 915 (C — 1 sheet ex herb. Schumacher, type of *A. tomentosa*, microf. 3: III. 5, 6); 1846 (C — 1 sheet ex herb. Hofman Bang, type of *A. tomentosa*, with field label "La. 14", microf. 3: II. 5, 6).
Forsskål s.n. (BM — 1 sheet ex herb. Banks, type of *A. tomentosa*).
Forsskål s.n. (LD — 1 sheet ex herb. Retzius with label on reverse "Bolghose").
LOCALITY. Uncertain.
MATERIAL. *Forsskål* 917 (C — 1 sheet ex herb. Hornemann, type of *A. tomentosa*, microf. 3: II. 3, 4).
NOTE. This is one of the cases where a new genus and species is described without an epithet in the 'Centuriae', and an epithet is provided in the 'Florae'.

Aerva lanata (*L.*) *Juss.* in Ann. Mus. Paris 2: 131 (1803); Deflers 1889: 194; Schweinfurth 1896: 165; Schwartz 1939: 44.
Achyranthes villosa Forssk. 1775: 48 (CVII No. 169; Cent. II No. 64); Christensen 1922: 16.
VERNACULAR NAME. *Schadjaret el athleb* (Arabic).
LOCALITY. Yemen: Al Hadiyah ("Hadiê"), Mar. 1763.
MATERIAL. *Forsskål* 203 (C — 1 sheet, type of *A. villosa*, microf. 2: II. 1, 2). *Forsskål* s.n. (BM — 1 sheet ex herb. Banks, type of *A. villosa*).

Alternanthera sessilis (*L.*) *DC.* in Cat. Hort. Monspel: 77 (1813); Melville in Kew Bull. 13: 171 (1958); Tackholm 1974: pl. 34B.
Alternanthera achyranth. Forssk. 1775: 28, LIX No. 17; Christensen 1922: 13 =
Alternanthera [sp. without epithet] Forssk. 1775: 28 (Cent. I No. 100).
A. repens J.F. Gmel., Syst. Nat. ed. 13, 2 (1): 106 (1791).
VERNACULAR NAME. *Hámel* (Arabic).
LOCALITY. Egypt: Rashid (*"Rosettae"*), 1762.
MATERIAL. *Forsskål* 200 (C — 1 sheet, type of *A. achyranth* and *A. repens*, microf. 5: I. 1, 2); 201 (C — 1 sheet, type of *A. achyranth* and *A. repens*, microf. 5: I. 3, 4).
NOTE. No. 200 named and labelled by R. Melville, 1958, who notes that R. Brown did not make the combination as indicated by Christensen (l.c.). Forsskål's abbreviated binomial *Alternanthera achyranth.* appears on p. LIX (No. 17) but the description is on p.28 (No. 100). The name *Alternanthera* also appears without description on p. CIII, No. 49 with the vernacular *Kavar el abid* from Yemen Wadi Mawr ('Môr').

Amaranthus aristatus *Forssk.* 1775: LXXV No. 486; Christensen 1922: 36.
LOCALITY. Egypt: Rosetta ("Rh = Rosettae hortenses"), 31 Oct. 1761.
MATERIAL. *Forsskål* no specimen found.
NOTE. The validating description "Parvus, cum aristis" is insufficient to identify this garden plant (? weed). This name should not be adopted if it upsets widely accepted nomenclature.

Celosia polystachia (*Forssk.*) *C.C. Townsend* in Hook. Icon. Pl. 38(2): 23 t. 3728 (1975).
Achyranthes polystachia Forssk. 1775: 48 (CVII No. 164; Cent. II No. 59); Christensen 1922: 15.
Celosia populifolia Moq. in DC. Prodr. 13(2): 239 (1849); Schweinfurth 1896: 163.
VERNACULAR NAME. *Suaed* (Arabic).
LOCALITY. Yemen: Wadi Surdud, ("Surdûd"), 1763.
MATERIAL. *Forsskål* s.n. (BM — 1 sheet, type of *A. polystachia*).
NOTE. Surprisingly, no type specimen has been found in the Copenhagen herbarium and it has been missing since before Christensen worked on the collection.

Celosia trigyna *L.*, Mant. Pl. Alt.: 212 (1771); Schwartz 1939: 40; F.T.E.A. Amaranth.: 12 (1985).
Achyranthes decumbens Forssk. 1775: 47 (CVII No. 168; Cent. II No. 58); Christensen 1922: 15.
VERNACULAR NAME. *Mehat abjad, Mehut abjat* (Arabic).
LOCALITY. Yemen: Wadi Sordud, Al Hadiyah ("Surdûd, Hadîe &c.), Mar. 1763.
MATERIAL. *Forsskål* 202 (C — 1 sheet with field label, type of *A. decumbens*, microf. 1: III. 5,6)
Forsskål (LD — 1 sheet ex herb. Retzius, type of *A. decumbens*).

Celosia trigyna *L.* var. **fasciculiflora** *Fenzl* in DC., Prodr. 13(2): 241 (1849); Schweinfurth 1896: 162; Schwartz 1939: 40.
? *Achyranthes paniculata* Forssk. 1775: 48 (CVII No. 165; Cent. II No. 62); Christensen 1922: 16.
Celosia caudata Vahl 1790: 21.
VERNACULAR NAME. *Suaed* (Arabic).
LOCALITY. Yemen: Milhan ("Melhân"), 1763.
MATERIAL. *Forsskål* 204 (C — 1 sheet with field label, type of *A. paniculata* and *C. caudata*, microf. 2: I. 1, 2); 208 (C — 1 sheet, type of *A. paniculata* and *C. caudata*, microf. 1: III. 7, 8);

912 (C — 1 sheet, type of *A. paniculata* and *C. caudata*, microf. nil, Kew negative 14002). *Forsskål* s.n. (BM — 2 sheets ex herb. Banks, types of *A. paniculata* and *C. caudata*, microf. nil).

Digera muricata (*L.*) *Mart.* in Nov. Act. Acad. Caes. Leop.-Carol., Nat. Curios 13(1): 285 (1826); F.T.E.A. Amaranth.: 36 (1985).
D. alternifolia (L.) Asch. in Schweinfurth, Beitr. Fl. Aeth. 180 (1867); Schweinfurth 1896: 164.
D. arvensis Forssk. 1775: 65 (CVI No. 141; Cent. III No. 7); Deflers 1889: 193; Christensen 1922: 17; Schwartz 1939: 42.
VERNACULAR NAMES. *Didjar, Dyddjer, Budjer, Buddjer* (Arabic).
LOCALITY. Yemen: Wadi Mawr and Wadi Surdud ("Môr, Surdûd"), Feb. 1763.
MATERIAL. *Forsskål* 1228 (C — 1 sheet, type of *D. arvensis*, microf. 39: III. 3, 4); 1229 (C —1 sheet ex herb. Vahl with original field label 'Mour', type of *D. arvensis*, microf. 39: III. 5, 6).
NOTE. On the original field label the new name is spelt "* Diddjera arvensis".

Saltia papposa (*Forssk.*) *Moq.* in DC., Prodr. 13(2): 325 (1849); Deflers 1889: 193; Schweinfurth 1896: 164; Ascherson in Oest. Bot. Zeitschr. 1899: 99–100; Tackholm 1974: 134.
Achyranthes papposa Forssk. 1775: 48 (CVII No. 167; Cent. II. No. 60); Christensen 1922: 15.
Axyris ceratoides sensu Vahl 1790: 76, non L.
VERNACULAR NAME. *Saenáam, Saelaam* (Arabic).
LOCALITY. Yemen: Zabid ("*Zebîd*"), 1763.
MATERIAL. *Forsskål* 205 (C — 1 sheet, type of *A. papposa*, microf. 2; I, 3, 4); 206 (C — 1 sheet, type of *A. papposa*, microf. 2: I. 5, 6).
Forsskål s.n. (BM — 1 sheet ex herb. Banks, type of *A. papposa*).

ANACARDIACEAE

Mangifera indica *L.* 1753: 200; Vahl 1790: 7.
M. amba Forssk. 1775: 205 (CVII No. 170; Cent. VIII No. 16); Christensen 1922: 33.
VERNACULAR NAME. *Amb* (Arabic).
LOCALITY. Yemen: Al Hadiyah ("Hadie illata hortis"), 1763.
MATERIAL. *Forsskål* no type specimen found.

Pistacia terebinthus *L.* 1753: 1025; Forssk. 1775: 219 (XXXV No. 435; Cent. VIII No. 95); Christensen 1922: 35; Fl. Turk. 2: 546 (1967).
VERNACULAR NAME. *Terebinthus*; *Schinos* (Greek).
LOCALITY. Turkey; Dardanelles ("in Natolia" "Dd = Dardanelli, Cph = Constantinople, Sm = Smirma"), July 1761.
MATERIAL. *Forsskål* 746 (C — 1 sheet, microf. 80: II. 5,6);
Forsskål s.n. (LD — 1 sheet ex herb. Retzius).

Rhus cf. **abyssinica** *Hochst. ex Oliv.*, Fl. Trop. Afr. 1: 438 (1868); Schwartz 1939: 148.
LOCALITY. Unspecified, prob. Yemen, 1763.
MATERIAL. *Forsskål* 747 (C — 1 sheet, microf. 80: II. 7, 8).
NOTE. The leaves of this species are almost glabrous, unlike those of *R. abyssinica* which usually have a hairy and glandular undersurface.

Rhus natalensis *Krauss* in Flora 27: 349 (1844); F.T.E.A. Anacard.: 28 (1986).
LOCALITY. Yemen, 1763.
MATERIAL. *Forsskål* s.n. (BM — 1 sheet ex herb. Banks).
NOTE. This fragment is tentatively identified.

ANNONACEAE

Annona squamosa *L.* 1753: 537; Deflers 1889: 108; Schweinfurth 1896: 177; Schwartz 1939: 60; Bircher 1960: 297. Fig. 11.
A. glabra Forssk. 1775: 102 (LXVII No. 286; CXIV No. 347; Cent. IV No. 16); Icones t.15 (1776); Christensen 1922: 20.
VERNACULAR NAME. *Keschta* (Arabic), *s'ferdjel hindi* (in Yemen).
LOCALITY. Egypt: Rashid, Damietta, Giza, Cairo, 1761–62.
MATERIAL. *Forsskål* no type specimen found.

APOCYNACEAE

Adenium obesum (*Forssk.*) *Roem. & Schult.*, Syst. 4: 411 (1819); Deflers 1889: 163; Blatter 1921: 293; Schwartz 1939: 186.
Nerium obesum Forssk. 1775: 205 (CVII No. 173; Cent. VIII No. 17); Vahl 1791: 45; Christensen 1922: 33.
VERNACULAR NAME. *Öddaejn, Öddein, Aden* (Arabic).
LOCALITY. Yemen: Milhan ("Melhân"), 1763.
MATERIAL. *Forsskål* 235 (C — 1 sheet, type of *N. obesum*, microf. 70: III. 7, 8); 236 (C — 1 sheet mixed with *N. salicinum* Vahl, flowers only, microf. 70: III. 3, 4); 237 (C — 1 sheet mixed with *N. salicinum*, flowers only, microf. 70: III. 5, 6); 1833 (C — 1 sheet ex herb. Vahl, type of *N. obesum*, microf. 70: I. 3, 4).
NOTE. Although *Forsskål* 235 is annotated by Plaizier (in 1980) as holotype, we presume this should be considered the lectotype since there are two sheets. The third sheet (microf.71: I. 1,2) in this type cover is in fact *Anisotes trisulcus* (Acanthaceae-q.v.) and is now removed to that cover.

Carissa edulis (*Forssk.*) *Vahl* 1790: 22; Deflers 1889: 163; Schwartz 1939: 185.
Antura edulis Forssk. 1775: CVI No. 137; Christensen 1922: 17. = *Antura* [*sp. without epithet*] Forssk. 1775: 63 (Cent. III No. 3)
VERNACULAR NAME. *Emîr jasir, Anthur, Antur, Arm* (Arabic).
LOCALITY. Yemen: Al Hadiyah ("In montibus *Hadîensibus* frequens"), 1763.
MATERIAL. *Forsskål* 234 (C — 1 sheet, type of *A. edulis*, microf. 8: II. 3, 4).
Forsskål s.n. (BM — 1 sheet Banks, type of *A. edulis*).

NOTE. This is one of the cases where a new genus and species was described in the 'Centuriae', while an epithet is provided in the 'Florae'.

Nerium foliis integris *Forssk.* 1775: 205 (CVIII No. 175; Cent. VIII No. 19); invalid trinomial.
VERNACULAR NAME. *Dharaf* (Arabic).
LOCALITY. Yemen: Kusma ("Kurmae"), 1763.
MATERIAL. *Forsskål* no specimen found.
NOTE. The brief description is insufficient to identify this species in the absence of a specimen.

FIG. 11. **Annona squamosa** L. — Icones, Tab. XV *Annona glabra*.

Apocynaceae gen. & sp. indet. "Tabernaemontana".
LOCALITY. Unspecified, prob. Yemen.
MATERIAL. *Forsskål* 238 (C — 1 sheet microf. 138: II. 6, 7).

ARALIACEAE

Hedera helix *L.* 1753: 202; Forssk. 1775: XIII No. 29; Fl. Europ. 2: 314 (1968).
LOCALITY. Malta, 12 June 1761.
MATERIAL. *Forsskål* 1313 (C — 1 sheet ex herb. Horneman, microf. 52: II. 1, 2).

ARISTOLOCHIACEAE

Aristolochia bracteolata *Lam.*, Encycl. Méth. Bot. 1: 258 (1783).
A. bracteata Retz., Obs. 5: 29 (1789); Deflers 1889: 196; Schweinfurth, 1896: 153; Schwartz 1939: 31.
A. sempervirens sensu Forssk. 1775: 156 (CXX No. 522; Cent. VI No. 4), non L.; Christensen 1922: 27.
VERNACULAR NAME. *Ghaga, Löaeja* (Arabic).
LOCALITY. Yemen: Al Luhayyah ("Lohajae", "Môr"), Jan. 1763.
MATERIAL. *Forsskål* no specimen found.

ASCLEPIADACEAE

Asclepias contorta *Forssk.* 1775: CVIII No. 188; Christensen 1922: 37.
VERNACULAR NAME. *Hommed, Ockas, Dagabis, Rodaa* (Arabic).
LOCALITY. Yemen: Al Hadiyah ("Hadie"), Wadi Sordud (Srd = Sordûd), 1763.
MATERIAL. *Forsskål* no type specimen found.
NOTE. The name is validated by the description "Teres dichotoma, aphylla, lactescens; edulis". However the editor of Flora Aeg.-Arab. gave a warning "An eadum cum 186.i)?" which is *A. aphylla*, here placed under *Sarcostemma* sp. A (q.v.). This name should not be adopted if it upsets widely accepted nomenclature.

Asclepias glabra *Forssk.* 1775: 51 (CVIII No. 185; Cent. II No. 74); Christensen 1922: 16.
Daemia? glabra (Forssk.) Schult., Syst. Veg. 6: 113 (1820); Deflers 1896: 280.
Pergularia glabra (Forssk.) Schwartz 1939: 196.
LOCALITY. Yemen: Taizz ("Taaes"), 1763.
MATERIAL. *Forsskål* no type specimen found.
NOTE. In the absence of a type specimen, Wood (l.c.) considers this to be possibly a glabrous form of *Pergularia daemia*. This name should not be adopted if it upsets widely accepted nomenclature.

Blyttia spiralis (*Forssk.*) *D.V. Field & J.R.I. Wood* in Kew Bull. 38: 219 (1983).
Asclepias spiralis Forssk. 1775: 49 (CVIII No. 179; Cent. II No. 66); Christensen 1922: 16.
Pentatropis spiralis (Forssk.) Decne. in Ann. Sci. Nat., Sér. 2, 9: 327 (1838); Deflers 1889: 164.
VERNACULAR NAME. *Schantob, Schuntob* (Arabic).
LOCALITY. Yemen: Al Luhayyah ("Lohajah", "Ad viam inter *Djaliae* and *Meneira*"), 1763.
MATERIAL. *Forsskål* 975 (C — 1 sheet, type of *A. spiralis*, microf. 138: III. 2, 3).
NOTE. When Field and Wood wrote their paper (q.v.) this type was missing so they designated a neotype (*Wood* 1423 K) which is now superfluous.

Caralluma dentata (*Forssk.*) *Blatter*, Rec. Bot. Serv. India 8: 303 (1921).
Boucerosia dentata (Forssk.) Deflers 1896: 273.
Stapelia dentata Forssk. 1775: CVIII No. 191.
VERNACULAR NAME. *Djadmel, Draat el kelb* (Arabic).
LOCALITY. Yemen: Al Hadiyah ("Hadîe"), Wadi Sordud ("Srd" = Surdûd), 1763.
MATERIAL. *Forsskål* ? 740 not seen.
NOTE. The name is validated by the description: "Exigua, longitudine digiti". Excluded by Wood (1988) who considers it to be probably a *Huernia*. Schwartz (1939: 193) entered it as a synonym of *C. sprengeri* (Damm.) N.E. Br.

Caralluma quadrangula (*Forssk.*) *N.E. Br.* in Gard. Chron. ser. 3, 12: 370 (1892); Gilbert in Bradleya 7: 20 (1989). Fig. 12.
Stapelia quadrangula Forssk. 1775: 52 (CVIII No. 190; Cent II No. 76), & Icones t.6 (1776); Christensen 1922: 16.
Desmidorchis forskalii Decne. in Ann. Sci. Nat. Sér. 2, 9: 265 (1838); Deflers 1889: 169.
Boucerosia forskalii (Decne.) Decne. in DC., Prodr. 8: 664 (1844).
B. quadrangula (Forssk.) Decne. in DC., Prodr. 664 (1844); Deflers 1889: 169.
VERNACULAR NAME. *Gholef, Gholaes, Gholak* (Arabic).
LOCALITY. Yemen: Sordud ("Surdûd"), Feb. 1763.
MATERIAL. *Forsskål* no type specimen found (see Note).
NOTE. Gilbert l.c. designated the plate as the lectotype, having rejected *Forsskål* 274 as the holotype (see *C. subulata*).

Caralluma subulata (*Forssk.*) *Decne.* in Ann. Sci. Nat. Sér. 2, 9: 267 (1838); Gilbert in Bradleya 7: 11 (1989). Fig. 13.
Stapelia subulata Forssk. 1775: CVIII No. 193, & Icones t.7 (1776); Christensen 1922: 37.
LOCALITY. Yemen ("Mm" = Montium regio media).
MATERIAL. *Forsskal* 274 (C — 1 sheet, microf. 103: II. 1, 2) (see Note).
NOTE. Gilbert l.c. designated the plate as the lectotype in preference to *Forsskål* 274 which is in fruit and hitherto has been identified as *C. quadrangula*.

Ceropegia variegata *Decne.* in Ann. Sci. Nat. Sér. 2, 9: 262, t.9A (1838), and in DC., Prodr. 8: 642 (1844); Deflers 1896: 262; Schwartz 1939: 191.
Stapelia variegata Forssk. 1775: 51 (CVIII No. 189; Cent. II No. 75), non L.; Christensen 1922: 16.
VERNACULAR NAME. *Drâet et kélbe* (Arabic).
LOCALITY. Sordud ("Surdûd"), 1763.
MATERIAL. *Forsskål* no type specimen found.

Echidnopsis multangula (*Forssk.*) *Chiov.* in Bull. Soc. Bot. Ital. 1923: 114 (1923).
Stapelia multangula Forssk. 1775: CVIII No. 192; Christensen 1922: 16. = *Stapelia* [*sp. without epithet*] Forssk. 1775: 52 (Cent. II No. 77); Christensen 1922: 16.
VERNACULAR NAME. *Sâk el ghorâb* (Arabic).
LOCALITY. Yemen: Wasab ("Uahfad"), 1763.
MATERIAL. *Forsskål* no type specimen found.
NOTE: Wood says that though probably an *Echidnopsis, E. multangula* ought not to be used as a name since it is impossible to prove its identity in the absence of a type specimen. It was, in any case, sterile.

Gomphocarpus fruticosus (*L.*) *Ait. f.* var. **setosus** (*Forssk.*) *Schwartz* 1939: 188.
Asclepias setosa Forssk. 1775: 51 (CVIII No. 181; Cent. II No. 70); Vahl 1790: 23, pl. 8; Christensen 1922: 16.
Gomphocarpus setosus (Forssk.) R.Br.in Mem. Wern. Soc. 1: 38 (1810); Deflers 1889: 164.

FIG. 12. **Caralluma quadrangula** (Forssk.) N.E.Br. — Icones, Tab. VI *Stapelia quadrangula*.

FIG. 13. **Caralluma subulata** (Forssk.) Decn. — Icones, Tab. VII *Stapelia subulata*.

VERNACULAR NAME. *Sabia, Dhraeba* (Arabic).
LOCALITY. Yemen: Al Hadiyah ("Hadîe"), Mar. 1763.
MATERIAL. *Forsskål* 264 (C — 1 sheet, type of *A. setosa*, microf. 11: III. 7, 8).
NOTE. This specimen is illustrated in Vahl 1790: pl. 8.

Kanahia laniflora (*Forssk.*) *R.Br.* in Mem. Wern. Soc. 1: 40 (1810); Bullock in Kew Bull. 7: 421 (1952).
Ascelpias laniflora Forssk. 1775: 51 (CVIII No. 180; Cent. II No. 72); Vahl 1790: 23 t.7; Christensen 1922: 16.
Kanahia kannah Roem. & Schult., Syst. Veg. 6: 94 (1820).
K. forskalii Decne. in DC. Prodr. 8: 537 (1844); Deflers 1889: 164.
VERNACULAR NAMES. *Kanah, Kanahh* (Arabic).
LOCALITY. Yemen: Jiblah ("Djöbla"), 1763.
MATERIAL. *Forsskål* 265 (C — 1 sheet, type of *A. laniflora*, microf. 11: II. 5, 6); 268 (C — 1 sheet type of *A. laniflora*, microf. 11: II. 7, 8); 973 (C —1 sheet ex herb. Schumacher, type *A. laniflora*, microf. 11: II. 1, 2); 974 (C — 1 sheet ex herb. Liebmann, type of *A. laniflora*, microf. 11: II. 3, 4).
Forsskål s.n. (BM — 1 sheet ex herb. Banks, type of *A. laniflora*).

Leptadenia arborea (*Forssk.*) *Schweinf.* 1912: 167; Schwartz 1939: 191; F.W.T.A. ed. 2, 2: 98 (1963).
Cynanchum arboreum Forssk. 1775: 53 (CVIII No. 177; Cent. II No. 80); Christensen 1922: 16.
VERNACULAR NAME. *Keranna, Kerenna, Kesch, Torah* (Arabic), *Dali* (Indian: Banjani).
LOCALITY. Yemen: Al Luhayyah ("Lohaja" according to label on reverse of the sheet), Jan. 1763.
MATERIAL. *Forsskål* 271 (C — 1 sheet with field label, type of *C. arboreum*, microf. 35: III. 7, 8); 273 (C — 1 sheet with field label "Mo.6", type of *C. arboreum*, microf. 35: III. 5, 6).
Forsskål s.n. (LD — 1 sheet ex herb. Retzius, type of *C. arboreum*).

Leptadenia pyrotechnica (*Forssk.*) *Decne.* in Ann. Sci. Nat. Sér. 2, 9: 270 (1838); Deflers 1889: 166; Schwartz 1939: 191; Tackholm 1974: 416.
Cynanchum pyrotechnicum Forssk. 1775: 53 (CVIII No. 176; Cent. II No. 79); Christensen 1922: 16.
VERNACULAR NAME. *March* (Arabic).
LOCALITY. Yemen: Wadi Mawr ("Môr"), 1763.
MATERIAL. *Forsskål* no specimen found.
NOTE. Although Forsskål says "Frequens ubique", he gives no locality in the text of the 'Descriptiones'; in the Flora Arabiae Felicis he cites 'Môr'.

Odontanthera radians (*Forssk.*) *D.V. Field* in Kew Bull. 37: 342 (1982).
Asclepias radians Forssk. 1775: 49 (CVIII No. 182; Cent. II No. 67); Christensen 1922: 16.
Steinheilia radians (Forssk.) Decne. in Ann. Sci. Nat. sér. 2, 9: 339, t.12E (1838) and in DC., Prodr. 8: 510 (1844); Deflers 1889: 163; Schwartz 1939: 187.
LOCALITY. Yemen: Bayt al Faqih ("In apricis circa Beit el fakíh rarius"), 1763.
MATERIAL. *Forsskål* 275 (C — 1 sheet with obscure field label, type, microf. 11: III 3, 4); 276 (C — 1 sheet with field label "Betelfaki", type of *A. radians*, microf. 11. III 5, 6).

Pentatropis nivalis (*J.F. Gmel.*) *D.V. Field & J.R.I. Wood* in Kew Bull. 38: 215 (1983).
Asclepias nivalis J.F. Gmel., Syst. Nat. ed. 13, 2: 444 (1791).
A. nivea Forssk. 1775: 51 (CVIII No. 183; Cent. II No. 73), non L.; Christensen 1922: 16.
A. forskalei Schult., Syst. Veg. 6: 85 (1820); Deflers 1896: 258; Schwartz 1939: 189.

VERNACULAR NAME. *Ghaschve* (Arabic).
LOCALITY. Yemen: Al Luhayyah ("Lohajae"), 1763.
MATERIAL. *Forsskål* 272 (C — 1 sheet, type of *A. nivea*, also type of *A. forsskåolii* Schult., microf. 11: III 1, 2); 267 (C — 1 sheet, type?, microf. 12: I 1, 2).
NOTE. Owing to confusion of the type specimens this has been known as *Pentatropis spiralis* but the epithet should be applied to the species now placed under *Blyttia*.

Pergularia daemia (*Forssk.*) *Chiov.* Res. Sci. Mus. Stefan-Paoli Somal. Ital. 1: 115 (1916).
Asclepias daemia Forssk. 1775: 51 (Cent. II No. 71); Christensen 1922: 16.
Daemia forskali Schult., Syst. Veg., ed. 15, 6: 113 (1820).
VERNACULAR NAME. *Dhraeba* (Arabic).
LOCALITY. Yemen: Zabid ("Zebïd"), 1763.
MATERIAL. *Forsskål* no type specimen found.
NOTE. See note under *Asclepias glabra*.

Pergularia tomentosa *L.*, Mant. Pl.: 53 (1767); Vahl 1790: 23; Schwartz 1939: 195; Tackholm 1974 pl. 143A.
LOCALITY. 'Arabia'.
MATERIAL. *Forsskål* no type in Herb. LINN (see Note).
Asclepias cordata Forssk. 1775: 49 (LXIII No. 148, CVIII No. 178; Cent. II No. 65); Christensen 1922: 65.
Daemia cordata (Forssk.) R.Br. in Mem. Wern. Soc. 1: 50 (1809); Deflers 1889: 165.
VERNACULAR NAME. *Daemia* (Arabic).
LOCALITY. Egypt: Cairo ("In desertis Káhirinis"), 1762.
MATERIAL. *Forsskål* 269 (C — 1 sheet with field label "Ca. 600" type of *A. cordata*, microf. 11: I. 3, 4); 971 (C — 1 sheet ex herb. Hornemann, type of *A. cordata*, microf. nil).
NOTE. Presumably Linnaeus grew Forsskål's seeds at Uppsala (H.U.), but no specimen is in his herbarium. The material named *P. tomentosa* came from Macao.
LOCALITY. Yemen: Taizz ("Taaes"), 1763.
MATERIAL. *Forsskål* 270 (C — 1 sheet, type of *A. cordata*, microf. 11: I. 5, 6); 972 (C — 1 sheet ex herb. Schumacher, type of *A. cordata*, microf. 11: I. 7, 8).
Forsskål s.n. (BM — 1 sheet ex herb. Banks, type of *A. cordata*).
NOTE. Although 270 is not annotated 'Taaes', this locality is cited on p. CVIII.

Sarcostemma viminale (*L.*) *R.Br.* in Mem. Wern. Soc. 1: 51 (1809); Deflers 1889: 164; Schwartz 1939: 190.
Asclepias stipitacea Forssk. 1775: 50 (CVIII No. 187; Cent. II No. 69); Christensen 1922: 16.
Sarcostemma stipitaceum (Forssk.) R.Br. in Mem. Wern. Soc. 1: 51 (1809); Schult., Syst. Veg. 6: 116 (1820); Deflers 1889: 164; Schwartz 1939: 190.
VERNACULAR NAME. *Rideh* (Arabic).
LOCALITY. Yemen: without location ("Yemen in sylvis", "Ubique"), 1763.
MATERIAL. *Forsskål* 277 (C — 1 sheet, type of *A. stipitacea*, microf. 12: I. 3, 4).

Sarcostemma sp. indet.
Asclepias aphylla Forssk. 1775: 50 (CVIII No. 186; Cent. II No. 68); Christensen 1922: 16.
Sarcostemma forskolianum Schult., Syst. Veg., ed. 15, 6: 117 (1820).
VERNACULAR NAME. *Milaeb* (Arabic).
LOCALITY. Yemen: Milhan ("In sylvis prope montem Melhân"), 1763.
MATERIAL. *Forsskål* no specimen found.

AVICENNIACEAE

Avicennia marina (*Forssk.*) *Vierh.* in Denkschr. Akad. Wien, Math. - Nat. 71: 435 (1907); Schwartz 1939: 217; Tackholm 1974: 454.
Sceura marina Forssk. 1775: 37 (CV No. 85; Cent. II No. 18); Christensen 1922: 14.
VERNACULAR NAME. *Schura* (Arabic), *Germ* (Muscat).
LOCALITY. Yemen: Al Luhayyah ("Lohaja", "Frequens in Insulis & littoribus *Maris rubri* ..."), 1763.
MATERIAL. *Forsskål* 317 (C — 1 sheet, type, microf. 96: II. 1,2); 318 (C — 1 sheet with field label "Sceura * marina. Insula Ghorab. Mocha. Loheja", type of *S. marina* microf. 96: II. 3, 4).
Forsskål s.n. (BM — 1 sheet, type of *S. marina*, Banks, MS).
NOTE. Vahl 1790: 47 misidentified this as *A. tomentosa* L. Deflers 1889: 185 and Christensen 1922: 14 considered it to be *A. officinalis* L.

BERBERIDACEAE

Berberis forskaliana *Schneider* in Bull. Herb. Boissier 5: 456 (1905); Ahrendt in J. Linn. Soc. 57: 108 (1961).
Berberis [*sp. without epithet*] — Forssk. 1775: CIX No. 230; Deflers 1889: 109.
VERNACULAR NAMES. *Tarat, Mösuk* (Arabic).
LOCALITY. Yemen: Mukhajah (or J. Barad) ("Moxhaja" on field label, "Kurma" on p. 1763.
MATERIAL. *Forsskål* 701 (C — 1 sheet, microf. 14: III. 7, 8).
NOTE. The type specimen, Yemen, *Schweinfurth* 1682 (K) has much smaller leaves than this sterile specimen, which seems to have been overlooked by workers on the genus. *B. forskaliana* may be a synonym of *B. holstii* Engl. (1894).

BETULACEAE

Alnus glutinosa (*L.*) *Gaertn.*, Fruct. Sem. Pl. 2: 54, t. 90 (1790); Fl. Turk. 7: 691 (1982).
Betula alnus L. 1753: 983; Forssk. 1775: XXXIII No. 407.
VERNACULAR NAME. *Skilithro* (Greek).
LOCALITY. Turkey: Belgrad, Istanbul ("Bg" = "Belgrad"), Aug 1761.
MATERIAL. *Forsskål* 793 (C — 1 sheet, microf. 15: I. 3, 4).

BORAGINACEAE

Alkanna lehmanii (*Tineo*) *A. DC.* in DC., Prodr. 10: 588 (1846); Meikle 1985: 1139.
Alkanna tinctoria (L.) Tausch (1824) — Tackholm 1974: 448, pl. 155B; Fl. Turk. 6: 425 (1978).
Anchusa tuberculata Forssk. 1775: 41 (LXII No. 112; Cent. II No. 32); Christensen 1922: 15.
Alkanna tuberculata (Forssk.) Meikle in Kew Bull. 34: 823 (1980), non Greuter (1972).
LOCALITY. Egypt: Alexandria ("Alexandriae"), Apr. 1761.
MATERIAL. *Forsskål* 283 (C — 1 sheet, type of *A. tuberculata*, microf. 7: I. 1, 2); 308 (C — 1 sheet, type of *A. tuberculata*, microf. 6: III. 7, 8); 940 (C —1 sheet ex herb. Schumacher, microf. 117: III. 5, 6).
Forsskål s.n. (LD — 1 sheet ex herb. Retzius).

Alkanna strigosa *Boiss. & Hohen.* in Boiss., Diagn. ser. 1, 4: 46 (1844).
LOCALITY. Unspecified.
MATERIAL. *Forsskål* s.n. (LD — 1 sheet ex herb. Retzius).
NOTES. If this identification is correct it is surprising to find an E. Mediterranean species on Forsskål's route.

Anchusa aegyptiaca (*L.*) *DC.*, Prodr. 10: 48 (1846); Tackholm 1974: 446.
A. flava Forssk. 1775: 40 (LXII No. 111; Cent. II No. 30); Christensen 1922: 15.
VERNACULAR NAME. *Sjubbaejta, Dabbûna* (Arabic).
LOCALITY. Egypt: Alexandria ("*Alexandriae* ad segetes"), 1761.
MATERIAL. *Forsskål* 292 (C — 1 sheet, type of *A. flava* microf. 6: II. 7, 8); 296 (C — 1 sheet, type of *A. flava*, microf. 6: III. 1, 2).
Forsskål s.n. (BM — 1 sheet ex herb. Banks, type of *A. flava*).
Forsskål s.n. (LIV — 1 sheet ex herb. Pallas, 1909. LBG. 3130, type of *A. flava*).

Anchusa bugloss *Forssk.* 1775: XXI No. 83.
LOCALITY. Turkey: Burgaz ("Brg" = Borghàs fons.).
MATERIAL. *Forsskål* no specimen found.
NOTE. The validating description is "Hispidissima: flore flavo". Its identity is unknown. This name should not be adopted if it upsets widely accepted nomenclature.

Anchusa undulata *L.* 1753: 133; Forssk. 1775: LXII No. 113.
LOCALITY. Egypt: Alexandria ("Alexandriae spontaneae"), 1 Apr. 1762.
MATERIAL. *Forsskål* 295 (C — 1 sheet, microf. 7: I. 3, 4); 309 (C — 1 sheet, microf. 7: I. 5, 6).
NOTE. This is probably what Tackholm (1974: 446) calls *A. hybrida* Ten. var. *pubescens* Gusul.

Arnebia tinctoria *Forssk.* 1775: 63 (Sub Cent. III No. 2); Riedl in Oesterr. Bot. Zeitschr. 109: 45–80 (1962); Fl. Palaest. 2: 69, pl. 111 (1978).
Lithospermum tinctorium (Forssk.) Vahl 1791: 33, t. 28.
Arnebia '*tetrastigma*' Forssk. 1775: 62 (LXII No. 110; Cent. III No. 2); Christensen 1922: 17; Tackholm 1974: 450.
Lithospermum tetrastigmum (Forssk.) Lam., Encycl. Méth. Bot. 3: 30 (1789).
VERNACULAR NAME. *Sagaret el arneb* (Arabic).
LOCALITY. Egypt: Cairo ("In desertis Kahirinis ad Caid Bey"), 1762.
MATERIAL. *Forsskål* 305 (C — 1 sheet, type of *A. tetrastigma* & *A. tinctoria*, microf. 9: II. 7, 8); 957 (C — 1 sheet ex herb. Schumacher, type of *A. tetrastigma* & *A. tinctoria*, microf. 9: III. 1, 2).
NOTE. The epithets *tetrastigma* and *tinctoria* were apparently published for the same species at the same time. Christensen favoured the use of the former, and was followed by Tackholm. Reidl argues in favour of *tinctoria* and was followed by Zohary in Flora Palaestina and by Index Nominum Genericorum. '*Tetrastigma*' was probably intended as part of the diagnosis.

Borago officinalis *L.* 1753: 137; Forssk. 1775: LXII No. 116; Bircher 1960: 658.
VERNACULAR NAME. *Lissân ettôr* (Arabic).
LOCALITY. Egypt: Cairo, garden ("Ch."), 1762.
MATERIAL. *Forsskål* 289 (C — 1 sheet, microf. 15: II. 3, 4).

Cordia sinensis *Lam.*, Illustr 1: 423 No. 1914 (1792).
Cordia myxa sensu Vahl 1790: 19, non L.

Cornus sanguinea sensu Forssk. 1775: 33 (CV No. 96; Cent. II No. 10), non L.; Christensen 1922: 14.
VERNACULAR NAME. *Gharaf, Onneb, Eschell, Sehaeli* (Arabic).
LOCALITY. Yemen: Al Luhayyah, Wadi Sordũd, Al Hadiyah ("Ubique per Yemen. Lohajae, Surdûd, Hadîe"), "in Hortu Beitelfaki" on field label, 1763.
MATERIAL. *Forsskål* 465 (C — 1 sheet, microf. 32: III. 1, 2); 688 (C — 1 sheet with field label, microf. 32: II. 7, 8).
Forsskål s.n. (BM — 1 sheet ex herb. Banks).
Forsskål s.n. (H — 1 sheet ex herb. Vahl).
NOTE. Formerly known as *C. gharaf* Ehr. ex Asch.

Cynoglossum dubium ? *Forssk.* 1775: CV No. 113; Christensen 1922: 37.
LOCALITY. Yemen: Barad ("Barah"), 1763.
MATERIAL. *Forsskål* no specimen found.
NOTE. The brief description is "Antheris corolla brevioribus". Mill & Miller in Notes. Roy. Bot. Gard. Edinb. 41: 474 (1984) dismiss it as "not intended as a name", *dubia* being repeatedly used for species which could not be identified. This view is in agreement with Art. 23 of the Code.

Cynoglossum lanceolatum *Forssk.* 1775: 41 (CV No. 111; Cent. II No. 33) Vahl 1791: 34; Deflers 1889: 173; Christensen 1922: 15; Schwartz 1939: 212.
VERNACULAR NAME. *Schenaf* (Arabic).
LOCALITY. Yemen: Al Hadiyah ("Hadîe"), Mar. 1763.
MATERIAL. *Forsskål* 312 (C — 1 sheet, type, microf. 36: I. 1, 2).

Cynoglossum linifolium sensu Forssk. 1775: 41 (CV No. 112; Cent. II No. 34), non L.; Christensen 1922: 15.
VERNACULAR NAME. *Hauscheb* (Arabic).
LOCALITY. Yemen: Kusma ("Kurmae"), 1763.
MATERIAL. *Forsskål* no type specimen found.
NOTE. An imperfectly known plant.

Echium angustifolium *Miller* subsp. **sericeum** (*Vahl*) *Klotz* in Wiss. Zeitschr. Univ. Halle 11: 298 (1962).
E. sericeum Vahl 1791: 35; Tackholm 1974: 451.
E. fruticosum ? sensu Forssk. 1775: LXII No. 118, non L.
VERNACULAR NAMES. *Sakhamam, Lesan el asal* (Arabic).
LOCALITY. Egypt: Alexandria ("Alexandriae"), 1 Apr. 1762.
MATERIAL. *Forsskål* 280 (C — 1 sheet, type of *E. sericeum*, microf. 40: III. 3, 4); 284 (C — 1 sheet, type of *E. sericeum*, microf. 40: III. 7, 8); 286 (C — 1 sheet with locality on reverse "inter Alexandriae et Cairo", type of *E. sericeum*, microf. 117: II. 5, 6); 291 (C — 1 sheet, type, microf. 40: III. 5, 6); 298 (C — 1 sheet, type of *E. sericeum*, microf. 40: III. 1, 2).

? Echium creticum *L.* 1753: 139; Forssk. 1775: 42 (CV No. 114; Cent. II No. 36); Vahl 1790: 15, t. 5; Christensen 1922: 15.
VERNACULAR NAME. *Kibedet el ard*? (Egyptian Arabic).
LOCALITY. Yemen: Taizz ("Prope Taaes"), 1763.
MATERIAL. *Forsskål* no specimen found.

Echium glomeratum *Poir.* in Lam., Encycl. 8: 670 (1808); Fl. Turk. 6: 321 (1978).
LOCALITY. ? Turkey, 1761.

MATERIAL. Forsskål 1240 (C — 1 sheet ex herb. Hornemann, microf. 117: II. 1, 2); 290 (C — 1 sheet, microf. 117: II. 7, 8).
NOTE. No. 1240 is named and labelled by F. Klotz 1961; No. 290 by B. Verdcourt, 1990.

Echium longifolium *Del.*, Fl. Egypte 184, t. 16 f. 3 (1812).
E. creticum sensu Forssk. 1775: XXI No. 79, non L.
LOCALITY. Turkey: Dardanelles (see note).
MATERIAL. *Forsskål* 287 (C — 1 sheet, microf. 40: II: 7, 8).
NOTE. This sheet, named by Klotz 1961, has on its reverse the locality Natolia. Forsskål also records it from "Borghas" and "Tenedos".

Echium cf. **rauwolfii** *Del.*, Fl. Egypte: 195 t. 19 f. 3 (1812); Tackholm 1974: 451, pl. 55A.
LOCALITY. Egypt, 1761-62.
MATERIAL. *Forsskål* 281 (C — 1 sheet, microf. 117: III. 1, 2).

Echium rubrum *Forssk.* 1775: 41 (LXII No. 119; Cent. 2: 36); Christensen 1922: 15; Klotz in Wiss. Zeitschr. Univ. Halle 11: 705 (1962).
E. setosum sensu Delile, Fl. Eg. 42, t. 17 (1812), non Vahl (1791).
VERNACULAR NAME. *El kahaeli* (Arabic).
LOCALITY. Egypt: Alexandria ("*Alexandriae*"), 1761-2.
MATERIAL. *Forsskål* 297 (C — 1 sheet, type, microf. 41: I. 3, 4); 1241 (C — 1 sheet ex herb. Schumacher, type, microf. 41: I. 1, 2).

Ehretia cymosa *Thonning* in Schum. & Thonn., Beskr. Guin. Pl.: 129 (1827); F.T.E.A. Boraginac.: 37 (1991); Friis in Forests & For. Trees N.E. Trop. Afr.: 259, map 155 (1992).
Sideroxylon inerme sensu Forssk. 1775: 204 (CVI No. 144; Cent. VIII No. 13), non L.; Christensen 1922: 33.
VERNACULAR NAME. *Uaraf* (Arabic).
LOCALITY. Yemen: Al Hadiyah ("Hadîe"), 1763.
MATERIAL. *Forsskål* 279 (C — 1 sheet with long description on field label and "Uaraf. Prope Ersch.", microf. 99: II. 5, 6).

Gastrocotyle hispida (*Forssk.*) *Bunge* in Mém. Sav. Etr. Pétersb. 7: 405 (1847); Tackholm 1974: 448.
Anchusa hispida Forssk. 1775: 40 (LXII No. 114; Cent. II No. 29); Vahl 1792: 33; Christensen 1922: 14.
LOCALITY. Egypt: Cairo ("In desertis *Káhirinis*"), 1761-62.
MATERIAL. *Forsskål* 310 (C — 1 sheet with field label "Ca. 148", type, microf. 6: III. 3, 4).

Heliotropium bacciferum *Forssk.* 1775: 38 (CV No. 106; Cent. II No. 22); Christensen 1922: 14; Tackholm 1974: 441.
VERNACULAR NAME. *Habbfa, Haschfa* (Arabic).
LOCALITY. Yemen: Al Luhayyah ("*Lohajae*"), Jan. 1763.
MATERIAL. *Forsskål* 1329 (C — sheet herb Vahl, type of *H. bacciferum*, microf. 53: II. 1, 2). *Forsskål* s.n. (BM — 1 sheet ex herb. Banks, type of *H. bacciferum*).
Lithospermum hispidum Forssk. 1775: 38 (LXII No. 106; Cent. II No. 24); Christensen 1922: 14.
Heliotropium undulatum Vahl 1790: 13.
LOCALITY. Egypt: Cairo ("In desertis Káhirinis"), 1761-62.
MATERIAL. *Forsskål* 300 (C — 1 sheet, type of *L. hispidum*, microf. 63: II. 3, 4); 1416 (C — 1 sheet ex herb. Hofman Bang, type of *L. hispidum* & *H. undulatum*, microf. 63: II. 5, 6); 1417 (C — 1 sheet ex herb. Hofman Bang, type of *L. hispidum* & *H. undulatum*, microf.

63: II. 7, 8); 1418 (C — 1 sheet ex herb. Horneman, type of *H. undulatum*, microf. 63: III. 1, 2); 1419 (C — 1 sheet ex herb. Schumacher, type of *L. hispidum* & *H. undulatum*, microf. 63: III. 3, 4); 1420 (C — 1 sheet ex herb Liebmann, type of *L. hispidum* & *H. undulatum*, microf. 63: III. 5, 6).

Heliotropium digynum (*Forssk.*) *Asch. ex C. Christensen* 1922: 14; Tackholm 1974: 441.
Lithospermum digynum Forssk. 1775: 40 (LXII No. 108; Cent. II No. 28).
VERNACULAR NAME. *Roghlae, Naetaefj* (Arabic).
LOCALITY. Egypt: Cairo ("In desertis montosis Káhirinis, rarius"), 1762.
MATERIAL. *Forsskål* 278 (C — 1 sheet, type, microf. nil); 285 (C — 1 sheet, type, microf. 117: II. 3, 4); 1415 (C — 1 sheet ex herb. Schumacher, microf. nil).

Heliotropium europaeum *L.* 1753: 130; Forssk. 1775: 38 (XXI No. 84; LXII No. 103; CV No. 108; Cent. II No. 19); Christensen 1922: 14; Fl. Turk. 6: 252 (1964).
VERNACULAR NAMES. *Sackrân* (Arabic, Egypt); *Kerîr Akrîr* (Arabic, Yemen).
LOCALITY. Turkey: Dardanelles, Aug. 1761, (also Eracliffe & Rhodes).
MATERIAL. *Forsskål* 301 (C — 1 sheet, microf. 53: II. 5, 6).
LOCALITY. Egypt.
MATERIAL. *Forsskål* no specimen found.
LOCALITY. Yemen: Wadi Mawr ("Môr"), 1763.
MATERIAL. *Forsskål* no specimen found.

Heliotropium longiflorum (*A. DC.*) *Steud. & Hochst. ex DC.*, Prodr. 9: 555 (1845).
H. curvassavicum sensu Forssk. 1775: 38 (CV No. 109; Cent. II No. 21), non L.; Christensen 1922: 14.
VERNACULAR NAME. *Kerîr, Akrîr* (Arabic).
LOCALITY. Yemen: Wadi Mawr ("Môr" according to cover), Feb. 1763.
MATERIAL. *Forsskål* 1330 (C — 1 sheet ex herb. Vahl, microf. 53: II. 3, 4).

Heliotropium ovalifolium *Forssk.* 1775: 38 (CV No. 110; Cent. II No. 23); Christensen 1922: 38.
H. coromandelianum Retz. — Vahl 1790: 13.
LOCALITY. Yemen: Al Hadiyah ("Hadîe"), Mar. 1763.
MATERIAL. *Forsskål* 299 (C — 1 sheet, type of *H. ovalifolium*, microf. nil).
Forsskål s.n. (BM — 1 sheet ex herb. Banks, type of *H. ovalifolium*).

Heliotropium supinum *L.* 1753: 130; Aschers. & Schweinf., Ill. Fl. Egypte 109 (1887); Tackholm 1974: 442.
Lithospermum heliotropoides Forssk. 1775: 39 (LXII No. 109; Cent. II No. 25); Christensen 1922: 14.
Heliotropium lineatum Vahl 1790: 13.
LOCALITY. Egypt: Cairo ("*Kahirae* florens initio Juni, inter Cucumeres"), June 1762.
MATERIAL. *Forsskål* 302 (C — 1 sheet, type of *L. heliotropoides*, microf. nil).
NOTE. Named and labelled by B L Burtt (1949), who recognised this as the missing type specimen not seen by Christensen.

Heliotropium sp. indet. A.
H. fruticosum sensu Forssk. 1775: 38 (CV No. 107, Cent. II. No. 20), non L; Schweinfurth 1912: 143; Christensen 1922: 14.
VERNACULAR NAME. *Haschfae, Haschfae* (Arabic).
LOCALITY. Yemen: Wadi Mawr ("Môr"), 1763.

MATERIAL. *Forsskål* 303 (C — 1 sheet, with field label "Heliotropium prope Taas" and notes by Vahl: "Myosotis an varietas M. fruticosi. Lithospermum fruticosum? Cent. 2. no. 20?").
NOTE. Christensen notes that "Vahl has with doubt referred a scanty specimen to *H. fruticosum* L.; it belongs probably to a species of *Myosotis*, but scarcely to *H. fruticosum* F. which was collected at Môr, while the specimen in question was collected near Taaes". The specimen in question belongs to the genus *Heliotropium*.

? **Heliotropium** sp. indet. B
LOCALITY. Unspecified.
MATERIAL. ? *Forsskål* s.n. (LD — 1 sheet ex herb. Retzius) see note.
NOTE. This specimen has not been identified and it is doubtful whether its attribution to Forsskål is correct.

Lappula spinocarpos (*Forssk.*) *Asch. ex Kuntze* in Acta Hort. Petrop. 10: 215 (1887); Tackholm 1974: 445.
Anchusa spinocarpos Forssk. 1775: 41 (LXII No. 115; Cent. II No. 31); Christensen 1922: 15.
Myosotis spinocarpos (Forssk.) Vahl 1791: 32.
Echinospermum vahlianum Lehm., Pl. Asperif.: 132 (1818).
E. spinocarpos (Forssk.) Boiss., Fl. Orient. 4: 249 (1875).
LOCALITY. Egypt: Alexandria ("*Alexandriae*"), Apr. 1762.
MATERIAL. *Forsskål* 304 (C — 1 sheet, type of *A. spinocarpos*, microf. 6: III. 5, 6).

Moltkiopsis ciliata (*Forssk.*) *I.M. Johnston* in Journ. Arn. Arb. 34: 2 (1953); Tackholm 1974: 449, pl. 154A.
Lithospermum ciliatum Forssk. 1775: 39 (LXII No. 105; Cent. II No. 26); Vahl 1790: 14; Christensen 1922: 14.
LOCALITY. Egypt: Cairo ("Ad *canales* Káhirensis"), 1762.
MATERIAL. *Forsskål* 282 (C — 1 sheet, type of *M. ciliatum*, microf. 117: I. 5, 6); 306 (C — 1 sheet, type, microf. 63: I. 5, 6); 311 (C — 1 sheet, type, microf. 63: I. 7, 8); 1414 (C — 1 sheet ex herb. Schumacher, type, microf. 63: II. 1, 2).

Lithospermum angustifolium Forssk. 1775: 39 (LXII No. 107; Cent. II No. 27); Christensen 1922: 14.
L. callosum Vahl 1790: 4.
VERNACULAR NAME. *Halamae* (Arabic).
LOCALITY. Egypt: Cairo ("In desertis Káhirinis"), 1762.
MATERIAL. *Forsskål* 307 (C — 1 sheet, type of *L. angustifolium*, microf. 63: I. 1, 2); 1412 (C — 1 sheet ex Hofman Bang with field label 'Ca. 178', type of *L. angustifolium*, microf. 63: I. 3, 4); 1413 (C — 1 sheet ex herb. Schumacher, type of *L. angustifolium*, microf. 117: III. 3, 4).
Forsskål s.n. (BM — 1 sheet ex herb. Banks, type of *L. angustifolium*).

Trichodesma africana (*L.*) *R. Br.*, Prodr. 496 (1810); Tackholm 1974: 444, pl. 154C.
Borago verrucosa Forssk. 1775: 41 (LXII No. 117; Cent. II No. 35); Christensen 1922: 15.
VERNACULAR NAME. *Lusseq, Höreig, Horraejg* (Arabic).
LOCALITY. Egypt: Cairo ("Káhirae") 1761–2.
MATERIAL. *Forsskål* 288 (C — 1 sheet, type of *B. verrucosa*, microf. 15: II. 5, 6).

BURSERACEAE

Amyris kafal *Forssk.* 1775: 80 (CX No. 256; Cent. III No. 50); Christensen 1922: 18.
VERNACULAR NAME. *Kafal* (Arabic).
LOCALITY. Yemen: Bayt al Faqih ("Beit el fakih"), 1763.
MATERIAL. *Forsskål* no type specimen found.
NOTE. Christensen (l.c.) notes that Schweinfurth (1912: 135) identifies this as *Commiphora erythraea* (Ehrbg.) Engl. He also mentions that "two other species by Forssk. named by their indigenous names [*Schadjaret el murr, Chadasch*] only under n. 50 are both [according to Schweinfurth] *Commiphora abyssinica* Engl. These identifications are likely as they are myrrh-producing species. Since the full description of *A. kafal* amply validates the name it would antedate either of the other species in *Commiphora*, but in the absence of a type specimen a transfer to that genus is not possible. These names should not be accepted if they upset widely accepted nomenclature.

Commiphora gileadensis (*L.*) *Christensen* 1922: 18; Schweinfurth 1899: 294; Schwartz 1939: 128.
Amyris gileadensis L., Dissent. Opobal.: 13 (1764); Mant.: 65 (1767) & Amoen. Acad. 7: 68 (1769); Vahl 1790: 28 t. 11.
C. opobalsamum (L.) Engl. in DC., Monogr. Phan. 4: 16 (1883) as var. *gileadensis*.
Balsamodendron opobalsamum (L.) Kunth (1824) — Deflers 1889: 120.
Amyris opobalsamum L., Dissert. Opobal. 14 (1764); Mant.: 65 (1767) & Amoen. Acad. 7: 69 (1769); Forssk. 1775: 79 (CX No. 254; Cent. III No. 48).
VERNACULAR NAME. *Abn scham* (Arabic).
LOCALITY. Yemen: Uday near Hays ("Öude, non longe ad urbe Haes"), 4 Apr. 1763.
MATERIAL. *Forsskål* 935 (C — 1 sheet ex herb. Schumacher, microf. 6: I. 1, 2); 936 (C — 1 sheet ex herb. Vahl, microf. 6: I. 3, 4); 937 (C — 1 sheet ex herb. Schumacher, microf. 6: I. 5, 6).
Forsskål s.n. (BM — 1 sheet ex herb. Nolte); s.n. (BM — 1 sheet ex herb. Banks).
NOTE. Although Linnaeus based *A. gileadensis* on a specimen sheet by Forsskål from Yemen with a letter dated 3 June 1763, no type specimen is in herb. LINN. This plant produces the balm or opobalsamum of the Bible and Forsskål wrote ecstatically about his discovery.

Commiphora kataf (*Forssk.*) *Engl.* in DC., Monogr. Phan. 4: 19 (1883); Schweinfurth 1899: 284.
Amyris kataf Forssk. 1775: 80 (CX No. 255; Cent. III No. 49); Vahl 1790: 28; Christensen 1922: 18.
VERNACULAR NAME. *Kataf* (Arabic).
LOCALITY. Yemen: Bayt al Faqih ("Beit el fakih plantatae erant nonnulle arbores"), Feb. 1763.
MATERIAL. *Forsskål* 934 (C — 1 sheet, holotype of *A. kataf*, microf. 5: III. 7, 8).
Forsskål s.n. (LD — 1 sheet ex herb. Retzius, type of *A. kataf*).
NOTE. This is one of the myrrh-producing species of the Bible.

CAESALPINIACEAE

(see Leguminosae)

CAMPANULACEAE

Campanula edulis *Forssk.* 1775: 44 (CVI No. 127; Cent. II No. 46); Christensen 1922: 15; Thulin in Bot. Notiser 128: 354–355 (1976) & in F.T.E.A. Campanulac: 35 (1976).
VERNACULAR NAME. *Chobs el okab, Rîam* (Arabic).
LOCALITY. Yemen: Kusma, Al Hadiyah ("Kurmae, Hadîe"), 1763.
MATERIAL. *Forsskål* 243 (C — 1 sheet, type, microf. 20: II. 5, 6); 245 (C — 1 sheet, type, microf. 20: II. 1, 2); 246 (C — 1 sheet, type, microf. 20: II. 3, 4).
Forsskål s.n. (BM — 1 sheet ex herb. Banks, type of *C. edulis*).

Legousia speculum-veneris (*L.*) *Chaix* in Vill., Hist. Pl. Dauph. 1: 338 (1786); Fl. Europ. 4: 94 (1976).
Campanula speculum(-veneris) L. (1753) — Forssk. 1775: V No. 57.
LOCALITY. France: Marseille ("Estac"), May 1761.
MATERIAL. *Forsskål* 244 (C — 1 sheet with field label "27", microf. 20: II. 7, 8).

CAPPARIDACEAE

Cadaba farinosa *Forssk.* 1775: 68 (CVI No. 140, Cent. III. 12); Deflers 1889: 110: Schweinfurth 1896: 193; Christensen 1922: 17; Schwartz 1939: 68; F.T.E.A. Cappar.: 75 (1964).
Stroemia farinosa (Forssk.) Vahl 1790: 20.
VERNACULAR NAMES. *Asal, El bejad, Korraeb, Saerah, Toraeb* (Arabic).
LOCALITY. Yemen: Surdud ("Elbej ad Surdûd" on field label), 1763.
MATERIAL. *Forsskål* 634 (C — 1 sheet with original field label, type of *C. farinosa*, microf. 18: III. 7, 8); 645 (C — 1 sheet, type of *C. farinosa*, microf. 18: III. 5, 6); 1034 (C — 1 sheet ex herb. Schumacher, type of *C. farinosa*, microf. 18: III. 3, 4).

Cadaba glandulosa *Forssk.* 1775: 68 (CVI No. 139; Cent. III No. 13); Schweinfurth 1896: 193; Christensen 1922: 17; Schwartz 1939: 67; F.T.E.A. Cappar.: 74 (1964).
Stroemia glandulosa (Forssk.) Vahl 1790: 20.
VERNACULAR NAME. *Taennaim* (Arabic).
LOCALITY. Yemen: Al Hadiyah ("Hadîe"), Mar. 1763.
MATERIAL. *Forsskål* 633 (C — 1 sheet, type of *C. glandulosa*, microf. 19: I. 1, 2); 636 (C — 1 sheet, type of *C. glandulosa*, microf. 19: I. 5, 6); 637 (C — 1 sheet, type of *C. glandulosa*, microf. 19: I. 3, 4); 646 (C — 1 sheet, type of *O. glandulosa*, microf. 19: I. 7, 8); 1035 (C — 1 sheet ex herb. Schumacher, type of *C. glandulosa*, microf. 19: II. 1, 2).

Cadaba rotundifolia *Forssk.* 1775: 68 (CVI No. 138; Cent. III No. 11); Deflers 1889: 110; Schweinfurth 1896: 192; Christensen 1922: 17; Schwartz 1939: 66; F.T.E.A. Cappar.: 75 (1964).
Stroemia rotundifolia (Forssk.) Vahl 1790: 20.
VERNACULAR NAME. *Kadhab* (Arabic).
LOCALITY. Yemen: Al Luhayyah ("Lohaja frequens"), Jan. 1763.
MATERIAL. *Forsskål* 635 (C — 1 sheet, type of *C. rotundifolia*, microf. 19: II. 7, 8); 644 (C —1 sheet, type of *C. rotundifolia*, microf. 19: II. 5, 6); 1037 (C — 1 sheet ex herb. Schumacher, type of *C. rotundifolia*, microf. 19: III. 1, 2); 1036 (C — 1 sheet ex herb. Vahl, type of *C. angustifolia*, microf. 19: II. 3, 4).
Forsskål s.n. (BM — 1 sheet ex herb. Pallas with original field label, type of *C. rotundifolia*).

Capparis cartilaginea *Decne.* in Ann. Sci. Nat. Sér. 2, 3: 273 (1835).
C. spinosa sensu Forssk. 1775: 99 (CXIII No. 332; Cent. IV No. 6); Christensen 1922: 20, non L.
C. galeata Fres. (1836) — Deflers 1889: 110; Schweinfurth 1896: 190; Schwartz 1939: 66.
VERNACULAR NAME. *Lasaf* (Arabic).
LOCALITY. Yemen: Taizz ("Taaes"), 1763.
MATERIAL. *Forsskål* no specimen found.
NOTE. The taxonomy of *Capparis* is so confused that it is impossible to decide the identity of Forsskål's plant without a specimen. His *C. inermis* (q.v.) seems to be similar except for the stipular species, as suggested by Forsskål.

Capparis dahi *Forssk.* 1775: 212 (CXIII No. 335; Cent. VIII No. 55); Christensen 1922: 34.
VERNACULAR NAME. *Dahi* (Arabic).
LOCALITY. Yemen: Milhan ("In monte Melhân"), 1763.
MATERIAL. *Forsskål* no specimen found.
NOTE. Unknown species in absence of specimen but validly described. It is most likely to be a form of *C. cartilaginea*. Dahi is listed as *Acacia orfota* by Schweinfurth 1912: 94. The name *C. dahi* should not be adopted if it upsets widely accepted nomenclature.

Capparis decidua (*Forssk.*) *Edgew.* in J. Linn. Soc. Bot. 6: 184 (1862); Schwartz 1939: 66; Fl. Palaest. 1: 244, pl. 361 (1966); Tackholm 1974: 162, pl. 48B.
Sodada decidua Forssk. 1775: 81 (CX No. 253; Cent. III No. 53); Christensen 1922: 19.
C. sodada R. Br. (1826) — Deflers 1889: 111.
VERNACULAR NAME. *Sodâd* (Arabic).
LOCALITY. Yemen: Kusma, Wadi Mawr, Bayt al Faqih ("Kurma, Môr, Beit el Fakih" on p. CX, "ubique per *Yemen*" on p. 82), 1763.
MATERIAL. *Forsskål* 641 (C — 1 sheet, type of *S. decidua*, microf. 100: III. 5, 6); 647 (C — 1 sheet, type of *S. decidua*, microf. 101: I. 1, 2); 1742 (C — 1 sheet, microf. 100: III. 3, 4); 1743 (C — 1 sheet ex herb. Hofman Bang, type of *S. decidua*, microf. 100: III. 7, 8).
NOTE. The new genus, *Sodada*, and species, *S. decidua*, are described in the 'Centuriae' and provided with an epithet there.

Capparis inermis *Forssk.* 1775: 100 (CXIII No. 333; Cent. IV No. 7); Christensen 1922: 20.
LOCALITY. Yemen: Taizz ("Taaes"), 1763.
MATERIAL. *Forsskål* no type specimen found.
NOTE. See note under *C. cartilaginea*. Schweinfurth 1896: 191 considered this to be *C. galeata* var. *montana* Schweinf. This name should not be adopted if it upsets widely accepted nomenclature.

Cleome ambylocarpa *Barr. & Murb.* in Lunds Univ. Arsskrift, n.s. 2, i No. 4: 25 (1905); Botschantzev in Not. Syst. Vasc. Pl. Acad. Sc. URSS 1964: 129–131 (1964) & 1968: 236–237 (1968). Fig. 19B.
Siliquaria glandulosa Forssk. 1775: 78 (LXV No. 194; Cent. III. No. 46); & Icones t.16B (1776); Christensen 1922: 18.
LOCALITY. Egypt: Cairo ('ad Birket el hádgi prope Káhiram'), 1762.
MATERIAL. *Forsskål* 640 (C — 1 sheet, type of *S. glandulosa*, microf. 100: I. 7–8); 1735 (C —1 sheet ex herb. Hofman Bang, type of *S. glandulosa*, microf. 117: III. 7, 8).
NOTE. Named and labelled by D. Hillcoat 1973. It has been named previously *C. arabica* L. which is *C. africana* Botsch. — Tackholm 1974: 169, pl. 50B. The new genus, *Siliquaria*, and the new species, *S. glandulosa*, are described in the 'Centuriae' and provided with an epithet there.

Cleome angustifolia *Forssk.* 1775: 120 (CXVI No. 404; Cent. IV No. 71); Christensen 1922: 23.
C. filifolia Vahl 1790: 48.
LOCALITY. Yemen: Taizz ("circa Taaes"), 1763.
MATERIAL. *Forsskål* no type specimen found at C.
Forsskål s.n. (BM — 2 sheets ex herb. Banks, types of *C. angustifolia*).
NOTE. Neither name is accounted for by Schwartz (1939).

Cleome digitata *Forssk.* 1775: 120 (CXVI No. 403; Cent. IV No. 70); Christensen 1922: 23.
VERNACULAR NAME. *Biss* (Arabic).
LOCALITY. Yemen: Surdud ("Surdûd in arvis"), 1763.
MATERIAL. *Forsskål* no type specimen found.
NOTE. Not accounted for by Schwartz 1939. This name should not be adopted if it upsets widely accepted nomenclature.

Cleome droserifolia (*Forssk.*) *Del.*, Fl. Egypte: 250 (1813); Tackholm 1974: 167.
Roridula droserifolia Forssk. 1775: LXII No. 101; Christensen 1922: 14. = *Roridula* [*sp. without epithet*] Forssk. 1775: 35 (Cent. II No. 16).
Cleome roridula R. Br. in Salt, Abyss. App. 65 (1814).
LOCALITY. Egypt: Suez ("Sués in valle Mosbaeha versus fontem Bîr"), Sept. 1762.
MATERIAL. *Forsskål* 619 (C — 1 sheet, type, microf. 88: I. 1, 2); 620 (C — 1 sheet, type, microf. 88: I. 3, 4); 1630 (C — 1 sheet ex herb. Schumacher, type, microf. 87: III. 7, 8).
Forsskål s.n. (BM — 1 sheet, ex herb Banks, type).
NOTE. This is one of the cases where a new genus and species are established with description but without epithet in the 'Centuriae', while the epithet is provided in the 'Florae'.

Maerua crassifolia *Forssk.* 1775: CXIII No. 330; Christensen 1921: 21; Schwartz 1939: 69; Fl. Palaest. 1: 241, pl. 357 (1966); Tackholm 1974: 165, pl. 49 B. = *Maerua* [*sp. without epithet*] Forssk. 1775: 104 (Cent. IV No. 22).
M. uniflora Vahl 1790: 36.
M. arabica J.F. Gmel., Syst. Nat. ed. 13, 2: 827 (1791).
VERNACULAR NAME. *Maeru, Meru* (Arabic).
LOCALITY. Yemen: without location ("Ps = planities argillacea, sicca"), 1763.
MATERIAL. *Forsskål* 649 (C — 1 sheet, type of *M. crassifolia, M. uniflora, M. arabia,* microf. 65(1–2): III. 7).
NOTE. This is again one of several new genera with one or more new species described without an epithet in the 'Centuriae'. The epithet is supplied in the 'Florae' only.

Maerua oblongifolia (*Forssk.*) *A. Rich.*, Tent. Fl. Abyss. 1: 32, t. 6 (1847); Deflers 1889: 110; Schweinfurth 1896: 194; Schwartz 1939: 69; F.T.E.A. Cappar.: 37 (1964).
Capparis oblongifolia Forssk. 1775: 99 (CXIII No. 334, Cent. IV No. 4); Christensen 1922: 20.
VERNACULAR NAME. *Redif, Asal* (Arabic).
LOCALITY. Yemen: Wadi Mawr ("Mór"), Jan. 1763.
MATERIAL. *Forsskål* 643 (C — 1 sheet, type of *C. oblongifolia*, microf. 20: III. 5, 6).
Forsskål s.n. (BM — 1 sheet ex herb Banks, type of *C. oblongifolia*).

Capparis mithridatica Forssk. 1775: 99 (CXIII No. 331, Cent. IV No. 5); Christensen 1922: 20.
VERNACULAR NAMES. *Schaegar, Oud essymm* (Arabic).
LOCALITY. Yemen: Surdud ("Surdûd"), 1763.
MATERIAL. *Forsskål* 638 (C — 1 sheet, type of *C. mithridatica*, microf. 20: III. 3, 4); 639 (C — 1 sheet, type of *C. mithridatica*, microf. 20: III. 1, 2).

Maerua racemosa *Vahl* 1790: 36; Christensen 1922: 39.
LOCALITY. "Habitat in Arabia".
MATERIAL. *Forsskål* no type specimen found.
NOTE. Unknown species, not necessarily from Arabia. This name should not be adopted if it upsets widely accepted nomenclature.

CARYOPHYLLACEAE

Agrostemma githago *L.* 1753: 435; Forssk. 1775: VI No. 100; Fl. Europ. 1: 157 (1964), 1: 190 (1993).
LOCALITY. France: Marseille ("Estac"), 9 May–3 June 1961.
MATERIAL. *Forsskål* 1733 (C — 1 sheet, microf. 118: I. 7, 8).

Cerastium procumbens *Forssk.* 1775: 211 (XXVI No. 214; Cent. VIII No. 50).
LOCALITY. Turkey: Istanbul, ("Ad pagum Belgrad prope Constantinop"), 1761.
MATERIAL. *Forsskål* no type specimen found.
NOTE. This is an unknown species which was validly published, yet not accounted for in the Flora of Turkey vol. 2 (1967). This name should not be adopted if it upsets widely accepted nomenclature.

Dianthus caryophyllus *L.* 1753: 410; Forssk. 1775: XXV No. 204; Fl. Europ. 1: 195 (1966); 1: 240 (1993).
LOCALITY. Malta, 12 June 1761.
MATERIAL. *Forsskål* 1225 (C — 1 sheet with field label "Malta", microf. 118: I. 1, 2).
LOCALITY. Turkey: Istanbul cult. ("Cph"; Imros also cited), 1761.
MATERIAL. *Forsskål* 1224 (C — 1 sheet ex herb. Hofman Bang, with field label "Co. 33", microf. 39: I. 5, 6).

Dianthus uniflorus *Forssk.* 1775: CXI No. 284; Schweinfurth 1896: 174; Christensen 1922: 37; Schwartz 1939: 58.
D. pumilus Vahl 1790: 32; Deflers 1889: 112.
VERNACULAR NAME. *Zabr es zirr* (Arabic).
LOCALITY. Yemen: Kusma ("Kurma"), Mar. 1763.
MATERIAL. *Forsskål* 549 (C — 1 sheet, type of *D. uniflorus* and *D. pumilus*, microf. 39: II. 7, 8); 554 (C — 1 sheet, type of *D. uniflorus* and *D. pumilus*, microf. 39: III. 1, 2); 1226 (C — 1 sheet, type of *D. uniflorus* and *D. pumilus*, microf. 39: II. 5, 6).
NOTE. The epithet is validated by the brief description: "Scapo brevissimo, unifloro; foliis linearibus".

Gymnocarpos decandrum *Forssk.* 1775: 65 (Cent. III No. 8); & Icones: 4, t. 10; Christensen 1922: 17; Fl. Palaest. 1: 130, pl. 183 (1966); Tackholm 1974: 101. = *G. deserti* Forssk. 1775: LXIII No. 144. Fig. 14.
Trianthema fruticosa Vahl 1790: 32.
Gymnocarpos fruticosus (Vahl) Pers., Syn. 1: 262 (1805).
VERNACULAR NAMES. *Syrr*, *Djarad* (Arabic).
LOCALITY. Egypt: Cairo ("In desertis Káhirinis orient"), 1762.
MATERIAL. *Forsskål* 537 (C — 1 sheet, type of *G. decandrum*, microf. 52: I. 5, 6); 1312 (C — 1 sheet ex herb. Schumacher, type of *G. decandrum*, microf. 128: III. 7, 8).
Forsskål s.n. (BM — 1 sheet ex herb. Banks, with field label, type of *G. decandrum*).
NOTE. This is an example of a new genus and species, both described and provided with epithet in the 'Centuriae'. A different epithet for the species is provided in the 'Florae'.

Tab. X.

Peter . Haas sculps

Gymnocarpus decandrum

FIG. 14. **Gymnocarpos decandrum** Forssk. — Icones, Tab. X.

Gypsophila capillaris (*Forssk.*) *Christensen* 1922: 19; Tackholm 1974: 82.
Rokejeka capillaris Forssk. 1775: LXVI No. 236. = *Rokeja [sp. without epithet]* Forssk. 1775: 90 (Cent. III No. 77).
R. deserti J.F. Gmel., Syst. 703 (1791).
Gypsophila Rokejeka Delile, Fl. Egypte 87 (1813); Fl. Palaest. 1: 101 (1966).
VERNACULAR NAME. *Syrr, Rokaejeka* (Arabic).
LOCALITY. Egypt: Cairo ("In desertis Kahiram").
MATERIAL. *Forsskål* 1629 (C — 1 sheet, type of all above names, microf. 87: III. 5, 6).
NOTE. This is one of the cases where a new genus and species are described without epithet in the 'Centuriae', while an epithet is provided in the 'Florae'.

Herniaria hirsuta *L.* 1753: 218; Forssk. 1775: XXII No. 131; Fl. Turk. 2: 250 (1967); Fl. Palaest. 1: 133, pl. 191 (1966); Tackholm 1974: 99.
LOCALITY. Turkey: Dardanelles ("Dd" = Dardanelli), July 1761.
MATERIAL. *Forsskål* 1331 (C — 1 sheet, microf. 53: II. 7, 8); 1332 (C — 1 sheet ex herb. Liebmann, microf. 53: III. 1, 2).
LOCALITY. Egypt: Ortakin and Alexandria (on reverse of sheet), 1761.
MATERIAL. *Forsskål* 1333 (C — 1 sheet, microf. 53: III. 3, 4).

Minuartia filifolia (*Forssk.*) *Mattfeld* in Fedde Rep., Beih. 15: 93 (1922); F.T.E.A. Caryophyllac.: 17 (1956); McNeill in Notes Roy. Bot. Gard. Edinb. 24: 32 (1963).
Arenaria filifolia Forssk. 1775: 211 (CXI 287; Cent. 8: 49); Vahl 1790: 33 t. 12; Christensen 1922: 34.
Alsine filifolia (Forssk.) Schweinfurth 1896: 175.
LOCALITY. Yemen: Bughah ("Boka"), Mar. 1763.
MATERIAL. *Forsskål* 956 (C — 1 sheet ex herb. Schumacher, type of *A. filifolia*, microf. 8: II. 7, 8).
Forsskål s.n. (BM — 1 sheet ex herb Banks, type of *A. filifolia*).
Arenaria fasciculata sensu Forssk. 1775: 211 (CXI No. 286; Cent. VIII No. 48), non L.
LOCALITY. Yemen: Kusma ("Kurmae"), Mar. 1763.
MATERIAL. *Forsskål* 955 (C — 1 sheet ex herb. Schumacher, microf. 8: II. 5, 6).
NOTE. Christensen doubted whether "the specimen so named [*A. fasciculata*] really belongs here". It seems to be correct, but we are not sure whether it really is a small form of *M. filifolia* as the sepals are shorter, broader and more strongly 3-nerved.
Another specimen, *Forsskål* 566 (C — 1 sheet, microf. 39: II. 3, 4) could be a *Minuartia* sp.

Minuartia geniculata (*Poir.*) *Thell.*, Fl. Adv. Montpellier 232 (1912); McNeill in Notes Roy. Bot. Gard. Edinb. 24: 323 (1963).
Arenaria geniculata Poir. (1789).
Arenaria procumbens Vahl 1791: 51, t. 32; Christensen 1922: 38.
Alsine procumbens (Vahl) Fenzl, Versuch Verbreit. Vertheil Alsin. 57: (1833).
Rhodalsine procumbens (Vahl) J. Gay in Ann. Sci. Nat. Ser. 3, 4: 25 (1845).
Minuartia procumbens (Vahl) Zohary in Beih. Bot. Centralbl. 52 Abt. B. 576 (1935); Tackholm 1974: 91, pl. 19B.
Cherleria sedoides sensu Forssk. LXVI No. 241, non L.
VERNACULAR NAME. *Sesau* (Arabic).
LOCALITY. Egypt: Cairo ("Cd"), 1761–2.
MATERIAL. *Forsskål* no specimen found.
NOTE. The short description "Foliis obovato-succulent." would validate *C. sedoides* if it were intended as a new species rather than a error of identification for one of Linnaeus'.

Paronychia arabica (*L.*) *DC.* in Lam., Encycl. 5: 24 (1804); Fl. Palaest. 1: 132, pl. 187 (1966); Tackholm 1974: 100.
Illecebrum arabicum L. 1767: 51.
LOCALITY. Egypt, 1761–2.
MATERIAL. *Forsskål* s.n. (LINN — 1 sheet 290.19 raised from seeds sent to Linnaeus, holotype of *I. arabicum*, microf. 160: II. 1.

Corrigiola albella Forssk. 1775: 207 (LXIII No. 146; Cent. VIII No. 31).
VERNACULAR NAME. *Libbaejt* (Arabic).
LOCALITY. Egypt: Cairo ("Cd = Káhirae"), 1762.
MATERIAL. *Forsskål* 558 (C — 1 sheet, type of *C. albella*, microf. 32: III. 5, 6); 1148 (C — 1 sheet ex herb. Schumacher, type of *C. albella*, microf. 32: III. 7, 8).
Forsskål s.n. (BM — 1 sheet ex herb. Banks, type of *C. albella*).
NOTE. This is var. *longiseta* (Bertol.) Aschers. & Schweinf.

Paronychia desertorum *Boiss.*, Diagn. ser. 1, 3: 11 (1843); Fl. Palaest. 1: 132 (1966); Tackholm 1974: 101.
Herniaria lenticulata Forssk. 1775: 52 (LXIII No. 149; Cent. II No. 78), non L.; Christensen 1922: 16.
Paronychia lenticulata (Forssk.) Aschers. & Schweinf. in Österr. Bot. Zeit. 39: 300 (1889), nom. illegit.
VERNACULAR NAME. *Makr* (Arabic).
LOCALITY. Egypt: Cairo ("In desertis circa Káhiram."), 1762.
MATERIAL. *Forsskål* 563 (C — 1 sheet, type of *H. lenticulata*, microf. 53: III. 7, 8); 1334 (C — 1 sheet, type of *H. lenticulata* microf. 53: III. 5, 6).

Petrorhagia prolifera (*L.*) *Ball & Heywood* in Bull. Brit. Mus. (Nat. Hist.) 3(4): 161 (1964); Fl. Turk. 2: 135 (1967).
Dianthus prolifera L. 1753: 410; Forssk. 1775: XXV No. 203.
LOCALITY. Turkey: Dardanelles ("Dardanelli"), July 1761.
MATERIAL. *Forsskål* 555 (C — 1 sheet with field label "K. 2", microf. 39: II. 1, 2); 556 (C —1 sheet with field label "K. 8", microf. 39: I. 7, 8).

Polycarpaea repens (*Forssk.*) *Aschers. & Schweinf.* in Österr. Bot. Zeit. 39: 126 (1889); Fl. Palaest. 1: 127, pl. 179 (1966); Tackholm 1974: 96.
Corrigiola repens Forssk. 1775: 207 (LXIII No. 145; Cent. VIII No. 30); Christensen 1922: 33.
LOCALITY. Egypt: Cairo ("In desertis Káhirinis"), 1762.
MATERIAL. *Forsskål* 557 (C — 1 sheet, holotype of *C. repens*, microf. 33: I. 1, 2).

Polycarpon prostratum (*Forssk.*) *Aschers. & Schweinf.* in Österr. Bot. Zeit. 39: 128 (1889); F.T.E.A. Caryophyllac: 5 (1956); Tackholm 1974: 97.
Alsine prostrata Forssk. 1775: 207 (LXIV No. 184; Cent. VIII No. 33); Christensen 1922: 33.
Polycarpia prostrata (Forssk.) Christensen 1922: 33.
VERNACULAR NAME. *Robbaire* (Arabic).
LOCALITY. Egypt: Cairo ('Káhirae'), 1762.
MATERIAL. *Forsskål* 551 (C — 1 sheet, type of *A. prostrata*, microf. 5: I. 7, 8): 559 (C — 1 sheet, type of *A. prostrata*, microf. 5: I. 5, 6); 924 (C — 1 sheet ex herb. Hofman Bang, type of *A. prostrata*, microf. 5: II. 1, 2); 925 (C — 1 sheet ex herb. ? Vahl, type of *A. prostrata*, microf. 5: II. 3, 4); 926 (C — 1 sheet ex herb. ? Hornemann, type of *A. prostrata*, microf. nil);
Forsskål s.n. (BM — 1 sheet ex herb Banks, type of *A. prostrata*).

Pteranthus dichotomus *Forssk.* 1775: LXII No. 100; Christensen 1922: 14; Fl. Palaest. 1: 136, t. 195 (1966); Tackholm 1974: 102; Meikle 1977: 284. = *Pteranthus* [*sp. without epithet*] Forssk. 1775: 36 (Cent. II No. 17).
Camphorosma pteranthus L. 1767: 41.
LOCALITY. Egypt, 1761–62.
MATERIAL. *Forsskål* s.n. (LINN — 1 sheet 1655.5 raised from seeds sent to Linnaeus, type of *C. pteranthus*, microf. 101: II. 2).

Pteranthus echinatus Desf., Fl. Atlant. 1: 144 (1798).
Pteranthus forskaolii Mirb. in Hist. Nat. Pl. ed. 2, 10: 130 (1806).
LOCALITY. Egypt: Cairo ("In desertis Káhirinis florens"), Mar. 1762.
MATERIAL. *Forsskål* 562 (C — 1 sheet with field label "Pteranthus nov. gen. dichotomous. Ca. 164", type of *P. forskaolii*, microf. 84: III. 7, 8); 553 (C — 1 sheet, type of *P. forskaolii* microf. 85: I. 1, 2); 1605 (C — 1 sheet ex herb. Schumacher, type of *P. forskaolii* microf. 85: I. 3, 4); 1606 (C — 1 sheet ex herb. Vahl, type of *P. forskaolii* microf. 85: I. 5, 6); 1607 (C — 1 sheet, type of *P. forskaolii*, microf. 85: I. 7, 8); 1608 (C — 1 sheet ex herb. Hofman Bang, type of *P. forskaolii*, microf. 85: II. 1, 2).
Forsskål s.n. (BM — 1 sheet, type of *P. forskaolii*.
NOTE. This is another case where a new genus and species without epithet is established in the 'Centuriae', while the epithet is provided in the 'Florae'.

Saponaria officinalis *L.* 1753: 408; Forssk. 1775: XXV No. 201; Fl. Turk. 2: 141 (1967).
LOCALITY. Turkey: Dardanelles ("Dd"), July 1761.
MATERIAL. *Forsskål* 1693 (C — 1 sheet, microf. 96: I. 8).

Silene biappendiculata *Rohrb.* in Bot. Zeit. 25: 82 (1867); Tackholm 1974: 89.
LOCALITY. Egypt, 1761–62.
MATERIAL. *Forsskål* 570 (C — 1 sheet, microf. 118: I. 3, 4).
NOTE. Named by Ascherson as *S. canopica* Del., a synonym of this species, which is an Egyptian endemic.

Silene colorata *Poir.*, Voy. Barb. 2: 163 (1789); Fl. Europ. 1: 180 (1964), 1: 217 (1993).
LOCALITY. Malta (on reverse of sheet), 12 June 1761.
MATERIAL. *Forsskål* 1731 (C — 1 sheet, plant no. 1, microf. 118: II. 3, 4).

Silene conica *L.* 1753: 418; Forssk. 1775: VI No. 96; Fl. Europ. 1: 180 (1964); 1: 218 (1993).
LOCALITY. France: Marseille ("Estac"), May 1761.
MATERIAL. *Forsskål* 571 (C — 1 sheet, field label "K 3", microf. 99: II. 7, 8).

Silene gallica *L.* 1753: 417; Fl. Turk. 2: 238 (1967).
LOCALITY. Turkey: Istanbul ("Belgrad prope Constantinop" on reverse), 30 July–8 Sept. 1761.
MATERIAL. *Forsskål* 1732 (C — 1 sheet, plant no. 1, microf. 118. II. 1, 2).
NOTE. Also a specimen of *S. noctiflora* on the same sheet.

Silene involuta *Forssk.* 1775: 210 (XXVI No. 213; Cent. VIII No. 47); Christensen 1922: 34.
LOCALITY. Turkey: Dardanelles ("Ad Dardanellos in arvis"), July 1761.
MATERIAL. *Forsskål* 569 (C — 1 sheet, type of *S. involuta*, microf. 99: III. 1, 2).
NOTE. This name was not accounted for in Fl. Turkey vol. 2.

Silene cf. **italica** (*L.*) *Pers.*, Syn. Pl. 1: 498 (1805); Fl. Europ. 1: 163 (1964).
LOCALITY. Not indicated.

MATERIAL. *Forsskål* 1734 (C — 1 sheet ex herb. Schumacher, microf. 118: II. 5, 6).
NOTE. Considered by Rohrbach to be near *S. italica.*

Silene cf. **paradoxa** *L.* 1763: 1673; Fl. Europ. 1: 164 (1964).
LOCALITY. Not indicated.
MATERIAL. *Forsskål* 568 (C — 1 sheet microf. 118: I. 5, 6).
NOTE. Named and labelled by Ascherson.

Silene nocturna *L.* 1753: 416; Fl. Turk. 2: 237 (1967).
LOCALITY. Turkey: Istanbul ("Belgrad prope Constaninope" on reverse), 30 July–8 Sept. 1761.
MATERIAL. *Forsskål* 1732 (C — 1 sheet specimen 2, microf. 118: II. 1, 2).
NOTE. Also a specimen of *S. gallica* on the same sheet.

Silene succulenta *Forssk.* 1775: 89 (LXVI No. 238; Cent. III No. 72); Christensen 1922: 19; Tackholm 1974: 86.
LOCALITY. Egypt: Alexandria ("Alexandriae ad Catacombas"), 1 Apr. 1762.
MATERIAL. *Forsskål* 552 (C — 1 sheet, type of *S. succulenta,* microf. 99: III. 3, 4); 561 (C — 1 sheet, type of *S. succulenta,* microf. 99: III. 7, 8); 567 (C — 1 sheet, type of *S. succulenta,* microf. 99: III. 5, 6).
Forsskål s.n. (LD — 1 sheet ex herb. Retzius, type of *S. succulenta).*
[? Syria attributed to *Forsskål* but another collector: C — 1 sheet, microf. 100: I. 1, 2].

Silene villosa *Forssk.* 1775: 88 (LXVI No. 239; Cent. III No. 71); Christensen 1922: 19; Fl. Palaest. 1: 93, pl. 119 (1966); Tackholm 1974: 86.
VERNACULAR NAME. *Kabbli* (Arabic).
LOCALITY. Egypt: Giza ('Pyramides Gizenses'), Jan. 1762.
MATERIAL. *Forsskål* 550 (C — 1 sheet, type of *S. villosa,* microf. 100: I. 5, 6); 560 (C — 1 sheet with field label 'Ca. 123', type of *S. villosa,* microf. 100: I. 3, 4).
NOTE. The name *Silene villosa* Forssk. is applied to two taxa in Forssk. (1775), from Egypt and Turkey respectively. The Egyptian plant carries a field label with 'Cairo' (Ca.). The Turkish collection was made at Belgrad ("prope Constantinop") and is described on p. 210 (Cent. VIII No. 46) where the description occurs, but no specimen has been found. Both type specimens are taken to be the Egyptian plant.

Silene vulgaris (*Moench*) *Garcke,* Fl. Nord Mittel-Deutschl. ed. 9, 64 (1869); Fl. Europ. 1: 169 (1964), 1: 204 (1993).
Cucubalus behen L. 1753: 414; Forssk. 1775: VI (No. 95), XXVI (No. 208).
LOCALITY. France: Marseille ("Estac"), May 1761, and Turkey: Gökçeada I. ("Imros"), July 1761.
MATERIAL. *Forsskål* 1177 (C — 1 sheet ex herb. Hornemann, plant no. 1, microf. 35: II. 3, 4).
NOTE. The other small specimen (2) on this sheet was collected in the Canaries by C. Smith, not by Forsskål.

Silene sp. indet. 1
LOCALITY. ? Egypt.
MATERIAL. *Forsskål* 572 (C — 1 sheet with original label "Cucubalus sp. fol. crassis", without locality, microf. 35: II. 5, 6).

Silene sp. indet. 2
LOCALITY. Without locality.
MATERIAL. *Forsskål* 1178 (C — 1 sheet, microf. 35: II. 1, 2).
NOTE. This sheet bears the determinations *Melandrium vespertinum* and *Silene behen*, but it is neither.

Spergularia marina (*L.*) *Griseb.*, Spicil. Fl. Rumel. 1: 213 (1843); Tackholm 1974: 95, pl. 20A.
Alsine segetalis sensu Forssk. 1775: 207 (LXIV No. 183; Cent. VIII No. 32), non L.; Christensen 1922: 33.
VERNACULAR NAME. *Girghair* (Arabic).
LOCALITY. Egypt: Cairo ("Kahirae frequens"), 1762.
MATERIAL. *Forsskål* 564 (C — 1 sheet, microf. 5: II. 5, 6).
Forsskål s.n. (BM — 1 sheet ex herb. Banks).

Velezia rigida *L.* 1753: 332; Fl. Turk. 2: 137 (1967).
Saponaria ocymoid? sensu Forssk. 1775: XXV No. 202.
LOCALITY. Turkey: Gökçeada I. ("Imros"), 7 July 1761.
MATERIAL. *Forsskål* 565 (C — 1 sheet, microf. 96: I. 6, 7); 1692 (C — 1 sheet, microf. 96: I. 5).

Caryophyllaceae gen. & sp. indet.
LOCALITY. Unspecified.
MATERIAL. *Forsskål* s.n. (LD — 1 sheet ex herb. Retzius).
NOTE. This specimen is subsucculent, rather densely pubescent; flowers large for size of plant; petals broadly ovate, entire; capsule glossy straw-coloured.

CELASTRACEAE

Catha edulis (*Vahl*) *Forssk. ex Endl.*, Enchiv. Bot.: 575 (1841); = *Catha edulis* Forssk. 1775: CVII No. 155, not validly published/non rite publ.; Deflers 1889: 121; Schweinfurth 1899: 337; Christensen 1922: 17; Schwartz 1939: 149; Fl. Zambes. 2: 381, t. 80 (1966); Krikorian in Journ. Ethnopharmacology 12: 115–178 (1984). = *Catha* [*sp. without epithet*] Forssk. 1775: 63 (Cent. III No. 4).
Celastrus edulis Vahl 1790: 21.
Catha Forskalii A. Rich., Fl. Abyss. Tent. 1: 134, t. 30 (1847).
VERNACULAR NAME. *Gat, Kat, Kath* (Arabic).
LOCALITY. Yemen: Al Hadiyah ("Bulgose sive Hadîe", "In *Yemen* colistur iisdem hortus cum *Coffea*"), 1763.
MATERIAL. *Forsskål* 504 (C — 1 sheet type of *C. edulis*, microf. 22: II. 1, 2); 505 (C — 1 sheet, type of *C. edulis*, microf. 22: I. 5, 6); 506 (C — 1 sheet type of *C. edulis*, microf. 22: I. 7, 8); 1066 (C — 1 sheet ex herb. Schumacher, type of *C. edulis*, microf. 22: II. 3, 4).
NOTE. The genus *Catha* is an example of a new genus with two new species being described in the 'Centuriae', only one species with epithet (both are provided in the 'Florae'). However, there is no validating description to the genus.

Maytenus parviflora (*Vahl*) *Sebsebe* in Symb. Bot. Upsal. 25: 63 (1985).
Celastrus parviflorus Vahl 1790: 21.

Catha spinosa Forssk. 1775: 64 (CVII No. 156; Cent. III No. 5), nom. inval. (see note to *Catha edulis*).
Gymnosporia spinosa (Forssk.) Fiori 1912: 225; Christensen 1922: 17.
G. senegalensis (Lam.) Loesener (1893) — Schwartz 1939: 149; var. *spinosa* (Forssk.) Engl. Bot. Jahrb. 17: 541 (1893); Schweinfurth 1899: 335.
LOCALITY. Yemen: Kusma ("Kurma"), Mar. 1763.
MATERIAL. *Forsskål* 503 (C — 1 sheet, type of *C. spinosa* and *C. parviflorus*, microf. 22: II. 7, 8); 507 (C — 1 sheet, lectotype of *C. spinosa* and *C. parviflorus*, microf. 22: II. 5, 6).
Forsskål s.n. (BM — 1 sheet ex herb. Banks, isolectotype).

CHENOPODIACEAE

Anabasis articulata (*Forssk.*) *Moq.* in DC., Prodr. 13, 2: 212 (1849) emend Aschers. & Schweinf. 1887: 128, 131: Fl. Palast. 1: 177 (1966); Tackholm 1974: 128; Freitag in Flora 183: 167 (1989). Fig. 15A
Salsola articulata Forssk. 1775: 55 (LXIV No. 160; Cent. II No. 87) & Icones t.8A.
VERNACULAR NAME. *Tartîr* (Arabic).
LOCALITY. Egypt: Giza ("In arena mobili circa Pyramides"), 1761–62.
MATERIAL. *Forsskål* 150 (C — 1 sheet, microf. 92: II. 3, 4); 188 (C — 1 sheet, type, microf. 92: II. 1, 2); 1664 (C — 1 sheet ex herb. Schumacher, type of *S. articulata*, microf. 92: I. 3, 4); 1665 (C — 1 sheet ex herb. Vahl, type of *S. articulata*, microf. 92: I. 5, 6); 1666 (C — 1 sheet ex herb Hofman Bang, with field label "Ca. 32", type of *S. articulata*, microf. 92: I. 7, 8).
Forsskål s.n. (BM — 1 sheet ex herb. Banks, type of *S. articulata*, microf. nil).

Arthrocnemum macrostachyum (*Moric.*) *K. Koch*, Hort. Dendrol.: 96 No. 3 (1853); Moris & Delponte in Ind. Sem. Hort. Reg. Taur.: 35 pl. 2 (1854); Fl. Palaest. 1: 156, pl. 226 (1966); Freitag in Flora 183: 152 (1989).
A. glaucum (Del.) Ung.-Sternb. — Tackholm 1974: 119.
Salicornia virginica sensu Forssk. 1775: 2 (LIX No. 3; Cent. I. No. 2), non L. (1753); Christensen 1922: 10.
VERNACULAR NAME. *Chraesi* (Arabic).
LOCALITY. Egypt: Alexandria, 1961–2.
MATERIAL. *Forsskål* 146 (C — 1 sheet, microf. 91: III. 5, 6); 174 (C — 1 sheet, microf. 91: III. 3, 4); 169 (C — 1 sheet, microf. 91: I. 7, 8).
Forsskål s.n. (BM — 1 sheet ex herb. Banks).

Atriplex coriacea *Forssk.* 1775: 175 (LXXVII No. 551; Cent. VI No. 79); Christensen 1922: 29; Tackholm 1974: 114.
VERNACULAR NAME. *Raetaem* (Arabic).
LOCALITY. Egypt: Alexandria ("Alexandriae ad Catacombas"), Dec. 1761.
MATERIAL. *Forsskål* 186 (C — 1 sheet, type of *A. coriacea*, microf. 13: III. 3, 4); 191 (C — 1 sheet, type of *A. coriacea*, microf. 13: III. 5, 6).
Forsskål s.n. (BM — 1 sheet ex herb. Banks, type of *A. coriacea*).

Atriplex farinosa *Forssk.* 1775: CXXIII No. 602; Deflers 1895: 367; Christensen 1922: 37; Schwartz 1939: 35; Tackholm 1974: 114; Boulos, Friis & Gilbert in Nord. J. Bot. 11(3): 311 (1991).
Atriplex hastata sensu Forssk. non L.; Forssk. 1775: 175 (CXXII No. 600; Cent. VI No. 78).

Tab. VIII.

P. Haas, sculp.

A. Salsola articulata
B. = inermis
C. = imbricata

FIG. 15. (A) **Anabasis articulata** (Forssk.) Moq. — Icones, Tab. VIII *Salsola articulata.*
(B) **Salsola inermis** Forssk — Icones Tab. VIII.
(C) **Salsola imbricata** Forssk. — Icones Tab. VIII.

VERNACULAR NAME. *Ösfai* (Arabic).
LOCALITY. Saudi Arabia: Jiddah ("Djidda" in book, "Insula prope Geddam" on reverse of sheet), 29 Oct. 1762.
MATERIAL. *Forsskål* 976 (C — 1 sheet ex herb. Hornemann, lectotype of *A. farinosa*, microf. 13: III. 7, 8).
NOTE. The name *Atriplex farinosa* in the list is accompanied by a description adequate to validate it "Foliis cordato-ovalibus, retusis, crassisculis, farinosis". The location is given as "Pm" = "Planities argillosa maritima" in the section headed 'Flora Arabico-Yemen'. This type was not seen by Christensen; it has been designated the lectotype by Boulos, Friis & Gilbert (l.c.). *Atriplex hastata* is in the list indicated with a star, meaning a new species described in the 'Centuriae' where such a description is found.

Atriplex halimus *L.* 1753: 1052 var. **schweinfurthii** *Boiss.*, Fl. Orient. 4: 916 (1879); Fl. Palaest. 1: 145 (1966); Tackholm 1974: 114.
A. glauca sensu Forssk. 1775: LXXVII No. 549, partly, non L.
VERNACULAR NAME. *Gataf*, *Ragbar* (Arabic).
LOCALITY. Egypt: Cairo, 1762.
MATERIAL. *Forsskål* 185 (C — 1 sheet, microf. 14: I. 3, 4).

Atriplex leucoclada *Boiss.*, Diagn. ser. 1, 12: 95 (1853); Fl. Palaest. 1: 146 pl. 206–208; (1966); Tackholm 1974: 111.
A. glauca sensu Forssk. 1775: LXXVII No. 549, partly, non L.
VERNACULAR NAME. *Gataf*, *Ragbat* (Arabic).
LOCALITY. Egypt: Cairo, 1762.
MATERIAL. *Forsskål* 187 (C — 1 sheet with original field label "Ca. 193", microf. 14: I. 5, 6); 192 (C — 1 sheet, microf. 14: I. 7, 8).

Atriplex tartarica *L.* 1753: 1053; Fl. Europ. 1: 96 (1964); Fl. Turk. 2: 309 (1967).
A. glauca sensu Forssk. 1775: XXXV No. 450, non L.
LOCALITY. Turkey: Dardanelles, July 1761.
MATERIAL. *Forsskål* 190 (C — 1 sheet plant 1, microf. 14: I. 1, 2, see Note); 977 (C — 1 sheet ex herb. Vahl, microf. 118: II. 7, 8).
NOTE. There are two specimens on No. 190: (1) is *A. tartarica*, but (2) is sterile and indeterminate. The description of *A. glauca* in p. XXV No. 450 is different from that of the plant under the same name on p. LXXII No. 549.

Atriplex sp.
LOCALITY. Egypt: Cairo, 1762.
MATERIAL. *Forsskål* 979 (C — 1 sheet ex herb. Hornemann, microf. 118: III. 1). NOTE. This sheet was labelled "Atriplex" by Dr A. Ludwigin, 1909.

Bassia muricata (*L.*) *Aschers.* in Schweinf., Beitr. Fl. Aethiop. 1: 187 (1867); Fl. Palaest. 1: 152, pl. 219 (1966); Tackholm 1974: 115; Freitag in Flora 183: 152 (1989).
Salsola muricata L. 1771: 54; Vahl 1790: 24.
Kochia muricata (L.) Schrad. (1809).
LOCALITY. Egypt ("Europa australi, Aegypto"), 1761–62.
MATERIAL. LINN 315.22 microf. 174: I. 6 (see Note).
NOTE. This type specimen may not be Forsskål's but Vaillant's (Vahl q.v.).
Salsola monobracteata Forssk. 1775: 55 (LXIV No. 161; Cent. II. No. 85); Christensen 1922: 16.
VERNACULAR NAME. *Aeraejam* (Arabic).

LOCALITY. Egypt: between Alexandria and Rosetta, Cairo ("Inter Alexandriam & Rosettam ad castellum Buckier copiose and in deserti Káhirinis"), 1761–62.
MATERIAL. *Forsskål* 159 (C — 1 sheet, type of *S. monobracteata,* microf. 93: III. 3, 4); 1675 (C — 1 sheet ex herb. Schumacher, type of *S. monobracteata,* microf. 93: III. 5, 6); 1676 (C —1 sheet ex herb. Liebmann, type of *S. monobracteata,* microf. 93: III. 7, 8); 1677 (C — 1 sheet, type of *S. monobracteata,* microf. 119: I. 3, 4).
Forsskål s.n. (BM — 1 sheet ex herb. Banks, type of *S. monobracteata*).

Beta vulgaris *L.* 1753: 222; Forssk. 1775: XXII No. 130, LXIII No. 154; Bircher 1960: 281; Fl. Turk. 2: 299 (1967); Tackholm 1974: 105, pl. 22.
VERNACULAR NAME. *Saelle* or *saelg* (Arabic).
LOCALITY. Turkey: Istanbul garden ("Constaninopoli"), Aug. 1761.
MATERIAL. *Forsskål* 1004 (C — 1 sheet specimens 1 & 2, microf. 15: I. 1, 2); 1005 (C — 1 sheet ex herb. Hornemann, microf. 119: I. 1, 2).
LOCALITY. Egypt: Rosetta (on sheet), Cairo (in book), 1762.
MATERIAL. *Forsskål* 1004 (C — 1 sheet specimen 3, microf. 15: I. 1, 2).

Chenopodium album *L.* 1753: 219; Forssk. 1775: XXII No. 125; Tackholm 1974: 107 pl. 23A.
VERNACULAR NAME. *Siritjam* (Turkish at Bozcaada = Tenedos).
LOCALITY. Egypt: Alexandria, 1761–2.
MATERIAL. *Forsskål* no specimen found.

Chenopodium murale *L.* 1753: 219; Schwartz 1939: 34.
C. triangulare Forssk. 1775: 205 (CVIII No. 197; Cent. VIII No. 22).
LOCALITY. Yemen: Taizz ("circa Taaes frequens"), 1763.
MATERIAL. *Forsskål* no specimen found.

Chenopodium schraderianum *Schultes,* Syst. Veg. 6: 260 (1820); F.T.E.A. Chenopodiac.: 12 (1954).
C. foetidum Schrad. (1808) — Deflers 1889: 195; Schweinfurth 1896: 156; Schwartz 1939: 35.
C. botrys sensu Forssk. 1775: CVIII No. 195, non L.
VERNACULAR NAME. *Schokr el bomâr* (Arabic).
LOCALITY. Yemen: J. Khadra ("Chadra"), 2–3 Apr. 1763.
MATERIAL. *Forsskål* 184 (C — 1 sheet with original field label, microf. 26: I. 5, 6).

Chenopodium serotinum *Forssk.* 1775: 205 (CVIII No. 196; Cent. VIII No. 21).
LOCALITY. Yemen: Taiz ("Taaes in cultis"), 1763.
MATERIAL. *Forsskål* no specimen found.
NOTE. This remains an obscure species. The name should not be adopted if it upsets widely accepted nomenclature.

Chenopodium viride sensu Forssk. 1775: 205 (CVIII No. 198; Cent. VIII No. 20), ? non L.
VERNACULAR NAME. *Rokeb ed djemmel, Rockeb el djämmel* (Arabic).
LOCALITY. Yemen ("In Yemen frequens"), 1763.
MATERIAL. *Forsskål* no specimen found.
NOTE. The identity of this plant remains obscure until a specimen can be found.

Halimione portulacoides (*L.*) *Aellen*, Verh. Naturf. Ges. Basel 49: 126 (1938); Fl. Palaest. 1: 151 pl. 216 (1966); Tackholm 1974: 115.
Atriplex portulacoides L. 1753: 1053; Forssk. 1775: 175 (LXXVII No. 550; Cent. VI No. 80).
LOCALITY. Egypt: Rashid ("Rosettae"), Nov. 1761.
MATERIAL. *Forsskål* 183 (C — 1 sheet with field label "Damiata", microf. 14: II. 1, 2); 978 (C — 1 sheet ? ex herb. Hornemann, microf. 119: I. 5, 6).
Forsskål s.n. (BM — 1 sheet ex herb. Banks).
Forsskål s.n. (LD — 1 sheet ex herb. Retzius).

Halocnemum strobilaceum (*Pallas*) *M. Bieb.*, Fl. Taur.-Cauc. 3: 3 (1819); Tackholm 1974: 118.
Salicornia cruciata Forssk. 1775: 2 (LIX No. 4; Cent. I No. 3); Christensen 1922: 10.
VERNACULAR NAME. *Sabta* (Arabic).
LOCALITY. Egypt: Alexandria ("*Alexandriae* ad salinas copiose ..."), Oct. 1761.
MATERIAL. *Forsskål* 147 (C — 1 sheet, type of *S. cruciata*, microf. 91: I. 3, 4); 168 (C — 1 sheet, type, microf. 91: I. 5, 6); 171 (C — 1 sheet, type, microf. 90: III. 6, 7); 172 (C — 1 sheet, type, microf. 91: I. 1, 2); 1661 (C — 1 sheet with field label "Salsola * cruciata", type, microf. 90: III. 4, 5); 1662 (C — 1 sheet ex herb. Hornemann, type, microf. 91: II. 1, 2); 1663 (C — 1 sheet ex herb. Hornemann, microf. 91: II. 7, 8).
NOTE. *Forsskål* 169 was identified with this species by Ascherson and Christensen but it has been re-identified as *Arthrocnemum strobilaceum* (Moric.) Koch by Freitag, 1987.

Halopeplis perfoliata (*Forssk.*) *Bunge* in Linnaea 28: 573 (1874); Schweinfurth 1896: 157; Schwartz 1939: 36. Fig. 16.
Salicornia perfoliata Forssk. 1775: 3 (CII No. 7; Cent. I No. 4), & Icones t. 1; Vahl 1804: 13; Christensen 1922: 10.
LOCALITY. Saudi Arabia: Jedda and Qunfudhah ("*Djiddae* & *Ghomfodae* copiose ad littora ..."), Dec. 1762.
MATERIAL. *Forsskål* 173 (C — 1 sheet, type of *S. perfoliata*, microf. 91: III. 1,2).

Noaea mucronata (*Forssk.*) *Aschers. & Schweinf.* 1887: 131; Fl. Palaest. 1: 175, pl. 237 (1966); Tackholm 1974: 127; Freitag in Flora 183: 167 (1989).
Salsola mucronata Forssk. 1775: 56 (LXIV No. 158; Cent. II No. 88); Christensen 1922: 16.
Anabasis spinosissima L.f. 1782 (1781): 173; Vahl 1790: 24.
VERNACULAR NAME. *Sjök ihannasch* (Arabic).
LOCALITY. Egypt: Alexandria ("*Alexandriae* ad catacombas"), Oct. 1761.
MATERIAL. *Forsskål* 182 (C — 1 sheet, type of *S. mucronata* & *A. spinosissima*, microf. 94: I. 3, 4); 1678 (C — 1 sheet ex herb. Schumacher, type of *S. mucronata* & *A. spinosissima*, microf. 94: I. 1, 2).

Salicornia europaea *L.* 1753: 3; Forssk. 1775: 2 (LIX No. 2a; Cent. I No. 1), partly; Christensen 1922: 10; Freitag in Flora 183: 152 (1989).
S. herbacea L. — Tackholm 1974: 119.
VERNACULAR NAME. *Chraesi*, *Hattab badâda* (Arabic).
LOCALITY. Egypt: Alexandria ("*Alexandriae* ad *Ras ettîn*"), Oct. 1761.
MATERIAL. *Forsskål* s.n. (BM — 1 sheet ex herb. Banks).
Forsskål s.n. (LD — 1 sheet ex herb. Retzius).
NOTE. The other specimens labelled *S. europaea* have been re-identified as *S. fruticosa*.

Salicornia fruticosa (*L.*) *L.* 1762: 5; Schwartz 1939: 36; Tackholm 1974: 119; Freitag in Flora 183: 153 (1989).
S. europaea sensu Forssk. 1775: 2 (LIX No. 26; CII No. 6, Cent. I No. I), partly.

Tab. I.

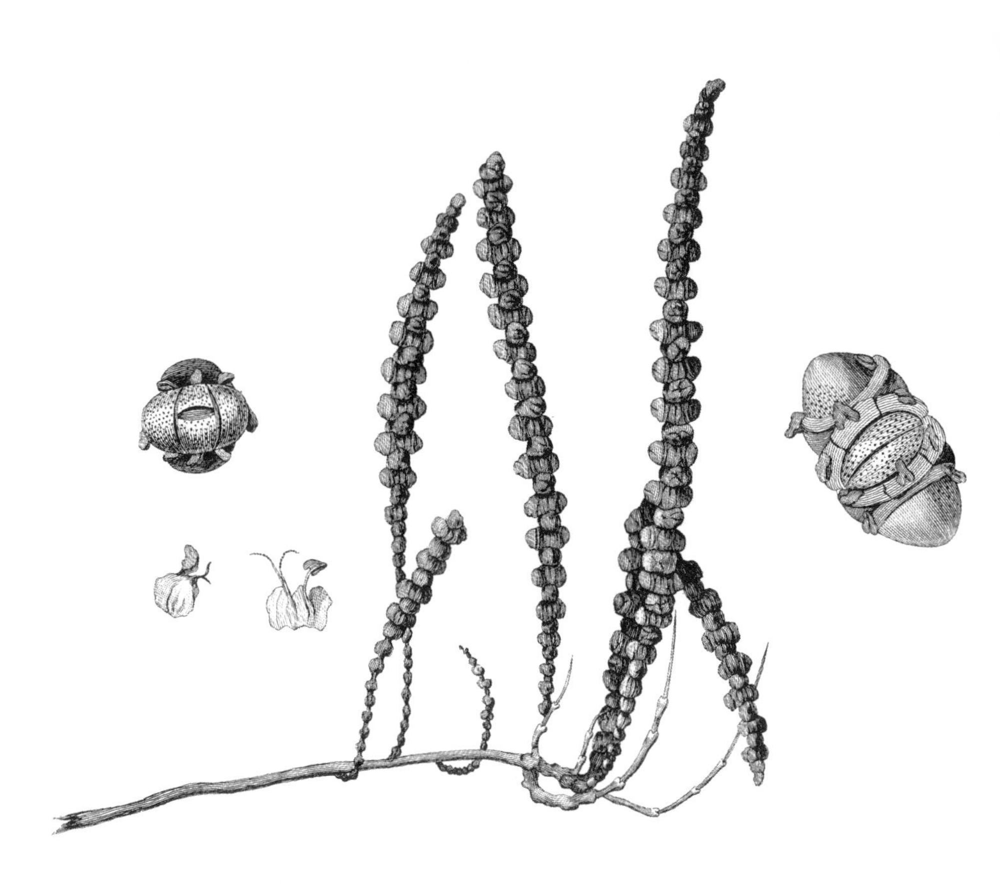

Salicornia perfoliata

Fig. 16. **Halopeplis perfolia** (Forssk.) Bunge — Icones, Tab. I *Salicornia perfoliata*.

VERNACULAR NAME. *Chraesi, Hattab badâde* (Arabic).
LOCALITY. Egypt: Alexandria ("*Alexandriae* ad *Ras ettîn*"), Oct. 1761.
Yemen: Mocha ("Mochha"), 1763.
MATERIAL. *Forsskål* 143 (C — 1 sheet, microf. 91: II. 3, 4); 170 (C — 1 sheet, microf. 91: II. 5, 6).
Forsskål s.n. (BM — 1 sheet ex herb. Banks).
NOTE. This was named as *S. fruticosa* by Schweinfurth in 1895 and separated from the *S. europaea*. The BM material could be from either Egypt or Yemen.
The specimen *Forsskål* 1663 (microf. 91: II. 7, 8) was previously placed here, but it has been re-identified by H. Freitag as *Halocnemum strobilaceum*.

Salsola imbricata *Forssk.* 1775: 57 (CXVIII No. 201; Cent. II No. 90), & Icones t.7 (1776); Deflers 1889: 195; Christensen 1922: 16; Schwartz 1939: 38. Fig. 15C.
Caroxylon imbricatum (Forssk.) Moq. in DC., Prodr. 13(2): 177 (1849).
Salsola forskalii Schweinfurth 1896: 160 nom. superfl.
VERNACULAR NAME. *Harm* (Arabic).
LOCALITY. Yemen: Al Luhayyah ("Lohajae in locis sterilibus"), Jan. 1763.
MATERIAL. *Forsskål* no type specimen found.
NOTE. An obscure species, although several authors have attempted to identify it:— Botshantzev in Novosti Sist. Vyssh. Rast 12: 165 (1975), Freitag in Flora 183: 158 (1989) and Boulos in Kew Bull. 46: 137 (1991). Schweinfurth (1896) and J.R.I. Wood (ined.) both think that it is possibly *S. spinescens* Moq., but it is impossible to be certain without a type specimen.

Salsola inermis *Forssk.* 1775: 57 (LXIV No. 157; Cent. II. No. 89), & Icones t. 8B (1776); Christensen 1922: 16; Fl. Palaest. 1: 171 pl. 249 (1966); Fl. Turk. 2: 330 (1967); Tackholm 1974: 125; Freitag in Flora 183: 160 (1989). Fig. 15B.
VERNACULAR NAMES. *Gummaeli, Naedaeva* or *Naedeuva* (Arabic).
LOCALITY. Egypt: Alexandria ("Alexandriae"), Oct. 1761.
MATERIAL. *Forsskål* 151 (C — 1 sheet, type of *S. inermis*, microf. 93: I. 3, 4); 158 (C — 1 sheet, type microf. 92: II. 7, 8); 163 (C — 1 sheet with field label No. "As 8", type, microf. 92: III. 3, 4); 181 (C — 1 sheet, type, microf. 92: III. 5, 6); 1668 (C — 1 sheet ex herb. Liebmann, type, microf. 92: III. 7, 8); 1669 (C — 1 sheet ex herb. Schumacher, type, microf. 93: I. 1, 2); 1670 (C — 1 sheet ex herb. Liebmann, type, microf. 93: I. 5, 6).
Forsskål s.n. (BM — 2 sheets ex herb. Banks, types of *S. inermis*).
Forsskål s.n. (LE — 1 sheet, type of *S. inermis*).
NOTE. The specimen No. 1667 has been re-identified as *S. volkensii* Aschers. & Schweinf. The identification of the remaining sheets has been confirmed.

Salsola kali *L.* 1753: 222; Fl. Palaest. 1: 170, pl. 246 (1966); Fl. Turk. 2: 329 (1967); Tackholm 1974: 125.
S. kali var. a) *glabra* Forssk. 1775: 54 (XXII No. 122; Cent. II No. 84); Christensen 1922: 16.
LOCALITY. Turkey ? (see b), 1761.
MATERIAL. *Forsskål* no specimen found.
S. kali var. b) *hispida* Forssk. 1775: 54 (XXII No. 122; Cent. II. No. 84); Christensen 1922: 16.
LOCALITY. Turkey: Sea of Marmora and elsewhere ("Copiose in Natoliae littore e regione Tenedos, glabra illa; quae deinde inventa hispidae mixta ad littora Maris Marmorae inter pageos Eraclissam & Merafte"), 1761.
MATERIAL. *Forsskål* 153 (C — 1 sheet, type, microf. 93: II. 3, 4); 156 (C — 1 sheet, microf. 93: I. 7, 8); 1671 (C — 1 sheet ex herb. Schumacher, type, microf. 93: II. 1, 2).
Forsskål s.n. (LD — 1 sheet ex herb. Retzius, ? type of *S. kali* var. *hispida*).

S. kali c) *hispida polygama* Forssk. 1775: 55 (XXII No. 122; Cent. II. No. 84C); Christensen 1922: 16.
LOCALITY. Egypt: Alexandria, 1761–62.
MATERIAL. *Forsskål* 1672 (C — 1 sheet with field label "Salsola kali hispida polygama. Ad salinas. As 17", microf. 93: II. 5).
NOTE. This taxon is not validly published.

Salsola longifolia *Forssk.* 1775: 55 (LXIV No. 159; Cent. II. No. 86); Fl. Palaest. 1: 173, pl. 253 (1966); Tackholm 1974: 125; Freitag in Flora 183: 161 (1989).
LOCALITY. Egypt: Between Alexandria & Rashid ("Buckier ..."), Oct. 1761.
MATERIAL. *Forsskål* 155 (C — 1 sheet, type of *S. longifolia,* microf. 93: III. 1, 2); 157 (C — 1 sheet, microf. 94: III. 5, 6); 166 (C — 1 sheet with field label "Salsola vermiculata. Alexandria. As. 24", microf. 94: III. 1, 2); 1674 (C — 1 sheet ex herb. Hofman Bang, with field label "Salsola vermiculata. Brikier. Bri. 12", microf. 94: III. 3, 4); 1673 (C — 1 sheet, ex herb. Liebmann, type of *S. longifolia,* microf. 93: II. 8).
Forsskål s.n. (BM — 2 sheets ex herb. Banks, type of *S. longifolia*).
NOTE. As appears from the field labels, Forsskål suggested that these specimens belonged to *S. vermiculata;* only one entry in the printed work refers to this name, and the associated specimen has not been found. All identifications have been confirmed by A. Freitag, 1989. Note that the micro edition of Herb. Forsskålii includes microf. 93: II. 6, 7 by mistake since it is a Roche specimen collected in Egypt.

Salsola soda *L.* 1753: 223; Forssk. 1775: V No. 67, XXII No. 124; Fl. Europ. 1: 105 (1964); Fl. Turk. 2: 330 (1967).
LOCALITY. France: Marseille ("Estac"), May 1761.
MATERIAL. *Forsskål* 154 (C — 1 sheet with field label "Salsola soda. Estac 52", microf. 94: I. 5, 6).
LOCALITY. Turkey: Izmia ("Sm" = Smyrna), June–July 1761.
MATERIAL. *Forsskål* 148 (C — 1 sheet microf. 94: II. 1, 2); 648 (C — 1 sheet microf. 94: I. 7, 8).

Salsola tetrandra *Forssk.* 1775: 58 (LXIV No. 163; Cent. II No. 92); Ascherson & Schweinfurth 1887: 129; Muschler 1912: 296; Christensen 1922: 16; Fl. Palaest. 1: 172, pl. 251 (1966); Tackholm 1974: 125; Freitag in Flora 183: 164 (1989).
LOCALITY. Egypt: Alexandria ("*Alexandriae*"), Oct. 1761.
MATERIAL. *Forsskål* 152 (C — 1 sheet, type of *S. tetrandra,* microf. 92: III. 1, 2); 176 (C — 1 sheet with field label "81 Salsola nov. tetrandra. Alex. Ras. ettim", type of *S. tetrandra,* microf. 94: II. 3, 4); 177 (C — 1 sheet, type of *S. tetrandra,* microf. 94: II. 5, 6); 1679 (C — 1 sheet ex herb. Schumacher, type of *S. tetrandra,* microf. 94: II. 7, 8).
Forsskål s.n. (BM — 1 sheet ex herb. Banks, type of *S. tetrandra*).

Salsola vermiculata ? sensu Forssk. 1775: 57 (CVIII No. 203; Cent. II No. 203), non L.?; Christensen 1922: 16.
LOCALITY. Saudi Arabia: Jiddah ("Djiddae in Insula Abu sáad"), 29 Oct. 1762.
MATERIAL. *Forsskål* no specimen found.
NOTE. *S. vermiculata* L. 1753: 223 was described from Spanish material, but in its numerous varieties the species has a wide distribution in dry countries including Saudi Arabia. In the absence of a Forsskål specimen confirmation is impossible.

Salsola volkensii *Aschers. & Schweinf.* in Mém. Inst. Egypt. 2: 130 (1887); Tackholm 1974: 125.
LOCALITY. Egypt: Alexandria, Oct. 1761.

MATERIAL. *Forsskål* 1667 (C — 1 sheet ex herb. Liebmann, microf. 92: II. 5, 6).
NOTE. This was formerly identified as *S. inermis* Forssk.

Suaeda aegyptiaca (*Hasselq.*) *Zohary* in J. Linn. Soc. 55: 635 (1957); Freitag in Flora 183: 153 (1989).
Schanginia aegyptiaca (Hasselq.) Aellen in Rechinger, Fl. Lowland Iraq: 195 (1964); Tackholm 1974: 122.
Suaeda baccata Forssk. 1775: 69 (LXIV No. 186, CIX No. 221; Cent. III No. 15); Christensen 1922: 17; Schwartz 1939: 36, not validly published/non rite publ.
VERNACULAR NAME. *Mullaeah* (Arabic).
LOCALITY. Egypt: Alexandria ("Alexandriae"), Sept. 1761.
MATERIAL. *Forsskål* 164 (C — 1 sheet, type of *S. baccata*, microf. 104: III. 5, 6).
NOTE. Seven new species of the new genus *Suaeda* are described in the 'Centuriae', one without epithet which is not provided in the 'Florae'. As no separate generic description is supplied, all names are invalid.

Suaeda fruticosa *Forssk. ex J.F. Gmel.*, Syst. Nat. ed. 13, 2: 503 (1791); (Forssk. 1775: 70 CIX No. 223; Cent. III No. 19), & Icones t. 9 (1776), not validly published/non rite publ.; Christensen 1922: 17; Schwartz 1939: 36; Fl. Palaest. 1: 159, pl. 130 (1966); Freitag in Flora 183: 156 (1989). Fig. 17.
LOCALITY. Egypt: Alexandria ("As"), Oct. 1761.
MATERIAL. *Forsskål* 161 (C — 1 sheet, microf. 105: II. 3, 4).
LOCALITY. Saudi Arabia: Jiddah ('Djddae' in book, 'Gomfudae' on field label), Nov. 1762.
MATERIAL. *Forsskål* 162 (C — 1 sheet with field label "Suaeda * Vera. Gomfodae", lectotype selected by Freitag, of *S. fruticosa*, microf. 104: II. 7, 8).
Forsskål s.n. (BM — 1 sheet with field label "As 10", type of *S. fruticosa*).
NOTE. See note to *S. aegyptiaca*.

Suaeda hortensis *Forssk. ex J.F. Gmel.*, Syst. Nat. ed. 13, 2: 503 (1791); Forssk. 1775: 71 (LXIV No. 188, CIX No. 222; Cent. III. No. 21), not validly published/non rite publ.); Christensen 1922: 18; Schwartz 1939: 37; Fl. Palaest. 1: 162 (1966).
Schanginia hortensis (Forssk. ex J.F. Gmel.) Moq., Chenopod. Monogr. Enum.: 119 (1840); Tackholm 1974: 122.
VERNACULAR NAME. *Mullah* (Arabic).
LOCALITY. Egypt: Cairo ("In Káhirinis hortis frequens"), 1762.
MATERIAL. *Forsskål* 165 (C — 1 sheet "Ca. 10. Suaeda baccata" on field label, microf. 105: I. 3, 4); 189 (C — 1 sheet, microf. 105: I. 5, 6).
LOCALITY. Yemen: Taizz ("In locis ruderatis arica agros urbis Taaes"), 1763.
MATERIAL. *Forsskål* 145 (C — 1 sheet lectotype of *S. hortensis*, microf. 105: I. 1, 2).
NOTE. See note to *S. aegyptiaca*.

Suaeda monoica *Forssk. ex J.F. Gmel.*, Syst. Nat. ed. 13, 2: 503 (1791); (Forssk. 1775: 70 (LXIV No. 189; CIX No. 220; Cent. III No. 20), not validly published/non rite publ.); Schwartz 1939: 37; Fl. Palaest. 1: 160, pl. 233 (1966); Tackholm 1974: 122, pl. 30B; Freitag in Flora 183: 154 (1989); Boulos in Kew Bull. 46: 296 (1991).
VERNACULAR NAME. *Asal* (Arabic).
LOCALITY. Egypt: ? Alexandria ('Alexandriae') see Note.
MATERIAL. *Forsskål* 180 (C — 1 sheet, lectotype of *S. monoica* selected by Freitag, microf. 105: II. 1, 2).
LOCALITY. Yemen: Al Luhayyah ('Lohajae'), Dec. 1762–Jan. 1763.

FIG. 17. **Suaeda fruticosa** Forssk. ex J.G. Gmel. — Icones, Tab. IX *Sueda fruticosa*?

MATERIAL. *Forsskål* 167 (C — 1 sheet with field label "Suaeda baccata ? vera ? Lohaja. Kudme", lectotype of *S. monoica* selected by Boulos, microf. 105: I. 7, 8).
NOTE. Forsskål cites both 'Alexandriae' and 'Lohajae' and one specimen is actually labelled 'Lohaja', but the other cannot have come from Alexandria since it is out of the range of this species. See also note to *S. aegyptiaca*.

Suaeda salsa (*L.*) *Pallas*, Illustr. Pl. 46, t. 39 (1803–06); Tackholm 1974: 123.
LOCALITY. Egypt, 1761–62.
MATERIAL. *Forsskal* 179 (C — 1 sheet, microf. 118: III. 2, 3).
NOTE. Named and labelled by G. Schweinfurth 1895.

Suaeda vera *Forssk. ex J.F. Gmel.*, Syst. Nat. ed. 13, 2: 503 (1791); (Forssk. 1775: 69 (LXIV No. 185; CIX No. 219; Cent. III No. 16) not validly published/non rite publ.); Christensen 1922: 17; Fl. Europ. 1: 105 (1964), 1: 124 (1993); Fl. Palaest. 1: 159, pl. 229 (1966); Tackholm 1974: 122.
Chenopodium fruticosum L. 1753: 221, non *Suaeda fruticosa* Forssk. ex J.F. Gmel.; Forssk. 1775: XIII No. 23.
Suaeda sp. — Forssk. 1775: 69 (Cent. III No. 17); Christensen 1922: 22.
VERNACULAR NAMES. *Suaed, Hömmam* (Arabic).
LOCALITY. Malta, 14–20 June 1761.
MATERIAL. *Forsskål* 149 (C — 1 sheet, type of *S. vera*, microf. 26: I. 7, 8).
LOCALITY. Egypt: Alexandria ("As"), Oct. 1761.
MATERIAL. *Forsskål* 161 (C — 1 sheet, type of *S. vera*, microf. 105: II. 3, 4).
NOTE. Schweinfurth 1912: 155 considered Forsskål's un-named *Suaeda* on p. 69 to be *S. vera*.

Suaeda vermiculata *Forssk. ex J.F. Gmel.*, Syst. Nat. ed. 13, 2: 503 (1791); (Forssk. 1775: 70 (LXIV No. 187; Cent. III No. 18), not validly published/non rite publ.); Christensen 1922: 17; Fl. Palaest. 1: 160, pl. 231 (1966); Tackholm 1974: 122; Freitag in Flora 183: 156 (1989). Fig. 20B.
LOCALITY. Egypt: Alexandria ('Alexandriae'), Dec. 1761.
MATERIAL. *Forsskål* 160 (C — 1 sheet with field label "Suaeda vermiculata. As 1", lectotype of *S. vermiculata* selected by Freitag, microf. 105: II. 5, 6).
NOTE. See note to *S. aegyptiaca*.

Tetragonia sp. indet.
LOCALITY. Unspecified.
MATERIAL. *Forsskål* s.n. (LD — 1 sheet ex herb. Retzius).

Traganum nudatum *Delile*, Fl. Aeg. Ill.: 60 (1813); Schwartz 1939: 37; Tackholm 1974: 123, pl. 32C.
LOCALITY. Egypt: Cairo, 1761–62.
MATERIAL. *Forsskål* 175 (C — 1 sheet, microf. 118: III. 6, 7); 178 (C — 1 sheet, microf. 118: III. 4, 5); 1840 (C — 1 sheet ex herb. Schumacher, microf. nil).

Chenopodiaceae gen. et sp. indet. LOCALITY. Egypt: Sinai, Mt. Horeb (on original label), Sept. 1762.
MATERIAL. *Forsskål* 1003 (C — 1 sheet ? ex herb. Hornemann with field label "Convolvulus cneor. M. Horeb.", microf. 31: I. 3, 4).

NOTE. This incomplete material was originally named *Convolvulus cneorum* by Hornemann, but it seems to belong to the Chenopodiaceae. H. Freitag has suggested that it perhaps is a species of *Bassia*, possibly *B. eriophora* (Schrad.) Aschers. & Schweinf. It has been associated with sheets of *Convolvulus lanatus* (*C. cneorum* sensu Forssk.) and it is possible that some kind of mix-up has occurred. Presumably it was collected by Niebuhr or von Haven on their trip to Sinai.

CISTACEAE

Helianthemum kahiricum *Delile*, Fl. Aeg. Ill.: 65 (1813), & Fl. Egypte: 93, 31, f. 2 (1813); Fl. Palaest. 2: 340, pl. 500 (1972); Tackholm 1974: 364.
Cistus stipulatus β Forssk. 1775: 101 (Cent. IV No. 11); Christensen 1922: 20.
LOCALITY. Egypt: Cairo ("vidi in desertis Káhirinis ad Liblab"), 1762.
MATERIAL. *Forsskål* 614 (C — 1 sheet, microf. 27: III. 7, 8).

Helianthemum stipulatum (*Forssk.*) *Christensen* 1922: 20; Fl. Turk. 1: 511 (1965); Verdcourt in Bol. Soc. Brot. 40: 58 (1966); Meikle 1977: 192; Tackholm 1974: 364.
Cistus stipulatus Forssk. 1775: 100 (Cent. IV No. 11); Christensen 1922: 20.
C. thymifol(ia)? — Forssk. 1775: LXVII No. 280.
VERNACULAR NAME. *Chosjaein* (Arabic).
LOCALITY. Egypt: Alexandria ("Alexandriae circa Salinas in collibus siccissimus"), April 1762.
MATERIAL. *Forsskål* 615 (C — 1 sheet, type of *C. stipulatus*, microf. 27: III. 3, 4); 616 (C — 1 sheet, type of *C. stipulatus*, microf. 27: III. 1, 2); 617 (C — 1 sheet, type of *C. stipulatus*, microf. 27: III. 5, 6); 618 (C — 1 sheet, type of *C. stipulatus*, microf. 27: II. 7, 8).
NOTE. Forsskål recognised var. *β* which is *Helianthemum kahiricum* (q.v.).

COMPOSITAE

Achillea fragrantissima (*Forssk.*) *Sch. Bip.* in Flora 38: 13 (1855); Fl. Palaest. 3: 343 pl. 579 (1978); Tackholm 1974: 577, pl. 206C.
Santolina fragrantissima Forssk. 1775: 147 (LXXIII No. 436; Cent. V No. 71); Vahl 1797: 70; Christensen 1922: 26.
VERNACULAR NAME. *Kaejsum, Keisûm gébeli* (Arabic).
LOCALITY. Egypt: Sinai, Mt Horeb ('Horeb' on field label; 'In desertis arenosis ad Káhiram & Suês' in book), Aug. 1762.
MATERIAL. *Forsskål* 1686 (C — 1 sheet, type of *S. fragrantissima*, microf. 95: III. 1, 2); 1687 (C — 1 sheet with field label "Santolina * fragrantissima. M. Horeb", type of *S. fragrantissima*, microf. 95: III. 3, 4); 1688 (C — 1 sheet, type of *S. fragrantissima*, microf. 95: III. 5, 6).

Achillea lobata *Forssk.* 1775: 218 (LXXIV No. 459 as 'lobatifolia'; Cent. VIII No. 93);
LOCALITY. Egypt: Alexandria ("Alexandriae rarior"), 1761.
MATERIAL. *Forsskål* no specimen found.
NOTE. An unknown species validly published and described: "foliis villosis, linearibus, subdentatis, apice lobatis. Planta annua. Corolla radio alba." This name should not be adopted if it upsets widely accepted nomenclature.

Achillea santolina *L.* 1753: 896; Tackholm 1974: 577.
A. falcata sensu Forssk. 1775: LXXIV No. 458.

VERNACULAR NAME. *Bastran, Kaejsûn* (Arabic).
LOCALITY. Egypt: Alexandria (As = Alexandriae spontanea), 1 Apr. 1762.
MATERIAL. *Forsskål* 908 (C — 1 sheet ex herb. Hornemann, on reverse "In itinere terrestri inter Alexandriam et Cairo", microf. 1: II. 7, 8); 909 (C — 1 sheet ex herb. Liebmann, microf. 1. III. 1, 2).

Aegialophila pumilio (*L.*) *Boiss.*, Diagn. ser. I, 10: 105 (1849); Tackholm 1974: 541, pl. 190 B (as 'A. pumila'); Fl. Palaest. 3: 391 pl. 658 (1978).
Centaurea mucronata Forssk. 1775: 151 (LXXIV No. 468 '*mucronulata*'; Cent. V No. 91); Christensen 1922: 26.
LOCALITY. Egypt: Alexandria ("Alexandriae"), April 1762.
MATERIAL. *Forsskål* 1082 (C — 1 sheet ex herb. Vahl, type of *C. mucronata*, microf. 24: I. 3, 4).

Aetheorhiza bulbosa (*L.*) *Cass.*, Dict. Sci. Nat. 48: 426 (1827); Fl. Europ. 4: 326 (1976).
LOCALITY. Unspecified — a species with a southern European distribution.
MATERIAL. *Forsskål* 919 (C — 1 sheet, microf. 121: I. 3, 4).

Ambrosia maritima *L.* 1753: 988; Tackholm, 1974: 568 pl. 202;
Ambrosia villosissima Forssk. 1775: 160 (LXXV No. 848; Cent. VI No. 20); Christensen 1922: 28.
VERNACULAR NAME. *Demsise* (Arabic).
LOCALITY. Egypt: Cairo ("In Insulis Niloticis circa Kâhiram frequens"), June 1762.
MATERIAL. *Forsskål* 928 (C — 1 sheet ex herb. Hofman Bang, type of *A. villosissima*, microf. 5: II. 7, 8); 929 (C — 1 sheet ex herb Schumacher, type of *A. villosissima*, microf. 5: III. 1, 2); 930 (C — 1 sheet ? ex herb. Vahl, type of *A. villosissima*, microf. nil); 931 (C — 1 sheet ? ex herb. Vahl, type of *A. villosissima*, microf. nil); 932 (C — 2 specimens on one sheet ex herb. Hornemann, type of *A. villosissima*, microf. nil).

Anacyclus monanthos (*L.*) *Thell.* in Mém. Soc. Nat. Sci. Cherbourg Sér. IV, 38: 518 (1911–12); Christensen 1922: 26; Humphries in Bull. Brit. Mus. (N.H.) 7: 116 fig. 14 (1979).
Tanacetum monanthos L. 1767: 111; Vahl 1790: 70.
LOCALITY. Egypt: Alexandria, 1761–62.
MATERIAL. Seeds sent by Forsskål to Linnaeus and grown in Hort. Uppsal. (as *T. monanthemum*) (LINN — 1 sheet 987: 8, type of *T. monanthos*, microf. 563: III. 6).

Santolina terrestris Forssk. 1775: 147 (LXXIII No. 435; Cent. V No. 72); Christensen 1922: 26.
LOCALITY. Egypt: Alexandria ('In loco inundatis Matareae'), Apr. 1762.
MATERIAL. *Forsskål* 1689 (C — 1 sheet, type of *S. terrestris*, microf. 95: III. 7, 8); 1690 (C — 1 sheet ex herb. Schumacher, type of *S. terrestris*, microf. 96: I. 1, 2); 1691 (C — 1 sheet with field label "Av. 122 * Santolina nova terrestris. Alex. in vicinia columna pompeyi", type of *S. terrestris*, microf. 96: I. 3, 4).

Anacyclus alexandrinus Willd., Sp. Pl. 3: 2173 (1803); Tackholm 1974: 577 pl. 205A.
LOCALITY. Egypt: Alexandria, 1761–62.
MATERIAL. Presumably Forsskål's sent to Willdenow either by Linnaeus or Vahl (B — WILLD — 3 sheets, 16307/1–3, types of *A. alexandrinus* microf. 1174: III. 9, 1175: I. 1, 2).

Anthemis melampodina *Delile*, Fl. Egypte: 268, t. 45 f. 1 (1813); Tackholm 1974: 575 pl. 204B.
Anthemis Frach om ali Forssk. 1775: LXXIV no. 457, nom. illegit.
LOCALITY. Egypt: Cairo (Cd = in desertis Kahirinis), 1762.
MATERIAL. *Forsskål* 944 (C — 1 sheet ex herb. Schumacher, microf. 7: III. 1, 2).
NOTE. Since *Frach om ali* does not conform to the binomial system it cannot be considered to be an epithet and a name is not therefore validly published for this species, in spite of there being a perfectly adequate brief description. The specimen is therefore not a type.

Anthemis tinctoria *L.* 1753: 896; Vahl 1790: 74; Fl. Turk. 5: 211 (1975).
Santolina flava Forssk. 1775: XXXI No. 356; Christensen 1922: 36.
LOCALITY. Turkey: Istanbul ('Belgrad'), Aug. 1761.
MATERIAL. *Forsskål* 1685 (C — 1 sheet, type of *S. flava*, microf. 95: II. 7, 8).
NOTE. R. Fernandes in 1975 annotated this sheet "perhaps the discoid form of Anthemis tinctoria L. subsp. tinctoria".

The name *S. flava* is validated by a rather full description: "Floribus flavis; pedunculis unifloris, longissimis; foliis inferioribus pinnatis; foliolis linearibus, serratis; paleis receptaculi pungentibus".

Artemisia judaica *L.* 1771: 281; Tackholm 1974: 581, pl. 206A.
VERNACULAR NAME. *Schiach, Schrâck* (?) (Arabic).
LOCALITY. Egypt: Suez, Sept./Oct. 1762.
MATERIAL. *Forsskål* 959 (C — 1 sheet with field label "iter ad Sues, Schiach", microf. 120: I. 1, 2); 960 (C — 1 sheet with field label "Schrâck", microf. nil).

Artemisia monosperma *Delile*, Fl. Egypte: 263, t. 43 f. 1 (1813); Tackholm 1974: 579, pl. 206B.
LOCALITY. Unspecified, probably Egypt, 1761–62.
MATERIAL. *Forsskål* 961 (C — 1 sheet, microf. 119: III. 5, 6).

Artemisia semsek *Forssk.* 1775: 218 (LXXIII No. 440 as "*A. abrotan*?"; Cent. VIII No. 87); Ascherson & Schweinfurth, Ill. Fl. Egypte: 762 (1889); Christensen 1922: 35.
VERNACULAR NAME. *Semsek, Msaeka, Semsaek* (Arabic).
LOCALITY. Egypt: Cairo ("Káhirae hortensis"), 1761–62.
MATERIAL. *Forsskål* no type specimen found.
NOTE. Although this is validly published, it appears to be the well known garden plant *A. abrotanum* L.

Artemisia sp. indet.
LOCALITY. Not specified.
MATERIAL. *Forsskål* 962 (C — 1 sheet, microf. 119: II. 7, 8).

Asteriscus aquaticus (*L.*) *Less.*, Syn. Gen. Comp.: 210 (1832); Fl. Europ. 3: 139 (1976).
LOCALITY. Uncertain.
MATERIAL. *Forsskål* 995 (C — 1 sheet ex herb. Schumacher, microf. 17: II. 3, 4).
NOTE. This sheet has been mixed with the material of *Nauplius graveolens*, but was excluded from that species by A. Wiklund, 1979.

Asteriscus spinosus (*L.*) *Sch. Bip.* in Webb & Benth., Hist. Nat. Iles Canar. 3 (2: 2): 230 (1844).
Pallenis spinosa (L.) Cass., Dict. Sci. Nat. 37: 276 (1825); Fl. Europ. 4: 139 (1976).
Buphthalmum ('*Buphtalmum*') *melitense* Forssk. 1775: 218 (XIV No. 73; Cent. VIII No. 94); Christensen 1922: 35.
LOCALITY. Malta ("Maltae in ruderatis frequens"), 12 June 1761.
MATERIAL. *Forsskål* no type specimen found.
NOTE. This synonymy is likely since the original description compares it with *A. spinosus* but with smaller flowers etc.

Atractylis carduus (*Forssk.*) *Christensen* 1922: 27; Tackholm 1974: 534; Fl. Palaest. 3: 370, pl. 621 (1978).
Centaurea carduus Forssk. 1775: 152 (LXXIV No. 467; Cent. V No. 93).
Atractylis humilis sensu Vahl 1790: 63, non L.
VERNACULAR NAME. *Akaesj*, *Akesch* (Arabic).
LOCALITY. Egypt: Alexandria ("Alexandriae"), 1762.
MATERIAL. *Forsskål* 1070 (C — 1 sheet, type of *C. carduus*, microf. 23: II. 3, 4); 1071 (C — 1 sheet, type of *C. carduus*, microf. 23: II. 5, 6).
NOTE. This species is often known as *A. flava* Desf. (1799). The sheet No. 1070 carries the inscription in Vahl's hand "ex Oriente ded. Broussonet" and may therefore not be —in spite of a probably later inscription "Coll. Forskal" —a Forsskål specimen after all.

Bellis flava *Forssk.* 1775: CXX No. 507, Christensen 1922: 26. = *Bellis* [*sp. without epithet*] Forssk.: 151 (Cent. V No. 87); Christensen 1922: 26.
LOCALITY. Yemen: Taizz ("Taaes"), 1763.
MATERIAL. *Forsskål* no type specimen found.
NOTE. This is not identifiable in the absence of a type specimen. The name should not be adopted if it upsets widely accepted nomenclature.

Bidens apiifolia *Forssk.* 1775: CXIX No. 481; Schweinfurth 1912: 118; Christensen 1922: 37.
VERNACULAR NAME. *Sis* (Arabic).
LOCALITY. Yemen: Al Hadiyah ("Hadîe"), 1763.
MATERIAL. *Forsskål* no type specimen found.
NOTE. This species was validated by the description: "Foliis laciniis late-ovatis, incisis, serratis; Floribus velutham; coacervatis in capitula". It is imperfectly known but was considered by Schweinfurth (l.c.) to be a *Chrysanthellum* sp. The name should not be adopted if it upsets widely accepted nomenclature.

Blumea axillaris (*Lam.*) *DC.*, Prodr. 5: 4434 (1836).
LOCALITY. Yemen: unlocalised, 1763.
MATERIAL. *Forsskål* 1007 (C — 1 sheet, microf. 120: II. 1, 2); 1008 (C — 1 sheet, microf. 120: II. 3, 4).
NOTE. This was formerly known as *Blumea solidaginoides* (Poir.) DC. (1836).

Buphthalmum ('*Buphtalmum*') **ramosum** *Forssk.* 1775: 151 (Cent. V No. 89); Vahl 1791: 92; Christensen 1922: 26.
LOCALITY. Yemen: Al Hadiyah ("Hadîe"), 1763.
MATERIAL. *Forsskål* no type specimen found.
NOTE. This is not identifiable in the absence of a type specimen. The name should not be adopted if it upsets widely accepted nomenclature.

Calendula arvensis *L.* 1763: 1303; Tackholm 1974: 585.
VERNACULAR NAMES. See note below.
LOCALITY. Not specified.
MATERIAL. *Forsskål* 1040 (C — 1 sheet, microf. 20: I. 1, 2).
NOTE. This specimen is in a folder named *Calendula officinalis*, a species thrice listed by Forsskål 1775: XXXIII No. 393 (Istanbul vern. name *Xamobyoreia)*; LXXIV No. 470 (Egypt: Cairo vern. names *Tubbaejni, Kaebbli)* and CXX No. 513 (Yemen vern. name *Zobejde*).

Cardopatium corymbosum (*L.*) *Pers.*, Synops. Pl. 2: 500 (1807); Fl. Turk. 5: 597 (1975).
Carthamus corymbosus L. — Vahl 1790: 69.
Cnicus horridus Forssk. 1775: 217 (XXXI No. 352 e; Cent. VIII No. 85);
VERNACULAR NAMES: *Mauraggathos* or *Chamolio* (Greek).
LOCALITY. Turkey: Dardanelles ("ad Dardanellos"), July 1761. Gökçeada I. ("Imr.").
MATERIAL. *Forsskål* 1124 (C — 1 sheet type of *C. horridus*, microf. 28: III. 1, 2); 1125 (C — 1 sheet type of *C. horridus*, microf. 28: III. 3, 4).

Carduus lanatus *Forssk.* 1775: XXXII No. 360; Christensen 1922: 36.
LOCALITY. Turkey: Anatolia: Burgaz ("Brg" = Borghas).
MATERIAL. *Forsskål* no specimen found.
NOTE. The description is hardly sufficient to validate this name: "spinosus. Rarisi an *Carlina*?" If considered valid, the name should not be adopted if it upsets widely accepted nomenclature.

Carlina corymbosa *L.* subsp. **graeca** (*Boiss.*) *Nyman*, Consp.: 400 (1879); Fl. Turk. 5: 598 (1975); Fl. Europ. 4: 209 (1976).
Carlina rubra Forssk. 1775: XXXI No. 346, nomen nudum.
VERNACULAR NAME. *Kokinagatho* (Greek).
LOCALITY. Turkey: Bozcaada I. ("Tenedos"), July 1761.
MATERIAL. *Forsskål* 1046 (C — 1 sheet, microf. 21: I. 1, 2); 1047 (C — 1 sheet ex herb. Vahl, with original field label "Te. 32", microf. 21: I. 3, 4).
NOTE. The specimens on these sheets present a very different appearance: one being elongated and the other very congested, but C. Jeffrey identifies them as the same species.

Carthamus dentatus (*Forssk.*) *Vahl* 1790: 69; Fl. Turk. 5: 592 (1975); Fl. Europ. 4: 302 (1976).
Kentrophyllum dentatum (Forssk.) DC., Prodr. 6: 611 (1838).
Cnicus dentatus Forssk. 1775: 217 (Cent. VIII No. 83); Christensen 1922: 35.
LOCALITY. Malta ("Maltae frequens in ruderatis"), 12 June 1761.
MATERIAL. *Forsskål* 1121 (C — 1 sheet, lectotype of *Cnicus dentatus*, microf. 28: II. 7, 8).
NOTE. P. Hanelt, 1959, has labelled this specimen as 'lectotype'. Two other specimens have been identified and separated as *C. lanatus* (q.v.).

Carthamus glaucus *Bieb.*, Tabl. Prov. Casp.: 118 (1798); Fl. Turk. 5: 593 (1975).
Carlina ch. lanatae Forssk. 1775: XXXI No. 347.
LOCALITY. Turkey: Burgaz ("Borghas"), 15–17 Sept. 1761.
MATERIAL. *Forsskål* 1045 (C — 1 sheet, microf. 20: III. 7, 8).

Carthamus lanatus *L.* subsp. **baeticus** (*Boiss. & Reuter*) *Nyman*, Consp.: 419 (1879); Fl. Turk. 5: 592 (1975); Fl. Europ. 4: 303 (1976).
LOCALITY. Turkey: Bozcaada I. ("Tenedos"), July 1761.

MATERIAL. *Forsskål* 1122 (C — 1 sheet, microf. 28: II. 3, 4); 1123; (C — 1 sheet, microf. 28: II. 5, 6).
NOTE. P. Hanelt, 1959, separated these sheets from *C. dentatus* with which they were confused.

Centaurea aspera *L.* 1753: 916; Forssk. 1775: XI No. 232; Fl. Europ. 4: 284 (1976).
LOCALITY. France: Marseille ("Estac"), May 1761.
MATERIAL. *Forsskål* 1073 (C — 1 sheet, with field label "W. 24", microf. 23: I. 7, 8).

Centaurea aegyptiaca *L.* 1767: 118; Tackholm 1974: 541; Fl. Palaest. 3: 398, pl. 671 (1978).
C. eriophora? Forssk. 1775: LXXXIV No. 465.
VERNACULAR NAME. *Jamrûr* (Arabic).
LOCALITY. Egypt: Cairo ("Cairi vel Káhirae loca deserta"), 1762.
MATERIAL. *Forsskål* 1074 (C — 1 sheet, microf. 23: II. 7, 8).

Centaurea alexandrina *Delile*, Fl. Egypte: 280, t. 49 f. 3 (1813); Tackholm 1974: 541.
LOCALITY. Egypt?
MATERIAL. *Forsskål* s.n. (LD — 1 sheet ex herb. Retzius).

Centaurea calcitrapa *L.* 1753: 917; Forssk. 1775: 152 (LXXIV No. 466; Cent. V No. 95); Christensen 1922: 27; Tackholm 1974: 541.
VERNACULAR NAME. *Schôk, Morrejr* (Arabic).
LOCALITY. Egypt: Alexandria ("Alexandriae"), 1762.
MATERIAL. *Forsskål* 1072 (C — 1 sheet with original label "Nr. 6", microf. 23: II. 1, 2); 1839 (C — sheet, microf. nil).
Forsskål s.n. (BM — 1 sheet mixed with other specimens).

Centaurea glomerata *Vahl* 1791: 94; Tackholm 1974: 539.
C. acaulis sensu Forssk. 1775: 152 (LXXIV No. 469; Cent. V No. 94); Christensen 1922: 27, non L.
VERNACULAR NAME. *Serrat enaghi, Sorrat en naghi* (Arabic).
LOCALITY. Egypt: Alexandria ("Alexandriae"), April 1762.
MATERIAL. *Forsskål* 1075 (C — 1 sheet, type of *C. glomerata*, microf. 23: I. 5, 6).
Forsskål s.n. (BM — 1 sheet ex herb. Banks).

Centaurea spinosa *L.* 1753: 912; Fl. Turk. 5: 499 (1975).
Serratula spinosa Forssk. 1775: 217 (XXXII No. 363; Cent. VIII No. 82); Christensen 1922: 35.
VERNACULAR NAME. *Djevann* (Turkish), *Iala sebia* (Greek).
LOCALITY. Turkey: Bozcaada and Anatolia ("Tenedos, frequens in ficcis", "In littore Natoliae inventa magis tomentosa"), July 1761.
MATERIAL. *Forsskål* 1724 (C — 1 sheet ex herb. Schumacher, type of *S. spinosa*, microf. 98: II. 1, 2).

Centaurea sp. indet. 1
LOCALITY. Not specified.
MATERIAL. *Forsskål* s.n (LD — 1 sheet ex herb. Retzius).
NOTE. This has been identified as *C. napifolia* L., but its stems are not winged with decurrent leaves, nor are the spines on the phyllaries exactly similar.

Centaurea sp. indet. 2
LOCALITY. Not specified.
MATERIAL. *Forsskål* s.n. (LD — 1 sheet ex herb. Retzius).
NOTE. A multi-headed specimen devoid of leaves.

Centaurothamnus maximus (*Forssk.*) *Wagenitz & Dittrich* in Candollea 37: 111 (1982).
Centaurea maxima Forssk. 1775: 152 (CXX No. 512; Cent. 5 No. 92); Deflers 1889: 157; Christensen 1922: 26; Schwartz 1939: 293.
C. verbascifolia Vahl 1790: 75 (based on *C. maxima*), nom. illeg. superfl.
VERNACULAR NAME. *Mokâr, Bôgad, Baejud, Baeruad* (Arabic).
LOCALITY. Yemen: Barad ("Barah" "in montibus Yemen"), 1763.
MATERIAL. *Forsskål* 1078 (C — 1 sheet, type of *C. maxima*, microf. 23: III. 5, 6); 1079 (C —1 sheet ex herb. Schumacher, type of *C. maxima*, microf. 23: III. 7, 8); 1080 (C — 1 sheet ex herb. Vahl, type of *C. maxima*, microf. 24: I. 1, 2).
Forsskål s.n. (BM — 1 sheet, type of *C. maxima*).

Ceruana pratensis *Forssk.* 1775: LXXIV No. 453; Christensen 1922: 27; Tackholm 1974: 547, pl. 192A. = *Cernana* [*sp. without epithet*] Forssk. 1775: 153 (Cent. V No. 99).
Buphthalmum pratense (Forssk.) Vahl 1790: 75.
VERNACULAR NAME. *Kaeruan* (Arabic).
LOCALITY. Egypt: Cairo ("Káhirae"), 1762.
MATERIAL. *Forsskål* 1084 (C — 1 sheet, type of *C. pratensis*, microf. 24: I. 5, 6).
Forsskål s.n. (BM — 1 sheet ex herb. Banks, type of *C. pratensis*).
Forsskål s.n. (LD — 1 sheet ex herb. Retzius, type of *C. pratensis*).
NOTE. This is one of the examples of a new genus and species described without epithet in the 'Centuriae', and provided with an epithet in the 'Florae'.

Chiliadenus montanus (*Vahl*) *Brullo* in Webbia 34: 301 (1980); Fl. Palaest. 4: 405 (1986).
Jasonia montana (Vahl) Botsch. in Nov. Syst. Pl. Vasc. Leningr. 1964: 365 (1964); Tackholm 1974: 562.
Varthemia montana (Vahl) Boiss., Fl. Orient. 3: 212 (1875).
Chrysocoma montana Vahl 1790: 70.
Linosyris montana (Vahl) DC., Prodr. 5: 352 (1836).
LOCALITY. Egypt: Sinai, Mt. Horeb, Sept. 1762.
MATERIAL. *Forsskål* 1111 (C — 1 sheet, type of *C. montana*, microf. 121: III. 1, 2); 1110 (C — 1 sheet, type of *C. montana*, microf. 121: II. 7, 8).
NOTE. These must have been collected by Niebuhr or von Haven.

Chrysocoma corymbosa *Forssk.* 1775: CXIX No. 488.
LOCALITY. Yemen: Al Hadiyah ("Hadîe"), 1763.
MATERIAL. *Forsskål* no type specimen found.
NOTE. The name is validated by the Latin description: "Receptaculo nudo; seminibus papposis; floribus flavis. Foliis & odore similis *Eupatorio barnuf. Aegypt.*" This name should not be adopted if it upsets widely accepted nomenclature.

Chrysocoma ovata *Forssk.* 1775: 147 (CXIX No. 486; Cent. V No. 69); Christensen 1922: 26; Schwartz 1939: 275.
LOCALITY. Yemen: Surdud ("Surdûd"), 1763.
MATERIAL. *Forsskål* no type specimen found.

Cichorium spinosum *L.* 1753; 813; Forssk. 1775: XIV No. 63; Fl. Europ. 4: 305 (1976).
LOCALITY. Malta, 12 June 1761.

MATERIAL. *Forsskål* 1113 (C — 1 sheet, microf. 27: I. 5, 6); 1112 (C — 1 sheet ex herb. Hofman Bang, microf. 27: I. 3, 4); 1114 (C — 1 sheet ex herb. Hofman Bang, microf. 27: I. 7, 8); 1115 (C — 1 sheet, microf. nil).

Cineraria abyssinica *Sch. Bip. ex A. Rich.*, Tent. Fl. Abyss. 1: 433 (1847); Deflers 1889: 153; Schwartz 1939: 289.
LOCALITY. Yemen ?
MATERIAL. *Forsskål* 1116 (C — 1 sheet, microf. 119: III. 1, 2).

Cnicus abortivus *Forssk.* 1775: XXXI No. 349; Christensen 1922: 36.
LOCALITY. Turkey: Istanbul (Bj = Bujuchtari, pagus"), July–Sept. 1761.
MATERIAL. *Forsskål* no specimen found.
NOTE. The epithet is validated by the description: "Seminae in radio: discus abortiens, tenacissime & copiose lanatus". It is not included in Fl. Turkey 5 (1975). This name should not be adopted if it upsets widely accepted nomenclature.

Cnicus spinosissmus sensu Forssk. 1775: 218 (LXXIII No. 424; Cent. VIII No. 86), non L.
LOCALITY. Egypt: Cairo ("In desertis Káhirinis"), 1761–62.
MATERIAL. *Forsskål* no specimen found.
NOTE. This must remain unknown until material can be traced to make an identification.

Conyza aegyptiaca (*L.*) *Ait.*, Hort. Kew. 3: 183 (1789); Vahl 1790: 72; Tackholm 1974: 549, pl. 193B.
Erigeron aegyptiacum L. 1767: 112.
E. serratum Forssk. 1775: 148 (LXXIII No. 445; Cent. V No. 76); Christensen 1922: 26.
LOCALITY. Egypt: Cairo ("prope Masr el atik (Káhiram veterem)", "Cs = Káhirae"), 1761–2.
MATERIAL. *Forsskål* 1139 (C — 1 sheet, microf. 120: I. 4, 5).
Forsskål s.n. (BM — 1 sheet ex herb. Banks, ?type of *E. serratum*).
NOTE. Although Linnaeus does not state the source of his seed from which he raised plants of *E. aegyptiacum*, they may have come from Forsskål.

Conyza incana (*Vahl*) *Willd.*, Sp. Pl. 3: 1937 (1804); Schwartz 1939: 274.
Erigeron incanum Vahl 1790: 72; Christensen 1922: 38.
LOCALITY. Yemen, 1763.
MATERIAL. *Forsskål* no type specimen found.

[**Conyza ivifolia** (*L.*) *Less.* in Linnaea 6: 138 (1831)].
NOTE. A sheet attributed to Forsskål (microf. 119: I. 7, 8) cannot have been collected by him since this is a South African species.

Conyza pyrrhopappa *Sch. Bip. ex A. Rich.* Tent. Fl. Abyss. 1: 389 (1847); Deflers 1889: 147; Schwartz 1939: 274.
VERNACULAR NAMES: (?) *Schadj'asch, elmâ* (on original label) (?Arabic).
LOCALITY. Yemen: Bolghose, 1763.
MATERIAL. *Forsskål* 1142 (C — 1 sheet, with field label "Bolghose", microf. 120: II. 5, 6).

Cotula anthemoides *L.* 1753: 891; Vahl 1790: 73.
Tanacetum humile Forssk. 1775: 148 (LXXIII No. 438; Cent. V No. 73); Christensen 1922: 26; Tackholm 1974: 579, pl. 205D.

LOCALITY. Egypt: Cairo ("Circa Káhiram"), Dec. 1762.
MATERIAL. *Forsskål* 1779 (C — 1 sheet, type of *T. humile*, microf. 106: II. 5, 6); 1780 (C — 1 sheet ex herb. Hornemann, type of *T. humile*, microf. 106: II. 7, 8); 1778 (C — 1 sheet, type of *T. humile*, microf. 106: II. 3, 4).
Forsskål s.n. (BM — 1 sheet ex herb. Banks, type of *T. humile*).

Crepis micrantha *Czerep.* in Bobrov & Tevelev, Fl. URSS 29: 684 (1964); Fl. Turk. 5: 837 (1975).
LOCALITY. Turkey?
MATERIAL. *Forsskål* 1151 (C — 1 sheet ex herb. Schumacher, microf. 121: II. 5, 6).

Echinops microcephalus *Sm.* in Sibth. & Sm., Prodr. Fl. Graec. 2: 209 (1813); Fl. Turk. 5: 611 (1975).
LOCALITY. Turkey: Bozcaada, July 1761.
MATERIAL. *Forsskål* 1239 (C — 1 sheet, microf. nil).
NOTE. In Flora Constantinopolitana there are three *Echinops* listed and they are described in Centuria VIII as 79 *E. sphaerocephalus*, 80 *E. strigosus* and 81 *E. ritro*. The specimens cited above appear to be additional.

Echinops ritro *L.* 1753: 815; Forssk. 1775: 217 (XXXII No. 365; Cent. VIII No. 81); Fl. Turk. 5: 610 (1975).
LOCALITY. Turkey: Dardanelles ("Ad Dardanellos"), July 1761.
MATERIAL. *Forsskål* no specimen found.

Echinops spinosissimus *Turra*, Farset. Nov. Gen. 13 (1765); Tackholm 1974: 532, pl. 184.
E. sphaerocephalus sensu Forssk. 1775: LXXIII No. 423, non L.
VERNACULAR NAMES. *Chasjîr, Sjôk edsjemmel* (Arabic).
LOCALITY. Egypt: Cairo ("Cd."), 1762.
MATERIAL. *Forsskål* 1238 (C — 1 sheet, microf. 40: II. 5, 6).
NOTE. *E. sphaerocephalus* sensu Forssk. 1775: 216 (XXXII No. 366; Cent. VIII No. 79) from Turkey: Bozcaada ("Tenedos") may be the same, but no specimen has been found.

[**Echinops strigosus** *L.* 1753: 815; Forssk. 1775: 216 (XXXII No. 364; Cent. VIII No. 80); Christensen 1922: 35].
VERNACULAR NAME. *Kathar aggatho* (Greek).
LOCALITY. Turkey: Bozcaada ("Tenedos"), July 1761.
MATERIAL. *Forsskål* no specimen found.
NOTE. Since *E. strigosus* is a Western Mediterranean species, this is unlikely to be the correct identification for a Turkish plant.

Eclipta prostrata (*L.*) *L.*, Mant. 2: 286 (1771).
Eclipta alba (L.) Hassk., Pl. jav. Rar.: 528 (1848); Tackholm 1974: 568, pl. 203A; Fl. Palaest. 3: 324, pl. 544 (1978).
E. erecta (L.) L. — Vahl 1790: 74.
Micrelium asteroides Forssk. 1775: LXXIV No. 460; Christensen 1922: 27. = *Micrelium* [*sp. without epithet*] Forssk. 1775: 152 (Cent. V No. 96).
LOCALITY. Egypt: Rashid ("Rosettae"), Nov. 1761.
MATERIAL. *Forsskål* 1470 (C — 1 sheet, type of *M. asteroides*, microf. 68: II. 1, 2).
Forsskål s.n. (BM — 1 sheet ex herb. Banks, type of *M. asteroides*).
Micrelium tolak Forssk. 1775: 153 (CXX No. 511; Cent. V No. 97); Schweinfurth 1912: 138; Christensen 1922: 27.

VERNACULAR NAME. *Tolak* (Arabic).
LOCALITY. Yemen: Al Hadiyah ("Hadie"), 1763.
MATERIAL. *Forsskål* no specimen found at C.
Forsskål s.n. (LD — 1 sheet ex herb. Retzius).
NOTE. The new genus *Micrelium* has two species described in the 'Centuriae', one is provided with an epithet in the 'Florae'; no separate generic diagnosis is provided, and none of the names is therefore valid.

Ethulia conyzoides *L.* 1763: 1171; Vahl 1790: 69; Tackholm 1974: 547.
Kahiria [*sp. without epithet*] Forssk. 1775: 153 (LXXIII No. 434; Cent. V, No. 98); Christensen 1922: 27.
LOCALITY. Egypt: Rashid ("Ad ripam Nili in limo", "Rs = Rosettae"), Nov. 1761.
MATERIAL. *Forsskål* 1384 (C — 1 sheet, type of *Kahiria*, microf. 60: II. 7, 8); 1385 (C — 1 sheet with field label 'Ni 14', type of *Kahiria*, microf. 60: III. 1, 2); 1386 (C — 1 sheet ex herb. Schumacher, type of *Kahiria*, microf. 60: III. 3, 4); 1387 (C — 1 sheet, type of *Kahiria*, microf. 60: III. 5, 6).
NOTE. This is an example of a new genus and a new species described in the 'Centuriae' without an epithet. Nowhere in the 'Flora Aegyptiaco-Arabica' is an epithet provided for this new species.

Filago pyramidata *L.* 1753: 1199; Fl. Europ. 4: 122 (1976); Fl. Turk. 5: 104 (1975).
LOCALITY. Unspecified.
MATERIAL. *Forsskål* s.n. (LD — 1 sheet ex herb. Retzius).
NOTE. Named and labelled by G. Wagenitz 1966. Synonym: *F. spathulata* C. Presl.

Gerbera piloselloides (*L.*) *Cass.*, Dict. Sci. Nat. 18: 461 (1820).
Piloselloides hirsuta (Forssk.) C. Jeffrey ex Cufod. in Bull. Jard. Bot. Nat. Belg. 37 Suppl. 1180 (1967).
Arnica hirsuta Forssk. 1775: 151 (CXX No. 506; Cent. V No. 86); Christensen 1922: 26.
Gerbera hirsuta (Forssk.) Less. in Linnaea 5: 298 (1830).
LOCALITY. Yemen: Barad & Bughah ("In monte Barah", "Boka"), Mar. 1763.
MATERIAL. *Forsskål* 958 (C — 1 sheet, type of *A. hirsuta*, microf. 9: III. 4, 5).
Forsskål s.n. (BM — 1 sheet ex herb. Banks, type of *A. hirsuta*).
NOTE. Microf. 9: III. 3 is of the copious notes made by J. Claussen, 1925.

Gnaphalium arabicum *J.F. Gmel.*, Syst. Nat. 2: 1216 (1791).
Gnaphalium [sp. 1 without epithet] Forssk. 1775: 218 (CXIX No. 492; Cent. VIII No. 88); Christensen 1922: 35.
LOCALITY. Yemen: Kusma ("Kurma"), 1763.
MATERIAL. *Forsskål* no specimen found.

Gnaphalium cinereum *J.F. Gmel.*, Syst. Nat. 2: 1216 (1791).
Gnaphalium [sp. 2 without epithet] Forssk. 1775: 218 (CXIX No. 493; Cent. VIII No. 89); Christensen 1922: 35.
LOCALITY. Yemen: Kusma ("Kurma"), 1763.
MATERIAL. *Forsskål* no specimen found.

Gnaphalium crispatulum *Delile*, Fl. Egypte: 267, t. 44 f. 3 (1813); Tackholm 1974: 559.
G. margaritaceum sensu Forssk. 1775: LXXIII No. 441, non L.
LOCALITY. Egypt: Cairo ("Cairi spontaneae"), 1762.
MATERIAL. *Forsskål* 1305 (C — 1 sheet ex herb. Hornemann, microf. 51: III. 6, 7).

Gnaphalium cuneifolium *J.F. Gmel.*, Syst. Nat. 2: 1216 (1791).
Gnaphalium [sp. 3 without epithet] Forssk. 1775: 218 (CXIX No. 494; Cent. VIII No. 90); Christensen 1922: 35.
LOCALITY. Yemen: Kusma ("Kurmae"), 1763.
MATERIAL. *Forsskål* no type specimen found.

Gnaphalium orientale sensu Forssk. 1775: 218 (CXIX No. 490; Cent. VIII No. 92), non L.; Christensen 1922: 35.
VERNACULAR NAME. *Adhaun el kelb* (Arabic).
LOCALITY. Yemen: Al Hadiyah ("Hadie"), 1763.
MATERIAL. *Forsskål* no type specimen found.
NOTE. If this is the correct identity, then the species should be called *Helichrysum orientale* (L.) Gaertn.

Gnaphalium pulvinatum *Delile*, Fl. Egypte: 266, t. 44 (1813); Tackholm 1974: 558.
LOCALITY. Egypt, 1761–62.
MATERIAL. *Forsskål* 1308 (C — 1 sheet, microf. 120: II. 7, 8); 1309 (C — 1 sheet, ex herb. Schumacher, microf. 120: III. 1, 2).

Hedypnois cretica (*L.*) *Dum.-Cours.*, Bot. Cult. 2: 339 (1802); Fl. Turk. 5: 686 (1975); Fl. Europ. 4: 307 (1976).
H. rhagadioloides (L.) F.W. Schmidt, Samml. Phys.-Ök. Anfs. 1: 279 (1795).
LOCALITY. Unspecified.
MATERIAL. *Forsskål* s.n. (LD — 1 sheet ex herb. Retzius).

Helichrysum forskahlii (*J.F. Gmel.*) *Hilliard & Burtt* in Notes Roy. Bot. Gard. Edin. 38: 146 (1980).
Gnaphalium fruticosum Forssk. 1775: 218 (CXIX No. 491; Cent. VIII No. 91); Christensen 1922: 35, non Miller (1768).
G. forskalii J.F. Gmel., Syst. Nat. 2: 1214 (1791).
Helichrysum fruticosum (Forssk.) Vatke in Linnaea 39: 491 (1875), nom. illegit.; Schwartz 1939: 280.
H. cymosum (L.) D. Don subsp. *fruticosum* (Forssk.) Hedberg in Symb. Bot. Ups. 15(1): 203 (1957).
Gnaphalium? kurmense Mart. in Denkschr. Akad. Muench. 6: 193 (1820); DC., Prodr. 6: 237 (1837).
VERNACULAR NAME. *Scheratat, Sinde, Synde* (Arabic).
LOCALITY. Yemen: Kusma ("Kurmae"), 1763.
MATERIAL. *Forsskål* 1306 (C — 1 sheet ex herb. Vahl, type of *G. fruticosum* & *G. forskahlii*, microf. 51: III. 4, 5).
Forsskål s.n. (LD — 1 sheet ex herb. Retzius, with "Rosettae", type of *G. fruticosum* & *G. forskahlii*).

Helichrysum schimperi (*Sch. Bip. ex A. Rich.*) *Moeser* in Engl. Bot. Jahrb. 44: 244 (1910); Schwartz 1939: 279.
LOCALITY. Unspecified.
MATERIAL. *Forsskål* 1326 (C — 1 sheet, microf. nil).

Hieracium hispidum *Forssk.* 1775: 216 (XXXI No. 342; Cent. VIII No. 78); Christensen 1922: 34).
H. forskohlei Froel. — DC., Prodr. 7: 223 (1838).

LOCALITY. Turkey: Istanbul ("Ad Constantinop. in sylvis rarius"), July–Sept. 1761.
MATERIAL. *Forsskål* no type specimen found.
NOTE. An unknown species not accounted for in Fl. Turkey 5 (1975). This name should not be adopted if it upsets widely accepted nomenclature.

Hieracium multiflorum *Forssk.* 1775: CXVIII No. 477; Christensen 1922: 37.
LOCALITY. Yemen: Bolghose ("Blg") 1763.
MATERIAL. *Forsskål* no type specimen found.
NOTE. An obscure species, probably not a *Hieracium*, with a brief description: "Foliis denticulatis". This name should not be adopted if it upsets widely accepted nomenclature.

Hieracium uniflorum *Forssk.* 1775: CXVIII No. 478; Christensen 1922: 37.
VERNACULAR NAME. *Morrejr* (Arabic).
LOCALITY. Yemen: Kusma ("Kurma"), 1763.
MATERIAL. *Forsskål* no type specimen found.
NOTE. An imperfectly known species with a brief validating description: "Scapo decumbente, squamato, unifloro". This name should not be accepted if it upsets widely accepted nomenclature.

Hieracium sp. indet.
LOCALITY. Unspecified.
MATERIAL. *Forsskål* s.n. (LD — 1 sheet ex herb. Retzius).

Hyoseris lucida *L.* 1767: 108; Vahl 1790: 66; Tackholm 1974: 587, pl. 211A.
LOCALITY. Egypt.
MATERIAL. (LINN — 1 sheet 957: 3, 4, cult. Hort. Upsala from seeds sent by Forsskål, type of *H. lucida*, microf. 544: II. 2, 3).
Forsskål 1348 (C — 1 sheet ex herb. Schumacher, microf. 121: I. 1, 2).
Lapsana taraxacoides Forssk. 1775: 145 (LXXII No. 419; Cent. V No. 63); Christensen 1922: 25.
LOCALITY. Egypt: Alexandria ("Alexandriae"), April 1762.
MATERIAL. *Forsskål* 1393 (C — 1 sheet, type of *L. taraxacoides*, microf. 61: II. 1, 2); 1394 (C — 1 sheet ex herb. Vahl, type of *L. taraxacoides*, microf. 61: II. 3, 4).

Ifloga spicata (*Forssk.*) *Sch. Bip.* in Webb & Berth., Phyt. Canar. 2: 310 (1845); Tackholm 1974: 552, pl. 196C.
Chrysocoma spicata Forssk. 1775: LXXIII No. 433; Christensen 1922: 36.
Gnaphalium spicatum (Forssk.) Vahl 1790: 70.
LOCALITY. Egypt: Cairo ("Cairi vel Káhirae loca deserta"), 1763.
MATERIAL. *Forsskål* 1107 (C — 1 sheet ex herb. Schumacher, type of *C. spicata*, microf. 26: III. 7, 8); 1108 (C — 1 sheet, type of *C. spicata*, microf. 27: I. 1, 2); 1109 (C — 1 sheet ex herb. Hornemann, type of *C. spicata*, microf. 120: III. 3, 4).

Inula crithmoides *L.* 1753: 883; Tackholm 1974: 560.
Senecio succulentus Forssk. 1775: 149 (LXXIII No. 448; Cent. V No. 78); Christensen 1922: 26.
Inula crithmifolia Vahl 1790: 73.
LOCALITY. Egypt: Alexandria ("Alexandriae ad catacombas in arena littorea copiose"), Sept–Oct. 1761.
MATERIAL. *Forsskål* 1723 (C — 1 sheet, type of *S. succulentus*, microf. 98: I. 7, 8).

Iphiona mucronata (*Forssk.*) *Aschers & Schweinf.* 1887: 186; Anderberg, Nord. Journ. Bot. 5: 190 fig. 16 D–H (1985).
Chrysocoma mucronata Forssk. 1775: 147 (LXXIII No. 432; Cent. V No. 68); Christensen 1922: 26.
Staehelina spinosa Vahl 1790: 69, nom. illeg. superfl.
Chrysocoma spinosa (Vahl) Delile, Fl. Egypte: 128 (1813).
VERNACULAR NAME. *Dafra, Dasri* (Arabic).
LOCALITY. Egypt: Cairo ("Káhirae"), 1761–2.
MATERIAL. *Forsskål* 1100 (C — 1 sheet, type of *C. mucronata* & *S. spinosa*, microf. 26: II. 1, 2); 1101 (C — 1 sheet, lectotype of *C. mucronata*, selected by Anderberg, and type of *S. spinosa*, microf. 26: II. 3, 4); 1102 (C — 1 sheet ex herb. Vahl, ? type of *C. mucronata* & *S. spinosa* (see Note), microf. 26: II. 5, 6); 1103 (C — 1 sheet ex herb. Schumacher, type of *C. mucronata* & *S. spinosa*, microf. 26: II. 7, 8); 1104 (C — 1 sheet ex herb. Hornemann, type of *C. mucronata* & *S. spinosa*, microf. 26: III. 1, 2).
Forsskål s.n. (BM — 1 sheet ex herb. Banks type of *C. mucronata*).
NOTE. The sheet No. 1102 at C may not represent a Forsskål collection as it is annotated "ded. Broussonet" by Vahl.

Kleinia odora (*Forssk.*) *DC.*, Prodr. 6: 339 (1838); Berger in Monatsschr. Kakt. 15: 38 (1905); Halliday in Kew Bull. 39: 825 (1984).
Cacalia odora Forssk. 1775: 146 (CXIX No. 483; Cent. V No. 66); Vahl 1794: 90; Christensen 1922: 25.
Senecio ?odorus (Forssk.) Sch. Bip. in Flora 28: 500 (1845); Deflers 1889: 158; Schwartz 1939: 289.
S. anteuphorbium (L.) Sch. Bip. var. *odorus* (Forssk.) Rowley in Nat. Cact. & Succ. Journ. 13: 78 (1958).
VERNACULAR NAME. *Edcher* (Arabic).
LOCALITY. Yemen: Al Hadiyah ("Hadîe", "Yemen in montibus frequens"), 1763.
MATERIAL. *Forsskål* 1030 (C — 1 sheet, holotype of *C. odora*, microf. 18: I. 3, 4).

Kleinia pendula (*Forssk.*) *DC.*, Prodr. 6: 339 (1838); Hook. f. in Bot. Mag. 125: t. 7659 (1899); Berger in Monatsschr. Kakt. 15: 39 (1905); Halliday in Kew Bull. 39: 822 (1984).
Cacalia pendula Forssk. 1775: 145 (CXIX No. 482; Cent. V No. 65); Vahl 1794: 90; Christensen 1922: 25.
Senecio pendulus (Forssk.) Sch. Bip. in Flora 28: 500 (1845); Schwartz 1939: 289.
Notonia pendula (Forssk.) Chiov. in Anna. Istit. Bot. Roma 8: 188 (1904).
VERNACULAR NAME. *Kaad* (Arabic).
LOCALITY. Yemen: Al Hadiyah ("Hadîe", "Yemen in montibus ... "), 1763.
MATERIAL. *Forsskål* 1031 (C — 1 sheet, type of *C. pendula*, microf. 18: I. 5, 6, 7); 1032 (C — 1 sheet, type of *C. pendula*, microf. 18: I. 8).

Kleinia semperviva (*Forssk.*) *DC.*, Prodr. 6: 339 (1838); Halliday in Bot. Mag. 183: t. 830 (1982) & in Kew Bull. 39: 820 fig. 2 (1984).
Cacalia semperviva Forssk. 1775: 146 (CXIX No. 484; Cent. V No. 67); Christensen 1922: 26.
C. sempervirens Vahl 1794: 92, nom. illeg. citing *C. semperviva* Forssk.
Senecio sempervivus (Forssk.) Sch. Bip. in Flora 28: 500 (1845).
Notonia semperviva (Forssk.) Aschers. in Schweinfurth, Beitr. Fl. Aeth.: 152 (1867); Deflers 1889: 153; Schweinfurth 1912: 147.
VERNACULAR NAME. *Tyflok* (Arabic).
LOCALITY. Yemen: Bughah to Kusma ("Inter Boka et Kurma in montibus"), Mar. 1763.
MATERIAL. *Forsskål* 1033 (C — 1 sheet, holotype of *C. semperviva*, microf. 18: II. 1, 2).

Lactuca inermis *Forssk.* 1775: 144 (CXVIII No. 475; Cent. V No. 58); Vahl 1791: 89; Christensen 1922: 25; Wood in Kew Bull. 39: 132 (1984).
VERNACULAR NAME. *Kat er raejan* (Arabic).
LOCALITY. Yemen: Kusmah ("Kurmae"), 1763.
MATERIAL. *Forsskål* no type specimen found.
NOTE. In the absence of the type Wood designated a neotype (*J.R.I. Wood* 75/833 (BM)).
Lactuca capensis Thunb. — Schwartz 1939: 297.

Lactuca serriola *L.*, Cent. Pl. 2: 29 (1756); Tackholm 1974: 608, pl. 221; Fl. Turk. 5: 778 (1975).
Lactuca sinuata Forssk. 1775: 215 (XXXI No. 339; Cent. VIII No. 75); Christensen 1922: 34.
L. virosa sensu Vahl 1791: 90, non L.;
VERNACULAR NAME. *Galat zida* (Greek).
LOCALITY. Turkey: Dardanelles ("Natoliae in arvis ad Dardanellos;"), July 1761, & 1762.
MATERIAL. *Forsskål* 1389 (C — 1 sheet ex herb. Vahl, type of *L. sinuata*, microf. 61: I. 3, 4). *Forsskål* s.n. (BM — 1 sheet ex herb. Banks, type of *L. sinuata*).

Laggera crispata (*Vahl*) *Hepper & Wood* in Kew Bull. 38: 83 (1983).
Conyza crispata Vahl 1790: 71; Christensen 1922: 38.
Conyza caule alato Forssk. 1775: CXIX No. 495; invalid name.
LOCALITY. Yemen: Al Hadiyah ("Bulgose sive Hadie"), Mar. 1763.
MATERIAL. *Forsskål* 1140 (C — 1 sheet, type of *C. crispata*, microf. 32: I. 5, 6).

Laggera decurrens (*Vahl*) *Hepper & Wood* in Kew Bull. 38: 84 (1983).
Erigeron decurrens Vahl 1790: 72.
Conyza arabica Willd., Sp. Pl. 3: 1949 (1804); Schwartz 1939: 274, non *C. decurrens* L.
Laggera arabica (Willd.) Deflers 1889: 149.
LOCALITY. Yemen, 1763.
MATERIAL. *Forsskål* no type specimen found.

[**Launaea cervicornis** (*Boiss.*) *Font Quer & Rothm.* in Sched. Fl. Iber. Select. Cent. I No. 99 (1934)]
MATERIAL. Two specimens at C (ex herb. Schumacher and unmarked) have been indicated as collected by Forsskål. The specimens are the basis of the statement by Christensen (1922: 25) that material of *Prenanthes spinosa* Forssk. (*Launaea spinosa*) is present in Herb. Forsskålii. However, *Launaea cervicornis* only occurs on the Balearic Islands which were not visited by Forsskål.

Launaea mucronata (*Forssk.*) *Muschl.* 1912: 1057; Schwartz 1939: 295; Tackholm 1974: 602; Fl. Europ. 4: 326 (1976); Fl. Palaest. 3: 431, pl. 728 (1978).
Leontodon mucronatum Forssk. 1775: 144 (LXXII No. 415; Cent. V No. 60); Christensen 1922: 25.
Zollikoferia mucronata (Forssk.) Boiss., Diagn., ser. 1, 7: 12 (1846).
Launaea resedifolia var. *mucronata* (Forssk.) Beauverd
Scorzonera resedifolia sensu Vahl 1791: 87, non L.
VERNACULAR NAME. *Jamrur* (Arabic).
LOCALITY. Egypt: Cairo ("In desertis Káhirinis"), 1762.
MATERIAL. *Forsskål* 1401 (C — 1 sheet, type of *L. mucronatum*, microf. nil); 1402 (C — 1 sheet ex herb. Schumacher, type of *L. mucronatum*, microf. nil).

Launaea nudicaulis (*L.*) *Hook.* f. in Fl. Brit. Ind. 3: 416 (1882); Tackholm 1974: 602; Schwartz 1939: 296.
Lactuca flava Forssk. 1775: 143 (CXVIII No. 476; Cent. V No. 57); Vahl 1791: 89; Christensen 1922: 25.
VERNACULAR NAME. *Hendibe, Hindibe* (Arabic).
LOCALITY. Yemen: Surdud ("Surdûd"), 1763.
MATERIAL. *Forsskål* no specimen found at C.
Forsskål s.n. (LD — 1 sheet ex herb. Retzius).

Scorzonera ciliata Forssk. 1775: 143 (LXXII No. 405; Cent. V No. 55); Christensen 1922: 55.
VERNACULAR NAME. *Nuar, Sjadaejd, Schadaeid, Hauve, Aliis Hauve* (Arabic).
LOCALITY. Egypt: Alexandria ("Alexandriae"), 1761–62.
MATERIAL. *Forsskål* 1711 (C — 1 sheet ex herb. Schumacher, type of *S. ciliata*, microf. nil); 1710 (C — 1 sheet, type of *S. ciliata.*, microf. nil).

Launaea spinosa (*Forssk.*) *Sch. Bip. ex O. Kuntze*, Revis. Gen. 1: 350 (1891); Tackholm 1974: 601; Fl. Palaest. 3: 432, pl. 732 (1978).
Prenanthes spinosa Forssk. 1775: 144 (LXXII No. 409; Cent. V No. 59); Vahl 1790: 661; Christensen 1922: 25.
Zollikoferia spinosa (Forssk.) Boiss., Fl. Or. 3: 826 (1875).
VERNACULAR NAME. *Kebbâd, Kaeddad* (Arabic).
LOCALITY. Egypt: Suez, "Ghobeibe" ("Sués", "Ghobeibe"), 16–18 Sept. 1762.
MATERIAL. *Forsskål* s.n. (BM — 1 sheet ex herb. Banks, type of *P. spinosa*).
NOTE. The material in Herb. Forsskålii at C stated to be this species is *Launaea cervicornis*, which could not have been collected by Forsskål — see note under that name.

Launaea sp. indet. 1
LOCALITY. Unspecified.
MATERIAL. *Forsskål* 1397 (C — 1 sheet, microf. 121: III. 5, 6).

Launaea sp. indet. 2
LOCALITY. Unspecified.
MATERIAL. *Forsskål* 1396 (C — 1 sheet ex herb. Schumacher, microf. 62: I. 7, 8).
NOTE. This sheet was in the folder containing *Leontodon asperum* Forssk. and this name appears on the reverse of the sheet, but it is not that species. Unfortunately the specimen is too deteriorated for accurate identification, though it is evidently a *Launaea.*

Leontodon hispidulus (*Del.*) *Boiss.*, Fl. Or. 3: 727 (1875); Tackholm 1974: 596, pl. 214.
LOCALITY. Egypt: Caid Bey, near Cairo, 1761–62.
MATERIAL. *Forsskål* 1400 (C — 1 sheet with original field label "Hierac" "Caid Bey", microf. 120: III. 7, 8).

Leontodon maritim[um] *Forssk.* 1775: X No. 211 short descr; Legré 1900: 5–27; Christensen 1922: 35.
LOCALITY. France: Marseille ("Estac"), 9 May–3 June 1761.
MATERIAL. *Forsskål* no specimen found.
NOTE. The expansion of the abbreviated epithet is not endorsed by the Code. It should not be adopted.

Leontodon tuberosus *L.* 1753: 799; Fl. Turk. 5: 673 (1975); Fl. Europ. 4: 315 (1976).
LOCALITY. Unspecified, ?Turkey.
MATERIAL. *Forsskål* 1403 (C — 1 sheet ex herb. Schumacher, microf. 121: I. 5, 6).

Macowania ericifolia (*Forssk.*) *Burtt & Grau* in Notes Roy. Bot. Gard. Edinb. 31(3): 376 (1972).
Aster ericifolius Forssk. 1775: 150 (CXIX No. 504; Cent. V No. 83), as '*ericaefolius*'; Christensen 1922: 26.
Arthrixia ericifolia (Forssk.) DC., Prodr. 6: 277 (1837).
VERNACULAR NAME. *Ansif* (Arabic).
LOCALITY. Yemen: Kusma ("Kurmae"), Mar. 1763.
MATERIAL. *Forsskål* 992 (C — 1 sheet ex herb. Schumacher, type of *A. ericifolius*, microf. 13: II. 5, 6).
Forsskål s.n. (BM — 1 sheet ex herb. Banks, type of *A. ericifolius*).

Nauplius graveolens (*Forssk.*) *A. Wicklund*, Some genera *Inula*-group. 80 (1986) & in Nordic J. Bot. 7: 16 (1987).
Asteriscus graveolens (Forssk.) Less., Syn.: 210 (1832); Tackholm 1974: 565 pl. 199C; Fl. Palaest. 3: 321, pl. 539 (1978).
Buphthalmum ('*Buphtalmum*') *graveolens* Forssk. 1775: 151 (LXXIV No. 461; Cent. V No. 90); Vahl 1790: 74, t. 19; Christensen 1922: 26.
Odontospermum graveolens (Forssk.) Sch. Bip. in Webb & Berth., Phyt. Canar. 2: 232 (1844).
VERNACULAR NAME. *Rabd* (Arabic).
LOCALITY. Egypt: Cairo ("Káhirae"), 1762.
MATERIAL. *Forsskål* 1021 (C — 1 sheet with original field label "Ca. 171", lectotype of *B. graveolens*, microf. 17: II. 1, 2); 1020 (C — 1 sheet, isotype of *B. graveolens*, microf. 17: I. 7, 8); 1022 (C — 1 sheet ex herb. Hornemann, isotype of *B. graveolens*, microf. 17: II. 7, 8); 1019 (C — 1 sheet ex herb. Schumacher, isotype of *B. graveolens*, microf. 17: I. 5, 6).
Forsskål s.n. (LD — 1 sheet ex herb. Retzius, isotype of *B. graveolens*.
NOTES. The C sheets were labelled by A. Wilklund, 1979, who selected the lectotype. The lectotype sheet, No. 1021, is the one which is illustrated by Vahl 1790, t. 19.

Notobasis syriaca (*L.*) *Cass.*, Dict. Sci. Nat. 35: 171 (1825); Fl. Turk. 5: 419 (1975); Fl. Europ. 4: 242 (1976).
LOCALITY. Unspecified.
MATERIAL. *Forsskål* 1497 (C — 1 sheet ex herb. Hofman Bang, microf. 119: II. 1, 2).

Onopordum tauricum *Willd.*, Sp. Pl. 3(3): 1687 (1803); Fl. Turk. 5: 366 (1975).
LOCALITY. Turkey, 1761.
MATERIAL. *Forsskål* 1515 (C — 1 sheet, microf. 119: II. 3, 4).

[**Osteospermum vaillantii** (*Decne.*) *T. Norl.* in Bot. Notiser 1939: 305 (1939).
LOCALITY. Unspecified.
MATERIAL. '*Forsskål*' 1538 (C — 1 sheet, microf. 120: I. 6, 7).
NOTE. Since this is a South African species its attribution to Herb. Forssk. must be an error].

Otanthus maritimus (*L*). *Hoffm. & Link*, Fl. Portug. 2: 364 (1809?); Fl. Turk. 5: 253 (1975).
LOCALITY. Turkey: Anatolia = Dardanelles ("E. Natoli" on sheet), 1761.
MATERIAL. *Forsskål* 1539 (C — 1 sheet, microf. nil).
Forsskål s.n. (LD — 1 sheet ex herb. Retzius).

Phagnalon harazianum *Deflers* 1889: 150; Schwartz 1939: 278.
Conzya tomentosa sensu Forssk. partly 1775: 148 (CXIX No. 496; Cent. V No. 75 pp. excl. descr.).

LOCALITY. Yemen: Kusma ("prope Kurma" on field label, "Blg" = Bulgose in book), 1763.
MATERIAL. *Forsskål* 1147 (C — 1 sheet with field label, microf. nil).

Phagnalon rupestre (*L.*) *DC.* Prodr. 5: 396 (1836); Schwartz 1939: 278; Tackholm 1974: 558.
Conyza rupestris L. 1767: 113; Vahl 1790: 71.
LOCALITY. Egypt: Alexandria?.
MATERIAL. *Forsskål* (Type: LINN 993: 5, 8 grown by Linnaeus at Uppsala HU from Forsskål's seeds).

C. tomentosa Forssk. 1775: 148 (LXXIII No. 444; Cent. V No. 75); Christensen 1922: 26.
VERNACULAR NAME. *Tom ernéb, Mottaej* (Arabic).
LOCALITY. Egypt: Alexandria ("Alexandriae"), April 1762.
Yemen: Al Hadiyah ("Hadîe"), 1763.
MATERIAL. *Forsskål* 1143 (C — 1 sheet with field label "Alex. Ras ettin" type of *C. tomentosa*, microf. 32: II. 1, 2); 1144 (C — 1 sheet ex herb. Schumacher, type of *C. tomentosa*, microf. 32: II. 3, 4); 1145 (C — 1 sheet, type of *C. tomentosa*, microf. 32: II. 5, 6); 1146 (C — 1 sheet, type of *C. tomentosa*, microf. nil).
Forsskål s.n. (BM — 1 sheet mixed, ex herb. Banks, type of *C. tomentosa*).
Forsskål s.n. (LD — 1 sheet ex herb. Retzius, type of *C. tomentosa*).

Phagnalon sordidum (*L.*) *Reichenb.*, Fl. Germ. Excurs.: 224 (1831); Fl. Europ. 4: 133 (1976).
LOCALITY. France: Marseille, May 1761 (see Note).
MATERIAL. *Forsskål* 1310 (C — 1 sheet, microf. nil).
NOTE. Although the specimen is not labelled as to its provenance, since this species occurs in south western Europe it is reasonable to assume it was collected near Marseille.

Picnomon acarna (*L.*) *Cass.*, Dict. Sci. Nat. 40: 188 (1826); Fl. Turk. 5: 413 (1975).
Cnicus acarna (L.) L. 1763: 1158; Forssk. 1775: 217 (XXXI No. 351; Cent. VIII No. 84).
VERNACULAR NAME. *Agri aggatho* (Greek).
LOCALITY. Turkey: Dardanelles ("Ad Dardanellos"), July 1761.
MATERIAL. *Forsskål* no specimen found.
NOTE. This may not be the species intended by Forsskål's reference since Linnaeus' new combination was not published until 1763, based on *Carduus acarna*.

Picris altissima *Delile*, Fl. Egypte: 260 t. 41 (1813); Fl. Turk. 5: 682 (1975).
P. laevis Forssk. 1775: XXXI No. 332.
LOCALITY. Turkey: Burgaz I. ("Brg" = Borghàs fons).
MATERIAL. *Forsskål* no specimen found.
NOTE. The epithet is validated by the description: "Calyculo non deciduo; pappo stipitato; seminib non transulcatis". *P. altissima* is better known as *P. sprengeriana* (L.) Chaix. The name *P. laevis* should not be adopted if it upsets widely accepted nomenclature.

Picris asplenioides *L.* 1753: 793; Lack, Dissert. Univ. Wien 16: 137 (1974).
Leontodon asperum Forssk. 1775: 145 (LXXII No. 416; Cent. V, No. 61); Christensen 1922: 25.
LOCALITY. Egypt: Alexandria ("Alexandriae"), 1761–62.
MATERIAL. *Forsskål* 1398 (C — 1 sheet, type of *L. asperum*, microf. 62: I. 3, 4); 1399 (C — 1 sheet ex herb. Schumacher, type of *L. asperum*, microf. 62: II. 1, 2).
Forsskål s.n. (BM — 1 sheet ex herb. Banks, type of *L. asperum*).

NOTE. Named by H.W. Lack 1985; a third Copenhagen sheet is uncertain: *Forsskål* 1581 (C — 1 sheet, microf. 62: I. 5, 6).

Crepis radicata Forssk. 1775: 145 (LXXII No. 410; Cent. V No. 62); Christensen 1922: 25.
Picris radicata (Forssk.) Less. — Tackholm 1974: 597, pl. 216.
VERNACULAR NAME. *Sraegha, Haudân* (Arabic).
LOCALITY. Egypt: Cairo ("Circa Pyramides"), 1762.
MATERIAL. *Forsskål* 1152 (C — 1 sheet, type of *C. radicata*, microf. 33: II. 7, 8).
NOTE. This specimen was not seen by Christensen (1922); it is named and labelled by H.W. Lack 1972.
MATERIAL. *Forsskål* 1154 (C — 1 sheet ex herb. Schumacher, microf. 121: II. 3, 4); 1153 (C — 1 sheet ex herb. Schumacher, microf. 119: II. 5, 6).

Picris scabra *Forssk.* 1775: 143 (CXVIII No. 472; Cent. V No. 56); Christensen 1922: 25; Schwartz 1939: 295.
VERNACULAR NAME. *Murreir, Murrejr* (Arabic).
LOCALITY. Yemen: Khadra ("In monte Châdra"), 1763.
MATERIAL. *Forsskål* no type specimen found.
NOTE. If this proved to be *P. abyssinica* Sch. Bip. it would provide the earlier name (Schwartz l.c.). However, Fosskål's name should not be adopted if it upsets widely accepted nomenclature.

Picris sulphurea *Delile*, Fl. Egypte: 258, t. 40 (1813); Schwartz 1939: 295; Tackholm 1974: 597.
LOCALITY. Unspecified.
MATERIAL. *Forsskål* 1580 (C — 1 sheet ex herb. Schumacher, microf. 120: I. 3).

Pluchea dioscoridis (*L.*) *DC.*, Prodr. 5: 450 (1836).
Conyza dioscoridis (L.) Desf., Tabl. ed. 2: 114 (1815). Tackholm 1974: 549, pl. 193A.
C. odora? (abbreviation of *odorata* L.) Forssk. 1775: 148 (LXXIII No. 443; Cent. V No. 74); Christensen 1922: 26, 37.
VERNACULAR NAME. *Barnûf* (Arabic).
LOCALITY. Egypt: Alexandria ("Ob odorem suavem culta in hortis Aegypti"), April 1762.
MATERIAL. *Forsskål* 1141 (C — 1 sheet with field label "Alex. hortensis Ae. 9", microf. 32: I. 7, 8).

Pseudognaphalium luteo-album (*L.*) *Hilliard & Burt* in Bot. J. Linn. Soc. 82: 205 (1981).
Gnaphalium luteo-album L. 1753: 851; Tackholm 1974: 558, pl. 197B.
G. obtusifolium sensu Forssk. 1775: LXXIII No. 442, non L.
LOCALITY. Egypt: Cairo ("Cairi spontaneae"), 1762.
MATERIAL. *Forsskål* 1307 (C — 1 sheet ex herb. Hornemann, microf. 51: III. 8, 52: I. 1, 2).

Psiadia punctulata (*DC.*) *Vatke* in Oestr. Bot. Zeitschr. 27: 196 (1877).
P. arabica Jaub. & Spach (1852) — Schwartz 1939: 273.
LOCALITY. Yemen: "Fâtha inter Ersch et Aludje" on reverse of sheet, 1763.
MATERIAL. *Forsskål* s.n. (LD — 1 sheet ex herb. Retzius).

[**Pterocaulon redolens** (*Forst.*) *Berol.*, Handl. Fl. Ned. Indië 2(1): 240 (1891).
MATERIAL. C — 1 sheet attributed to Forsskål but a species from India outside his collecting area. Included here because it is housed in Herb. Forsskålii and photographed in the IDC set microf. 17: II. 5, 6].

Pulicaria arabica (*L.*) *Cass.*, Dict. Sci. Nat. 44: 94 (1826); Tackholm 1974: 562, pl. 198A; Fl. Palaest. 3: 318, pl. 532 (1978).
Inula arabica L. 1767: 114.
LOCALITY. Egypt, 1761–62.
MATERIAL. *Forsskål* (LINN — 1 sheet 999: 20, cult. hort. Uppsala from seeds sent to Linnaeus; microf. 595: I. 4, 5).
NOTE. Although the above sheet has Hort. Cliff. written on the reverse, it is presumed that the specimen was raised from Forsskål's seeds.

Inula dysenterica sensu Forssk. 1775: 150 (LXXIII No. 450; Cent. V No. 84), non L.; Christensen 1922: 26.
VERNACULAR NAME. *Rara Ejub, Raràa eijub* (Arabic).
LOCALITY. Egypt: Cairo ("Káhirae"), 1762.
MATERIAL. *Forsskål* 1360 (C — 1 sheet ex herb. Schumacher, microf. 57: I. 1, 2). *Forsskål* s.n. (BM — 1 sheet ex herb. Banks).

Pulicaria crispa (*Forssk.*) *Benth. ex Oliver* in Trans. Linn. Soc. 29: 96 (1873) & in Fl. Trop. Afr. 3: 366 (1877).
Aster crispus Forssk. 1775: 150 (LXXIII No. 449; Cent. V No. 82); Christensen 1922: 26.
Inula crispa (Forssk.) Pers., Syn. Pl. 2: 450 (1807).
Francoeuria crispa (Forssk.) Cass. in Dict. Sci. Nat. 38: 374 (1825); Tackholm 1974: 563, pl. 198B.
VERNACULAR NAME. *Sabat* (Arabic).
LOCALITY. Egypt: Cairo ("Circa Káhiram ad Pyramides Gisenses"), 1762.
MATERIAL. *Forsskål* 987 (C — 1 sheet, type of *A. crispus*, microf. 13: I. 3, 4); 988 (C — 1 sheet, type of *A. crispus*, microf. 13:I. 5, 6); 989 (C — 1 sheet ex herb. Vahl, type of *A. crispus*, microf. 13: I. 7, 8); 990 (C — 1 sheet, lectotype — see note — of *A. crispus*, microf. 13: II. 1, 2); 991 (C — 1 sheet ex herb. Schumacher, type of *A. crispus*, microf. 13: II. 3, 4). *Forsskål* s.n. (BM — 1 sheet ex herb. Banks, type of *A. crispus*).
NOTE. Most of the sheets were labelled by E. Gamal-Eldin, 1981, who chose the lectotype and indicated that this is subsp. *crispa*.

Pulicaria grandidentata *Jaub. & Spach*, Illustr. 4: 71, t. 345 (1852); Schwartz 1939: 283.
LOCALITY. Yemen, 1763.
MATERIAL. *Forsskål* 1609 (C — 1 sheet, microf. 120: III. 5, 6).

Pulicaria incisa (*Lam.*) *DC.*, Prodr. 5: 479 (1836).
P. undulata sensu auct., non (L.) DC.; Deflers 1889: 151; Schwartz 1939: 283.
Inula odora sensu Forssk. 1775: 150 (CXIX No. 505; Cent. V No. 84), non L.; Christensen 1922: 26.
VERNACULAR NAMES. *Munis, Neschusch, Cháa* (Arabic).
LOCALITY. Yemen: Wadi Mawr & Al Luhayyah & Surdud ("Môr, Lohaja, Surdûd"), 1763.
MATERIAL. *Forsskål* s.n. (BM — 1 sheet, ex herb Banks).

Reichardia tingitana (*L.*) *Roth*, Bot. Abh.: 35 (1787); Tackholm 1974: 604, pl. 219.
LOCALITY. Unspecified.
MATERIAL. *Forsskål* 1611 (C — 1 sheet ex herb. Schumacher, microf. nil).

Santolina ambigua *Forssk.* 1775: XXXII No. 357; Christensen 1922: 36.
LOCALITY. Turkey: Tekirdag ("Ecl." = Eraclissa), 1761.
MATERIAL. *Forsskål* no specimen found.
NOTE. The epithet is validated by the description: "Flava; recept. nudo; pappo nullo; polygamia aequali, discoidea". However, the name should not be adopted if it upsets widely accepted nomenclature.

Scariola viminea (*L.*) *F. W. Schmidt*, Samml. Phys.-Oekon. Anfsätzl: 270 (1795); Fl. Turk. 5: 783 (1975); Meikle 1985: 1029.
Lactuca decorticata Forssk. 1775: 216 (XXXI No. 341; Cent. VIII No. 76).
LOCALITY. Turkey: Golzcehada I. & Istanbul ("In insula Imros & circa Constantinop."), July–Aug. 1761.
MATERIAL. *Forsskål* 1388 (C — 1 sheet, type of *L. decorticata*, microf. 61: I. 1, 2).

Scolymus hispanicus *L.* 1753: 813; Forssk. 1775: 145 (LXXIII No. 426; Cent. V No. 64); Christensen 1922: 25; Tackholm 1974: 585.
VERNACULAR NAME. *Laehlech*, *Laehlah* (Arabic).
LOCALITY. Egypt: Cairo ("Káhirae"), 1961–62.
MATERIAL. *Forsskål* no specimen found.

Scolymus maculatus *L.* 1753: 813; Fl. Europ. 4: 304 (1976); Tackholm 1974: 885; Fl. Turk. 5: 624 (1975).
LOCALITY. Unspecified.
MATERIAL. *Forsskål* s.n. (LD — 1 sheet ex herb. Retzius).

Scorzonera hispida *Forssk.* 1775: 215 (Cent. VIII No. 73); Christensen 1922: 34.
LOCALITY. France: Marseille ("Ad Estac prope Massiliam"), May 1761.
MATERIAL. *Forsskål* no type specimen found.
NOTE. An unknown species with a validly published name. Probably a synonym of another species, but it is not accounted for in Flora Europaea. The name should not be adopted if it upsets widely accepted nomenclature.

? **Scorzonera tingitana** *L.* 1753: 791; Forssk. 1775; 143 (LXXII No. 404; Cent. V No. 54); Christensen 1922: 25.
VERNACULAR NAME. *Nuggd* (Arabic).
LOCALITY. Egypt: Cairo ("In desertis arenosis Káhirinis"), 1761–62.
MATERIAL. *Forsskål* no specimen found.
NOTE. This species was described from Tangier and is not represented in Egypt, so it is unlikely that Forsskål's determination was correct.

Senecio aegyptius *L.* var. **discoideus** *Boiss.*, Fl. Orient. 3: 388 (1875); Tackholm 1974: 583.
S. arabicus L. 1767: 114; Vahl 1790: 72.
LOCALITY. Egypt, 1761–62.
MATERIAL. (LINN 996: 8, 9, cultivated in Uppsala by Linnaeus from Forsskål's seeds, type of *S. arabicus*; microf. 583: I. 7; II. 1).
S. hieracifolius sensu Forssk. 1775: LXXIII No. 446, non L.
VERNACULAR NAME. *Kus* (Arabic).
LOCALITY. Egypt: Cairo & Rashid ("Cs" in list, "Rosetto" on field label), 1761–62.
MATERIAL. *Forsskål* 1715 (C — 1 sheet with field label "Senecio novus. Rosette. Ro. 21", microf. 97: III. 1, 2); 1716 (C — 1 sheet with field label "Senecio novus. Rosette", microf. 97: III. 3, 4); 1717 (C — 1 sheet microf. 97: III. 5, 6); 1718 (C — 1 sheet "circa Cairo" on back, microf. 119: III. 3, 4).
NOTE. The annotation 'novus' on the field label does not agree with Forsskål's identification of the material with a Linnaean species.

Senecio aquaticus *L.* var. **erraticus** (*Bertol.*) *Matthews* in Fl. Turk. 5: 150 (1975).
S. jacobaea sensu Forssk. 1775: XXXII No. 374, non L.
LOCALITY. Turkey: Burgaz I. ("Borghas"), July 1761.
MATERIAL. *Forsskål* 1719 (C — 1 sheet with field label "Co. 29", microf. 97: III. 7, 8).

Senecio biflorus *Vahl* 1790: 72; Christensen 1922: 39; Schwartz 1939: 291.
S. linifolius radio nullo Forssk. 1775: CXIX No. 502, invalidly published name.
LOCALITY. Yemen: Jiblah ("Djöbla"), 1763.
MATERIAL. *Forsskål* no specimen found.

Senecio glaucus *L.* 1753: 868; Fl. Palaest. 3: 356, pl. 601 (1978).
S. squalidus sensu Forssk. 1775: 150 (LXXIII No. 447; Cent. V No. 81); Christensen 1922: 26.
VERNACULAR NAME. *Korraejr, Korreis* (Arabic).
LOCALITY. Egypt: Cairo ("Káhirae"), 1761.
MATERIAL. *Forsskål* 1722 (C — 1 sheet, microf. 98: I. 5, 6).
S. coronopifolius Desf., Fl. Atl. 2: 273 (1799), non Burm.f.
S. desfontainei Druce, Brit. Pl. List. ed. 2, 61 (1928); Tackholm 1974: 581, pl. 207A.

Senecio hadiensis *Forssk.* 1775: 149 (CXIX No. 498; Cent. V No. 79), & Icones t. 19; Vahl 1790: 73; Deflers 1889: 153; Christensen 1922: 26; Schwartz 1939: 290. Fig. 18.
VERNACULAR NAME. *Saelá abjad; Saelá el bákar, Chodrab, Oud el Karah,* (Arabic).
LOCALITY. Yemen: Bolghose ("Yemen in montibus", "Blg" = Bulgose), 1763.
MATERIAL. *Forsskål* 1714 (C — 1 sheet, type of *S. hadiensis*, microf. 97: II. 7, 8).
Forsskål s.n. (BM — 1 sheet ex herb. Banks, type of *S. hadiensis*).

Senecio linifolius *Forssk.* 1775: 150 (CXIX No. 501; Cent. V No. 80), nom. illegit., non L. (1759), nec (L.) L. (1763); Christensen 1922: 26.
LOCALITY. Yemen: Al Hadiyah ("Hadie"), 1763.
MATERIAL. *Forsskål* no specimen found.

Senecio lyratus *Forssk.* 1775: 148 (CXIX No. 497; Cent. V No. 77); Christensen 1922: 26; non *S. lyratus* L.f. (1781).
S. auriculatus Vahl 1790: 72, t. 18; Schwartz 1939: 291, nom. illegit.
VERNACULAR NAME. *Mekátkat, Hörimrim* (Arabic).
LOCALITY. Yemen: Hadie & Barad ("In montibus Hadîensibus", "Barah"), 1763.
MATERIAL. *Forsskål* 1720 (C — 1 sheet, type of *S. lyratus* & *S. auriculatus*, microf. 98: I. 1, 2); 1721 (C — 1 sheet, type of *S. lyratus* & *S. auriculatus*, microf. 98: I. 3, 4).
Forsskål s.n. (BM — 1 sheet ex herb. Banks with field label "Senecio * lyratus. Mekatkat. Bolghose", type of *S. lyratus* & *S. auriculatus*).

[**Senecio** sp. indet. 1
MATERIAL. *Unknown collector* attributed to *Forsskål* s.n. (C — 1 sheet ex herb. Schumacher, microf. 121: II. 1, 2).
NOTE. Not a Forsskål collection as this may be a South African plant].

Senecio sp. indet. 2
LOCALITY. Unspecified.
MATERIAL. *Forsskål* s.n. (LD — 1 sheet ex herb. Retzius).

Serratula ciliata *Vahl* 1790: 67; Christensen 1922: 39, non Bieb. (1808).
LOCALITY. Turkey: Büyukdere ("Bujuchtari legit in hortis"), 1761.
MATERIAL. *Forsskål* no type specimen found.
NOTE. The type was not seen by Christensen; an unknown species.

FIG. 18. **Senecio hadiensis** Forssk. — Icones, Tab. XIX.

Sigesbeckia orientalis *L.* 1753: 900; Forssk. 1775: 151 (CXX No. 510; Cent. V No. 88); Christensen 1922: 26.
LOCALITY. Yemen: Al Hadiyah, Wasab ("Hadîe", "Uhf" = Uahfad), 1763.
MATERIAL. *Forsskål* no specimen found.

Solanecio angulatus (*Vahl*) *C. Jeffrey* in Kew Bull. 41: 922 (1986).
Cacalia sonchifolia sensu Forssk. 1775: CXIX No. 485, non L.
Cacalia angulata Vahl 1794: 92; Christensen 1922: 38.
Kleinia angulata (Vahl) Willd., Sp. Pl. 3: 1739 (1803).
Emilia angulata (Vahl) DC., Prodr. 6: 303 (1837).
VERNACULAR NAME. *Oud el kârab, Saela* (Arabic).
LOCALITY. Yemen: Al Hadiyah ("Hadîe"), Mar. 1763.
MATERIAL. *Forsskål* 1026 (C — 1 sheet with field label "Bulghose", type of *C. angulata*, microf. 18: II. 3, 4); 1027 (C — 1 sheet, type of *C. angulata*, microf. 18: II. 5, 6); 1028 (C — 1 sheet, type of *C. angulata*, microf. 18: II. 7, 8); 1029 (C — 1 sheet ex herb. Schumacher, type of *C. angulata* microf. 18: III. 1, 2).

Sonchus oleraceus *L.* 1753: 794; Forssk. 1775: 215 (CXVIII No. 473; Cent. VIII No. 74); Christensen 1922: 34; Schwartz 1939: 297; Tackholm 1974: 608.
VERNACULAR NAME. *Myrrejr* (Arabic).
LOCALITY. Yemen: Jiblah ("Djöblae"), 30–31 Mar. 1763.
MATERIAL. *Forsskål* 1754 (C — 1 sheet, ex herb. Schumacher, microf. 102: III. 3, 4).

Sonchus tenerrimus *L.* 1753: 794; Fl. Turk. 5: 692 (1975).
VERNACULAR NAME. *Myrrejr* (Arabic).
LOCALITY. Yemen: Bolghose, 1763.
MATERIAL. *Forsskål* 1755 (C — 1 sheet with field label "Sonchus tenerrimus. Myrrejr. Bolghose", microf. 102: III. 1, 2).
NOTE. Although this mainly Mediterranean species is known from Saudi Arabia, this appears to be the only record from Yemen.

Sphaeranthus suaveolens (*Forssk.*) *DC.*, Prodr. 5: 370 (1836); Tackholm 1974: 552, pl. 195.
Polycephalos suaveolens Forssk. 1775: LXXII No. 422; Christensen 1927: 27. = *Polycephalos* [*sp. without epithet*] Forssk. 1775: 154 (Cent. V No. 100).
VERNACULAR NAME. *Habagbâg* (Arabic).
LOCALITY. Egypt: Rashid ("Roseitae in littore Nili"), Nov. 1761.
MATERIAL. *Forsskål* 1588 (C — 1 sheet with field label "Syngenes. Capitata n. gen. Ro. 12", type of *P. suaveolens*, microf. 82: II. 7, 8); 1589 (C — 1 sheet, type of *P. suaveolens*, microf. 82: III. 1, 2); 1590 (C — 1 sheet ex herb. Vahl, type of *P. suaveolens*, microf. 82: III. 3, 4); 1591 (C — 1 sheet ex herb. Liebmann, type of *P. suaveolens*, microf. 82: III. 5, 6). *Forsskål* s.n. (BM — 1 sheet ex herb. Banks, type of *P. suaveolens*).
NOTE. This is one of the cases with the description of a new genus and an unnamed species in the 'Centuriae' and a specific epithet in the 'Florae'.

Synedrella nodiflora (*L.*) *Gaertn.*, Fruct. et Sem. Pl. 2: 456, t. 171 (1791).
LOCALITY. ? Yemen, 1763.
MATERIAL. *Forsskål* 1773 (C — 1 sheet, microf. 121: III. 3, 4).

Tanacetum parthenium (*L.*) *Sch. Bip.*, Tanacet.: 55 (1844).
Matricaria parth.? — Forssk. 1775: CXX No. 514.
VERNACULAR NAME. *Möniât* (Arabic).

LOCALITY. Yemen: Kusma ("Kurma"), Mar. 1763.
MATERIAL. *Forsskål* 1451 (C — 1 sheet, microf. 66: III. 1, 2); 1452 (C — 1 sheet with field label "Kurma", microf. 66: III. 3, 4).
NOTE. This is listed as a wild plant but it is a cultivated garden herb in Europe and Yemen is outside its normal range, yet the label and published list correspond for the Yemen. On the label in Forsskål's hand is: "Matricaria parthe. Mböniat."

Taraxacum megalorrhizon (*Forssk.*) *Hand.-Mazz.*, Monogr. Tarax.: 35 (1907); Fl. Turk. 5: 798 (1975).
Leontodon megalorrhizon Forssk. 1775: 216 (XXXI No. 334; Cent. VIII No. 77); Christensen 1922: 35.
LOCALITY. Turkey: Dardanelles ("Ad Dardanellos"), July 1761.
MATERIAL. *Forsskål* 1406 (C — 1 sheet, microf. nil).
NOTE. Named and labelled by H. Dahlstedt 1914, as "T. macrorrhizum". (Also old annotations "an Scorzonera orientalis" and "potius Leontodon").

Tolpis virgata (*Desf.*) *Bertol.*, Rar. Lig. Pl. Dec. 1: 135 (1803); Fl. Turk. 5: 630 (1975).
LOCALITY. Unspecified, ?Turkey.
MATERIAL. *Forsskål* 1786 (C — 1 sheet ex herb. Schumacher, microf. 121: I. 7, 8).

Urospermum picroides (*L.*) *Scop. ex F.W. Schmidt*, Samml. Phys. Aufs. Naturk.: 276 (1795); Fl. Europ. 4: 308 (1976).
Tragopogon picroid. — Forssk. 1775: X No. 198.
LOCALITY. France: Marseille ("Estac"), May–June 1761.
MATERIAL. *Forsskål* 1787 (C — 1 sheet, microf. 107: III. 3, 4).

Vernonia spatulata (*Forssk.*) *Hochst. ex Sch. Bip.* in Schweinfurth 1867: 162.
Chrysocoma spatulata Forssk. 1775: 147 (CXIX No. 487; Cent. V No. 70); Christensen 1922: 26.
Staehelina hastata Vahl 1790: 70, nom. illeg. superfl.
Linosyris hastata (Vahl) DC., Prodr. 5: 352 (1836).
LOCALITY. Yemen: Taizz ("prope Taaes"), 2 Apr. 1763.
MATERIAL. *Forsskål* 1105 (C — 1 sheet, type of *C. spatulata* and *S. hastata*, microf. 26: III. 3, 4); 1106 (C — 1 sheet, type of *C. spatulata* and *S. hastata*, microf. 26: III. 5, 6).
V. cinerascens sensu auct., non Sch. Bip. — Deflers 1889: 146; Schwartz 1939: 270.

Volutaria lippii (*L.*) *Cass.* in Bull. Soc. Philom. Paris 1816: 200 (1816).
Amberboa lippii (L.) DC., Prodr. 6: 559 (1838); Tackholm 1974: 542; Fl. Palaest. 3: 390, pl. 656 (1978).
Centaurea lippii L. 1753: 910; Forssk. 1775: LXXIV No. 464.
VERNACULAR NAME. *Chaisarân* (Arabic).
LOCALITY. Egypt: Cairo ("Cairi vel Káhirae loca deserta"), 1762.
MATERIAL. *Forsskål* 1076 (C — 1 sheet, microf. 23: III. 1, 2); 1077 (C — 1 sheet, microf. 23: III. 3, 4).
Forsskål s.n. (LD — 1 sheet ex herb. Retzius).

Compositae (tribe Anthemideae) gen. & sp. indet. 1
LOCALITY. Unspecified.
MATERIAL. *Forsskål* 994 (C — 1 sheet ex herb. Vahl, microf. 120: I. 8).

Compositae (tribe Anthemideae) gen. & sp. indet. 2
LOCALITY. Unspecified.
MATERIAL. *Forsskål* 993 (C — 1 sheet, ex herb. Hornemann, microf. 119: III. 7, 8).

CONVOLVULACEAE

Calystegia soldanella (*L.*) *R. Br.*, Prodr. Fl. Nov. Holl.: 484 (1810); Fl. Europ. 3: 78 (1972).
LOCALITY. Unspecified.
MATERIAL. *Forsskål* s.n. (LD — 1 sheet ex herb. Retzius).

Convolvulus althaeoides *L.* subsp. **tenuissimus** (*Sibth. & Sm.*) *Stace* in Bot. Jour. Linn. Soc. 64: 59 (1971); Fl. Europ. 3: 82 (1972); Fl. Turk. 6: 214 (1978).
C. sericeus sensu Forssk. 1775: 204 (XXI No. 90; Cent. VIII No. 11), non L.
LOCALITY. Turkey: Sea of Marmara ("Ad littora Maris Marmorae in sepibus"), July–Aug 1761.
MATERIAL. *Forsskål* 439 (C — 1 sheet, microf. 31: III. 5, 6); 449 (C — 1 sheet, microf. 31: III. 7, 8).

Convolvulus arvensis *L.* 1753: 153; Forssk. 1775: IV No. 55; Tackholm 1974: 430, pl. 149 B (C).
LOCALITY. France: Marseille ("Estac"), May–June 1761.
MATERIAL. *Forsskål* 1130 (C — 1 sheet with field label "E.4", microf. 30: III. 2).
C. hastatus Forssk. 1775: 203 (LXII No. 121; Cent. VIII No. 10).
VERNACULAR NAME. *Öllaeik* (Arabic).
LOCALITY. Egypt: Cairo ("Kåhirae"), 1762.
MATERIAL. *Forsskål* 1131 (C — 1 sheet with field label "No. 13", ex herb. Hofman Bang, type of *C. hastatus*, microf. 31: II. 1, 2); 1132 (C — 1 sheet ex herb. Hornemann, type of *C. hastatus*, microf. 31: II. 5, 6); 1133 (C — 1 sheet, ex herb. Horneman, microf. 122: II. 3, 4).
Forsskål s.n. (LD — 1 sheet ex herb. Retzius).
NOTE. The sheet No. 445, placed with these on the microfiches and annotated as *C. hastatus* by Ascherson, 1881, is *Ipomoea eriocarpa* R. Br.

Convolvulus biflorus sensu Forssk. 1775: 203 (LXII No. 122; Cent. VIII No. 9), non L.
LOCALITY. Egypt: Cairo ("In desertis Kåhirinis"), 1761–62.
MATERIAL. *Forsskål* no specimen found.

Convolvulus hystrix *Vahl* 1790: 16; Schwartz 1939: 201; Tackholm 1974: 426.
Convolvulus spinosus Forssk. 1775: CVI No. 121, non Burm. (1768); Christensen 1922: 37.
VERNACULAR NAME. *Dahhi* (?) (Arabic).
LOCALITY. Yemen: Bayt al Faqih ("Beit el fakih"), 1763.
MATERIAL. *Forsskål* 440 (C — 1 sheet, type of *C. hystrix* and *C. spinosus*, microf. 32: I. 3, 4); 454 (C — 1 sheet with field label, type of *C. hystrix* and *C. spinosus*, microf. 32: I. 1, 2).

Convolvulus lanatus *Vahl* 1790: 16; Christensen 1922: 38; Tackholm 1974: 426, pl. 149A.
C. cneorum sensu Forssk. 1775: LXIII No. 124, CVI No. 120, non L.
VERNACULAR NAME. *Baejâd* (Arabic).
LOCALITY. Egypt: Cairo? & Sinai, 1762.
MATERIAL. *Forsskål* 455 (C — 1 sheet, type of *C. lanatus*, microf. 30: III. 3, 4); 456 (C — 1 sheet, type of *C. lanatus*, microf. 31: I. 5, 6); 1134 (C — 1 sheet ex herb. Hofman Bang,

type of *C. lanatus*, microf. 30: III. 5, 6); 1136 (C — 1 sheet, type of *C. lanatus*, microf. 31: I. 7, 8); 1135 (C — 1 sheet ex herb. Hornemann, type of *C. lanatus*, microf. 31: I. 1, 2); 1137 (C — 1 sheet ex herb. Horneman, type of *C. lanatus*, microf. 122: II. 7, 8).
Forsskål s.n. (BM — 1 sheet ex herb. Banks, type of *C. lanatus*).
Forsskål s.n. (LD — 1 sheet ex herb. Retzius, type of *C. lanatus*).
NOTE. F. Saad, 1965, labelled the Copenhagen sheets and designated 456 as holotype, but it should be considered to be a lectotype with all the other sheets syntypes).
See also Chenopod. gen. et sp. indet (1003, microf. 31: I. 3, 4) which has an original field label bearing the locality "Mt Horeb" (Sinai). The collections from Sinai have probably been made by Niebuhr or von Haven.

Convolvulus prostratus *Forssk.* 1775: 203 (CVI No. 119; Cent. VIII No. 8); Christensen 1922: 33; Tackholm 1974: 429.
C. pentapetaloides sensu Vahl 1791: 36, non L.
LOCALITY. Yemen: Wadi Mawr ("Môr"), Jan. 1763.
MATERIAL. *Forsskål* 438 (C — 1 sheet, type of *C. prostratus*, microf. 31: III. 1, 2); 446 (C — 1 sheet, type of *C. prostratus*, microf. 31: II. 7, 8); 1138 (C — 1 sheet ex herb. Schumacher, type of *C. prostratus*, microf. 31: III. 3, 4).
NOTE. F. Saad, 1965, designated 438 as the holotype, but it should correctly be the lectotype.

Convolvulus siculus *L.* subsp. **agrestis** (*Schweinf.*) *Verdc.* in Kew Bull. 12: 344 (1957); F.T.E.A., Convolvulac. 41 (1963).
LOCALITY. Unspecified, probably Yemen.
MATERIAL. *Forsskål* 457 (C — 1 sheet, microf. 122: I. 7, 8); 458 (C — 1 sheet, microf. 122: II. 1, 2).

Cressa cretica *L.* 1753: 223; Tackholm 1974: 435, pl. 151D.
LOCALITY. Egypt: El Tûr ("Tor"), Sept. 1762.
MATERIAL. *Forsskål* 441 (C — 1 sheet, microf. 34: I. 3, 4); 442 (C — 1 sheet, microf. 34: I. 1, 2); 443 (C — 1 sheet, microf. 33: III. 7, 8); 1157 (C — 1 sheet ex herb. Hofman Bang, microf. 33: III. 1, 2); 1158 (C — 1 sheet ex herb. Hofman Bang, with field label "Rai .. â juxta Tor" microf. 33: III. 3, 4); 1159 (C — 1 sheet ex herb. Hofman Bang, with field label "ad salinas P.V." microf. 33: III. 5, 6); 1155 (C — 1 sheet ex herb. Hornemann, microf. 34: I. 5); 1156 (C — 1 sheet ? ex herb. Hornemann, microf. 34: I. 6, 7).
Forsskål s.n. (BM — 1 sheet ex herb. Banks).
Forsskål s.n. (LD — 1 sheet ex herb. Retzius).

Hildebrandtia africana *Vatke* in Journ. Bot. 14: 313 (1876); Schwartz 1939: 198.
LOCALITY. Yemen ("Arabia" on sheet), 1763.
MATERIAL. *Forsskål* 1337 (C — 1 sheet, microf. 132: II. 3, 4).
NOTE. There is some doubt as to whether the Arabian plant is taxonomically the same as the Somali one known as *H. africana*.

Ipomoea aquatica *Forssk.* 1775: 44 (CVI No. 124; Cent. II: 44); Christensen 1922: 15; F.T.E.A. Convolvulac.: 120 (1963).
LOCALITY. Yemen: Zabid ("Ad Zebîd"), 5–6 Apr. 1763.
MATERIAL. *Forsskål* 447 (C — 1 sheet, type, microf. 57: I. 3, 4).
NOTE. This name was for long displaced in favour of *I. reptans* (Schwartz 1939: 200), but Linnaeus's type of that species is a *Merremia*.

Ipomoea cairica (*L.*) *Sweet*, Hort. Brit.: 287 (1827); F.T.E.A., Convolvulac.: 125 (1963); Tackholm 1974: 432.
I. palmata Forssk. 1775: 43 (LXIII No. 125; Cent. II: 40); Christensen 1922: 15; Schwartz 1939: 200.
VERNACULAR NAME. *Sett el bösn, Sett el hösch, Öllaech, Ollaea* (Arabic).
LOCALITY. Egypt: Rashid ("*Rosettae* in hortis scandens ..."), Nov. 1761.
MATERIAL. *Forsskål* 432 (C — 1 sheet, type of *I. palmata*, microf. 57: II. 1, 2); 448 (C — 1 sheet with field label "Ipomoea tuberosa ? Rossette", type of *I. palmata*, microf. 57: I. 7, 8).
NOTE. *Forsskål* 448 is named and labelled by van Oostroom 1949.

Ipomoea eriocarpa *R. Br.*, Prodr. Fl., Nov. Holl.: 484 (1810); Tackholm 1974: 432.
LOCALITY. Egypt: Cairo?
MATERIAL. *Forsskål* 445 (C — 1 sheet, microf. 31: II. 3, 4).
NOTE. Forsskål 445 was written up by Ascherson, 1881, as "Convolvulus hastatus".

Ipomoea nil (*L.*) *Roth*, Cat. Bot. 1: 36 (1797); F.T.E.A. Convolvulac.: 113 (1963).
I. scabra Forssk. 1775: 44 (CVI No. 122; Cent. II: 43); Christensen 1922: 15; Schwartz 1939: 199.
VERNACULAR NAME. *Scherdjedja* (Arabic).
LOCALITY. Yemen: Al Hadiyah ("Hadîe"), Mar. 1763.
MATERIAL. *Forsskål* 450 (C — 1 sheet, type, microf. 57: II. 3, 4).

Ipomoea obscura (*L.*) *Ker-Gawl.* in Bot. Reg. 3, t. 239 (1817); Deflers 1889: 175; Schwartz 1939: 199; F.T.E.A. Convolvulac.: 116 (1963).
LOCALITY. Yemen: Wadi Mawr ("Môr", "Mour"), Feb. 1763.
MATERIAL. *Forsskål* 444 (C — 1 sheet with field label "Ipomoea * pallidiflora. Mour", microf. 122: II. 5, 6).
NOTE. Although Forsskål on the field label suggested the name "Ipomoea pallidiflora" with an asterisk indicating a new species, the names does not appear in Flora Aegyptiaco-Arabica.

Ipomoea pes-caprae (*L.*) *R. Br.* in Tuckey, Narrative Exped. R. Zaire: 477 (1818); Schwartz 1939: 200; F.T.E.A. Convolvulac.: 121 (1963).
I. biloba Forssk. 1775: 44 (CVI No. 123; Cent. II: 42); Christensen 1922: 15.
VERNACULAR NAME. *Sokar* (Arabic).
LOCALITY. Yemen: Zabid ("*Zebîd*"), 4–5 Apr. 1763.
MATERIAL. *Forsskål* 451 (C — 1 sheet, type, microf. 57: I. 5, 6).
Forsskål s.n. (BM — 1 sheet, designated lectotype of *I. biloba* by Verdcourt).
NOTE. Verdcourt (F.T.E.A.: 122) writes: "Subsp. *pes-caprae*, with more deeply bilobed leaves, has not been recorded from Africa. Prof. H. St. John pointed out to me after examining material in the East African Herbarium that both subsp. *pes-caprae* and subsp. *brasiliensis* occurred in Arabia. It is therefore of interest to me to which subsp. the type of Forsskål's *I. biloba* belongs. The type material is in my opinion rather too poor for a firm decision, but it is probable that it belongs to subsp. *b.* and not subsp. *p-c* as previously thought".

Ipomoea quamoclit *L.* 1753: 159.
LOCALITY. ? Egypt.
MATERIAL. *Forsskål* 1361 (C — 1 sheet ex herb. Hofman Bang, microf. 122: III. 1, 2).
NOTE. A native of tropical America long-since cultivated in warm countries of the Old World as *Quamoclit pennata* Bojer (e.g. Bircher 1960: 647).

Ipomoea triflora *Forssk.* 1775: 44 (CVI No. 125; Cent. II: 45); F.T.E.A. Convolvulac.: 133 (1963).
VERNACULAR NAME. *Sotar, Gaschue, Gaschve* (Arabic).
LOCALITY. Yemen: Al Hadiyah ("Hadîe"), Mar. 1763.
MATERIAL. *Forsskål* 452 (C — 1 sheet with original field label, type of *I. triflora*, microf. 57: II. 5, 6).
NOTE. The original ticket on the reverse bears the name, the vernacular name "sotar" and the locality "Bölgholi", presumably Bolghose near Al Hadiyah.
Schwartz 1939: 199 placed this under *I. obscura.*

Ipomoea verticillata *Forssk.* 1775: 44 (CVI No. 126; Cent. II No. 41); Vahl 1794: 33; Christensen 1922: 15; Schwartz 1939: 198.
Convolvulus forskahlii Sprengel, Syst. 1: 596 (1825), non Delile 1809: t. 18, 3.
VERNACULAR NAMES. *Teraeba, Toraeha, Sebelli, Shelli* (Arabic).
LOCALITY. Yemen: Wadi Mawr ("Môr", "Cudmiae", a locality near Mor, on field label), Jan. 1763.
MATERIAL. *Forsskål* 453 (C — 1 sheet with original field label "Ipomoea, Toraeha ... Cudmiae", type, microf. 57: II. 7, 8).

Quamoclit coccinea (*L.*) *Moench* Meth.: 453 (1794).
LOCALITY. ? Yemen.
MATERIAL. *Forsskål* 1361 (C — 1 sheet, microf. 122: III. 1, 2).

Seddera arabica (*Forssk.*) *Choisy* in DC., Prodr. 9: 441 (1845); Schwartz 1939: 197; F.T.E.A. Convolvulac.: 27 (1963).
Cressa arabica Forssk. 1775: 54 (CVIII No. 206; Cent. II. No. 81); Christensen 1922: 16.
LOCALITY. Yemen: Taizz ("In locis montosis ad *Taaes* semel lecta".), 1763.
MATERIAL. *Forsskål* no specimen found.

CORNACEAE

Cornus sanguinea *L.* susp. **australis** (*C.A. Meyer*) *Jáv.* in Soó & Jávorka, Magyar Növ. Kéz. 1: 398 (1951); Fl. Turk. 4: 340 (1972).
LOCALITY. Turkey: Istanbul ("Constp. hortense"), 1761.
MATERIAL. *Forsskål* 1149 (C — 1 sheet, ex herb. Hofman Bang, with field label "Co. 74", microf. 32: III. 3, 4).

CRASSULACEAE

Crassula alba *Forssk.* 1775: 60 (CIX No. 228; Cent. II. No. 100); Christensen 1922: 16; Schwartz 1939: 78; Tolken in Contrib. Bolus Herb. 8: 362 (1977); F.T.E.A. Crassulac.: 17, fig. 2 (1987).
LOCALITY. Yemen: Al Hadiyah & Bughah ("In montibus *Hadîe* & *Boka*"), 1763.
MATERIAL. *Forsskål* 691 (C — 1 sheet, Tolken's lectotype of *C. alba*, microf. 33: I. 8, II. 1), 1150 (C — 1 sheet ex herb. Liebmann, microf. 33: II. 2).

Kalanchoe alternans (*Vahl*) *Pers.*, Syn. 1: 446 (1805); DC., Prodr. 3: 395 (1828); Schweinfurth 1896: 201; Schwartz 1939: 79 (except *Deflers* 444); Raadts in Willdenowia 11: 329 (1981).

Cotyledon alternans Vahl 1791: 51 non Haw (1819); Christensen 1922: 38.
Vereia alternans (Vahl) Spreng., Syst. Veg. ed. 16, 2: 260 (1825).
Cotyledon orbiculata sensu Forssk. 1775: CXII No. 292, non L. (1753).
Kalanchoe rosulata Raadts in EBJ 91: 480, fig. 2 (1972).
VERNACULAR NAME. *Chodardar* (Arabic).
LOCALITY. Yemen: Hays ("Hadie" on label, Haes), 1763.
MATERIAL. *Forsskål* 690 (C — 1 sheet with field label, microf. 33: I. 3, 4).

Kalanchoe deficiens (*Forssk.*) *Asch. & Schweinf.* in Mem. Inst. Egypte 2: 79 (1887); Raadts in Willdenowia 11: 327 (1981).
Cotyledon deficiens Forssk., 1775: 89 (CXI No. 290; Cent. III No. 73).
C. aegyptiaca Lam., Encycl. 2: 142 (1786), nom. illegit.
Kalanchoe aegyptiaca (Lam.) DC., Pl. Hist. Succ.: t. 64 (1801).
K. glaucescens Britten var. *deficiens* (Forssk.) Senni in Boll. Reale Orto Bot. Giardino Colon. Palermo 4: 12 (1905).
VERNACULAR NAME. *Odejn* (Yemen, Arabic).
LOCALITY. Egypt: "Colitur in hortus Aegypti". Yemen: Milhan ("Melhan"), 1763.
MATERIAL. *Forsskål* no type specimen found.
NOTE. *Deflers* 572 (P) was designated the lectotype by Raadts, but her conclusions are disputed by J.R.I. Wood.

Kalanchoe lanceolata (*Forssk.*) *Pers.*, Synops. Pl. 1: 446 (1805); Schwartz 1939: 79; R. Fernandes in Bol. Soc. Brot. ser. 2, 53: 381 (1980); F.T.E.A. Crassulac.: 49 fig. 7 (1987).
Cotyledon lanceolata Forssk. 1775: 89 (CXI No. 291, Cent. III No. 74); Christensen 1922: 19.
VERNACULAR NAME. *Hömed errobat, Homedet er robah* (Arabic).
LOCALITY. Yemen: Kusma ("Kurmae"), 1763.
MATERIAL. *Forsskål* 689 (C — 1 sheet, holotype of *C. lanceolata*, microf. 33: I. 6, 7).

CRUCIFERAE

Anastatica hierochuntica *L.* 1753: 641; Forssk. 1775: 117 (LXIX No. 309; Cent. IV No. 57); Schwartz 1939: 73; Fl. Palaest. 1: 276, pl. 406 (1966); Tackholm 183 pl. 54A.
VERNACULAR NAME. *Kaf marjam* (Arabic), *manus Mariae* (Latin translation of Arabic name).
LOCALITY. Egypt: Cairo ("Káhirae in desertis rarior"), 1761–62.
MATERIAL. *Forsskål* 718 (C — 1 sheet, microf. 6: II. 3, 4); 939 (C — 1 sheet ex herb. Hofman Bang, microf. 6: II. 5, 6).
Forsskål s.n. (BM — 1 sheet ex herb. Banks).

Bunias orientalis *L.* 1953: 670; Forssk. 1775: 120 (CXVI No. 400; Cent. IV. 69).
VERNACULAR NAME. *Doraema, Chodejva,* (*Chodeira*), *Fussa* (Arabic).
LOCALITY. Yemen: Al Luhayyah ("Lhj"), Wadi Mawr ("Môr"), Bayt al Faqih ("Btf").
MATERIAL. *Forsskål* no specimen found.

Cakile maritima *Scop.*, Fl. Carn. ed. 2, 2: 35 (1772); Fl. Turk. 1: 275 (1965) ssp. **aegyptiaca** (*Willd.*) *Nyman*, Consp. 29 (1878); Fl. Europ. 1: 343 (1964), 1: 414 (1993); Tackholm 1974: 197.
Isatis aegyptia(ca) L. - Forssk. 1775: 121 (XXIX No. 292; LXIX No. 329; Cent. IV No. 72); Christensen 1922: 23
Bunias cakile L. *β* - Vahl 1792: 78.
VERNACULAR NAME. *Fidjl el djemal* (Arabic).

LOCALITY. Egypt: Alexandria ("Alexandriae, vernalis"), 1 Apr. 1761.
MATERIAL. *Forsskål* 721 (C — 1 sheet, microf. 57: III. 5, 6); 1363 (C — 1 sheet ex herb. Vahl, microf. 57: III. 3, 4); 1364 (C — 1 sheet ex herb. Vahl, microf. 57: III. 7, 8).

Isatis pinnata Forssk. 1775: 121 (XXIX No. 293; LXIX No. 330; Cent. IV No. 73); Christensen 1922; 23.
LOCALITY. Turkey: Anatolia without locality ("Natoliae"), & Egypt: Alexandria ("Alexandriae"), 1761.
MATERIAL. *Forsskål* 1365 (C — 1 sheet ex herb Liebmann, type of *I. pinnata*, microf. 58: I. 1, 2); 1366 (C — 1 sheet ex herb Vahl, type of *I. pinnata*, microf. 58: I. 3, 4); 1367 (C — 1 sheet ex herb. Schumacher, type of *I. pinnata*, microf. 58: I. 5, 6).
Forsskål s.n. (BM — ½ sheet, microf. nil).

Cardamine africana *L.* 1753: 655; Vahl 1791: 77; F.T.E.A. Crucif.: 38 (1982).
LOCALITY. Yemen, 1763.
MATERIAL. *Forsskål* 1044 (C — 1 sheet ex herb. Vahl, microf. 122: III. 3, 4).

Coronopus squamatus (*Forssk.*) *Aschers.*, Fl. Prov. Brandenb. 1: 62 (1860); Fl. Palaest. 1: 303, t. 449 (1966).
Lepidium squamatum Forssk. 1775: 117 (LXIX No. 310; Cent. IV No. 59); Christensen 1922: 22.
VERNACULAR NAME. *Raschât* (Arabic).
LOCALITY. Egypt: Alexandria & Cairo ("Alexandriae in desertis vicinis" "Cs Cd"), 1 Apr. 1762.
MATERIAL. *Forsskål* 1407 (C — 1 sheet ex herb. Vahl, type of *L. squamatum*, microf. 62: II. 7, 8).
Forsskål s.n. (BM — 1 sheet ex herb. Banks, type of *L. squamatum*).

Diplotaxis acris (*Forssk.*) *Boiss.*, Fl. Orient. 1: 389 (1867); Schwartz 1939: 70; Fl. Palaest. 1: 306, t. 453 (1966); Tackholm 1974: 188.
Hesperis acris Forssk. 1775: 118 (LXIX No. 320; Cent. IV No. 62); Delile 1809: t. 35, 2; Christensen 1922: 22.
VERNACULAR NAME. *Sphaeri* (Arabic ex Latin).
LOCALITY. Egypt: Cairo ("In convallibus circa Káhiram frequens"), 1762.
MATERIAL. *Forsskål* 1335 (C — 1 sheet ex herb Vahl, type of *H. acris*, microf. 54: I. 1, 2).

Diplotaxis harra (*Forssk.*) *Boiss.*, Fl. Orient. 1: 388 (1867); Schwartz 1939: 70; Fl. Palaest. 1: 306, t. 454 (1966); Tackholm 1974: 188, pl. 56A.
Sinapis harra Forssk. 1775: 118 (LXIX No. 326; Cent. IV No. 64); Christensen 1922: 22.
Sisymbrium hispidium Vahl 1792: 77.
VERNACULAR NAME. *Harra* (Arabic).
LOCALITY. Egypt: Cairo ('Káhirae'), 1762.
MATERIAL. *Forsskål* 722 (C — 1 sheet, type of *S. harra*, microf. 100: III. 1, 2); 1739 (C — 1 sheet with 2 specimens; ex herb. Vahl with field label "Harra, Ca. 91", type of *S. harra*; "1" ex herb. Liebmann, type of *S. harra*; microf. 100: II. 8); 1738 (C — 1 sheet with 2 specimens; "2" ex herb. Schumacher, type of *S. harra*; "3" ex herb. Vahl, type of *S. harra*; microf. 100: II. 7).

Enarthrocarpus lyratus (*Forssk.*) *DC.*, Syst. 2: 661 (1821); Schwartz 1939: 71; Fl. Palaest. 1: 328 (1966); Tackholm 1974: 197, pl. 56B.
Raphanus lyratus Forssk. 1775: 119 (LXIX No. 328; Cent. IV No. 65); Christensen 1922: 22.
VERNACULAR NAME. *Rsjâd el barr*, *Rschâd el barr* (Arabic).
LOCALITY. Egypt: Giza ("Pyramides Gizensis ... insula Djesiret ed dáhab"), Jan. 1762.

a. Lunaria scabra
B. Siliquaria glandulosa.

Fig. 19. (A) **Farsetia aegyptia** Turra — Icones, Tab. XVI *Lunaria scabra*.
(B) **Cleome amblyocarpa** Barr & Murb. — Icones, Tab. XVI *Siliquaria glandulosa*.

MATERIAL. *Forsskål* 710 (C — 1 sheet, type of *R. lyratus*, microf. 86: II. 5, 6); 724 (C — 1 sheet, type of *R. lyratus*, microf. 86: II. 3, 4).
Forsskål s.n. (LIV — 1 sheet labelled "Siberia, Pallas", 1909. LBG.6717, type of *R. lyratus*).
Forsskål s.n. (BM — 1 sheet ex herb. Banks, type of *R. lyratus*).

Eremobium aegyptiacum (*Spreng.*) *Hochr.* in Ann. Conserv. & Jard. Genèv. 7–8: 159 (1904); Tackholm 1974: 178.
LOCALITY. ? Egypt.
MATERIAL. *Forsskål* 712 (C — 1 sheet, microf. 122: III. 7, 8).

Erucaria crassifolia (*Forssk.*) *Delile*, Fl. Egypte: 100 = 244, t. 34, 1 (1813); Tackholm 1974: 194.
Brassica crassifolia Forssk. 1775: 118 (LXIX No. 323; Cent. IV: 63); Christensen 1922: 22.
VERNACULAR NAME. *Krumb bissabra* (Arabic).
LOCALITY. Egypt: Cairo ("Ad Pyramides frequens"), Jan. 1762.
MATERIAL. *Forsskål* 709 (C — 1 sheet, type of *B. crassifolia*, microf. 15: III. 5, 6); 711 (C — 1 sheet, type of *B. crassifolia*, microf. 15: III. 1, 2); 713 (C — 1 sheet with original field label "Ca. 119", type of *B. crassifolia*, microf. 15: III. 3, 4); 1011 (C — 1 sheet ex herb. Schumacher, type of *B. crassifolia*, microf. 15: II. 7, 8).

Farsetia aegyptia *Turra*, Farset. Nov. Gen.: 5, t. 1 (1765); Fl. Palaest. 1: 281, pl. 415 (1966); Tackholm 1974: 183 pl. 55B. Fig. 19A.
Cheiranthus farsetia L. 1767: 94; Vahl 1790: 48.
LOCALITY. Egypt: Cairo, 1761–62.
MATERIAL. *Forsskål* seeds cult. Hort. Upsala (LINN 839: 29, type of *C. farsetia*, microf. 443: III. 1, 2).

Lunaria scabra Forssk. 1775: 117 (LXIX No. 313; Cent. IV No. 60), & Icones t. 16A; Christensen 1922: 22.
VERNACULAR NAME. *Djarba* (Arabic).
LOCALITY. Egypt: Cairo ("In desertis Káhirinis ad Liblab"), Mar. 1762.
MATERIAL. *Forsskål* 708 (C — 1 sheet, type of *L. scabra*, microf. 65 (1–2): I. 5, 6); 1435 (C — 1 sheet, type of *L. scabra*, microf. 65 (1–2): I. 7, 8); 1436 (C — 1 sheet ex herb. Vahl, type of *L. scabra*, microf. 64: III. 7, 8, 65 (1–2): II. 3, 4); 1437 (C — 1 sheet ex herb. Schumacher, type of *L. scabra*, microf. 65 (1–2): II. 1, 2); 1089 (C — 1 sheet ex herb. Hofman Bang, microf. 24: III. 7, 8); 1090 (C — 1 sheet ex herb. Hofman Bang, microf. 25: I. 1, 2); 1091 (C — 1 sheet ex herb. Drejer, microf. 25: I. 3, 4).
Forsskål s.n. (BM — 1 sheet, type of *L. scabra*).

Farsetia linearis (*Forssk.*) *Decne. ex Boiss.* in Ann. Sci. Nat., Sér. 2, 17: 150 (1842).
Cheiranthus linearis Forssk. 1775: 120 (CXVI No. 398; Cent. IV No. 67); Christensen 1922: 22.
LOCALITY. Yemen: Bayt al Faqih ("*Beit el fakíh*"), 1763.
MATERIAL. *Forsskål* 723 (C — 1 sheet, type of *C. linearis*, microf. 25: I. 5–6); 1088 (C — 1 sheet ex herb. Vahl, type of *C. linearis*, microf. nil).
NOTE. This is *F. longisiliqua* Decne. in Ann. Sci. Nat. Sér. 2, 4: 69 (1835); Schweinfurth 1896: 185.

Iberis pinnata *L.*, Cent. Pl. 1: 18 (1755); Fl. Europ. 1: 325 (1964), 1: 393 (1993); Fl. Turk. 1: 312 (1965).
Iberis — Forssk. 1775: XXIX No. 286.

LOCALITY. Turkey: Bozcaada I. ("Imr." = Imros), July 1761.
MATERIAL. *Forsskål* 714 (C — 1 sheet, microf. 56: II. 3, 4).

Lepidium armoracia *Fisch. & Mey.* in Index Sem. hort. Petrop. 1842: 77 (1842); Schweinfurth 1896: 182; Schwartz 1939: 72.
Lepidium alpinum ? — Forssk. 1775: CXVI No. 395.
LOCALITY. Yemen: Al Hadiyah ("Ms" = montum regio superior), Mar. 1763.
MATERIAL. *Forsskål* 715 (C — 1 sheet, microf. 62: II. 3, 4).

Lepidium graminifolium *L.*, Syst. Nat. ed. 10, 2: 1127 (1759); Forssk. 1775: XXIX No. 287; Fl. Europ. 1: 333 (1964), 1 : 402 (1993); Fl. Turk. 1: 285 (1965).
LOCALITY. Turkey: Dardanelles ("Dd"), July 1761.
MATERIAL. *Forsskål* 717 (C — 1 sheet with field label, microf. 62: II. 5, 6).

Matthiola livida (*Del.*) *DC.*, Syst. 2: 174 (1821); Fl. Palaest. 1: 273, pl. 400 (1966); Tackholm 1974: 181, pl. 53A.
Cheiranthus tristis sensu Forssk. 1775: 119 (LXIX No. 317; Cent. IV No. 66), non L.; Christensen 1922: 22.
VERNACULAR NAMES. *Naegeisi, Schudjara* (Arabic).
LOCALITY. Egypt: Cairo ("Káhiriae ad pagum Caid Bey"), Mar. 1762.
MATERIAL. *Forsskål* 720 (C — 1 sheet, microf. 25: I. 7, 8); 1092 (C — 1 sheet ex herb. Schumacher, microf. 25: II. 1, 2); 1093 (C — 1 sheet ? ex herb. Hornemann, microf. 25: II. 3, 4); 1094 (C — 1 sheet ex herb. Vahl, microf. 25: II. 5, 6); 1095 (C — 1 sheet ex herb. Schumacher, microf. 25: II. 7, 8); 1096 (C — 1 sheet ex herb. Liebmann, microf. 25: III. 1, 2).
Forsskål s.n. (LD — 1 sheet ex herb. Retzius).

Matthiola tricuspidata (*L.*) *R. Br.* in Aiton, Hort. Kew. ed. 2, 4: 120 (1812); Fl. Palaest. 1: 271, pl. 396 (1966).
Cheiranthus villosus Forssk. 1775: 120 (LXIX No. 318; Cent. IV No. 68); Christensen 1922: 22.
VERNACULAR NAME. *Mantur* (Arabic).
LOCALITY. Egypt: Alexandria ("Alexandriae"), Oct. 1761.
MATERIAL. *Forsskål* 1097 (C — 1 sheet ex herb. Vahl with original field label, type of *C. villosus*, microf. 25: III. 3, 4).
NOTE. Tackholm does not mention this species in her Students' Flora (1974), yet several 19th century specimens at Kew confirm its presence at Alexandria from where Forsskål collected his, so it may now be extinct in that location.

Schouwia purpurea (*Forssk.*) *Schweinf.* 1896: 183; Schwartz 1939: 71.
Subularia purpurea Forssk. 1775: 117 (CXVI No. 393; Cent. IV No. 58); Christensen 1922: 22.
VERNACULAR NAME. *Bökel, Bockel* (Arabic).
LOCALITY. Yemen: Wadi Mawr ("Môr"), Jan.–Feb. 1763.
MATERIAL. *Forsskål* 716 (C — 1 sheet type of *S. purpurea*, microf. 105: III. 3, 4); 719 (C — 1 sheet, type of *S. purpurea*, microf. 105: III. 5, 6); *Forsskål* 1771 (C — 1 sheet ex herb. Vahl type of *S. purpurea*, microf. 105: II. 7, 8); 1772 (C — 1 sheet ex herb. Schumacher, type of *S. purpurea*, microf. 105: III. 1, 2).

Sinapis allionii *Jacq.*, Hort. Bot. Vindob. 2: 79, t. 168 (1773); Tackholm 1974: 192, pl. 58.
S. arvensis sensu Forssk. 1775: LXIX No. 325, non L.
VERNACULAR NAME. *Karilli, Chardel* (Arabic).

FIG. 20. (A) **Zilla spinosa** (L.) Prantl — Icones, Tab. XVIII *Zilla*
(B) **Suaeda vermiculata** Forssk. ex J.G. Gmel. — Icones, Tab. XVIII.

FIG. 21. (A) **Zilla spinosa** (L.) Prantl — Icones, Tab. XVII *Zilla myagroides*.
(B) **Andrachne telephioides** L. — Icones, Tab. XVII *Eraclissa hexagyna*

LOCALITY. Egypt: El Giza ("Guise" on field label, "As" = Alexandria in book); 1762.
MATERIAL. *Forsskål* 1740 (C — 1 sheet ex herb. Vahl, with field label "Sinapis arvensis", microf. 100: II. 5, 6).
NOTE. This specimen - with broad leaf segments - matches material of *S. turgida* Delile, now considered to be synonymous with *S. allionii* which has narrow leaf segments.

Sisymbrium irio *L.* 1753: 659; Vahl 1791: 77; Fl. Europ. 1: 264 (1964), 1: 319 (1993).
LOCALITY. Unknown, probably France.
MATERIAL. *Forsskål* s.n. (LD — 1 sheet ex herb. Retzius).
NOTE. The sheet has "S. monense F." written on the reverse.

Sisymbrium pinnatifidum Forssk. 1775: 118 (CXVI. 396; Cent. IV 61); Christensen 1922: 22; Schwartz 1939: 74.
LOCALITY. Yemen: Al Hadiyah ("In montibus Hadiensibus"), Bolghose ("Blg."), 1763.
MATERIAL. *Forsskål* 1741 (C — 1 sheet ex herb. Vahl, type of *s. pinnatifidum*, with field label "Sisymbrium * pinnatifidum. Bolghose", microf. nil).
NOTE. Christensen (1922) did not see this specimen.

Zilla spinosa (*L.*) *Prantl* in Engl. & Prantl, Nat. Pflanzenfam. 3, 2: 174, t. 112 (1891); Fl. Palaest. 1: 324, pl. 474 (1966); Tackholm 1974: 197, pl. 59A. Fig. 20A, 21A.
Zilla myagrioides Forssk. 1775: 121 (Cent. IV No. 74); Christensen 1922: 23 = *Zilla myagroides* Forssk., Icones 5, t. 17A (1776). = *Zilla* [*sp. without epithet*] Forssk. 1775: 121 (Cent. IV No. 75), Icones 5, t. 18A (1776).
VERNACULAR NAMES. *Zilla, Zillae* (Arabic).
LOCALITY. Egypt: Cairo ("In desertis Kahirinis"), Mar. 1762.
MATERIAL. *Forsskål* 725 (C — 1 sheet with field label "Myagr. n. Zilla ... Cairi des. ad Esbodna" (?), type of *Z. mayagrioides*, microf. 113: I. 5, 6); 1826 (C — 1 sheet ex herb. Schumacher, type of *Z. mayagrioides*, microf. 113: I. 1, 2); 1827 (C — 1 sheet ex herb. Vahl, type of *Z. myagrioides*, microf. 113: I. 3, 4).
Forsskål s.n. (BM — 1 sheet ex herb. Banks, type of *S. myragrioides*).
NOTE. Another sheet (C — ex herb. Hofman Bang, microf. 122:III. 5, 6) is attributed to *Forsskål*, but there is some doubt as to whether he collected the specimen in view of the note on the reverse of the sheet "Jardin de la Malmaison". The nomenclatural complications of *Zilla myagrioides* have been discussed by Friis in Taxon 33: 659 (1984).

CUCURBITACEAE

Citrullus colocynthis (*L.*) *Schrad.* in Linnaea 12: 414 (1838); Schwartz 1939: 267; F.T.E.A. Cucurbitac.: 46 (1967).
Cucumis colocynthis L. 1753: 1011; Forssk. 1775: CXXII No. 575.
VERNACULAR NAMES. *Hamdal, Dabak* (Arabic).
LOCALITY. Yemen: Al Luhayyah ("Lohoja" on field label, "Môr" in book), Jan. 1763.
MATERIAL. *Forsskål* 664 (C — 1 sheet with original field label, microf. 35: III. 1, 2).

Citrullus lanatus (*Thunb.*) *Matsum. & Nakai*, Cat. Sem. Sp. Hort. Bot. Univ. Imp. Tokyo: 30 (1916); F.T.E.A. Cucurbitac.: 46 (1967).
Citrullus battich Forssk. 1775: 167 (CXXII No. 572; Cent. VI, No. 43); Christensen 1922: 28.
VERNACULAR NAME. *Dubba farakis* (Arabic).
Cucurbita citrullus Forssk. 1775: CXXII No. 570; Christensen 1922: 28 = *Citrullus* [*sp. without epithet*] Forssk. 1775: 165 (Cent. VI No. 44).
VERNACULAR NAME. *Schurredj* (Arabic).
Cucurbita citrullus Forssk. 1775: CXXII No. 571; Christensen 1922: 28 = *Citrullus* [*sp. 2 without epithet*] Forssk. 1775: 167 (Cent. VI No. 45).

VERNACULAR NAME. *Kasch* (Arabic).
MATERIAL. *Forsskål* no specimens found.
NOTE. Christensen points out that these are cultivated forms ie. cultivars.

Coccinia grandis (*L.*) *Voigt*, Hort. Suburb. Calc.: 59 (1845); F.T.E.A. Cucurbitac.: 68 (1867).
Turia moghadd Forssk. 1775: 166 (CXXI No. 554; Cent. VI No. 39); J.F. Gmel., Syst. Nat. ed. 13, 2: 403 (1791); Christensen 1922: 28.
Coccinia moghadd (Forssk.) Schweinf. 1867: 251.
VERNACULAR NAME. *Moghadd* (Arabic).
LOCALITY. Yemen: Al Luhayyah and Wadi Mawr ("Lohajae", "Môr"), Feb. 1763.
MATERIAL. *Forsskål* 662 (C — 1 sheet, type of *T. moghadd*, microf. 109: III. 7, 8); 663 (C — 1 sheet, type of *T. moghadd*, microf. 110: I. 1, 2); 666 (C — 1 sheet, type of *T. moghadd*, microf. 110: I. 3, 4).
NOTE. These sheets were formerly included in the *Turia sativa* folder and Christensen considered them to be that species. Now named by C. Jeffrey 1986.
See also note under *Luffa cylindrica*.
Cucumis inedulis Forssk. 1775 (CXXII No. 580) nomen nudum; Schweinfurth 1896: 136.
VERNACULAR NAME: *Arakis* (Arabic on field label).
LOCALITY. Yemen: Wadi Mawr ("Mour" on field label), 1763.
MATERIAL. *Forsskål* 660 (C — 1 sheet with original field label, microf. 35: II. 7, 8).
NOTE. Although this sheet has the original field label stuck on the back bearing the vernacular name *Arakis*, it does not match the description of *Cucumis sativa* 'Arakis' in Forssk. 1775: 169 (Cent. VI No. 61), according to C. Jeffrey who identified the specimen.

Ctenolepis cerasiformis (*Stocks*) *Hook.f. ex Oliv.* in Fl. Trop. Afr. 2: 558 (1871); F.T.E.A. Cucurbit.: 92, fig. 14 (1967).
Cucumis angulatus Forssk. 1775: 168 (LXXVI No. 510; Cent. VI No. 51); Christensen 1922: 28.
VERNACULAR NAME. *Kauûn* (Arabic).
LOCALITY. Egypt: Cairo ("Káhirae"), 1761–2.
MATERIAL. *Forsskål* no specimen found.

Cucumis melo *L.* 1753: 1011; Forssk. 1775: 168 (LXXVI No. 507; Cent. VI No. 52); Christensen 1922: 28; Schwartz 1939: 268.
VERNACULAR NAME. *Dummeiri, Dummaejri* (Arabic).
LOCALITY. Egypt: Cairo ("Káhirae"), 1761–62.
MATERIAL. *Forsskål* 661 (C — 1 sheet, microf. 109: III. 5, 6, mixed with *Kedrostis leloja*).

C. melo *L.* var. **chate** (*L.*) *Sageret* in Mém. Agric. Soc. Roy. Centr. Aric. 58: 488 (1825).
C. chate L. — Forssk. 1775: 168 (LXXVI No. 508; Cent. VI No. 53); Christensen 1922: 28; Schwartz 1939: 268.
VERNACULAR NAME. *Abdellavi, Adjûr* (Egyptian Arabic).
LOCALITY. Egypt: Cairo garden ("Ch = Káhirae").
MATERIAL. *Forsskål* no specimen found.

Cucumis sativus *L.* 1753: 1012; Schwartz 1939: 269.
The following cultivars with descriptions are listed by Forsskål (1775: 169). No material has been found but the following identifications are the most likely.
No. 54 *fakus - Fakûs* = **Cucumis melo** L.
No. 55 *smillii - Smilli* (Arabic, Al Luhayyah) — identity doubtful.
No. 56 *chiar - Chiâr* (Arabic, Cairo) = **Cucumis sativus** L.

No. 57 *chatte* - *Qatte* (Arabic, Cairo) = prob. **Cucumis melo** L.
No. 58 *battich djebbal* - *Battich djebbeli* (Arabic, Yemen mountains) = **Citrullus lanatus** (Thunb.) Matsum. & Nakai.
No. 59 *brullos* - *Battich brullosi* (Arabic, Egypt Delta) = **Citrullus lanatus** (Thunb.) Matsum. & Nakai.
No. 60 *ennemis* - *Battich ennemis* (Arabic, Cairo) = **Citrullus lanatus** (Thunb.) Matsum. & Nakai.
No. 61 *arakis* - *Arakîs* (Arabic, Al Luhayyah, see note under *Coccinia grandis*) —identity doubtful.
No. 62 *schemmam* - *Schemmâm* (Arabic; Cairo) = **Cucumis melo** L.

Cucumis prophetarum *L.* Cent. Pl. 1: 33 (1755); Schwartz 1939: 269; Tackholm 1974: 376, pl. 128B.
C. anguria sensu Forssk. 1775: 168 (CXXII No. 576; Cent. VI No. 48), non L.; Christensen 1922: 28.
LOCALITY. Yemen: Al Mukha ("Mochhae"), Musa, 1763.
MATERIAL. *Forsskål* 657 (C — 1 sheet, microf. 109: III. 1, 2); 658 (C — 1 sheet, microf. 109: II. 7, 8); 659 (C — 1 sheet, microf. 109: III. 3, 4).
Forsskål s.n. (LD — 1 sheet ex herb. Retzius).

Cucumis trilobatus sensu Forssk. 1775: 168 (CXXII No. 577; Cent. VI No. 49), non L.; Christensen 1922: 28.
LOCALITY. Yemen: Taizz ("Taaes"), Musa, 1763.
MATERIAL. *Forsskål* no specimen found.
NOTE. Its identity is obscure.

Cucumis [*sp. without epithet*] Forssk. 1775: 168 (Cent. VI No. 50); Christensen 1922: 28. = *Cucumis m'haeimta* Forssk. 1775: CXXII No. 578.
VERNACULAR NAME. *M'haeimta* (Arabic).
LOCALITY. Yemen: Surdud ("Surdûd), 1763.
MATERIAL. *Forsskål* no specimen found.
NOTE. Imperfectly known.

Cucurbita pepo *L.* 1753: 1010.
C. maxima Forssk. 1775: LXXVI & 198.
Pepo longa Forssk. 1775: 168 (Cent VI. No. 46) = *Pepo p. longa* Forssk. 1775: LXXVI No. 501. = *Cucurbita pepo longa* Forssk. 1775: CXXII No. 573.
VERNACULAR NAME. *Kara, Garna* (Egyptian Arabic).

Pepo qara stambuli Forssk. 1775: LXXVI No. 502. = *Pepo* [*sp. without epithet*] Forssk. 1775: 168 (Cent. VI. No. 47); Christensen 1922: 28.
VERNACULAR NAME. *Qarà stambuli* (Egyptian Arabic).
MATERIAL. *Forsskål* no specimens found.
NOTE. Cultivars.

Kedrostis gijef (*Forssk.*) *C. Jeffrey* in Kew Bull. 15: 354 (1962); F.T.E.A., Cucurbitac.: 136 (1967).
Turia gijef Forssk. 1775: 166 (CXXI No. 553; Cent. VI No. 38); J.F. Gmel., Syst. Nat. ed. 13, 2: 403 (1791); Christensen 1922: 28.
Corallocarpus gijef (Forssk.) Hook. f. in Fl. Trop. Afr. 2: 566 (1871); Schwartz 1939: 266.
VERNACULAR NAMES. *Gijef* (Arabic).
LOCALITY. Yemen: Al Luhayyah ("Lohajae"), Jan. 1763.
MATERIAL. *Forsskål* no specimen found.
NOTE. See note under *Luffa cylindrica*.

Kedrostis leloja (*Forssk.*) *C. Jeffrey* in Kew Bull. 15: 354 (1962), non sensu F.T.E.A., Cucurbitac.: 134 (1967).
Turia leloja Forssk. 1775: 165 (CXXI No. 552; Cent. VI No. 36); Christensen 1922: 36.
VERNACULAR NAME. *Lua, Hack el omja, Leloja* (Arabic).
LOCALITY. Yemen: Wadi Mawr ("Mour" on field label), Al Luhayyah ("Lhj" = Lohaja p. CXXI), Jan. 1763.
MATERIAL. *Forsskål* 661, partly (C — 1 sheet with field label "NB. est Gijef araborum. Mour", type of *T. leloja*, microf. 109: III. 5, 6, mixed with *Cucumis melo*); 665 (C — 1 sheet, microf. 123: I. 5, 6).

K. hirtella (Naud.) Cogn. in DC., Monogr. Phan. 3: 644 (1881); C. Jeffrey in F.T.E.A. Cucurbitac.: 133 (1967).
NOTE. See note under *Luffa cylindrica*. The name should not be adopted if it upsets widely accepted nomenclature.

Lagenaria siceraria (*Molina*) *Standley* in Publ. Field Mus. Nat. Hist. Chicago, Bot., Ser. 3: 435 (1930); F.T.E.A., Cucurbitac.: 51 (1967).
Cucurbita lagenaria L. — Forssk. 1775: 167 (LXXV No. 495, 496, 497, CXXII No. 569; Cent. VI Nos. 40, 41, 42); Christensen 1922: 28.
VERNACULAR NAMES. *Qará m'aauer, Dubba dybbe, Qará tauvil* (Arabic).
LOCALITY. Egypt, Yemen (cultivated), 1761–63.
MATERIAL. *Forsskål* 652 (C — 1 sheet with field label "Karda Fauvil. Cairi culta", microf. 123: I. 7, 8).
NOTE. Three cultivars, each with a distinct Arabic name. This sheet was not seen by Christensen.

Luffa cordata (*J.F. Gmel.*) *Meissn.* in M.J. Roem., Syn. Pepon: 67 (1846).
Turia cordata J.F. Gmel., Syst. Nat. ed. 13: 403 (1791).
Turia foliis cordatis Forssk. 1775: CXXI No. 551. = *Turia* [*sp. 1 without epithet*] Forssk. 1775: 166 (Cent. VI No. 37); Christensen 1922: 28.
VERNACULAR NAME. *Turia* (Arabic).
LOCALITY. Yemen: Wadi Mawr ("Môr"?) 1763.
MATERIAL. *Forsskål* no specimen found.
NOTE. Its identify is obscure. See note under *Luffa cylindrica*. The name should not be adopted if it upsets widely accepted nomenclature.

Luffa cylindrica (*L.*) *M.J. Roem.*, Syn. Monogr. 2: 63 (1846); Schwartz 1939: 266; F.T.E.A., Cucurbit. 76 (1967).
Turia Forssk. 1775: 165 (Cent. VI No. 35) generic name; Christensen 1922: 28.
T. sativa Forssk. 1775: CXXI No. 550; Christensen 1922: 28 = *Turia* [*sp. 2 without epithet*] Forssk. 1775: 165 (Cent. VI No. 35).
VERNACULAR NAME. *Turia* (Arabic).
LOCALITY. Yemen: Wadi Mawr ("Môr", cult.), Feb. 1763.
MATERIAL. *Forsskål* no specimen found.
NOTE. The new genus *Turia* was described with a short diagnosis and five new species in the 'Centuriae'. Three new species (*T. leloja, T. gijef* and *T. moghadd*) were provided with epithets there; *T. sativa* was provided with an epithet in the 'Florae'. See discussion by Friis in Taxon 33: 665–666 (1984).

Sicyos angulatus *L.* 1753: 1013.
LOCALITY. Unspecified.
MATERIAL. *Forsskål* 1725 (C — 1 sheet ex herb. Vahl & Zucagni, microf. 123: II. 1, 2).

Zehneria thwaitesii (*Schweinf.*) *C. Jeffrey* in Kew Bull. 15: 371 (1962); F.T.E.A. Cucurbitac.: 128 (1967).
LOCALITY. Unspecified, probably Yemen.
MATERIAL. *Forsskål* 656 (C — 1 sheet, microf. 123: I. 1, 2).

DIPSACACEAE

Cephalaria transsylvanica (*L.*) *Schrader*, Syst. Veg. 3: 45 (1818); Fl. Europ. 4: 58 (1976).
LOCALITY. Not stated.
MATERIAL. *Forsskål* s.n. (LD — 2 sheets ex herb. Retzius).

Cephalaria sp.
LOCALITY. Not stated.
MATERIAL. *Forsskål* 247 (C — 1 sheet, microf. 124: II. 1, 2).

Scabiosa arenaria *Forssk.* 1775: LXI No. 89; Christensen 1922: 36; Tackholm 1974: 519 pl. 181A.
LOCALITY. Egypt: Rashid ("Rs"), 1761.
MATERIAL. *Forsskål* no type specimen found.
NOTE. The name is validated by the description: "Flore albo; calyce longiore". The name has been taken up by Tackholm (l.c.).

Scabiosa argentea *L.* 1753: 100; Fl. Turk. 4: 613 (1972).
LOCALITY. Unspecified: (? Turkey).
MATERIAL. *Forsskål* 1694 (C — 1 sheet ex herb. Schumacher, microf. nil).
Forsskål s.n. (LD — 1 sheet ex herb. Retzius).

Scabiosa atropurpurea *L.* 1753: 100; Forssk. 1775: XIII No. 14; Fl. Europ. 4: 71 (1976).
LOCALITY. Malta (garden), 12 June 1761.
MATERIAL. *Forsskål* 1695 (C — 1 sheet ex herb. Hornemann, microf. nil).

Scabiosa columbaria *L.* 1753: 99; Forssk. 1775: IV No. 34, CV No. 86?; Fl. Europ. 4: 73 (1976).
LOCALITY. France ('Estac'); Yemen?.
MATERIAL. *Forsskål* s.n. (LD — 1 sheet ex herb. Retzius).
NOTE. Two more LD specimens may be this species.

Scabiosa sicula *L.* 1767: 196; Fl. Turk. 4: 617 (1972).
S. lyrata Forssk. 1775: 203 (XX No. 59; Cent. VIII No. 6); Vahl 1791: 27; Christensen 1922: 33.
LOCALITY. Turkey: Dardanelles ("Bujuchtari, Eracliffe", "Dardanelli"), July 1761.
MATERIAL. *Forsskål* no type specimen found.

Scabiosa sp. indet.
LOCALITY. Unspecified.
MATERIAL. *Forsskål* s.n. (LD — 1 sheet ex herb. Retzius).

EBENACEAE

Diospyros lotus *L.* 1753: 1057; Vahl 1790: 82.
Dactylus trapezuntius Forssk. 1775: XXXVI No. 481; Christensen 1922: 36.
VERNACULAR NAME. *Trebizon chormasi* (Turkish).
LOCALITY. Turkey: Istanbul ('Cph.').
MATERIAL. *Forsskål* no specimen found.
NOTE. The ample description validates *D. trapezuntius*. The name is not likely to upset any well known name.

Diospyros mespiliformis *Hochst. ex A.DC.*, Prodr. 8: 672 (1844); F.W.T.A. ed. 2, 2: 12 (1963).
LOCALITY. Yemen: near 'Alujah ("Besulf inter Erset & Amdje" on label), 1763.
MATERIAL. *Forsskål* s.n. (BM — 1 sheet ex herb. Banks, with label "ignota arbor").

ELAEAGNACEAE

Elaeagnus angustifolia *L.* 1753: 121; Forssk. 1775: XX No. 75; Fl. Europ. 2: 261 (1968).
VERNACULAR NAME. *Idae* (Turkish).
LOCALITY. Turkey: Bozcaada I. ("Tenedos, Eraclissa"), July 1761.
MATERIAL. *Forsskål* 529 (C — 1 sheet, microf. 41: I. 5, 6); 531 (C — 1 sheet, microf. 41: I. 7, 8).
LOCALITY. Unspecified.
MATERIAL. *Forsskål* 530 (C — 1 sheet, microf. 124: II. 3, 4); 1242 (C — 1 sheet ex herb. Schumacher, microf. 124: II. 5).

ERICACEAE

Erica arborea *L.* 1753: 353; Fl. Turk. 6: 96 (1978).
E. scoparia sensu Forssk. 1775: XXV No. 186, non L.
LOCALITY. Turkey: Belgrad forest, Istanbul ("in sylvis Belgrad" on field label), Aug. 1761.
MATERIAL. *Forsskål* 726 (C — 1 sheet with field label, microf. 41: III. 5, 6).

Erica manipuliflora *Salisb.* in Trans. Linn. Soc. Bot. 6: 344 (1802); Fl. Turk. 6: 96 (1978).
E. verticillata Forssk. 1775: 210 (XXV No. 187; Cent. VIII No. 43), non Bergius (1767).
VERNACULAR NAME. *Phonta* (Greek).
LOCALITY. Turkey: Büyükdere near Istanbul ("Bujuchtari"), July–Sept. 1761.
MATERIAL. *Forsskål* 727 (C — 1 sheet, type of *C. verticillata*, microf. 41: III. 7, 8); 728 (C — 1 sheet with field label "Co. 65", type of *C. verticillata*, microf. 42: I. 1, 2).

EUPHORBIACEAE

Acalypha ciliata *Forssk.* 1775: 162 (CXXI No. 561; Cent. VI No. 25); Vahl 1790: 77 t. 20; Schweinfurth 1899: 310; Christensen 1922: 28; Schwartz 1939: 137; F.T.E.A. Euphorb. 1: 197 (1987).
LOCALITY. Yemen: Taizz ("Yemen in montibus inferioribus ad Taaes"), Apr. 1763.
MATERIAL. *Forsskål* 902 (C — 1 sheet ex herb. Schumacher, holotype of *A. ciliata*, microf. 1: I. 1, 2).
Forsskål s.n. (BM — 1 sheet ex herb. Banks; isotype of *A. ciliata*).

Acalypha decidua *Forssk.* 1775: 161 (CXXI No. 558; Cent. VI No. 22); Christensen 1922: 28.
VERNACULAR NAME. *Bortom soghaier, Bortom saghajar* (Arabic).
LOCALITY. Yemen: Wadi Surdud ("Srd = Uadi Surdûd"), 1763.
MATERIAL. *Forsskål* no type specimen found.
NOTE. See note under *A. indica.*

Acalypha fruticosa *Forssk.* 1775: 161 (CXXI No. 557; Cent. VI. No. 21); Schweinfurth 1899: 309; Deflers 1889: 204; Christensen 1922: 28; Schwartz 1939: 138; F.T.E.A. Euphorb. 1: 206 (1987).
Acalypha betulina Retz., Observ. Bot. 5: 30 (1788); Vahl 1790: 77.
VERNACULAR NAME. *Borton, Bortam, Schohat, Anschat, Daefrân* (Arabic).
LOCALITY. Yemen: Al Hadiyah ("In Hadîe"), 1763.
MATERIAL. *Forsskål* 903 (C — 1 sheet ex herb. Schumacher, type of *A. fruticosa,* microf. 1: I. 3, 4).
Forsskål s.n. (BM — 1 sheet ex herb. Banks, type of *A. fruticosa*).
Forsskål s.n. (LD — 1 sheet ex herb. Retzius, type of *A. fruticosa* and *A. betulina*).

Acalypha indica *L.* 1753: 1003; Schwartz 1939: 137; F.T.E.A. Euphorbiac. 1: 199 (1987).
A. spicata Forssk. 1775: 161 (CXXI No. 559; Cent. VI No. 23); Schweinfurth 1899: 309; Christensen 1922: 28.
LOCALITY. Yemen: Wadi Surdud ("Srd" = Uadi Surdûd), 1763.
MATERIAL. *Forsskål* no type specimen found at C.
Forsskål s.n. (LD — 1 sheet ex Retzius, type of *A. spicata*).
NOTE. On the reverse there is the name *A. decidua* not *A. spicata.*

Acalypha supera *Forssk.* 1775: 162 (CXXI No. 560; Cent. VI No. 24); Christensen 1922: 28.
LOCALITY. Yemen: Wadi Surdud ("Srd" = Uadi Surdûd), 1763.
MATERIAL. *Forsskål* no type specimen found.
NOTE. An imperfectly known species, possibly *A. brachystachya* Hornem. — Schweinfurth 1899: 309; Christensen 1922: 28; Pax in Pflanzenr. IV. 147. xvi: 101 (1924); Schwartz 1939: 138.

Andrachne telephioides *L.* 1753: 1014; Fl. Turk. 7 (1982). Fig. 21B.
Eraclissa hexagyna Forssk. 1775: 208 (XXIV No. 161; Cent. VIII No. 35), & Icones t. 13B (1776).
LOCALITY. Turkey: Tekirdag (Marmaraereglisa), ("Eraclissae ... in littore occidentali Maris Marmorae").
MATERIAL. *Forsskål* no type specimen found.

Chrozophora obliqua (*Vahl*) *A. Juss. ex Spreng.*, Syst. Veg. 3: 850 (1826); Fl. Palaest. 2: 267, pl. 386 (1972).
Croton obliquum Vahl 1790: 78.
Croton argenteum sensu Forssk. 1775: LXXV No. 491, non L.
Chrozophora verbascifolia (Willd.) A. Juss. — Tackholm 1974: 316.
LOCALITY. Egypt: Cairo and Alexandria ("Cairi spontaneae" in book, "Alexandriae spontaneae" on field label), 1761–2.
MATERIAL. *Forsskål* 1162 (C — 1 sheet with original field label "As. 32", type of *C. obliqua,* microf. 34: I. 8–34, II. 1).

Chrozophora oblongifolia (*Del.*) *A. Juss. ex Spreng.*, Syst. Veg. 3: 850 (1826); Schwartz 1939: 136; Fl. Palaest. 2: 268, pl. 387 (1972).
C. obliqua sensu Tackholm 1974: 316, non (Vahl) A. Juss. ex Spreng.

Croton tinctorium ? sensu Forssk. 1775: CXXI No. 563, non L.
VERNACULAR NAME. *Hadak* (Arabic).
LOCALITY. Yemen: Al Luhayyah ("Lohaja"), Al Mukha ("Mokha" on field label), Jan. 1763.
MATERIAL. *Forsskål* 1166 (C — 1 sheet ex herb. Schumacher, microf. 34: II. 6, 7); 1167 (1 sheet, microf. 34. II, 8–34. III. 1); 1165 (C — 1 sheet with field label, microf. 34: II. 4, 5).
Forsskål s.n. (LD — 1 sheet ex herb. Retzius).

Chrozophora plicata (*Vahl*) *A. Juss. ex Spreng.*, Syst. Veg. 3: 850 (1826); Prain in Bull. Misc. Inf. Kew 1918: 66, 92 (1918); Schwartz 1939: 136; Tackholm 1974: 316, pl. 102A.
Croton tinctorium ? - Forssk. 1775: 162 (LXXV No. 490; Cent. VI No. 26), non L. (1753); Christensen 1922: 28.
Croton plicatum Vahl 1790: 78.
VERNACULAR NAME. *Ghobbaejre, Battich el malaike* (Arabic).
LOCALITY. Egypt: Giza ("Ad *Gize*, pagum e regione *Masr el atîk*"), 1762.
MATERIAL. *Forsskål* 1164 (C — 1 sheet, microf. 34: III. 4, 5); 1163 (C — 1 sheet ex herb. Hofman Bang, microf. 34: III. 2, 3).
Forsskål s.n. (BM — 1 sheet ex herb. Banks, type of *C. plicatum*).
NOTE. There is also a *Croton tinctor*[*ium*] sensu Forssk. 1775: XXXIV No. 420 from Dardanelles and Tenedos; in absence of material this remains obscure.

Clutia lanceolata *Forssk.* 1775: 170 (CXXII No. 589; Cent. VI No. 65); Vahl 1791: 101; Christensen 1922: 29; Schwartz 1939: 141.
VERNACULAR NAME. *Alloh, Lûch* (Arabic).
LOCALITY. Yemen: Bughah ("*Boka*"), 1763.
MATERIAL. *Forsskål* 1120 (C — 1 sheet ex herb. Vahl, holotype of *C. lanceolata*, microf. 28: II. 1, 2).
Forsskål s.n. (BM — 1 sheet, isotype of *C. lanceolata*).
Forsskål s.n. (M — 1 sheet Herb. Schreberianum ex Herb. Vahl, isotype of *C. lanceolata*).

Croton lobatus *L.* 1753: 1005; Hutchinson in Fl. Trop. Afr. 6(1): 751 (1912); Schwartz 1939: 135.
Croton trilobatus Forssk. 1775: 163 (CXXII No. 567; Cent. VI No. 31); Vahl 1790: 78; Christensen 1922: 28.
LOCALITY. Yemen: Wadi Surdud, ("Srd = Uadi Surdûd"), 1763.
MATERIAL. *Forsskål* no type specimen found.

Croton sp. indet.
LOCALITY. See note.
MATERIAL. ? *Forsskål* s.n. (LD — 1 sheet ex herb. Retzius).
NOTE. This specimen has not been matched with any species in the areas visited by Forsskål and there is a possiblity that the collection has been erroneously attributed to him.

Euphorbia aculeata *Forssk.* 1775: 94 (CXII No. 317; Cent. VI No. 91); Christensen 1922: 20; Schwartz 1939: 144.
VERNACULAR NAME. *Kerth, Kerâth, Sâl* (Arabic).
LOCALITY. Yemen: Taizz ("Taaes"), 1763.
MATERIAL. *Forsskål* no type specimen found.
NOTE. This validly published name remains obscure.

Euphorbia belgradica *Forssk.* 1775: 211 (XXVI No. 225; Cent. VIII No. 52); Fl. Turk. 7: 593 (1982).
LOCALITY. Turkey: Belgad, Istanbul ("Ad pagum Belgrad prope Constantinop.), Aug. 1761.
MATERIAL. *Forsskål* no type specimen found.
NOTE. Radcliffe-Smith (Fl. Turk. 7: 593) notes that this is probably referable to either *E. platyphyllos* L. or *E. pubescens* Vahl, according to Boissier in DC., Prodr. 15(2): 177 (1862).

Euphorbia cuneata *Vahl* 1791: 53; Christensen 1922: 38; Carter in Kew Bull. 35: 423 (1980); F.T.E.A. Euphorbiac.: 466 fig. 88 (1988).
LOCALITY. Yemen, 1763.
MATERIAL. *Forsskål* 1249 (C — 1 sheet, type of *E. cuneata*, microf. 124: II. 6, 7).

Euphorbia prob. **exigua** *L.* 1753: 456; Fl. Europ. 2: 222 (1968).
LOCALITY. Prob. France: Marseille, May–June 1761.
MATERIAL. *Forsskål* s.n. (LD — 1 sheet ex herb. Retzius).
NOTE. This may be a late-season branched specimen.

Euphorbia falcata *L.* 1753: 456; Fl. Turk. 7: 607 (1982).
E. polygonifolia sensu Forssk. 1775: XXVI No. 223, non L.
LOCALITY. Turkey: Dardanelles ("Dardanell." on field label), July 1761.
MATERIAL. *Forsskål* 1275 (C — 1 sheet, microf. 43: I. 3, 4).

[**Euphorbia glaucophylla** Poir.]
LOCALITY. A West African species not occurring on the route of this expedition.
MATERIAL. Attributed in error to *Forsskål* "191" (C — 1 sheet, microf. 124: III. 2, 3).

[**Euphorbia glomerifera** (Millsp.) Wheeler]
LOCALITY. A species from West Africa or West Indies not on Forsskål's route.
MATERIAL. Attributed in error to *Forsskål* "196" (C — 1 sheet, microf. 124: II. 8; III.1).

Euphorbia granulata *Forssk.* 1775: 94 (CXII No. 312; Vahl 1791: 54; Deflers 1889: 198; Schweinfurth 1899: 314; Christensen 1922: 20; Schwartz 1939: 143; F.T.E.A. Euphorbiac.: 424 (1988).
var. **granulata**
E. forskalei Gay var. *β* partly (the other part is *E. aegyptiaca* Boiss.).
VERNACULAR NAME. *Lebbejde, Lebbejn, Melaebene* (Arabic).
LOCALITY. Yemen: Mawr, Al Luhayyah, Al Hadiyah ("Môr, Lohajae, Hadîe"), Feb. 1763.
MATERIAL. *Forsskål* 1276 (C — 1 sheet, type of *E. granulata*, microf. 42: II. 5, 6); 1277 (C —1 sheet with original field label "Mor", type of *E. granulata*, microf. 42: II. 7, 8); 1278 (C — 1 sheet ex Schumacher, type of *E. granulata*, microf. 42: III. 1, 2).

var. **glabrata** (Gay) Boiss. in DC., Prodr. 15(2): 34 (1862); F.T.E.A. Euphorbiac.: 424 (1988).
LOCALITY. Yemen, 1763.
MATERIAL. *Forsskål* 1271 (C — 1 sheet, microf. nil).
NOTE. Named and labelled by S. Carter, 1985. It was previously named *E. chamaesyce* by J. Lange 1856.

Euphorbia hirta *L.* 1753: 454; Tackholm 1974: 323.
LOCALITY. Egypt or ? Yemen.
MATERIAL. *Forsskål* s.n. (LD — 1 sheet ex herb. Retzius).

Euphorbia inarticulata *Schweinfurth* 1899: 324; Schwartz 1939: 144.
E. antiquorum β minor inarticulata Forssk. 1775: 94 (CXII No. 303; Cent. III No. 88); Christensen 1922: 20.
VERNACULAR NAME. *Chorraesch* (Arabic).
LOCALITY. Yemen: Al Hadiyah ("Hadie"), 1763.
MATERIAL. *Forsskål* no specimen found.
NOTE. This species is based on the isotypes *Schweinfurth* s.n., 7111, 859, 1001.

Euphorbia indica *Lam.*, Encycl. Méth. Bot. 2: 423 (1786); Deflers 1889: 198; F.T.E.A. Euphorbiac: 471 (1988).
E. decumbens Forssk. 1775: CXII No. 313, nomen nudum.
VERNACULAR NAME. *Melaebene* (Arabic).
LOCALITY. Yemen: Wadi Surdid ("Surdud" on field label), 1763.
MATERIAL. *Forsskål* 1267 (C — 1 sheet with original field label "ex valle Surdud", microf. 42: III. 5, 6); 1266 (C — 1 sheet with original field label "Euphorbia. Melabene. Surdud", microf. 42: III. 3, 4).

Euphorbia fruticosa *Forssk.* 1775: 94 (CXII No. 306; Cent. VI No. 90); Schweinfurth 1899: 327; Christensen 1922: 20; Schwartz 1939: 145.
E. officinalis β caespitosa Forssk. 1775: CXII No. 305.
VERNACULAR NAME. *Schôrur* (Arabic).
LOCALITY. Yemen: J. Khadrá ("Chadrae"), 1763.
MATERIAL. *Forsskål* no type specimen found.

Euphorbia peplis *L.* 1753: 455; Vahl 1791: 54; Tackholm 1974: 323.
E. dichotoma Forssk. 1775: 93 (LXVII No. 260; Cent. III No. 84); Christensen 1922: 19.
LOCALITY. Egypt: Alexandria ("Alexandriae in ficetis *Peninsulae Ras-ettîn*"), 1 Apr. 1762.
MATERIAL. *Forsskål* 1269 (C — 1 sheet ex herb *Vahl* with field label "At. 13", type of *E. dichotoma*, microf. 42: II. 1, 2); 1268 (C — 1 sheet, type of *E. dichotoma*, microf. 42: I. 7, 8).

Euphorbia peplus L. 1753; Forssk. 1775: 94 (LXVII No. 255; CXII No. 314; Cent. III No. 92); Christensen 1922: 20; Schwartz 1939: 146; Tackholm 1974: 332.
VERNACULAR NAME: *Maelaeke* (Arabic-Egypt), *Subbejb*, *Sabía* (Arabic-Yemen).
LOCALITY. Egypt: Cairo ("Ca. 52" on field label), 1762.
MATERIAL. *Forsskål* 1273 (C — 1 sheet ex herb. Hofman Bang with field label, microf. 43: I. 1, 2).
LOCALITY. Yemen: Al Hadiyah ("Hadîe), 1763.
MATERIAL. *Forsskål* (no specimen).
LOCALITY. Unlocalised.
MATERIAL. *Forsskål* 1274 (C — 1 sheet with 2 plants ex herb. Hornemann, microf. nil).
NOTE. On *Forsskål* 1274 one of the two plants was collected by Forsskål while the other was grown in the Botanical Garden of Copenhagen.

Euphorbia cf. **platyphyllos** *L.* 1753: 460.
E. esula sensu Forrsk. 1775: 94 (CXII No. 315; Cent. III No. 94), non L.; Christensen 1922: 20.
E. monticola sensu Schweinfurth 1896: 330, non Hochst.
VERNACULAR NAMES. *Subaesib*, *Sauseh* (Arabic).
LOCALITY. Yemen: Al Hadiyah ("in montibus Hadiensibus"), Mar. 1763.
MATERIAL. *Forsskål* 1270 (C — 1 sheet ex herb. Hornemann, microf. 42: II. 3, 4).
NOTE. Tentatively named by S. Carter 1985, as the specimen is sterile.

Euphorbia retusa *Forssk.* 1775: 93 (LXVII No. 257; Cent. III No. 85) & Icones t. 13 (1776); Christensen 1922: 19; Tackholm 1974: 320, pl. 108. Fig. 22.
E. serrata sensu Vahl 1791: 55, non L.
VERNACULAR NAMES. *Melbaegju, Noömanîe* (Arabic).
LOCALITY. Egypt: Cairo ("In desertis *Káhirinis* rarius"), 1762.
MATERIAL. *Forsskål* 1280 (C — 1 sheet, type *E. retusa*, microf. 43: I. 7, 8); 1279 (C — 1 sheet ex herb. Hofman Bang, type of *E. retusa*, microf. 43: I. 5, 6); 1281 (C — 1 sheet, type of *E. retusa*, microf. 43: II. 1, 2); 1282 (C — 1 sheet, type of *E. retusa*, microf. 43: II. 3, 4).

Euphorbia schimperi *Presl* in Bemerk.: 109 (1844); Schwartz 1939: 144.
E. tirucalli α simplex sensu Forssk. 1775: CXII No. 308, non L.
VERNACULAR NAME. *Dahan* (Arabic).
LOCALITY. Yemen: Jiblah on field label "inter Djobla & Mekansch" ?), 1763.
MATERIAL. *Forsskål* 1283 (C — 1 sheet with field label "Euphorb. tirucalli. * Inter Djobla & Makaush", microf. nil); 1284 (C — 1 sheet, microf. nil).

Euphorbia schimperiana *Scheele* in Linnaea 17: 344 (1843); Schwartz 1939: 146; F.T.E.A. Euphorbiac.: 433 fig. 80 (1988).
LOCALITY. Yemen presumably.
MATERIAL. *Forsskål* s.n. (LD — 2 sheets ex herb. Retzius).

Euphorbia scordifolia *Jacq.*, Collect. 5: 113 (1786); Schwartz 1939: 142.
E. thymifolia Forssk. 1775: 94 (CXII No. 310; Cent. VI No. 93), non Burm. (1768); Schweinfurth 1899: 315.
VERNACULAR NAME. *Rummid* (Arabic).
LOCALITY. Saudi Arabia: Al Qunfudhah ("Ghunfude" in book); Yemen: Wadi Surdud ("ad via in Surdud" on sheet), 1763.
MATERIAL. *Forsskål* 1285 (C — 1 sheet, microf. nil).
NOTE. See comment under *E. thymifolia.*

Euphorbia terracina *L.* 1762: 654; Vahl 1791: 55; Tackholm 1974: 334.
E. obliquata Forssk. 1775: 93 (LXVII No. 258, Cent. III No. 86).
LOCALITY. Egypt: Alexandria ("*Alexandriae*"), 1 Apr. 1762.
MATERIAL. *Forsskål* 1272 (C — 1 sheet, type of *E. obliquata*, microf. 42: III. 7, 8).

Euphorbia thymifolia *L.* 1753: 454; F.T.E.A. Euphorbiac.: 420 (1988).
LOCALITY. Unspecified.
MATERIAL. *Forsskål* 1286 (C — 1 sheet, microf. nil).
NOTE. The *E. thymifolia* sensu Forssk. 1775: 94 (CXII No. 310; Cent. VI No. 93) is *E. scordifolia* (q.v.) as confirmed by Forsskål's description of the leaves being "serrulata" and the flowers "umbellis axillaribus", neither character applying to the true *E. thymifolia.*

Euphorbia triaculeata *Forssk.* 1775: 94 (CXII No. 311; Cent. III No. 96); Vahl 1791: 53, Schweinfurth 1899: 328; Deflers 1889: 201; Christensen 1922: 20; Schwartz 1939: 145.
LOCALITY. Yemen: Musa ("Ad urbem Musa frequens ..."), May 1763.
MATERIAL. *Forsskål* 1287 (C — 1 sheet, type of *E. triaculeata*, microf. 43: II. 5, 6).

Euphorbia sp. indet.
E. suffruticosa Forssk. 1775: XXVI No. 224.
LOCALITY. Turkey: Tekirdag ("Ecl" = Eraclissa).

FIG. 22. **Euphorbia retusa** Forssk. — Icones, Tab. XIII.

MATERIAL. *Forsskål* no specimen found.
LOCALITY. Rhodes ("Rh").
MATERIAL. *Forsskål* no specimen found.
NOTE. The name is validated by the description: "Foliis lanceolatis, sparsis, sessilibus, pungentibus, pollic. involucellis semi-orbiculatis, cum acumine".

? **Flueggea virosa** (*Willd.*) *Voigt*, Hort. Suburb. Calc.: 152 (1845); F.T.E.A. Euphorbiac.: 68, fig. 7 (1987).
? *F. obovata* (Willd.) Wall. ex F. Vill. — Schweinfurth 1886: 303.
Phyllanthus hamrur Forssk. 1775: 159 (CXXI No. 537; Cent. VI No. 13).
VERNACULAR NAME. *Hamrûr* (Arabic).
LOCALITY. Yemen: Milhan ("Melhân"), 1763.
MATERIAL. *Forsskål* no type specimen found.
NOTE. Unresolved in absence of type specimen. Schweinfurth tentatively identified *P. hamrur* with this species. The name should not be adopted.

Jatropha curcas *L.* 1753: 1006; Schwartz 1939: 141; F.T.E.A. Euphorbiac.: 356 (1987).
LOCALITY. ? Yemen, without locality, 1763.
MATERIAL. *Forsskål* 1369 (C — 1 sheet, microf. 124: III. 4, 5).

Jatropha glauca *Vahl* 1790: 78; Tackholm 1974: 318 pl. 103.
Croton lobatum sensu Forssk. 1775: 162 (CXXI No. 562; Cent. VI No. 27), non L.; Christensen 1922: 28.
Jatropha lobata Muell. Arg., nom. illeg. — Schweinfurth 1899: 311.
VERNACULAR NAME. *Mdjersche, Medjersehe, Ohâh* (Arabic).
LOCALITY. Yemen: Wadi Mawr ("Môr"), 1763.
MATERIAL. *Forsskål* 1370 (C — 1 sheet ex herb. Schumacher, type of *C. lobatum* and *J. glauca*, microf. 34: II. 2, 3).
Forsskål s.n. (BM — 1 sheet ex herb. Pallas, type of *J. glauca*).

Jatropha pelargoniifolia *Courbon* in Ann. Sci. Nat. sér. 4, 18: 150 (1862); F.T.E.A. Euphorbiac.: 357 (1987).
J. villosa (Forssk.) Muell. Arg. (1866), non Wight — Schweinfurth 1899: 311; Schwartz 1939: 140.
Croton villosum Forssk. 1775: 163 (CXXII No. 566; Cent. VI No. 30); Christensen 1922: 30.
Jatropha glandulosa Vahl 1790: 80.
VERNACULAR NAME. *Öbab, Bocka* (Arabic).
LOCALITY. Yemen: Wadi Mawr ("Môr").
MATERIAL. *Forsskål* 1371 (C — 1 sheet ex herb. Schumacher, type of *C. villosum* and *J. glandulosa* microf. 34: III. 6, 7).

Jatropha spinosa *Vahl* 1790: 79; Schweinfurth 1899: 311; Schwartz 1939: 140.
Croton spinosum sensu Forssk. 1775: 163 (CXXII No. 565; Cent VI No. 29), non L.; Christensen 1922: 28.
LOCALITY. Yemen: Wasab ("Uhf" = Uahfad), 1763.
MATERIAL. *Forsskål* no specimen found.

Jatropha variegata *Vahl* 1790: 79, t. 21; Schwartz 1939: 140.
Croton variegatum sensu Forssk. 1775: 163 (CXXI No. 564; Cent. VI No. 28), non L.; Christensen 1922: 28.
VERNACULAR NAME. *Dundul* (Arabic).

LOCALITY. Yemen: Wadi Zabid ("Uadi Zebîd"), 1763.
MATERIAL. *Forsskål* no type specimen found.

Phyllanthus niruri *L.* 1753: 981; Forssk. 1775: 159 (CXXI No. 534; Cent. VI No. 14); Schwartz 1939: 134.
VERNACULAR NAME. *Mekatkata, Meneckete* (Arabic).
LOCALITY. Yemen: Al Hadiyah ("Hadîe"), 1763.
MATERIAL. *Forsskål* no specimen found.

Phyllanthus ovalifolius *Forssk.* 1775: 159 (CXXI No. 536; Cent. VI No. 12).
VERNACULAR NAME. *Hömaemer* (Arabic).
LOCALITY. Yemen: Al Hadiyah ("*Hadîe*"), Mar. 1763.
MATERIAL. *Forsskål* 1834 (C — 1 sheet with field label "Phyllanthus * ovalifolius. Hömamer. Bölghose", type of *P. ovalifolius*, microf. 80: I. 5, 6); 1835 (C — 1 sheet, with field label "Phyllanthus emblica [?] inter Bolghose and Mockaja", type of *P. ovalifolius*, microf. 124: III. 6, 7).

Phyllanthus tenellus *Muell. Arg.* var. **arabicus** *Muell. Arg.* in DC., Prodr. 15(2): 339 (1866); Schwartz 1939: 134.
P. maderaspatens(is) sensu Forssk. 1775: 159 (CXXI No. 535; Cent. VI No. 11), non L.; Christensen 1922: 27.
LOCALITY. Yemen: Al Hadiyah ("Hadîe"), Mar. 1763.
MATERIAL. *Forsskål* 1574 (C — 1 sheet, type of *P. tenellus* var. *arabicus*, microf. 80: I. 1, 2); 1575 (C — 1 sheet, type of *P. tenellus* var. *arabicus*, microf. 80: I. 3, 4).
Forsskål s.n. (BM — 1 sheet ex herb. Banks, type of *P. tenellus* var. *arabicus*).

Ricinus communis *L.* 1753: 1007; Schwartz 1939: 139; Fl. Palaest. 2: 269, pl. 389 (1972).
R. medicus Forssk. 1775: 164 (LXXV No. 492; Cent. VI No. 33); Christensen 1922: 28.
VERNACULAR NAME. *Charua* (Arabic).
LOCALITY. Egypt: Rashid ("Rosettae"), 1761–2.
MATERIAL. *Forsskål* no specimen found.

Tragia pungens (*Forssk.*) *Muell. Arg.* in DC., Prodr. 15(2): 941 (1862); Schweinfurth 1899: 310; Schwartz 1939: 138.
Jatropha pungens Forssk. 1775: 163 (CXXI No. 555; Cent. VI No. 32); Christensen 1922: 28.
Tragia cordifolia Vahl 1790: 76.
VERNACULAR NAME. *Hörekrek, Meherkaka, Humejta, Mehaerkeka* (Arabic).
LOCALITY. Yemen: Al Hadiyah ("Hadîe"), 1763.
MATERIAL. *Forsskål* no type specimen found.

FAGACEAE

Quercus cerris *L.* 1753: 997; Forssk. 1775: XXXIV No. 415, as *Cervis*; Fl. Turk. 7: 674 (1982).
VERNACULAR NAME. *Mérre* (Greek).
LOCALITY. Turkey: Tekirdag ("Ecl." = Eraclissa), Burgaz ("Brg." = Borghas), 30 July–8 Sept. 1761.
MATERIAL. *Forsskål* 792 (C — 1 sheet, microf. 125: I. 5, 6).

Quercus coccifera *L.* 1753: 995; Fl. Europ. 1: 62 (1964).
Q. ilex c sensu Forssk. 1775: XI No. 243.
LOCALITY. France: Marseille ("Estac"), May 1761.
MATERIAL. *Forsskål* 790 (C — 1 sheet, microf. 86: I. 1, 2); 788 (C — 1 sheet, microf. 125: I. 3, 4); 789 (C — 1 sheet, microf. 124: III. 8; 125: I. 1, 2).
NOTE. All three collections have spiny leaf margins, although the leaves are most similar to those of *Q. ilex* in No. 790.

Quercus ilex *L.* 1753: 995; Fl. Europ. 1: 62 (1964), 1 : 74 (1993).
Q. ilex (a) foliis subintegris, acutis sensu Forssk. 1775: XI No. 243.
LOCALITY. France: Marseille ("Estac"), May 1761.
MATERIAL. *Forsskål* 786 (C — 1 sheet, microf. 85: III. 7, 8).

Q. ilex (b) integris obtusis sensu Forssk. 1775: XI No. 243.
LOCALITY. France: Marseille ("Estac"), May 1761.
MATERIAL. *Forsskål* 791 (C — 1 sheet, microf. 86: I. 3, 4).
NOTE. The typography indicates here that the apparent phrase names of var. (a) and (b) are descriptions.

Quercus pubescens *Willd.*, Berl. Baumz.: 279 (1796); Fl. Europ. 1: 64 (1964).
Q. robur; an *cervis* (sic) sensu Forssk. 1775: XI No. 242.
LOCALITY. France: Marseille ("Estac"), May 1761.
MATERIAL. *Forsskål* 787 (C — 1 sheet, microf. 86: I. 5, 6).
NOTE. This specimen has remained unidentified since it was collected, but in the absence of flowers or fruits, the leaves match *Q. pubescens* which is common in Southern France.

FLACOURTIACEAE

Oncoba spinosa *Forssk.* 1775: 103 (CXIII No. 337); Christensen 1922: 21; Schwartz 1939: 171; F.T.E.A. Flacourtiac.: 16, t. 6 (1975). = *Oncoba [sp. without epithet]* Forssk. 1775: 103 (Cent. IV No. 21).
VERNACULAR NAME. *Onkob, Korkor* (Arabic).
LOCALITY. Yemen: Bolghose ("Bulghose"), mar. 1763.
MATERIAL. *Forsskål* 626 (C — 1 sheet with field label "Bölghosi", type of *O. spinosa*, microf. 72: II. 3, 4); 627 (C — 1 sheet, type of *O. spinosa*, microf. 72: II. 5, 6).
NOTE. This is an example of a new genus and species described without an epithet in the 'Centuriae'. The epithet is provided in the 'Florae'.

FRANKENIACEAE

Frankenia pulverulenta *L.* 1753: 332; Tackholm 1974: 369, pl. 127B.
LOCALITY. Egypt ("Aegypto" on sheet in Vahl's hand, see Note), 1761–2.
MATERIAL. *Forsskål* 625 (C — 1 sheet, microf. 125: I. 7, 8).
NOTE. This species is listed on p. XIII No. 31 for Malta and not in the flora of Egypt although the sheet is so marked.

Frankenia revoluta *Forssk.* 1775: 75 (LXV No. 210; Cent. III No. 39); Christensen 1922: 18; Tackholm 1974: 371, pl. 127A.
F. hirsuta var. *revoluta* (Forssk.) Boiss., Fl. Or. 1: 780 (1867).

F. laevis L. var. *revoluta* (Forssk.) Dur. & Bar.
VERNACULAR NAMES. *Haejscheb, Nemaesje* (Arabic).
LOCALITY. Egypt: Alexandria ("Alexandriae"), 1 Apr. 1762.
MATERIAL. *Forsskål* 623 (C — 1 sheet, type of *F. revoluta*, microf. 45: III. 5, 6); 624 (C — 1 sheet, type of *F. revoluta*, microf. 45: III. 3, 4); 1264 (C — 1 sheet ex herb. Schumacher, type of *F. revoluta*, microf. 45: III. 7, 8); 1265 (C — 1 sheet, type of *F. revoluta*, microf. 46: I. 1, 2).

FUMARIACEAE

Fumaria densiflora *DC.*, Cat. Pl. Hort. Monsp.: 113 (1813); Tackholm 1974: 157.
F. officinalis sensu Forssk. 1775: LXX No. 348.
VERNACULAR NAME. *Sjaebtaredj* (Arabic).
LOCALITY. Egypt: Cairo ("Cairi spontaneae"), 1762.
MATERIAL. *Forsskål* 707 (C — 1 sheet with field label "27", microf. 49: I. 5, 6).

GENTIANACEAE

Centaurium pulchellum (*Swartz*) *Druce*, Fl. Berks.: 342 (1898); Fl. Turk. 6: 180 (1978).
LOCALITY. Turkey: Belgrad, Istanbul, 1761.
MATERIAL. *Forsskål* 1083 (C — 1 sheet specimen 3a, microf. nil).
NOTE. The second specimen (3b) on this sheet also attributed by Hornemann to Forsskål is *C. tenuiflorum*, but since it appears to be var. *tenuiflorum*, and not var. *acutiflorum* which occurs around Istanbul, it is unlikely to have been collected by Forsskål. Still further specimens were gathered by other collectors. Named and labelled by A. Melderis, 1951.
Gentiana centaur(ium) sensu Forssk. 1775: LXIV No. 165, non L.
VERNACULAR NAME. *Kantariân* (derived from Latin).
LOCALITY. Egypt: Cairo ("Cairi spontaneae"), 1762.
MATERIAL. *Forsskål* 241 (C — 1 sheet, microf. 49: III. 7, 8); 242 (C — 1 sheet, microf. 50: I. 1, 2); 1081 (C — 1 sheet ex herb. Hofman Bang, microf. 50: I. 3, 4).

Centaurium spicatum (*L.*) *Fritsch*, Mitt. Naturw. Ver. Wien 5: 97 (1907); Fl. Europ. 3: 59 (1972); Tackholm 1974: 407 pl. 141B; Fl. Turk. 6: 182 (1978).
LOCALITY. Not indicated.
MATERIAL. *Forsskål* s.n. (LD — 1 sheet ex herb. Retzius).

Enicostema axillare (*Lam.*) *A. Raynal* in Adansonia, n.s., 9: 75 (1969).
LOCALITY. Yemen ?
MATERIAL. *Forsskål* 239 (C — 1 sheet, microf. 125: II. 5, 6).
Forsskål s.n. (LD — 1 sheet ex herb. Retzius).
NOTE. This tropical species must have been collected in Yemen.

Swertia polynectaria (*Forssk.*) *Gilg* in Engl. & Prantl, Naturl. Pflanzenfam. IV. 2: 88 (1895). Fig. 24B.
Parnassia polynectaria Forssk. 1775: 207 (CIX No. 224; Cent. VIII No. 34); Icones t. 5B (1776); Christensen 1922: 33.
Swertia decumbens Vahl 1790: 24; Deflers 1889: 171; Schwartz 1939: 185.
LOCALITY. Yemen: Al Hadiyah ("Hadîe"), Mar. 1763.

MATERIAL. *Forsskål* 240 (C — 1 sheet, type of *P. polynectaria* and *S. decumbens*, microf. 77: II. 1, 2); 1556 (C — 1 sheet ex herb. Vahl, type of *P. polynectaria* and *S. decumbens*, microf. 77: II. 3, 4); 1557 (C — 1 sheet ex herb. Schumacher, type of *P. polynectaria* and *S. decumbens*, microf. nil).

GERANIACEAE

Erodium ciconium (*L.*) *L'Hérit.* in Aiton, Hort. Kew. ed. 1, 2: 415 (1789); Fl. Europ. 2: 201 (1968); Tackholm 1974: 299.
Geranium ciconium L., Cent. 1: 21 (1755); Forssk. 1775: 123 (LXIX No. 335, Cent. IV: 80); Christensen 1922: 23.
LOCALITY. Egypt: Cairo, 1761–62.
MATERIAL. *Forsskål* no specimen found.

Erodium crassifolium *L'Hérit.* in Aiton, Hort. Kew ed. 1, 2: 414 (1789); Meikle 1977: 344.
Geranium hirtum Forssk. 1775: 123 (LXIX No. 334; Cent. IV No. 77), non Burm. f. 1759; Vahl 1790: 49; Christensen 1922: 23.
Erodium hirtum (Forssk.) Willd., Sp. Pl. 3: 632 (1800); Tackholm 1974: 296, pl. 96A.
LOCALITY. Egypt: Cairo ("Cairi loco deserta"), 1762.
MATERIAL. *Forsskål* 731 (C — 1 sheet, type of *G. hirtum*, microf. 50: II. 7, 8); 738 (C — 1 sheet, type of *G. hirtum*, microf. 50: III. 1, 2).
Forsskål s.n. (LD — 1 sheet ex herb. Retzius, type of *G. hirtum*).

Erodium glaucophyllum (*L.*) *L'Hérit.* in Aiton, Hort. Kew. ed. 1, 2: 416 (1789); Fl. Palaest. 1: 234, pl. 333 (1966); Tackholm 1974: 296.
Geranium crassifolium Forssk. 1775: 123 (LXIX No. 332; Cent. IV No. 79); Christensen 1922: 23.
VERNACULAR NAME. *Tummaejr, Kabsjie* (Arabic).
LOCALITY. Egypt: Cairo ("Cairi loca deserta"), 1762.
MATERIAL. *Forsskål* 732 (C — 1 sheet with field label "Ca. 93. Geranium mihi crassifolium", type of *G. crassifolium*, microf. 50: II. 1, 2).; 739 (C — 1 sheet, type of *G. crassifolium*, microf. 50: II. 3, 4); 1299 (C — 1 sheet ex herb. Vahl, type of *G. crassifolium*, microf. 50: II. 5, 6).

Erodium laciniatum (*Cav.*) *Willd.*, Sp. Pl. 3: 633 (1800); Burtt & Lewis in Kew Bull. 9: 404 (1954); Fl. Palaest. 1: 240, pl. 345, 345a (1972); Tackholm 1974: 297; Meikle 1977: 341, 343.
Geranium triangulare Forssk. 1775: 123 (LXIX No. 336; Cent. IV No. 81), non L.; Christensen 1922: 23.
Erodium triangulare (Forssk.) Muschler 1912: 558; see Burtt & Lewis in Kew Bull. 9: 404 (1954) for discussion on rejection of this name.
LOCALITY. Egypt: Cairo ("Cairi loca deserta"), 1762.
MATERIAL. *Forsskål* 733 (C — 1 sheet with field label "Geranium marit.? Ca 121a Caid Bey", microf. 50: III. 3, 4); 734 (C — 1 sheet with field label "Ca. 121 Geranium maritim. ad Pyramides", microf. 50: III. 5, 6); 737 (C — 1 sheet, microf. 50: III. 7, 8).
NOTE. Christensen (l.c.) comments "Muschler 558 identified this with *Erodium laciniatum* (Cav.) Willd. which he therefore called *E. triangulare* (Forsk.) Muschler, but Forsk. has named a specimen belonging to *E. laciniatum G. maritimum* (p. LXIX), and there does not exist any specimen in hb. Forsk. which agrees with the description of *G. triangulare*; still it is possible that this really is a glabrescent form of *E. laciniatum*".

Zohary (l.c.) comments: "According to Christensen there is no specimen in Herbarium Forsskål which agrees with the description of *G. triangulare.* Instead a specimen named by Forsskål *G. maritimum* agrees with *E. laciniatum*". Zohary regards *G. triangulare* as *nomen ambiguum*. The name should not be taken up if it upsets widely accepted nomenclature.

Erodium malacoides (*L.*) *L'Hérit.* in Aiton, Hort. Kew. ed. 1, 2: 415 (1789); Tackholm 1974: 296.
Geranium malacoides L. 1753: 680; Forssk. 1775: 123 (LXIX No. 331; Cent. IV: 78); Christensen 1922: 23.
VERNACULAR NAME. *Garna, Djarma* (Arabic).
LOCALITY. Egypt: Cairo ("Cd"), 1761–62.
MATERIAL. *Forsskål* no specimen found.

Geranium arabicum *Forssk.* 1775: 124 (CXVI No. 407 without epithet; Cent. IV No. 82); Christensen 1922: 23; F.T.E.A. Geraniac.: 8 (1971).
VERNACULAR NAME. *Ghasl, Talab, Chada* (Arabic).
LOCALITY. Yemen: Kusma ("Kurma" on field label), 1763.
MATERIAL. *Forsskål* 735 (C — 1 sheet with field label, type of *G. arabicum,* microf. 50: I. 5, 6); 736 (C — 1 sheet, type of *G. arabicum,* microf. 50: I. 7, 8).
NOTE. Schwartz 1939: 117 reduces *G. arabicum* to *G. simense* Hochst., a later name.

GUTTIFERAE

Hypericum bithynicum *Boiss.*, Diagn. Pl. Or. Nov. 2(8): 112 (1849); Fl. Europ. 2: 266 (1968); Fl. Turk. 2: 36 (1967).
LOCALITY. Turkey: Istanbul (see note).
MATERIAL. *Forsskål* 494 (C — 1 sheet, microf. 128: III. 5, 6).
NOTE. Although there is no locality cited on this sheet, the species occurs only around Istanbul. Determined by N. Robson, 1988.

Hypericum calycinum *L.* 1767: 106; Fl. Turk. 2: 365 fig. 13/16 (1967).
H. asc[yron] sensu Forssk. 1775: XXX No. 320, non L.
LOCALITY. Turkey: Istanbul ("Bg" = Belgrad), Aug. 1761.
MATERIAL. *Forsskål* 495 (C — 1 sheet, microf. 56: I. 3, 4).

Hypericum perforatum *L.* 1753: 785; Fl. Turk. 2: 400 fig. 12, 13 (1967).
Hypericum c) *foliis linearibus* — Forssk. 1775: XXX No. 322.
LOCALITY. Turkey; 1761.
MATERIAL. *Forsskål* 493 (C — 1 sheet with field label "Co. 49", microf. 56: I. 5, 6); 1352 (1 sheet ex herb. Schumacher, microf. nil); 1353 (C — 1 sheet ex herb. Schumacher, microf. nil).
Forsskål s.n. (LD — 1 sheet ex herb. Retzius).
NOTE. Nos. 1352 and 1353 were named by R. Keller, 1896, as *H. perforatum* f. *veronense* Schr., a taxon not recognised by the current specialist N. Robson.

Hypericum pubescens *Boiss.*, Elenchus 26 (1838); Fl. Europ. 2: 266 (1968).
LOCALITY. Malta (see note), 6 June 1761.
MATERIAL. *Forsskål* s.n. (LD — 1 sheet ex herb. Retzius).
NOTE. Since this species occurs in Malta and not in other places visited by Forsskål, it is assumed to have been collected on that island. Named and labelled by N. Robson 1987.

Hypericum revolutum *Vahl* 1790: 66; Christensen 1922: 39.
H. kalmii sensu Forssk. 1775: CXVIII No. 469, non *H. kalmianum* L.
VERNACULAR NAME. *Ebaes* (Arabic).
LOCALITY. Yemen: mountains ("Ms"), 1763.
MATERIAL. *Forsskål* 492 (C — 1 sheet, type of *H. revolutum*, microf. 56: II. 1, 2); 496 (C — 1 sheet, type of *H. revolutum*, microf. 56: I. 7, 8).
Forsskål s.n. (LD — 1 sheet ex herb. Retzius, type of *H. revolutum*).

LABIATAE

Acinos arvensis (*Lam.*) *Dandy* in Journ. Ecol. 33: 326 (1946); Fl. Europ. 3: 166 (1972).
LOCALITY. Unspecified, prob. France.
MATERIAL. *Forsskål* s.n. (LD — 1 sheet ex herb Retzius).

Ajuga iva (*L.*) *Schreb.*, Pl. Vert. Unilab.: 25 (1773); Tackholm 1974: 471 pl. 162B.
Teucrium iva L. (1753) — Vahl 1790: 40.
Moscharia asperifolia Forssk. 1775: LXXIV No. 472; Christensen 1922: 27. = *Moscharia* [*sp. without epithet*] Forssk. 1775: 158 (Cent. VI No. 10).
VERNACULAR NAME. *Missaeka* (Arabic).
LOCALITY. Egypt: Alexandria ("In desertis circa Alexandriam"), Apr. 1762.
MATERIAL. *Forsskål* 330 (C — 1 sheet, type of *M. asperifolia*, microf. 70: III. 5, 6); 351 (C —1 sheet, type of *M. asperifolia*, microf. 70: II. 7, 8); 1493 (C — 1 sheet ex herb. Schumacher; type of *M. asperifolia*, microf. 70: III. 1, 2). *Forsskål* s.n. (LD — 1 sheet ex herb. Retzius, type of *M. asperifolia*).
NOTE. This is one of the cases where a new genus and species without epithet is described in the 'Centuriae', while an epithet is provided in the 'Florae'.

[**Anisomeles indica** (*L.*) *O. Kuntze*, Rev. Gen. 512 (1891)].
LOCALITY. Unspecified.
MATERIAL. ? *Forsskål* s.n. (C — 1 sheet ex herb. Hornemann, microf. 130: II. 3, 4).
NOTE. This specimen is attributed to Forsskål but probably in error since its distribution is mainly Asian. Det R.M. Harley, 1988.

Ballota acetabulosa (*L.*) *Benth.*, Lab. Gen. Sp.: 595 (1834); Fl. Turk. 7: 157 (1982).
Molluccella fruticosa Forssk. 1775: XXVIII No. 267 with brief validating description.
VERNACULAR NAMES. *Chazabo, Al thaschia* (Greek).
LOCALITY. Turkey: Bozcaada I. ("Td" = Tenedos), July 1761.
MATERIAL. *Forsskål* 223 (C — 1 sheet with field label 'Tenedos', type of *M. fruticosa*, microf. 70: II. 1, 2); 1492 (C — 1 sheet ex herb. Schumacher, type of *M. fruticosa*, microf. 70: II. 3, 4).

Becium filamentosum (*Forssk.*) *Chiov.* in Nuov. Giorn. Ital. 26: 162 (1919); Harley in Kew Bull. 38: 56 (1983).
Ocymum filamentosum Forssk. 1775: 108 (CXIV No. 365; Cent. IV No. 31); Deflers 1896: 227; Christensen 1922: 21; Schwartz 1939: 231.
LOCALITY. Yemen: Milhan ("Mlh."), 1763.
MATERIAL. *Forsskål* 394 (C — 1 sheet, type of *O. filamentosum*, microf. 71: II. 7, 8).
Forsskål s.n. (BM — 1 sheet ex herb. Banks, type of *O. filamentosum*).

Becium serpyllifolium (*Forssk.*) *J.R.I. Wood* in Kew Bull. 37: 602 (1983).
Ocymum serpyllifolium Forssk. 1775: 110 (CXIV No. 366; Cent. IV No. 36); Christensen 1922: 21; Schwartz 1939: 231.
LOCALITY. Yemen: between 'Dorebât' and Taizz ("inter Dorebât & Taaes"), 1763.
MATERIAL. *Forsskål* 213 (C — 1 sheet, type of *Ocimum serpyllifolium*, microf. 72: I. 3, 4); 326 (C — 1 sheet, type of *Ocimum serpyllifolium*, microf. 72: I. 5, 6); 325 (C — 1 sheet, microf. 72: I. 7, 8); 1502 (C — 1 sheet ex herb. Schumacher, microf. 72: I. 1, 2). (1503 (C — 1 sheet ex herb. Schumacher, microf. nil, identity uncertain, inflorescence condensed).
Forsskål s.n. (BM — 1 sheet ex herb. Banks, type of *O. serpyllifolium*).
Forsskål s.n. (LD — 1 sheet ex herb. Retzius, type of *O. serpyllifolium*).
NOTE. The present spelling of the locality 'Dorebât' is unknown (see also *Plectranthus arabicus*).

Leucas alba (*Forssk.*) *Sebald* in Stuttg. Beit. Naturk. 308: 38 (1978).
Ballota forskahlei Benth., Lab. Gen. et Sp.: 599 (1834); Schwartz 1939: 225.
Phlomis alba Forssk. 1775: 107 (CXIV No. 355; Cent. IV No. 26); Vahl 1790: 43; Christensen 1922: 21.
Elbunis alba (Forssk.) Raf., Fl. tell. 3: 88 (1837).
VERNACULAR NAME. *Schokab* (Arabic).
LOCALITY. Yemen: Al Hadiyah ("Hadie"), 1763.
MATERIAL. *Forsskål* 222 partly (C — 1 sheet with field label "Phlomis * alba. iter ad Taaês", type of *B. forskhalei* and *P. alba*, microf. 79: II. 5, 6); 1571 5, 6); 1571 (C — 1 sheet, type of *B. forskhalei* and *P. alba*, microf. nil).
NOTE. *Forsskål* 222 includes a specimen of *Leucas glabrata* (q.v.).

Leucas glabrata (*Vahl*) *R. Br.*, Prodr.: 504 (1810); Schwartz 1939: 223.
Phlomis glabrata Vahl 1790: 42; Christensen 1922: 39.
LOCALITY. Yemen, 1763.
MATERIAL. *Forsskål* 217 (C — 1 sheet, type of *P. glabrata*, microf. 79: II. 7, 8); 222 partly (C — 1 sheet, type of *P. glabrata*, microf. 79: II. 5, 6), the other plant on this sheet is *Leucas alba* (Forssk.) Sebald.
NOTE. Named and labelled by O. Sebald, 1977.
Forsskål s.n. (BM — 1 sheet ex herb. Banks, type of *P. glabrata*).

Leucas urticifolia (*Vahl*) *R.Br.*, Prodr.: 504 (1810); Schwartz 1939: 223.
Phlomis urticifolia Vahl 1794: 76; Christensen 1922: 39.
LOCALITY. Yemen?
MATERIAL. *Forsskål* 1573 (C — 1 sheet ex herb. Vahl, microf. 130: II. 7, 8).
Forsskål s.n. (BM — 1 sheet ex herb. Banks).

Marrubium alysson *L.* 1753: 582; Vahl 1791: 62; Tackholm 1974: 464 pl. 161B.
M. plicatum Forssk. 1775: 213 (LXVIII No. 294; Cent. VIII No. 62); Christensen 1922: 34.
VERNACULAR NAME. *Frasîun* (Arabic).
LOCALITY. Egypt: Alexandria ("Alexandriae"), Apr. 1762.
MATERIAL. *Forsskål* 220 (C — 1 sheet with field label "Marrubium nov. plicatum. Av. 123", type of *M. plicatum*, microf. 66: II. 3, 4); 221 (C — 1 sheet, type of *M. plicatum*, microf. 66: II. 5, 6); 346 (C — 1 sheet, type of *M. plicatum*, microf. 66: II. 7, 8).

Marrubium peregrinum *L.* 1753: 582; Fl. Turk. 7: 175 fig. 7, map 28 (1982).
M. candid(issimum) sensu Forssk. 1775: XXVIII No. 266, non L.

LOCALITY. Turkey: Dardanelles ("Dd." = Dardanelli), July 1761.
MATERIAL. *Forsskål* 219 (C — 1 sheet, microf. 66: II. 1, 2); 1449 (C — 1 sheet ex herb. Hofman Bang, microf. 66: I. 7, 8).

Marrubium vulgare *L.* 1753: 583; Forssk. 1775: VIII No. 141; Fl. Europ. 3: 138 (1972).
LOCALITY. France: Marseille ("Estac"), May–June 1761.
MATERIAL. *Forsskål* 1450 (C — 1 sheet ex herb. Hornemann, microf. nil).

Mentha aquatica *L.* 1753: 576; Fl. Europ. 3: 185 (1972).
LOCALITY. Unspecified.
MATERIAL. *Forsskål* 1463 (C — 1 sheet ex herb. Hornemann, microf. 130: II. 6).
NOTE. Named and labelled by J. Briquet, 1899, as var. *capitata* Briq. forma.

Mentha longifolia (*L.*) *L.*, Amoen. Acad. 4: 485 (1759); Hudson, Fl. Angl.: 221 (1762); Forssk. 1775: LXVIII No. 289; Fl. Turk. 7: 388 (1982); Fl. Palaest. 3: 158 pl. 261 (1978).
LOCALITY. Egypt: Rashid ("Rs." = Rosette), Oct. 1761.
MATERIAL. *Forsskål* 334 (C — 1 sheet with field label "Ni. 19", microf. 67: III. 1, 2).

M. spicata sensu Forssk. 1775: XXVIII No. 257, non L.
VERNACULAR NAME. *Agriodiosmos* (Greek).
LOCALITY. Turkey: Dardanelles ("Dd." = Dardanelli), July 1761.
MATERIAL. *Forsskål* 333 (C — 1 sheet with field label "Da. 41", microf. 67: III. 3, 4).
NOTE. Named and labelled by R.M. Harley, 1985.

Mentha longifolia var. **schimperi** (*Briq.*) *Briq.* in Bull. Herb. Boiss 4: 686 (1896).
LOCALITY. Egypt: Sinai, Mt. Horeb ("M. Horeb" on field label), Sept. 1762.
MATERIAL. *Forsskål* 336 (C — 1 sheet, microf. 131: II. 1, 2).
NOTE. This would have been collected by Niebuhr or von Haven, the only members of the expedition to visit Sinai.

Mentha × piperita *L.* 1753: 576; Bentham, Lab. Gen. & Sp.: 175 (1833); Bircher 1960: 675.
Mentha kahirina Forssk. 1775: 213 (LXVIII No. 291 without epithet; Cent. VIII No. 61); Christensen 1922: 34.
M. glabrata Vahl 1791: 75.
VERNACULAR NAME. *Lmam, Nmâme* (Arabic).
LOCALITY. Egypt: Cairo ("Káhirae"), 1961–62.
MATERIAL. *Forsskål* 335 (C — 1 sheet, type of *M. kahirina* and *M. glabrata*, microf. 67: II. 7, 8).
NOTE. Labelled by R.M. Harley, 1985.

Mentha pulegium *L.* 1753: 577; Forssk. 1775: XIII No. 48; Fl. Europ. 3: 184 (1972).
VERNACULAR NAME. *Poleg* (Maltese).
LOCALITY. Malta (cult.); June 1761.
MATERIAL. *Forsskål* 1464 (C — 1 sheet ex herb. Hornemann, microf. 130: II. 5).
NOTE. J. Briquet, 1899, named and labelled this var. *erectum*.

Mentha × verticillata *L.*, Syst. Nat. ed. 10, 2: 1099 (1759); Fl. Europ. 3: 184 (1972).
LOCALITY. Unlocalised.
MATERIAL. *Forsskål* 1465 (C — 1 sheet, microf. nil).
NOTE. J. Briquet in 1899 labelled this specimen as a form intermediate between var. *ovalifolium* and *atrorienta.*

Mentha sp. indet.
Mentha gentilis sensu Forssk. 1775: 213 (LXVIII No. 290; Cent. VIII No. 60), non L.
VERNACULAR NAME. *Naenàa, Naenae* (Arabic).
LOCALITY. Egypt: Cairo ("Kahirae hortensis"), 1761–62.
MATERIAL. *Forsskål* 1466 (C — 1 sheet, microf. 67: II. 6).
NOTE. R.M. Harley, 1985, annotated this as "Mentha-material too fragmentary for certain determination (unlikely to be *M. gentilis*!)". In 1897 Briquet had named it *M. gentilis* L. var. *tenuiceps* Briq.

Micromeria graeca (*L.*) *Benth. ex Reichenb.*, Fl. Germ. Excurs. 859 (1832); Fl. Europ. 3: 169 (1972); Fl. Turk. 7: 342 (1982).
Satureia graeca L. 1753: 568; Forssk. 1775: 212 (XXVIII No. 277; Cent. VIII No. 58); Christensen 1922: 34.
VERNACULAR NAME. *Phloskoni* (Greek).
LOCALITY. Turkey: Istanbul ("circa Constantinop in ruderatis").
MATERIAL. *Forsskål* no specimen found.

Micromeria imbricata (*Forssk.*) *C. Christensen* 1922: 21; Schwartz 1939: 228.
Satureja imbricata (Forssk.) Briq. in Engl. & Prantl, Pflanzenf. IV, 3A: 299 (1896).
Thymus imbricatus Forssk. 1775: 108 (CXIV No. 360; Cent. IV No. 29).
T. piperella Vahl 1791: 65, non L.
Micromeria forskahlei Benth., Lab. Gen. & Sp.: 379 (1834).
LOCALITY. Yemen: Kusma ("Kurmae"), 1763.
MATERIAL. *Forsskål* 210 (C — 1 sheet, type of *T. imbricatus*, microf. 107: I. 7, 8); 230 (C — 1 sheet, type of *T. imbricatus*, microf. 130: III. 6, 7); 231 (C —1 sheet with field label "Thymus * imbricatus. Kurma", type of *T. imbricatus*, microf. 107: I. 5, 6).
Forsskål s.n. (BM — 1 sheet ex herb. Banks).
Forsskål s.n. (LD — 1 sheet ex herb. Retzius).

Nepeta nepetellae *Forssk.* 1775: 212 (CXIV No. 352; Cent. VIII No. 59); Christensen 1922: 34.
LOCALITY. Yemen; Taiz ("circa Taaes in subhumidis"), 1763.
MATERIAL. *Forsskål* 216 (C — 1 sheet, ? type of *N. nepetellae*, microf. 131: I. 7, 8).

Ocimum forskolei *Benth.*, Lab. Gen. & Sp.: 6 (1832); Wood in Kew Bull. 37: 601 (1989); Paton in Kew Bull. 47: 427 (1992).
O. gratissimum Forssk. 1775: 110 (CXIV No. 363; Cent. IV No. 35), non L. (1753); Christensen 1922: 21).
[*O. hadiense* sensu E.A. Bruce in Bull. Misc. Inf. Kew 1935: 323, non *O. hadiense* Forssk. (1775)].
VERNACULAR NAME. *Höbokbok* (Arabic).
LOCALITY. Yemen: Al Hadiyah ("Hadie"), Wadi Surdud ("Srd."), 1763.
MATERIAL. *Forsskål* 211 (C — 1 sheet, type of *O. forskolei*, microf. 71: III. 4, 5); (cover —microf. 71: III. 3); 1500 (C — 1 sheet ex herb. Schumacher, microf. 71: III. 6, 7); 1501 (C — 1 sheet ex herb. Schumacher, microf. 71: III. 8).
Forsskål s.n. (BM — 1 sheet ex herb. Banks, holotype of *O. forskalei*).

Ocimum tenuiflorum *L.* 1753: 597; Paton in Kew Bull. 47: 432 (1992).
LOCALITY. Yemen: Jebel Milhan ("Mon Melhân" on original label), 1763.
MATERIAL. *Forsskål* 328 (C — 1 sheet with field label, microf. 131: I. 5, 6).

Ocimum vaalae *Forssk.* 1775: 111 (CXV No. 370; Cent. IV. No. 37); Christensen 1922: 21; Schwartz 1939: 232; Wood in Kew Bull. 37: 602 (1983).
VERNACULAR NAME. *Vaalae* (Arabic).
LOCALITY. Yemen: cult. Bayt al Faqih ("Btf."), 1763.
MATERIAL. *Forsskål* no type specimen found.
NOTE. Wood (l.c.) notes "that this cultivated, aromatic species which rarely flowered was known by the Arabic name Walah. It seems probable that the plant in question is *Plectranthus amboinicus* (Lour.) Spreng." which is still occasionally cultivated in Yemen and known by that Arabic name. Forsskål's name should not be adopted if it upsets widely accepted nomenclature.

Ocimum *α* **zatarhendi** *Forssk.* 1775: 109 (Cent. IV No. 33); Wood in Kew Bull. 37: 600 (1983).
O. aegyptiacum Forssk. 1775: CXV No. 368.
VERNACULAR NAME. *Medân* (Yemen Arabic), *Zatarhendi* (Egyptian Arabic).
LOCALITY. Yemen: Bayt al Faqih ('Beit el Fakih'), 1763.
MATERIAL. *Forsskål* no type specimen found.
NOTE. Wood (l.c.) rejected the name *O. α zatarhendi*, suggesting it might be *Plectranthus tenuiflorus* (Vatke) Agnew although it would be unwise to change names in view of the ambiguities. On p. CXV No. 368–370 are tentatively suggested to belong to a new genus '*Zatarhendi*'. It is doubtful if Forsskål intended to describe a new species called *O. zatarhendi*. Neither of Forsskål's names under this entry should be adopted if they upset widely accepted nomenclature.

Origanum majorana *L.* 1753: 590; Forssk. 1775: LXVIII No. 295; Fl. Turk. 7: 308 (1982).
VERNACULAR NAME. *Mardakvsj* (Arabic).
LOCALITY. Egypt: Cairo garden ("Ch"), 1762.
MATERIAL. *Forsskål* 229 (C — 1 sheet, microf. 74: III. 1, 2).

Origanum vulgare *L.* subsp. **hirtum** (*Link*) *Ietswaart*, Tax. Rev. Origanum: 112 (Leiden Bot. Ser.) (1980); Fl. Turk. 7: 310 (1982).
O. heracleot[icum] L. — Forssk. 1775: XXVIII No. 273.
LOCALITY. Turkey: Izmir, 30 June–10 July 1761; Gokçeada I. & Istanbul ("Imr., Cp."), July–Sept. 1761.
MATERIAL. *Forsskål* 225 (C — 1 sheet with field label "Co. 75 Belgrad", microf. 74: II. 3, 4); 226 (C — 1 sheet with field label "Co. 79", microf. 74:II. 5, 6); 227 (C — 1 sheet with field label "Sm. 18" = Smyrna = Ismir., microf. 74: II. 7, 8).
Forsskål s.n. (LD — 1 sheet ex herb. Retzius).

Otostegia fruticosa (*Forssk.*) *Schweinf. ex Penzig* in Atti Congr. Bot. Geneva 356 (1893); Sebald in Stuttg. Bietr. Naturk. Ser. A, 263: 58 (1973); Tackholm 1974: 466.
Clinopodium fruticosum Forssk. 1775: 107 (CXIV No. 357; Cent. IV No. 27); Christensen 1922: 21.
Phlomis molluccoides Vahl 1790: 42.
LOCALITY. Yemen: Al Hadiyah ("Hadîe"), Mar. 1763.
MATERIAL. *Forsskål* 1119 (C — 1 sheet ex herb. Schumacher, type of *C. fruticosum* and of *P. molluccoides*, microf. nil); 218 (C — 1 sheet, type of *C. fruticosum* and *P. molluccoides*, microf. 28: I. 7, 8); 224 (C — 1 sheet, type of *C. fruticosum* and *P. molluccoides*, microf. 28: I. 3, 4); 345 (C — 1 sheet, type of *C. fruticosum* and *P. molluccoides*, microf. 28: I. 5, 6).
Forsskål s.n. (BM — 1 sheet ex herb. Banks, type of *C. fruticosum*); s.n. (BM — 1 sheet ex herb. Pallas with field label, type of *C. fruticosum*).
Forsskål s.n. (LD — 1 sheet ex herb. Retzius, type of *C. fruticosum*).
NOTE. Named and labelled by M. Wasson, 1981.

Phlomis herba-venti *L.* 1775: 586; Forssk. 1775: VIII No. 140; Fl. Europ. 3: 144 (1972).
LOCALITY. France: Marseille ("Estac"), May 1761.
MATERIAL. *Forsskål* 228 (C — 1 sheet, microf. 79: III. 1, 2); 1572 (C — 1 sheet ex herb. Schumacher, microf. 79: III. 3, 4).

Plectranthus arabicus *E.A. Bruce* in Bull. Misc. Inf. Kew 1935: 324; Wood in Kew Bull. 37: 601 (1983).
Ocimum β zatarhendi ? Forssk. 1775: 110 (Cent. IV No. 34).
O. villosum Forssk. 1775: CXV No. 369, non *Plectranthus villosus* Sieb. ex Benth. (1832).
Plectranthus quadridentatus sensu O. Schwartz 1939: 229, non Schweinf. ex Baker (1900).
LOCALITY. Yemen: between Dorebat and Taizz ("inter Dorebat & Taaes"), 1763.
MATERIAL. *Forsskål* no type specimen found.
NOTE. The lectotype of *P. arabicus* is *Schweinfurth* 690 (K, isolectotype BM, P). The present spelling of the locality 'Dorebat' is unknown (see also *Becium serpyllifolium*). On p. CXV No. 368–370 are tentatively suggested to belong to a new genus '*Zatarhendi*'. It is, as mentioned under *Ocimum zatarhendi*, doubtful whether Forsskål intended to describe a new species with that epithet.

Plectranthus hadiensis (*Forssk.*) *Schweinf. ex Spreng.* in Wein Ill. Gart. Zeitung 19: 2 (1894); Christensen 1992: 21; Schwartz 1939: 229; Wood in Kew Bull. 37: 599 (1983).
Ocimum hadiense Forssk. 1775: 109 (CXV No. 367; Cent. IV No. 32).
Plectranthus forskalaei Vahl 1790: 44.
P. zatarhendi sensu E.A. Bruce in Bull. Misc. Inf. Kew 1935: 590, non *P. zatarhendi* (Forssk.) E.A. Bruce.
VERNACULAR NAME. *Medân* (Arabic).
LOCALITY. Al Hadiyah ("In montibus altioribus Hadiensibus"), 1763.
MATERIAL. *Forsskål* 348 (C — 1 sheet with field label, type of *O. hadiense*, microf. 72: II. 1, 2).
Forsskål s.n. (1 sheet ex herb. Banks, type of *O. hadiense*).
NOTE. Although the application of *P. zatarhendi* (Forssk.) E.A. Bruce is in doubt, it cannot refer to this species the identity of which is documented with a field label. The name should not be adopted.

Plectranthus ovatus Benth., Lab. Gen. & Sp.: 709 (1834); Schwartz 1939: 229; Wood in Kew Bull. 37: 602 (1989).
LOCALITY. Yemen: Ibb, July 1763.
MATERIAL. *Forsskål* 350 (C — 1 sheet type of *P. ovatus*, microf. 71: III. 1, 2).
NOTE. Wood comments: "It was doubtless snatched from the roadside [near Ibb where it is endemic] by one of Forsskål's companions, but Forsskål himself would have had no chance to write up the plant before his death a few days later at Yarim."

Salvia aegyptiaca *L.* 1753: 23, non L. (1767); Vahl 1790: 7; Tackholm 1974: 462 pl. 158A.
Melissa perennis Forssk. 1775: 108 (LXVIII No. 296; Cent. IV No. 30); Christensen 1922: 21.
VERNACULAR NAME. *Raàle, Sadjaret* (Arabic).
LOCALITY. Egypt: Cairo ("In desertis Káhirinis"), Apr. 1762.
MATERIAL. *Forsskal* 341 (C — 1 sheet, type of *M. perennis*, microf. 67: I. 7, 8); 1459 (C — 1 sheet, type of *M. perennis*, microf. 67: II. 1); 1460 (C — 1 sheet ex herb. Schumacher, type of *M. perennis*, microf. 67: II. 2, 3); 1461 (C — 1 sheet ex herb. Vahl, type of *M. perennis*, microf. 67: II. 4, 5).
LOCALITY. ? Egypt, without locality, 1761–2.
MATERIAL. *Forsskål* 1680 (C — 1 sheet ex herb. Hofman Bang, microf. 95: II. 3, 4).
NOTE. The specimen No. 1680 was determined by D. Hillcoat, 1973.

Salvia forskahlei *L.* 1767: 26; Fl. Turk. 7: 451 (1982).
MATERIAL. *Forsskål* s.n. (LINN — 1 sheet 42: 56, plants grown at Uppsala from Forsskål's Turkish seeds, microf. 19: III. 1).

S. bifida Forssk. 1775: 202 (XVIII No. 13; Cent. VIII No. 2); Christensen 1922: 33.
LOCALITY. Turkey: Istanbul ("Ad pagum Belgrad prop Constantinop. frequens in pratis"), July–Sept. 1761.
MATERIAL. *Forsskål* 344 (C — 1 sheet, microf. 95: I. 1, 2); 1681 (C — 1 sheet ex herb. Liebmann "HH", microf. 94: III. 7, 8); 1682 (C — 1 sheet ex herb. Schumacher "HH. 1810", microf. 95: I. 3, 4); 1683 (C — 1 sheet ex herb. Hofman Bang ex herb. Goetting. D. Mertens. 1810", microf. 95: I. 5, 6); 1684 (C — 1 sheet ex herb. Liebmann "HH", microf. 95: I. 7, 8).
NOTE. All the specimens cited here, with the possible exception of No. 344, were cultivated from Forsskål's seeds at Copenhagen or Goettingen, some as late as 1810. "HH" stands for Hortus Hafniensis, the Botanical Garden of the University of Copenhagen.

Salvia lanigera *Poir.* in Lam., Encycl. Meth. Bot. Suppl. 5: 49 (1817); Tackholm 1974: 462.
S. ceratophyt. Forssk. 1775: LIX No. 13, nomen nudum.
S. merjamie sensu Forssk. 1775: 10 (Cent. I. 29), partly (Egyptian specimen); Christensen 1922: 11.
VERNACULAR NAME. *Merjamîe* (Arabic).
LOCALITY. Egypt: Alexandria ("*Alexandriae*"), 1761–2.
MATERIAL. *Forsskål* 342 (C — 1 sheet, type, microf. 95: II. 1, 2).
NOTE. See also *S. merjamie* and *S. aegyptiaca* for other material included under *S. merjamie.*

Salvia merjamie *Forssk.* 1775: 10 (CIII No. 33; Cent. I. 29); Christensen 1922: 11.
S. nudicaulis Vahl 1804: 166.
S. nudicaulis var. *nubia* (Ait.) Baker — Schwartz 1939: 227.
VERNACULAR NAME. *Dharu* (Arabic).
LOCALITY. Yemen: Kusma ("Kurmae"), 1763.
MATERIAL. *Forsskål* 343 (C — 1 sheet with field label "Salvia aegyptiaca, inter Boka et Kurma", type, microf. 95: II. 5, 6).
NOTE. Of the three specimens labelled *S. merjamie* this must be considered the lectotype, according to Hedge in Notes R.B.G. Edinb. 33: 101 (1974); it is also the holotype of *S. nudicaulis.* (See *S. lanigera* and *S. aegyptiaca* for the other specimens previously included with *S. merjamie*).

[**Salvia phlomoides** *Asso*, Intr. Oryctogr. Arag.: 158 (1779); Fl. Europ. 3: 190 (1972).
MATERIAL. ? *Forsskål* s.n. (C — 1 sheet, microf. 130: III. 3).
NOTE. Since this is a Spanish/N.W. African species and Forsskål did not visit there, the specimen must have been attributed to him in error. Named by R.M. Harley, 1988].

Salvia spinosa *L.* 1771: 511; Tackholm 1974: 462 pl. 159.
S. aegyptiaca L. 1767: 26, non L. (1753).
MATERIAL. Egypt, 1761–2, *Forsskål* (LINN — 42.43 isolecto. probably grown from Forsskål's seed sent to Uppsala, microf. 18: III. 7; LINN — 42: 44 marked as Hort. Uppsala from Forsskål's seed, lectotype selected by Nahid, microf. 19: I. 1).
NOTE. Linnaeus gave two taxa the same name, *S. aegyptiaca,* so he re-named the second one *S. spinosa.*

Salvia virgata *Jacq.*, Hort. Vindob. 1: 14, t. 37 (1770); Fl. Turk. 7: 454 (1982).
LOCALITY. Unspecified, probably Turkey: Istanbul (see Note).

MATERIAL. *Forsskål* s.n. (BM — 1 sheet ex herb. Banks).
NOTE. This was tentatively identified from a Cibachrome photograph by I. Hedge, 1993, who commented: "it is a very widespread species ... and there are some forms of it in the Istanbul area which bear some similarity to this plant."

Satureja abyssinica (*A. Rich.*) *Briq.* in Engl. & Prantl, Pflanzenf. IV 3A: 301 (1896); Sebald in Stuttg. Beitr. Naturk., ser. A: 421: 15 (1988).
LOCALITY. Unspecified.
MATERIAL. *Forsskål* s.n. (BM — 1 sheet ex herb. Banks).
NOTE. Named by Harley, 1987.

Scutellaria arabica *Jaub. & Spach*, Ill. Pl. Or. 4: 114, t. 376 (1853).
S. peregrina sensu Schwartz 1939: 220, non L.
S. galeric[ulata] sensu Forssk. 1775: CXV No. 371, non L.
LOCALITY. Yemen: Kusma ("Kurma"), Mar. 1763.
MATERIAL. *Forsskål* 396 (C — 1 sheet, microf. 97: II. 1, 2); 399 (C — 1 sheet, microf. 97: II. 3, 4).

Sideritis montana *L.* 1753: 575; Fl. Europ. 3: 143 (1972).
LOCALITY. Unspecified, probably France.
MATERIAL. *Forsskål* s.n. (LD — 1 sheet ex herb. Retzius).

Sideritis romana *L.* 1753: 575; Fl. Europ. 3: 143 (1972).
LOCALITY. Malta.
MATERIAL. *Forsskål* s.n. (LD — 1 sheet ex herb. Retzius, "Maltha" on reverse).

Solenostemon latifolius (*Hochst. ex Benth.*) *Morton* in J. Linn. Soc. 58: 271 (1962).
LOCALITY. Yemen: Bolghose ("Bolghosi" on field label), 1763.
MATERIAL. *Forsskål* 349 (C — 1 sheet with field label "Ocymum ?", microf. 131: I. 1, 2).

Stachys aegyptiaca *Pers.*, Syn. Pl. 2: 124 (1806); Tackholm 1974: 464, pl. 161A.
S. oriental? Forssk. 1775: LXVIII No. 293, non L.
VERNACULAR NAME. *Raghat* (Arabic).
LOCALITY. Egypt: Cairo ("Cd"), 1762.
MATERIAL. *Forsskål* 339 (C — 1 sheet, microf. 103: I. 5, 6).

Stachys annua (*L.*) *L.* 1763: 813; Fl. Europ. 3: 157 (1972).
LOCALITY. Unspecified.
MATERIAL. *Forsskål* s.n. (BM — 1 sheet ex herb. Banks).

Stachys cf. **cretica** *L.* 1753: 581; Fl. Europ. 3: 153 (1972); Fl. Turk. 7: 216 (1982).
LOCALITY. Unspecified, probably Turkey.
MATERIAL. *Forsskål* s.n. (LD — 1 sheet ex herb. Retzius).

Stachys cf. **maritima** *Gouan*, Fl. Monsp.: 91 (1764); Fl. Europ. 3: 156 (1972); Fl. Turk. 7: 261 (1982).
LOCALITY. ?
MATERIAL. *Forsskål* 338 (C — 1 sheet, microf. 130: III. 4, 5).
NOTE. Named by R.M. Harley, 1988.

Stachys obliqua *Waldst. & Kit.*, Pl. Rar. Hung. 2: 142, t. 134 (1805); Fl. Turk. 7: 213 (1982).
S. cretica sensu Forssk. 1775: XXVIII No. 264, non L.
S. orientalis Vahl 1791: 64, non L.
LOCALITY. Turkey: Dardanelles ("Dd"), July 1761.
MATERIAL. *Forsskål* 340 (C — 1 sheet, microf. 103: I. 3, 4); 1759 (C — 1 sheet ex herb. Schumacher, microf. nil).

Stachys recta *L.* subsp. **subcrenata** (*Vis.*) *Briq.*, Lab. Alp. Marit.: 257 (1893); Fl. Turk. 7: 44 (1982).
S. sylvestris Forssk. 1775: XXVIII No. 263 (with brief validating description).
LOCALITY. Turkey: Burgaz ("Brg" = Borghas), Aug. 1761.
MATERIAL. *Forsskål* 337 (C — 1 sheet with field label "0.4", type of *S. sylvestris*, microf. 103: I. 7, 8).

Teucrium polium *L.* 1753: 566; Forssk. 1775: XXVIII No. 260; Fl. Turk. 7: 69 (1982).
VERNACULAR NAME. *Agapesbotane* or *Asperochorto* (Greek).
LOCALITY. Turkey: Gotzcehada, Dardanelles, Izmia ("Imros, Dardanelles, Smyrna"), July–Aug. 1761.
MATERIAL. *Forsskål* 332 (C — 1 sheet, microf. 106: III. 5, 6); 347 (C — 1 sheet with field label "Bickier Bi. 2", microf. 106: III. 1, 2); 1782 (C — 1 sheet ex herb. Liebmann, microf. 106: III. 3, 4); 1783 (C — 1 sheet ex herb. Liebmann, microf. nil).
Forsskål s.n. (LD — 1 sheet ex herb. Retzius).

Teucrium scordium *L.* subsp. **scordioides** (*Schreber*) *Maire & Petitmengin* in Bull. Soc. Bot. Sci. Nancy III, 9: 411 (1908); Fl. Turk. 7: 62 (1982).
T. scordioides Schreber — Forssk. 1775: XXVIII No. 259.
VERNACULAR NAME. *Skordochorto* (Greek).
LOCALITY. Turkey: Dardanelles, July 1761.
MATERIAL. *Forsskål* 331 (C — 1 sheet with field label "Da. 18", microf. 106: III. 7, 8); 1841 (C — 1 sheet ex herb. Schumacher, microf. nil).
Forsskål s.n. (LD — 1 sheet ex herb. Retzius with "Dardan" on reverse).

Thymus laevigatus *Vahl* 1791: 65.
T. serpyllum sensu Forssk. 1775: 107 (CXIV No. 359; Cent. IV No. 28), non L; Christensen 1922: 21; Schwartz 1939: 228.
VERNACULAR NAME. *Saatar* (Arabic).
LOCALITY. Yemen: Jebel Khadra ("Chadra"), Mar. 1763.
MATERIAL. *Forsskål* 233 (C — 1 sheet type of *T. laevigatus*, microf. 107: II. 1, 2).
Forsskål s.n. (BM — 1 sheet ex herb. Banks, type of *T. laevigatus*).
Forsskål s.n. (LD — 1 sheet ex herb. Retzius, type of *T. laevigatus*).
NOTE. This species is also known as *T. bovei* Benth.

Thymus vulgaris *L.* 1753: 591; Fl. Europ. 3: 176 (1972).
LOCALITY. Unspecified (see Note).
MATERIAL. *Forsskål* 1784 (C — 1 sheet ex herb. Hornemann, microf. nil).
NOTE. This specimen, labelled '2' in Hornemann's hand, is more likely to be the '*Thymus vulgaris*' of Forsskål 1775: VII No. 134 from Estac, although the *Forsskål* 209 (q.v. under *Thymus* sp.) has been so labelled on its folder by Christensen.

Thymus sp. indet. 1.
T. ciliatus Forssk. 1775: XXVIII No. 269, with validating description.
LOCALITY. Turkey: Tekirdag ("Ecl" = Eraclissa), Aug. 1761.

MATERIAL. *Forsskål* 232 (C — 1 sheet with field label "03", or "9.3", microf. 107: I. 3, 4).
NOTE. Forsskål's name should not be adopted if it upsets widely accepted nomenclature.

Thymus sp. indet. 2.
LOCALITY. Without locality (see note below).
MATERIAL. *Forsskål* 209 (C — 1 sheet, microf. 107: II. 3, 4).
NOTE. The inflorescence of this specimen has been eaten by insects so determination is difficult. Its folder bears '*Thymus vulgaris*' and locates it as Forsskål VII No. 134 from Estac but it is not that species.

Labiatae gen. & sp. indet.
LOCALITY. Unspecified.
MATERIAL. *Forsskål* 329 (C — 1 sheet, microf. 131: I. 3, 4).
NOTE. The specimen is too young for certain identification.

LAURACEAE

Cassytha filiformis *L.* 1753: 35; Vahl 1790: 29; Deflers 1889: 196.
Volutella aphylla Forssk. 1775: 84 (CXI No. 263; Cent. III No. 56); Christensen 1922: 19.
VERNACULAR NAMES. *Hadg mödeg, Djâha Örâk* (fruits?) (Arabic).
LOCALITY. Yemen: Wadi Mawr and Al Hadiyah ("Môr and Hadîe non infrequens"), 1763.
MATERIAL. *Forsskål* 535 (C — 1 sheet with field label "Tricatula strigens. Djaha Hadie", type of *V. aphylla*, microf. 112: III. 5, 6); 1825 (C — 1 sheet ex herb. Schumacher, type of *V. aphylla*, microf. 112: III. 3, 4).

LEGUMINOSAE

Abrus precatorius *L.*, Syst. Nat. ed. 12, 2: 472 (1767); Christensen 1922: 25; Schwartz 1939: 112; F.T.E.A., Legum.-Papil.: 114 (1971).
Glycine abrus L. 1753: 753; Forssk. 1775: 138 (CXVIII No. 458; Cent. V No. 30).
VERNACULAR NAME. *Byllia* (Arabic).
LOCALITY. Yemen: Al Hadiyah ('Hadie'), 1763.
MATERIAL. *Forsskål* no specimen found.

Acacia asak (*Forssk.*) *Willd.*, Sp. Pl. 4: 1077 (1806); Schweinfurth 1896: 215; Schwartz 1939: 83.
Mimosa asak Forssk. 1775: 176 (CXXIII No. 611; Cent. VI No. 85); Christensen 1922: 29.
VERNACULAR NAME. *Asak* (Arabic).
LOCALITY. Yemen: Al Hadiyah ("Hadîe"), Mar. 1763.
MATERIAL. *Forsskål* 1473 (C — 1 sheet, type of *M. asak*, microf. 68: II. 7, 8).

Acacia ehrenbergiana *Hayne*, Getreue Darstell Gew. 10 t. 29 (1827); Schwartz 1939: 87.
Mimosa flava Forssk. 1775: 176 (CXXIII No. 612; Cent. VI No. 84); Christensen 1922: 29.
Acacia flava (Forssk.) Schweinfurth 1896: 214, non DC.
VERNACULAR NAME. *Syllîm* (Arabic).
LOCALITY. Yemen: Surdud ("Surdûd"), 1763.
MATERIAL. *Forsskål* no type specimen found.

Acacia farnesiana (*L.*) *Willd.*, Sp. Pl. 4: 1083 (1806); Vahl 1790: 81; Schwartz 1939: 85; F.T.E.A.: Legum.-Mimos.: 111 (1959); Bircher 1960: 338.
Mimosa scorpioides sensu Forssk. 1775: XXXV No. 448 & LXXVII No. 553, non L.
LOCALITY. Rhodes: "Rhodos in horto" on field label.
MATERIAL. *Forsskål* 1480 (C — 1 sheet with field label "Mimosa scorpioides It. 12", microf. 69: I. 3, 4).
VERNACULAR NAME. *Faetue* (Arabic).
LOCALITY. Egypt: Cairo, cult. ("Ch"), 1762.
MATERIAL. *Forsskål* 1481 (C — 1 sheet, microf. 69: I. 5, 6); 1482 (C — 1 sheet, microf. 69: I. 7, 8); 1483 (C — 1 sheet, microf. 69: II. 1, 2).

Acacia hamulosa *Benth.* in Hook., Lond. Journ. Bot. 1: 509 (1842).
Mimosa senegalensis Forssk. 1775: 176 (CXXIII No. 605; Cent. VI No. 81); probably error for *M. senegal* L.; Lamarck, Encyc. Méth. Bot. 1: 19 (1783); Christensen 1922: 29.
A. senegal sensu Schwartz 1939: 84 partly, non (L.) Willd.
VERNACULAR NAME. *Ketât* (Arabic).
LOCALITY. Yemen: Hays ("Ad Haes"), 4 Apr. 1763.
MATERIAL. *Forsskål* 1484 (C — 1 sheet ex herb. Schumacher, type of *M. senegalensis*, microf. 69: II. 3, 4); 1485 (C — 1 sheet, type of *M. senegalensis*, microf. 69: II. 5, 6).
NOTE. Forsskål's name should not be taken up if it upsets widely accepted nomenclature.

Acacia hockii *De Wild.*, in Fedde, Repert. 11: 502 (1913); Wood in Kew Bull. 37: 454 (1982).
A. oerfota sensu Brenan in Kew Bull. 8: 101–102 (1953).
Mimosa gummifera Forssk. 1775: CXXIV, nomen nudum.
VERNACULAR NAME. *Talab* (Arabic).
LOCALITY. Yemen; Kusmah to Alujah ("inter Kurma & Aludje"), 1763.
MATERIAL. *Forsskål* 1474 (C — 1 sheet, microf. 68: III. 1, 2).
NOTE. Another specimen 1490 (C — 1 sheet, microf. 69: I. 1, 2) was identified by Christensen as *A. oerfota* and by Vahl as "*A. horrida*". See the cited note by Wood for a discussion of *A. oerfota*.

Acacia mellifera (*Vahl*) *Benth.* in Hook., Lond. Journ. Bot. 1: 507 (1842); Schweinfurth 1896: 217; Schwartz 1939: 84; F.T.E.A. Legum.-Mimos.: 84 (1959); Tackholm 1974: 289, pl. 93A.
Mimosa mellifera Vahl 1791: 103.
Mimosa unguis cati sensu Forssk. 1775: 176 (CXXIII No. 613; Cent. VI. No. 83); non L.; Christensen 1922: 29.
VERNACULAR NAME. *Dhoba, Dobb, Smurr* (Arabic).
LOCALITY. Yemen: Surdud ("Surdûd alibique"), Feb. 1763.
MATERIAL. *Forsskål* 1477 (C — 1 sheet with field label 'Do66 Mimosa ungu. cati?, Surdud', type of *M. mellifera*, microf. 70: I. 1, 2); 1476 (C — 1 sheet, type of *M. mellifera*, microf. 69: III. 7, 8).

Acacia nilotica (*L.*) *Willd. ex Delile*, Fl. Aegypt. Ill.: 79 (1812); F.T.E.A. Legum.-Mimos.: 109 (1959); Tackholm 1974: 290, pl. 93C.
Mimosa nilotica L. 1753: 521; Forssk. 1775: LXXVII No. 554.
VERNACULAR NAMES. *Sant, Qarad* (Arabic).
LOCALITY. Egypt: Cairo, cult. ("Ch"), 1761–62.
MATERIAL. *Forsskål* 1478 (C — 1 sheet, microf. 68: III. 5, 6); 1479 (C — 1 sheet, microf. 68: III. 7, 8).
Forsskål s.n. (LD — 1 sheet ex herb. Retzius).

Acacia arabica (Lam.) Willd., Sp. Pl. 4: 1085 (1806); Deflers 1889: 135; Schweinfurth 1896: 209.

Acacia oerfota (*Forssk.*) *Schweinf.* 1896: 213; Schwartz 1939: 85; Wood in Kew Bull. 37: 451–454 (1982) for discussion and synonymy.
Mimosa oerfota Forssk. 1775: 177 (CXXIII No. 607; Cent. VI No. 86); Christensen 1922: 29.
VERNACULAR NAME. *Örfota* (Arabic).
LOCALITY. Yemen: Ad Dahi ("Ad Dahhi"), 1763.
MATERIAL. *Forsskål* (no type specimen, neotype selected by Wood as *Radcliffe-Smith & Henchie* 4932 (K)).
NOTE. This has been confused with *A. nubica* Benth. in F.T.E.A. Legum.-Mimos.: 129 (1959).

Acacia seyal *Delile, Fl. Aegypte* 286, t. 52, f. 2 (1813); Schwartz 1939: 87.
Mimosa sejal Forssk. 1775: 177 (CXXIV No. 614; Cent. VI No. 90); Christensen 1922: 29.
Acacia hadiensis DC., Prodr. 2: 472 (1825).
VERNACULAR NAME. *Sejâl* (Arabic).
LOCALITY. Yemen: Al Hadiyah ("Hadîe"), 1763.
MATERIAL. *Forsskål* no type specimen of *M. sejal* and *A. hadiensis* found. The name *M. sejal* should not be taken up as it would upset widely accepted nomenclature.

Acacia tortilis (*Forssk.*) *Hayne*, Arzneyk. 10, t. 31 (1827) subsp. **tortilis** — Brenan in F.T.E.A. Legum.-Mimos.: 117 (1959); Tackholm 1974: 289.
Mimosa tortilis Forssk. 1775: 176 (CXXIII No. 606; Cent. VI No. 82); Christensen 1922: 29.
VERNACULAR NAME. *Hares* (Arabic).
LOCALITY. Yemen: Hays ("Haes"), 4–5 Apr. 1763.
MATERIAL. *Forsskål* 1488 (C — 1 sheet with field label "Mimosa tortilis. Hares. Mons Soudân", holotype of *A. tortilis*, microf. 69: III. 3, 4); 1489 (C — 1 sheet ex herb. Schumacher, syntype of *A. tortilis*, microf. 69: III. 5, 6).
A. spirocarpa Hochst. — Schweinfurth 1896: 207; Schwartz 1939: 86.

Albizia julibrissin *Durazz.* in Mag. Tosc. 3(4): 11 (1772); Fl. Turk. 3: 10 (1970).
Mimosa arborea sensu Forssk. 1775: 177 (XXXV No. 447; Cent. VI No. 89 note), non L.
VERNACULAR NAME. *Djul ibrzim* (Turkish).
LOCALITY. Turkey: Istanbul, cult. ("Cph"), July–Sept. 1761.
MATERIAL. *Forsskål* 1471 (C — 1 sheet, microf. 68: II. 3, 4); 1472 (C — 1 sheet with field label "Co. 42", microf. 68: II. 5, 6).

Albizia lebbeck (*L.*) *Benth.* in Hook. Lond. Journ. Bot. 3: 87 (1844); Schwartz 1939: 83; F.T.E.A. Legum.-Mimos.: 147 (1959); Bircher 1960: 332.
Mimosa lebbeck L. 1753: 516; Forssk. 1775: 177 (LXXVII No. 552, CXXIII No. 603; Cent. VI No. 89).
VERNACULAR NAME. *Laebach* (Arabic), *Dujl Ibrzim* (Turkish), *Serisch* (Indian), *Seiseban* (Yemen).
LOCALITY. Egypt: Cairo ("Káhirae hortensis illa"), 1762.
MATERIAL. *Forsskål* 1475 (C — 1 sheet, microf. 68: III. 3, 4).

Alhagi maurorum *Medikus*, Vorl. Churpf. Phys. — Oken. Ges. 2: 397 (1787) pro parte; Fl. Palaest. 2: 112, pl. 165 (1972); Tackholm 1974: 272, pl. 86C.
Hedysarum alhagi L. 1753: 745 pro parte; Forssk. 1775: 136 (LXXI No. 374; Cent. V No. 21; Christensen 1922: 24.

VERNACULAR NAME. *Aghûl, Azûb* (Arabic).
LOCALITY. Egypt: Cairo and Rashid ("Cs = Kâhirae, Rs = Rosettae", "In ruderatis locis Aegypti frequens"), June 1762.
MATERIAL. *Forsskål* 1317 (C — 1 sheet ex herb. Hornemann, on reverse 'ad Alexandr. prope Nilum', microf. 52. III. 1, 2).
Forsskål s.n. (BM — 1 sheet ex herb. Banks "Spartium scorpiurus" MS).
Forsskål s.n. (LD — 1 sheet ex herb. Retzius).

Alysicarpus glumaceus (*Vahl*) *DC.*, Prodr. 2: 353 (1825); F.T.E.A. Legum.-Papil. 497 (1971).
Hedysarum glumaceum Vahl 1791 (Add. & Corrig.): 106.
H. violaceum Forssk. 1775: 136 (CXVII No. 449; Cent. No. 23), non L.; Vahl 1790: 54; Christensen 1922: 23.
Alysicarpus rugosus sensu Schwartz 1939: III, non DC.
LOCALITY. Yemen: Surdud ('Surdûd'), 1763.
MATERIAL. *Forsskål* 1321 (C — 1 sheet, type of *H. violaceum* & *H. glumaceum*, microf. 52: III. 7, 8); 1322 (C — 1 sheet ex herb. Schumacher, type of *H. violaceum* & *H. glumaceum*, microf. 53: I. 1, 2); 1323 (C — 1 sheet ex herb. Schumacher, type of *H. violaceum* & *H. glumaceum*, microf. 53: I. 3, 4); 1324 (C — 1 sheet, type of *H. violaceum* & *H. glumaceum*, microf. 53: I. 5, 6).

Astragalus annularis *Forssk.* 1775: 139 (LXXI No. 378; Cent. V No. 32); Vahl 1790: 58; Christensen 1922: 25; Fl. Palaest. 2: 64, pl. 84 (1972); Tackholm 1974: 265.
LOCALITY. Egypt: Alexandria ('Alexandriae'), Mar., Apr. 1762.
MATERIAL. *Forsskål* 996 (C — 1 sheet ex herb. Hornemann, type of *A. annularis*, microf. 13: II. 7, 8).

Astragalus fruticosus *Forssk.* 1775: 139 (LXXI No. 379 as *fruticulosus*; Cent. V. No. 33); Christensen 1922: 33; Fl. Palaest. 2: 76, pl. 107 (1972).
A. christianus sensu Vahl 1790: 57, non L., based on Forsskål's specimen.
LOCALITY. Egypt: Rashid ("Rosettae"), 2–6 Nov. 1761.
MATERIAL. *Forsskål* 997 (C — 1 sheet ex herb. Schumacher, holotype of *A. fruticosus*, microf. 13: III. 1, 2).
A. tomentosus Lam. (1783) — Tackholm 1974: 268.

Astragalus peregrinus *Vahl* 1790: 57; Fl. Palaest. 2: 68, pl. 92 (1972); Tackholm 1974: 265, pl. 82C.
A. trimestris sensu Forssk. 1775: 138 (LXXI No. 376; Cent. V No. 31), non L; Christensen 1922: 25.
VERNACULAR NAMES. *Chamsaret el arûsi, Chamsarat el aruse* (Arabic).
LOCALITY. Egypt: Cairo ("Kâhirae"), 1761–2.
MATERIAL. *Forsskål* 998 (C — 1 sheet ex herb. Hornemann, type of *A. peregrinus*, microf. nil); 999 (C — 1 sheet ex herb. Schumacher, type of *A. peregrinus*, microf. nil).
Forsskål s.n. (BM — 1 sheet ex herb. Banks, type of *A. peregrinus*.
NOTE. Christensen did not see these sheets in 1922, nor were they photographed by IDC.

Astragalus spinosus (*Forssk.*) *Muschl.*, Verh. Bot. Ver. Prov. Brandenb. 49: 98 (1907); Fl. Palaest. 2: 78, pl. 112 (1972); Tackholm 1974: 268, pl. 83A.
Colutea spinosa Forssk. 1775: 131 (LXX No. 359; Cent. V No. 7); Christensen 1922: 24.
Astragalus rauwolfii Vahl 1790: 63.
A. forskalei Boiss., Diagn. ser. 1, 9: 101 (1849).
VERNACULAR NAME. *Keddâd* (Arabic).

LOCALITY. Egypt: Cairo ("In desertis Káhirinis ad Liblab), Jan–Mar. 1962.
MATERIAL. *Forsskål* 1127 (C — 1 sheet type of *C. spinosa*, microf. 29: I. 1, 2); 1128 (C — 1 sheet ex herb. Schumacher, type of *C. spinosa*, microf. 29: I. 3, 4).
Forsskål s.n. (BM — 1 sheet ex herb. Banks, type of *C. spinosa*).

Astragalus sp.
LOCALITY. Yemen ? ("in Arabia" on C sheet).
MATERIAL. *Forsskål* 1000 (C — 1 sheet ex herb. Schumacher, microf. 132: II. 5, 6).
Forsskål s.n. (LD — 1 sheet ex herb. Retzius).

Bauhinia tomentosa *L.*, 1753: 375; F.T.E.A. Legum.-Caesalp.: 209 (1967).
B. inermis Forssk. 1775: 85 (CXI No. 264; Cent. III. No. 57); Deflers 1889: 134; Schweinfurth 1896: 213; Christensen 1922: 19; Schwartz 1939: 89.
VERNACULAR NAME. *Hénn embas, Henn el bagar, Tummâr, Athbir* (Arabic).
LOCALITY. Yemen: Milhan ("prope montem Melhân"), 1763.
MATERIAL. *Forsskål* no type specimen found.
NOTE. We are grateful for Dr B. Verdcourt's opinion as to the identity of *B. inermis*.

Cadia purpurea (*Picciv.*) *Aiton*, Hort. Kew. 3: 492 (1792); F.T.E.A. Legum.-Papil.: 33, fig. 4 (1971).
Cadia [*sp. without epithet*] Forssk. 1775: 90 (CXI No. 276; Cent. III. No. 76); Christensen 1922: 19.
C. varia L'Hérit., Diss. in Mag. Enc. 5: 22, t. 5 (1795); Deflers 1889: 133; Schweinfurth 1896: 223; Schwartz 1939: 93.
C. arabica Raeusch., Nom. bot. ed. 3: 117 (1797).
VERNACULAR NAME. *Kadi* (Arabic).
LOCALITY. Yemen: Al Hadiyah ("*Hadîe*"), Mar. 1763.
MATERIAL. *Forsskål* 1038 (C — 1 sheet ex herb. Vahl, type of genus and each of above species, microf. 19: III. 3, 4).
NOTE. This sheet was annotated by G. Bentham, 1869. Only the generic name was proposed by Forsskål so epithets were provided by other authors.

Caesalpinia bonduc (*L.*) *Roxb.*, Fl. Ind., ed. 2, 2: 362 (1832); F.T.E.A. Legum.-Caesalp.: 37 (1967).
C. bonducella (L.) Fleming (1810).
C. crista sensu Schwartz 1939: 92, non L.
Guilandina bonducella L. (1762) — Vahl 1790: 30; Christensen 1922: 24.
Glycyrrhiza aculeata Forssk. 1775: 135 (CXVII No. 448; Cent. V. No. 20).
VERNACULAR NAME. *Sirs* (Indian "Banjan").
LOCALITY. Al Hadiyah & Zabid (cult.) ("Hadîe", "Zebîd"), Mar. 1763.
MATERIAL. *Forsskål* 1304 (C — 1 sheet with field label "Glycyrrhiza * aculeata. Hadîe ad templum. ex India", holotype of *G. aculeata*, microf. 51: III. 2, 3).

Canavalia africana *Dunn* in Piper & Dunn in Kew Bull. 1922: 135 (1922); Verdcourt in Kew Bull. 42: 658 (1987).
C. virosa (Roxb.) Wight & Arn., Prodr. Fl. Pen. Ind. Or. 1: 253 (1834); F.T.E.A. Legum.-Papil.: 573 (1971).
Dolichos polystachyos Forssk. 1775: 134 (CXVII No. 437; Cent. V. No. 17), non L.; Christensen 1922: 24.
Canavalia polystachyos (Forssk.) Schweinfurth 1896: 254; Schwartz 1939: 114.

VERNACULAR NAME. *Sjef, Syjef, Syjef er robacb* (Arabic).
LOCALITY. Yemen: Wadi Mawr ("Mor"), Al Hadiyah ("Hadie"), 1763.
MATERIAL. *Forsskål* no type specimen found.

Canavalia gladiata *DC.*, Prodr. 2: 404 (1825); F.T.E.A. Legum.-Papil.: 572 (1971).
Dolichos faba indica Forssk. 1775: 133 (CXVII No. 442; Cent. V. No. 10); Christensen 1922: 24.
VERNACULAR NAME. *Ful hendi, Didjre* (Arabic).
LOCALITY. Saudi Arabia: Jiddah ("Djiddae"), 29 Oct. 1762.
MATERIAL. *Forsskål* no specimen found.
NOTE. Forsskål's epithet should not be hyphenated and taken up.

Chamaecrista nigricans (*Vahl*) *Greene* in Pittonia 4: 30 (1899); Lock in Kew Bull. 43: 337 (1988).
Cassia nigricans Vahl 1790: 30; Schwartz 1939: 91; F.T.E.A. Legum.-Caesalp.: 81 (1967).
Cassia procumbens sensu Forssk. 1775: CXI No. 272.
VERNACULAR NAME. *Houmer* (Arabic).
LOCALITY. Yemen: Wadi Surdud ("Sordûd" on original label), 1763.
MATERIAL. *Forsskål* 1057 (C — 1 sheet with field label, type of *C. nigricans*, microf. 21: III. 5, 6); 1058 (C — 1 sheet, type of *C. nigricans*, microf. 21: III. 7, 8).

Clitorea ternatea *L.* 1753: 753; Vahl 1790: 53; Deflers 1889: 132; Schweinfurth 1896: 252; Schwartz 1939: 113; F.T.E.A., Legum.-Papil.: 515 (1971).
Lathyrus spectabilis Forssk. 1775: 135 (CXVII No. 445; Cent. V. No. 18); Christensen 1922: 24.
LOCALITY. Yemen: Surdud ("Surdûd in sylvis passim"), 1763.
MATERIAL. *Forsskål* no specimen found at C.
Forsskål s.n. (BM — 1 sheet ex herb. Banks, annotated by Frantz as the "probable type of *Lathyrus spectabilis*").
Forsskål s.n. (LD — 1 sheet ex herb. Retzius, ? type of *L. spectabilis*).

Orobus volubilis Forssk. 1775: CXVII No. 436; Vahl 1790: 53; Christensen 1922: 37.
LOCALITY. Yemen: Kusma ("Kurma"), 1763.
MATERIAL. *Forsskål* no type specimen found.
NOTE. The brief description ("Stylo apice plano") probably validates the name *Orobus volubilis*.

[**Crotalaria albida** *Heyne ex A.W. Roth*, Nov. Pl. Sp.: 333 (1821)].
NOTE. These two sheets in Forsskål's herbarium are unlikely to be of his material since this species occurs in tropical Asia not visited by him: (C — 1 sheet ex herb. Vahl, microf. 131: II. 7, 8); (C — 1 sheet ex herb Vahl, microf. 131: III. 1, 2).

Crotalaria incana *L.* subsp. **purpurascens** (*Lam.*) *Milne-Redh.* in Kew Bull. 15: 159 (1961); F.T.E.A., Legum.-Papil.: 870 (1971).
LOCALITY. ? Yemen.
MATERIAL. *Forsskål* s.n. (LD — 1 sheet ex herb. Retzius).

Crotalaria microphylla *Vahl* 1790: 52; Schweinfurth 1896: 223; Schwartz 1939: 97.
LOCALITY. ? Yemen.
MATERIAL. *Forsskål* 1160 (C — 1 sheet ex herb. Vahl, type of *C. microphylla*, microf. 113: II. 3, 4); 1161 (C — 1 sheet ex herb. Vahl, type of *C. microphylla*, microf. 113: II. 5, 6).

Crotalaria retusa *L.*, 1753: 715; Vahl 1790: 52; Schweinfurth 1896: 223; Schwartz 1939: 94; F.T.E.A., Legum.-Papil.: 958 (1971).
Dolichos cuneifolius Forssk. 1775: 134 (CXVII No. 438; Cent. V. No. 16); Christensen 1922: 24.
VERNACULAR NAME. *Kolkol, Kalakel* (Arabic).
LOCALITY. Yemen: Wadi Mawr ("Môr"), Feb. 1763.
MATERIAL. *Forsskål* 1233 (C — 1 sheet ex herb. Schumacher, type of *D. cuneifolius*, microf. 40: I. 5, 6).
Forsskål s.n. (LD — 1 sheet ex herb. Retzius, type of *D. cuneifolius*).

Crotalaria thebaica (*Del.*) *DC.*, Prodr. 2: 128 (1825); Tackholm 1974: 224; pl. 70B.
LOCALITY. ? Egypt.
MATERIAL. *Forsskål* s.n. (LD — 1 sheet ex herb. Retzius).

Cullen corylifolia (*L.*) *Medikus* in Vorles, Churf. Phys.-Oek. Ges. 2: 380 (1787); Stirton in Bothalia 13: 317 (1981).
Psoralea corylifolia L. 1753: 764; Schwartz 1939: 106.
Trifolium unifolium Forssk. 1775: 140 (CXVIII No. 460; Cent. V No. 36); Vahl 1790: 65; Christensen 1922: 25.
VERNACULAR NAME. *Löbab el abîd* (Arabic).
LOCALITY. Yemen: Taiz ("Taaes ad aquas"), 1763.
MATERIAL. *Forsskål* no specimen found at C.
Forsskål s.n. (BM — 1 sheet ex herb. Banks MS "Psoralea corylifolia, Trifolium unifolium", type of *T. unifolium*).

Delonix elata (*L.*) *Gamble*, Fl. Madras 1(3): 396 (1919); F.T.E.A. Legum.-Caesalp.: 23 (1967).
Poinciana elata L., Cent. II. Pl.: 16 (1756); Forssk. 1775: 86 (CXI No. 274; Cent. III. No. 63); Christensen 1922: 19; Schwartz 1939: 92.
VERNACULAR NAME. *Ranf, Mschillech, Mschillaech* (Arabic).
LOCALITY. Yemen: Milhan ("Ad *Melhân*"), Feb. 1763.
MATERIAL. *Forsskål* 1587 (C — 1 sheet ex herb. Hornemann with field label "Ranf. Poinciana elata via in Melhân", microf. 82: II. 5, 6).

Desmodium gangeticum (*L.*) *DC.* Prodr. 2: 327 (1825); Schwartz 1939: 111; F.T.E.A., Legum.-Papil.: 467 (1971).
LOCALITY. ? Yemen.
MATERIAL. *Forsskål* s.n. (LD — 1 sheet ex herb. Retzius).

Desmodium ospriostreblum *Chiov.* in Ann. Ist. Bot. Roma 8: 428 (1908); F.T.E.A., Legum.-Papil.: 475 (1971).
LOCALITY. ? Yemen.
MATERIAL. *Forsskål* s.n. (LD — 1 sheet ex herb. Retzius).
NOTE. This early record seems to confirm the plant as an Old World native species.

Desmodium repandum (*Vahl*) *DC.*, Prodr. 2: 334 (1825); F.T.E.A., Legum.-Papil.: 465 (1971).
Hedysarum repandum Vahl 1791: 82; Christensen 1922: 38.
LOCALITY. Yemen, 1763.
MATERIAL. *Forsskål* 1320 (C — 1 sheet ex herb. Schumacher, type of *H. repandum*, microf. Vahl 36: II. 6, 7).

NOTE. There is another early sheet of this species in Herb. Vahl (microf. Vahl 36: III. 1, 2); however, this is not indicated as being a Forsskål collection.

Desmodium triflorum (*L.*) *DC.*, Prodr. 2: 334 (1825); F.T.E.A., Legum.-Papil.: 459 (1971).
LOCALITY. Yemen: Mukhajah, Jebbel Barad ("Moxheja").
MATERIAL. *Forsskal* 1219 (C — 1 sheet ex herb. Hofman Bang, microf. 132: I. 5, 6). *Forsskål* s.n. (BM — 1 sheet ex herb. Nolte with field notes "ignota diadelphia miniata. Moxheja".

Dichrostachys cinerea (*L.*) *Wight & Arn.*, Prodr. Fl. Ind. Or.: 271 (1834) var. **cinerea** — F.T.E.A. Legum.-Mimosoid.: 36 (1959).
D. nutans (Pers.) Benth. (1841) — Schwartz 1939: 88.
Mimosa glomerata Forssk. 1775: 177 (CXXIII No. 610: Cent. VI No. 88); Christensen 1922: 29.
LOCALITY. Yemen.
MATERIAL. *Forsskål* no type specimen found.

Dolichos faba nigrita *Forssk.*, 1775: 132 (CXVII No. 441; Cent. V No. 9); Christensen 1922: 24.
VERNACULAR NAME. *Ful Djellâbe, Ful barabra* (Egypt); *Habb el kulae* (Syria & Greece); *Didjre* (Yemen Arabic).
LOCALITY. Egypt: Cairo ("Káhiram"), 1761–2. Yemen: Al Luhayyah ('Lhj'), Jan. 1763.
MATERIAL. *Forsskål* no specimen found.
NOTE. This climbing bean is not yet positively identified. The tri-nomial is invalid and should not be hyphenated.

Dorycnium graecum (*L.*) *Ser.* in DC., Prodr. 2: 208 (1825); Fl. Europ. 2: 173 (1968); Fl. Turk. 3: 514 (1970).
Lotus graecus L. (1767) — Vahl 1790: 65.
Lotus belgradica Forssk. 1775: 215 (XXX No. 318; Cent. VIII No. 71); Christensen 1922: 34.
LOCALITY. Turkey: Istanbul ("Ad Belgrad, pagum prope Constantinop."), Aug. 1761.
MATERIAL. *Forsskål* 1422 (C — 1 sheet, type of *L. belgradica*, microf. 64: I. 3, 4); 1423 (C — 1 sheet, type of *L. belgradica*, microf. 64: I 5, 6).

Flemingia cf. **grahamiana** *Wight & Arn.*, Prodr. Fl. Pen. Ind. Or.: 242 (1834).
Pyrus hadiensis Forssk. 1775: 212 (CXII No. 325; Cent. VIII No. 54); Christensen 1922: 34.
VERNACULAR NAME. *S'faerdjel* (Arabic).
LOCALITY. Yemen: Al Hadiyah ("Hadîe"), Mar. 1763.
MATERIAL. *Forsskål* 764 (C — 1 sheet, type of *P. hadiensis*, microf. 85: III 3, 4).
NOTE. This sterile specimen has been identified by Schweinfurth 1896: 203 as *F. rhodocarpa* Bak. which is a synonym of *F. grahamiana* Wight & Arn. but the specimen does not exactly match that species. Unfortunately Forsskål's epithet is earlier than either of the above and it should not be taken up.

Genista scorpius (*L.*) *DC.* in Lam. & DC., Fl. Franc. ed. 3, 4: 498 (1805); Fl. Europ. 2: 98 (1968).
LOCALITY. Unspecified, probably France.
MATERIAL. *Forsskål* s.n. (LD — 1 sheet ex herb. Retzius).
NOTE. Since this specimen is in fruit the determination is tentative.

Hedysarum immaculatum *Forssk.* 1775: 214 (CXVII No. 450; Cent. VIII No. 69); Christensen 1922: 34.
LOCALITY. Yemen: Al Hadiyah ("Hadîe"), 1763.
MATERIAL. *Forsskål* no type specimen found.
NOTE. This is a validly published name but in the absence of the type specimen it has not been possible to recognise the species. The name should not be taken up if it upsets widely accepted nomenclature.

Hippocrepis unisiliquosa *L.* 1753: 744; Tackholm 1974: 252.
var. **bisiliqua** (*Forssk.*) *Bornm.* in Notizbl. 10: 438 (1928); Fl. Palaest. 2: 104; pl. 153 (1972).
H. bisiliqua Forssk. 1775: LXXI No. 371; Christensen 1922: 36.
LOCALITY. Egypt: Alexandria ("As"), 1761–2.
MATERIAL. *Forsskål* 1338 (C — 1 sheet ex herb. Horneman, type of *H. bisiliqua*, microf. nil).
NOTE. There is a very short description which may be sufficient to validate the name. Christensen did not see this specimen.

Indigofera articulata *Gouan*, Illus.: 49 (1773); Gillett in Kew Bull., Addit. Ser. I: 101 (1958); Tackholm 1974: 255.
I. argentea sensu Vahl 1790: 56, non Burm. f. (1768); Deflers 1889: 129.
I. tinctoria sensu Forssk. 1775: 138 (LXXI No. 375; Cent. V No. 28), non L.; Christensen 1922: 24.
VERNACULAR NAME. *Nile* (Arabic).
LOCALITY. Egypt: Cairo, cult. ("Ch"), 1761–2.
MATERIAL. *Forsskål* 1359 (C — 1 sheet, microf. 56: III. 7, 8).

Indigofera oblongifolia *Forssk.* 1775: 137 (CXVIII No. 455; Cent. V No. 26); Deflers 1889: 130; Schweinfurth 1896: 240; Christensen 1922: 24; Schwartz 1939: 102; Gillett in Kew Bull., Addit. Ser. I: 116 (1958).
VERNACULAR NAME. *Hasar* (Arabic).
LOCALITY. Yemen: Al Luhayyah ("Lohajae"), Wadi Mawr ("Môr"), Al Hadiyah ("Hadie"), 1763.
MATERIAL. *Forsskål* no specimen found.

Indigofera semitrijuga *Forssk.* 1775: 137 (CXVIII No. 454; Cent. V No. 25); Vahl 1790: 56; Christensen 1922: 24; Schwartz 1939: 103; Gillett in Kew Bull., Addit. Ser. 1: 117 (1958).
LOCALITY. Yemen: Al Luhayyah ('Lohajae'), Jan. 1963.
MATERIAL. *Forsskål* 1354 (C — 1 sheet ex herb. Schumacher, type of *I. semitrijuga*, microf. 56: II. 5, 6); 1355 (C — 1 sheet, type of *I. semitrijuga*, microf. 56: II. 7, 8).
Forsskål s.n. (BM — 1 sheet ex herb. Banks MS 'Indigofera semitrijuga Forssk.', type of *I. semitrijuga*).

Indigofera spicata *Forssk.* 1775: 138 (CXVIII No. 456; Cent. V No. 29); Vahl 1790: 56; Christensen 1922: 25; Schwartz 1939: 105; Gillett in Kew Bull., Addit. Ser. 1: 119 (1958); Du Puy, Labat & Schrire in Kew Bull. 48: 727 (1993).
VERNACULAR NAME. *Mscheter*; *M'scheter*; *Schiter* (Arabic).
LOCALITY. Yemen: Bolghose ("Bolgôse"), Mar. 1763.
MATERIAL. *Forsskål* 1356 (C — 1 sheet ex herb. Schumacher with field label "Indigofera* ad Bolghose", type of *I. spicata*, microf. 56: III. 1,2).

Indigofera spinosa *Forssk.* 1775: 137 (CXVIII No. 457; Cent. V No. 27); Vahl 1790: 55; Deflers 1889: 129; Schweinfurth 1896; Christensen 1922: 24; Schwartz 1939: 101; Gillett in Kew Bull., Addit. Ser. 1: 51 (1958).
VERNACULAR NAME. *Haell* (Arabic).
LOCALITY. Yemen: Al Luhayyah ("Lohajae"), Surdud ("Surdûd"), Al Hadiyah ("Hadie"), 1763.
MATERIAL. *Forsskål* 1357 (C — 1 sheet, type of *I. spicata*, microf. 56: III. 3, 4); 1358 (C — 1 sheet ex herb. Schumacher, type of *I. spicata*, microf 56: III. 5, 6).
Forsskål s.n. (BM — 1 sheet ex herb. Banks MS 'Indigofera spinosa Forsk.', type of *I. spinosa*).
NOTE. On the BM sheet is mounted another African specimen (*Schimper* 772), not Forsskål's.

[**Indigofera tetrasperma** *Vahl ex Pers.* — A specimen at C from Ghana collected by *Thonning*, included in error in Forsskål collection and microf. 132: II. 7, 8].

Indigofera tinctoria *L.* 1753: 751; Schwartz 1939: 104; Gillett in Kew Bull., Addit. Ser. I: 106 (1958); F.T.E.A., Legum.-Papil.: 308 (1971).
I. houer Forssk. 1775: 137 (CXVIII No. 453; Cent. V No. 24); Vahl 1790: 56; Christensen 1922: 24.
VERNACULAR NAME. *Houer* (Arabic).
LOCALITY. Yemen: Surdud ("Surdûd in arvis spontanea"), 1763.
MATERIAL. *Forsskål* no type specimen found.

[**Kummerowia striata** (*Thunb.*) *Schindl.*
LOCALITY. Unspecified (see note below).
MATERIAL. *Unknown collector* s.n. (C — 1 sheet ex herb. Hofman Bang, microf. 132: I. 1, 2); *Unknown collector* (C — 1 sheet ex herb. Hofman Bang, microf. 132: I. 3, 4).
NOTE. This species occurs in Central Asia and the specimens cannot have been collected by Forsskål to whom they are attributed.]

Lablab purpureus (*L.*) *Sweet*, Hort. Brit. ed. 1: 481 (1827); F.T.E.A., Legum.-Papil.: 696 (1971).
L. vulgaris Savi (1824) — Schwartz 1939: 116.
Dolichos lablab L. 1753: 725; Schweinfurth 1896: 263.
D. cultratus Forssk. 1775: 134 (CXVII No. 439; Cent. V No. 15), non Thunb; Christensen 1922: 24.
VERNACULAR NAME. *Didjre, Kescht, Keschd* (Arabic).
LOCALITY. Yemen: Al Luhayyah ('Lohajae'), Kusma ('Kurma'), Jan. 1763.
MATERIAL. *Forsskål* no specimen found.

Lotus arabicus *L.*, 1767: 104; Vahl 1790: 65; Deflers 1889: 128; Schweinfurth 1896: 234; F.T.E.A., Legum.-Papil.: 1048 (1971); Tackholm 1974: 248, pl. 77D.
LOCALITY. Egypt: Cairo (not "Arabia"), 1761–2.
MATERIAL. *Forsskål* s.n. (LINN — 2 sheets 931.9 & 10, types of *L. arabicus* raised in Uppsala from Forsskål's seeds, microf. 523: I. 1, 2).

L. rosea Forssk. 1775: 140 (LXXI No. 387; Cent. V No. 38); Christensen 1922: 25.
LOCALITY. Egypt: Cairo ("Cs = Káhirae", "In insula Djesîret ed dáhalo prope Masr el atîk (Kâhiram veterem,)"), Jan. 1762.

MATERIAL. *Forsskål* 1429 (C — 1 sheet, type of *L. rosea*, microf. 64: II. 7, 8); 1430 (C — 1 sheet, with field label "Ca. 127 Lotus rosea, Gebel Eddahab", microf. 64: III. 1, 2). *Forsskål* s.n. (BM — 1 sheet ex herb. Banks MS 'Lotus rosea Forsk.', type of *L. rosea*).

Lotus corniculatus *L.* 1753: 755; Fl. Europ. 2: 174 (1968), var. **ciliatus** *Koch*, syn. ed. 1, 154 (1835).
L. hirsuta sensu Forssk. 1775: IX No. 176, non L.
LOCALITY. France: Marseille ("Estac"), May 1761.
MATERIAL. *Forsskål* 1428 (C — 1 sheet with field label "Lotus hirsuta? Estac", microf. 64: II. 5, 6).

Lotus corniculatus *L.* var. **tenuifolius** *L.* 1753: 776; Fl. Turk. 3: 526 (1970).
L. corniculatus L. — Forssk. 1775: XXX No. 319.
L. tenuis Waldst. & Kit. ex Willd., Enum. Pl. Hort. Berol.: 797 (1809); Fl. Europ. 2: 174 (1968).
LOCALITY. Turkey: Izmir ("Sm" = Smyrna), June–July 1761.
MATERIAL. *Forsskål* 1431 (C — 1 sheet, microf. nil).

Lotus creticus *L.*, 1775: 775; Tackholm 1974: 246, pl. 77A.
LOCALITY. Egypt: Cairo, 1761–62.
MATERIAL. *Forsskål* 1427 (C — 1 sheet, microf. 132: III. 5, 6).

Lotus halophilus *Boiss. & Sprun.* in Boiss., Diagn. ser. 1, 2: 37 (1843); Fl. Palaest. 2: 92, pl. 134 (1972); Tackholm 1974: 247.
L. peregrinus sensu Vahl 1790: 65, non L.
L. villosus Forssk. 1775: LXXI No. 386, non Burm. f.
VERNACULAR NAME. *Karn el gasal* (Arabic).
LOCALITY. Egypt: Alexandria ("As"), Apr. 1762.
MATERIAL. *Forsskål* 1432 (C — 1 sheet, type of *L. villosus*, microf. 64: III. 3, 4; 65(1–2): I. 1, 2); 1433 (C — 1 sheet with field label "Lotus novus villosus Alex. No. 15 ad salinas"; type of *L. villosus*, microf. 64: III. 5, 6, 65(1–2): I. 3, 4).

Lotus hirsutus sensu Forssk. 1775: 140 (CXVIII No. 462; Cent. V No. 39), non L.; Christensen 1922: 25.
LOCALITY. Yemen: Al Hadiyah, Bolghose ("Hadîe", "Blg = Bulgose"), 1763.
MATERIAL. *Forsskal* no specimen found.
NOTE. This remains obscure.

Lotus polyphyllus *Clarke*, Trav. 2(2): 41 (1814); Tackholm 1974: 247.
L. cretica sensu Forssk. 1775: 215 (LXXI No. 385; Cent. VIII No. 72), non L.; Christensen 1922: 34.
L. argenteus (Del.) Webb & Berth., Phyt. Canar. 2: 87 (1836); Ascherson & Schweinfurth 1887: 63.
VERNACULAR NAME. *Aesibe, Aesjbae* (Arabic).
LOCALITY. Egypt: Alexandria ("Alexandriae"), April 1762.
MATERIAL. *Forsskål* 1424 (C — 1 sheet with field label "Av. 48 Alex. Ras attin", microf. 64: I, 7, 8); 1425 (C — 1 sheet ex herb. Schumacher, microf. 64: II. 1, 2); 1426 (C — 1 sheet, microf. 64: II. 3, 4).

Lotus quinatus (*Forssk.*) *Gillett* in Kew Bull. 13: 373 (1959).
Dorycnium quinatum (Forssk.) Christensen 1922: 24, non *D. argenteum* Del. nec *D. latifolium* Willd. — Schwartz 1939: 100.
Ononis quinata Forssk. 1775: 130 (CXVII No. 433; Cent. V No. 4).
Lotus dorycnium Vahl 1790: 65, non L.
LOCALITY. Yemen: Khadra ("In monte Chadra"), 1763.
MATERIAL. *Forsskål* 1516 (C — 1 sheet, type of *O. quinata*, microf. 73: III. 4, 5); 1517 (C —1 sheet, type of *O. quinata*, microf. 73: III. 6, 7); 1518 (C — 1 sheet ex herb. Schumacher, type of *O. quinata*, microf. 73: III. 8).
Forsskål s.n. (BM — 1 sheet ex herb. Banks "Ononis Cherleri Forsk.", type of *O. quinata*); *Forsskål* s.n. (BM — 1 sheet ex herb. Banks "Lotus", type of *O. quinata*).

Lotus sp. indet.
LOCALITY. Not indicated.
MATERIAL. *Forsskål* 1434 (C — 1 sheet, microf. 132: III. 3, 4).
NOTE. A sterile specimen.

Lupinus albus *L.* 1753: 721; Fl. Palaest. 2: 42, pl. 55 (1972); Tackholm 1974: 225; J.S. Gladstones in W. Austral. Dept. Agric. Tech. Bull. 26: 5 (1974).
Lupinus termis Forssk. 1775: 131 (LXX No. 357; Cent. V No. 5); Christensen 1922: 24.
L. albus var. *termis* (Forssk.) Alef.
VERNACULAR NAME. *Termis* (Arabic).
LOCALITY. Egypt: Cairo ("Ch = Káhirae", "Colitur in Aegypto"), 1761-2.
MATERIAL. *Forsskål* 1439 (C — 1 sheet ex herb. Vahl, lectotype of *L. termis,* microf. 65(1-2): II. 5, 6); 1440 (C — 1 sheet ex herb. Vahl, isolectotype of *L. termis*, microf. nil).

Lupinus digitatus *Forssk.*, 1775: 131 (LXX No. 358; Cent. V No. 6); Christensen 1922: 24; J.S. Gladstones in W. Austral. Dept. Agric. Tech. Bull. 26: 24 (1974).
L. pilosus Murr. subsp. *digitatus* (Forssk.) Maire, Fl. Afr. Nord. 16: 111 (1987).
L. forskahlei Boiss., Diagn. Ser. I, 9: 10 (1849).
LOCALITY. Egypt: ("In Delta ad pagum Nedjil"), Mar. 1762.
MATERIAL. *Forsskål* 1438 (C — 1 sheet ex herb. Vahl with field label 'Av.100 it. Alex. ad Nedgile', lectotype of *L. digitatus*, microf. 65(1-2): II. 7, 8).
NOTE. This lectotype was selected by P. Hanelt op. cit. (1960).

Medicago littoralis *Rohde ex Loisel.*, Not. Pl. Fr.: 118 (1810); Fl. Europ. 2: 156 (1968); Tackholm 1974: 237.
LOCALITY. Egypt: Alexandria to Cairo ("in itenere inter Alexandriam ad Cairo" MS on reverse), 1761-62.
MATERIAL. *Forsskål* 1453 (C — 1 sheet ex herb. Horneman, microf. 67: I. 3, 4).

Medicago marina *L.* 1753: 779; Fl. Europ. 2: 156 (1968); Fl. Turk. 3: 501 (1970); Tackholm 1974: 234.
LOCALITY. Turkey: Dardanelles, July 1761.
MATERIAL. *Forsskål* 1455 (C — 1 sheet, ex herb. Horneman, microf. nil); 1454 (C — 1 sheet specimen 2 with field label "Natolia ante Dard.", microf. 66: III. 5, 6).
LOCALITY. Egypt: Alexandria to Rashid ("via inter Alex. et Rosette" on label), Oct. 1761.
MATERIAL. *Forsskål* 1455 (C — 1 sheet specimen 1 with field label "Av.147 via inter Alex et Rosetta", microf. 66: III. 5, 6).
NOTE. *Forsskål* 1455 has a MS name "Lotus creticus Linn" but the specimen is not the same as that on another sheet called *L. creticus* by Forsskål which is *L. polyphyllus* Clarke

and presumably is the one referred to in Forssk. 1775: 215 (LXXI No. 385; Cent. VIII no. 72).
Forsskål s.n. (LD — 1 sheet ex herb. Retzius).
Forsskål s.n. (LIV — 1 sheet 1909.LBG.7536).

Medicago polymorpha *L.* 1753: 779; Forssk. 1775: 141 (LXXI No. 391; Cent. V No. 42); Christensen 1922: 25; Fl. Europ. 2: 156 (1968); Tackholm 1974: 234, pl. 75A.
VERNACULAR NAME. *Nefl* (Arabic).
LOCALITY. Egypt: Cairo ("In Aegypto vulgaris), 1761–62.
MATERIAL. *Forsskål* 1456 (C — 1 sheet ex herb. Hornemann, microf. 66: III. 7, 8); *Forsskål* s.n. (LIV /1909 LBG.7554).
Forsskål s.n. (BM — 1 sheet ex herb. Banks MS 'Medicago polymorpha Forsk.').

Melilotus alba *Medikus*, Vorl. Churpf. Phys.-Ökon. Ges. 2: 382 (1787); Fl. Europ. 9: 149 (1968).
LOCALITY. ? ("Ad Kada" on sheet).
MATERIAL. *Forsskål* 1458 (C — 1 sheet, ex herb. Hornemann, microf. nil).

Melilotus indica *L.* 1753: 765; Tackholm 1974: 238.
Trifolium melilothus Forssk. 1775: 140 (LXXI Nos. 380, 381, CXVIII No. 459; Cent. V No. 37).
T. melilotus indicus L. 1753: 765; Christensen 1922: 25.
VERNACULAR NAME. *Gurt, Djulbân* (Egypt Arabic) *Rijam, Ryjam, Reinâm* (Yemen Arabic).
LOCALITY. Yemen: Kudmiyah and Egypt: Cairo ('Kudmîe', 'Frequens circa Káhiram'), 1761–62.
MATERIAL. *Forsskål* 1794 (C — 1 sheet ex herb. Hornemann, microf. 67: I. 1, 2); 1795 (C — 1 sheet ex herb. Hofman Bang, microf. 108: II. 3, 4).
Forsskål s.n. (BM — 1 sheet ex herb. Banks MS 'Trifolium m. indica Forsk.').

Neonotonia wightii (*Wight & Arn.*) *Lackey* var. **longicauda** (*Schweinf.*) *Lackey* in Iselya 2: 11–12 (1981).
Glycine javanica auctt., non L.; Schwartz 1939: 112.
LOCALITY. Yemen: ? Mukhajah, J. Barad ("Moxhaja, Barak" on field label), Mar. 1763.
MATERIAL. *Forsskål* 1302 (C — 1 sheet with field label, microf. 40: II. 1, 2); 1303 (C — 1 sheet, microf. 40: II. 3, 4).
NOTE. At C enclosed in the folder marked "Dolichos sinensis" (Cent. V No. 8).

Ononis natrix *L.* 1753: 717; Forssk. 1775: IX No. 169; Fl. Europ. 2: 144 (1968).
LOCALITY. France: Marseille ("Estac"), May 1761.
MATERIAL. *Forsskål* 1513 (C — 1 sheet ex herb. Vahl, microf. 73: II. 8, 73: III. 1); 1514 (C — 1 sheet ex herb. Hornemann, microf. 73: III. 2, 3).

Ononis minutissima *L.* 1753: 717; Forssk. 1775: IX No. 168; Fl. Europ. 2: 146 (1968).
LOCALITY. France: Marseille ("Estac"), May 1761.
MATERIAL. *Forsskål* 1511 (C — 1 sheet ex herb. Hornemann, microf. 73: II. 4, 5); 1512 (C — 1 sheet ex herb. Hornemann, microf. 73: II. 6, 7).

Ononis serrata *Forssk.* 1775: 130 (LXX No. 355; Cent. V No. 2); Christensen 1922: 24; Fl. Turk. 3: 383 (1970); Fl. Palaest. 2: 120, pl. 177 (1972); Tackholm 1974: 229, pl. 73A; Meikle 1977: 397.

LOCALITY. Egypt: Alexandria ("Alexandriae"), Mar. 1762.
MATERIAL. *Forsskål* 1519 (C — 1 sheet ex herb. Hornemann, with field label "18 Alex. Ras. ettin", type of *O. serrata*, microf. 74: I. 3, 4); 1520 (C — 1 sheet, type of *O. serrata*, microf. 74: I. 1, 2).
Forsskål s.n. (BM — 1 sheet ex herb. Banks, type of *O. serrata*).

Ononis spinosa *L.* 1753: 716; Fl. Turk. 3: 381 (1970).
O. alternifolia Forssk. 1775: XXX No. 305.
LOCALITY. Turkey: Dardanelles ("Dd."), July 1761.
MATERIAL. *Forsskål* 1504 (C — 1 sheet, with field label "Bo. 3", type of *O. alternifolia*, microf. 72: II. 7, 8); 1505 (C — 1 sheet ex herb. Schumacher, type of *O. alternifolia*, microf. 72: III. 1, 2); 1506 (C — 1 sheet, type of *O. alternifolia*, microf. 72: III. 3, 4); 1507 (C — 1 sheet ex herb. Vahl, type of *O. alternifolia*, microf. 72: III. 5, 6); 1508 (C — 1 sheet ex herb. Schumacher, type of *O. alternifolia*, microf. 72: III. 7, 8).
NOTE. These sheets named by J. Lange as *O. antiquorum* L., now considered to be *O. spinosa* ssp. *antiquorum* (L.) Briq. — Fl. Turk. 3: 381 (1970), but it could equally be ssp. *leiosperma* (Boiss.) Sirj. which differs only in seed characters not available on these specimens.

Ononis vaginalis *Vahl* 1790: 53; Tackholm 1974: 229.
O. cherleri sensu Forssk. 1775: 130 (LXX No. 354; Cent. V No. 1), non L.; Christensen 1922: 24.
LOCALITY. Egypt: Alexandria ("Frequens in collibus peninsulae Râs et tîn, Alexandriae"), Mar. 1762.
MATERIAL. *Forsskål* 1521 (C — 1 sheet with field label "As. 7", type of *O. vaginalis*, microf. 73: I. 1, 2); 1522 (C — 1 sheet ex herb. Hornemann with field label "As. 7 Alex", type of *O. vaginalis*, microf. 73: I. 5, 6); 1523 (C — 1 sheet type of *O. vaginalis*, microf. 73: I. 3, 4); 1524 (C — 1 sheet ex herb. Schumacher, type of *O. vaginalis*, microf. 73: I. 7).
Forsskål s.n. (BM — 1 sheet ex herb. Banks 'Ononis alternifolia Forsk. O. flava', with other specimens, type of *O. vaginalis*).

Ormocarpum yemenense *Gillett* in Kew Bull. 20: 339 (1966).
LOCALITY. Yemen: field label indistinct "inter Ersch & Amadja?" probably between Örs and 'Alujah ('Aludje') in Hadie Mts, 1763.
MATERIAL. *Forsskål* 1530 (C — 1 sheet with field label, microf. 132: II. 1, 2).

Ornithopus compressus *L.* 1753: 744; Forssk. 1775: IX No. 185; Fl. Europ. 2: 182 (1968).
LOCALITY. France: Marseille ("Estac"), May 1761.
MATERIAL. *Forsskål* 1531 (C — 1 sheet ex herb. Hornemann with old label, microf. 74: III. 5, 6).

Ornithopus sativus *Brot.*, Fl. Lusit. 2: 160 (1804); Fl. Europ. 2: 182 (1968).
LOCALITY. France: Montpellier, May 1761.
MATERIAL. *Forsskål* 1532 (C — sheet with field label "Gonan", microf. 132: III. -, 8).

Pterolobium stellatum (*Forssk.*) *Brenan* in Mem. N.Y. Bot. Gard. 8: 425 (1954); F.T.E.A. Legum.-Caesalp.: 42 (1967).
Mimosa stellata Forssk. 1775: 177 (CXXIIIVI, No. 609; Cent. VI No. 87), non Lour. (1790); Vahl 1790: 81; Christensen 1922: 29.
Acacia stellata (Forssk.) Willd., Sp. Pl. 4: 1078 (1806).
LOCALITY. Yemen: Kusma ("Kurmae"), Mar. 1763.

MATERIAL. *Forsskål* 1486 (C — 1 sheet ex herb. Viborg, type of *M. stellata*, microf. 69: II. 7, 8); 1487 (C — 1 sheet, type of *M. stellata*, microf. 69: III. 1, 2).

Retama raetam (*Forssk.*) *Webb* in Webb & Berth., Phyt. Canar. 2: 56 (1842); Fl. Palaest. 2: 47, pl. 63 (1972).
Genista raetam Forssk. 1775: 214 (VIII No. 66); Christensen 1922: 34.
Genista spartium sensu Forssk. 1775: LXX No. 353.
Lygos raetam (Forsk.) Heywood in Fedde, Repert. 79: 53 (1968); Tackholm 1974: 227.
VERNACULAR NAME. *Raetaem behâm* (Arabic).
LOCALITY. Egypt: Rashid & Suez ("Rosettae, Sués"), 1761–2.
MATERIAL. *Forsskål* 1339 (C — 1 sheet ex herb. Horneman, type of *G. raetam*, microf. 49: III. 3, 4); 1340 (C — 1 sheet ex herb. Hornemann with original field label "Ca. 223 Ignota Arab. Raetam", type of *G. raetam*, microf. 49: III. 5, 6); 1341 (C — 1 sheet ex herb. Hornemann with original field label "Sues", type of *G. raetam*, microf. nil); 1342 (C — 1 sheet ex herb. Hofman Bang, type of *G. raetam*, microf. 133: I. 1, 2).
Forsskål s.n. (LD — 1 sheet ex herb. Retzius, type of *G. raetam*).

Rhynchosia flava (*Forssk.*) *Thulin* in Nordic. Journ. Bot. 1: 37 (1981).
Ononis flava Forssk. 1775: 130 (CXVII No. 432; Cent. V No. 3); Christensen 1922: 24.
LOCALITY. Yemen: Jiblah ("Djöblae", Djöbla"), Apr. 1763.
MATERIAL. *Forsskål* 1509 (C — 1 sheet, type of *O. flava*, microf. 73: I. 8, II. 1); 1510 (C — 1 sheet, type of *O. flava*, microf. 73: II. 2, 3).

Rhynchosia malacophylla (*Sprengel*) *Bojer*, Hort. Maurit.: 104 (1837).
LOCALITY. Yemen: unspecified.
MATERIAL. *Forsskål* 1622 (C — 1 sheet, microf. nil).

Rhynchosia minima (*L.*) *DC.*, Prodr. 2: 385 (1825); Schwartz 1939: 114.
LOCALITY. Yemen: without locality, 1763.
MATERIAL. *Forsskål* 1624 (C — 1 sheet, microf. nil); 1625 (C — 1 sheet, microf. nil); 1626 (C — 1 sheet, microf. 131: III. 3, 4); 1627 (C — 1 sheet, microf. 131: III. 5, 6).

Scorpiurus muricatus *L.* 1753: 745; Fl. Europ. 2: 185 (1968).
S. subvillosa (sic) L. — Forssk. 1775: IX No. 181.
LOCALITY. France: Marseille ("Estac"), May 1761.
MATERIAL. *Forsskål* 1712 (C — 1 sheet ex herb. Hornemann with field label "Scorpiur. Subvil. Estac", microf. 97: I. 7, 8).

Senna alexandrina *Miller*, Gard. Dict., ed. 8 No. 1 (1768); Lock in Kew Bull. 43: 338 (1988).
Cassia senna L. 1753: 377; Brenan in Kew Bull. 13: 243 (1958); F.T.E.A. Legum.-Caesalp.: 65 (1967).
Cassia ligustrina sensu Forssk. 1775: 86 (CXI No. 268; Cent. III. No. 59), non L.; Christensen 1922: 19.
Cassia medica Forssk. 1775: CXI No. 271 nomen nudum; Christensen 1922: 27.
Cassia angustifolia Vahl 1790: 29; Schwartz 1939: 91.
LOCALITY. Yemen: Jiblah ("Ad Djöblam frequens"), 31 Mar. 1763.
MATERIAL. *Forsskål* 1049 (C — 1 sheet, type of *C. angustifolia*, microf. 21: II. 7, 8); 1048 (C — 1 sheet ex herb. Schumacher, type of *C. angustifolia*, microf. 21: II. 5, 6); 1050 (C — 1 sheet, type of *C. angustifolia*, microf. 21: III. 1, 2); 1051 (C — 1 sheet, type of *C. angustifolia*; microf. 21: III. 3, 4).

Forsskål s.n. (BM — 1 sheet ex herb. Banks 'Cassia ligustrum Forsk' MS).
Forsskål s.n. (LD — 1 sheet ex herb. Retzius).
NOTE. Sheet No. 1050 is marked "Niebuhr" on the back. It was collected on Forsskål's and Niebuhr's joint trip to Jiblah and Taizz in March–April 1763.

Senna italica *Miller*, Gard. Dict., ed. 8 Senna No. 2 (1768); Lock in Kew Bull. 43: 339 (1988).
Cassia italica (Mill.) Spreng., Bot. Gart. Univ. Halle: 21 (1800); F.T.E.A. Legum.-Caesalp.: 65 (1967); Tackholm 1974: 284, pl. 90B.
Cassia senna sensu Forssk. 1775: LXVI No. 222, non L.
VERNACULAR NAME. *Senna mecki, Hedjazi* (Arabic).
LOCALITY. Egypt: Cairo, 1761–2.
MATERIAL. *Forsskål* 1059 (C — 1 sheet, microf. 21: I. 5, 6); 1061 (C — 1 sheet ex herb. Hofman Bang, microf. nil); 1062 (C — 1 sheet ex herb. Hofman Bang, microf. nil); 1060 (C — 1 sheet ex herb. Vahl, microf. 132: I. 7, 8).
Forsskål s.n. (LD — 1 sheet ex herb. Retzius).
Forsskål s.n. (BM — 1 sheet ex herb. Banks).

Cassia aschrek Forssk. 1775: CXI No. 265; Schweinfurth 1896: 220; Christensen 1922: 19. = *Cassia* [*sp. without epithet*] Forssk. 1775: 86 (Cent. III No. 62).
VERNACULAR NAME. *Aschrek* (Arabic).
LOCALITY. Yemen: Wadi Mawr ("Môr"), Feb. 1763.
MATERIAL. *Forsskål* 1052 (C — 1 sheet with field label "Movr", holotype of *C. aschrek*, microf. nil).
Forsskål s.n. (BM — 1 sheet ex herb. Banks).

Senna obtusifolia (*L.*) *Irwin & Barneby* in Mem. N.Y. Bot. Gard. 35: 252 (1982); Lock in Kew Bull. 43: 340 (1988).
Cassia obtusifolia L. 1753: 377; F.T.E.A. Legum.-Caesalp.: 77 (1967).
Cassia tora sensu Forssk. 1775: 86 (CXI No. 266; Cent. III No. 60), non L.; Christensen 1922: 19.
VERNACULAR NAME. *Kolkol, Didjer el akbar* (Arabic).
LOCALITY. Yemen: Wadi Surdud & Wasab ("Ad *Surdud*"), Mar. 1763.
MATERIAL. *Forsskål* 1063 (C — 1 sheet, microf. 22: I. 1, 2); 1064 (C — 1 sheet with "Uahfâd" on field label, microf. 22: I. 3, 4).

Senna occidentalis (*L.*) *Link*, Handbuch 2: 140 (1831); Lock in Kew Bull. 43: 340 (1988).
Cassia occidentalis L. 1753: 377; F.T.E.A. Legum.-Caesalp.: 23, fig. 14 (1967); Schwartz 1939: 89.
LOCALITY. ? Yemen, not specified.
MATERIAL. *Forsskål* s.n. (LD — 1 sheet ex herb. Retzius).

Senna sophera (*L.*) *Roxb.*, Fl. Ind., ed. 2, 2: 347 (1832); Lock in Kew Bull. 43: 340 (1988).
Cassia sophera L. 1753: 379; F.T.E.A. Legum.-Caesalp.: 78 (1967); Schwartz 1939: 89.
Cassia lanceolata Forssk. 1775: 85 (CXI No. 270; Cent. III. No. 85); Christensen 1922: 19.
VERNACULAR NAME. *Súna* (Arabic).
LOCALITY. Yemen: Wadi Surdud & Wadi Mawr ("*Surdud.* Etiam circa *Môr* frequens"), Feb. 1763.
MATERIAL. *Forsskål* 1053 (1 sheet ex herb. Schumacher, type of *C. lanceolata*, microf. 21: I. 7, 8); 1054 (C — 1 sheet with original ticket bearing "Surdud", type of *C. lanceolata*, microf. 21: II. 1, 2); 1055 (C — 1 sheet, type of *C. lanceolata*, microf. 21: II. 3, 4); 1056 (C — 1 sheet with field label, microf. nil).
Forsskål s.n. (BM — 1 sheet ex herb. Banks with ms 'Cassia lanceolata').

Senna tora (*L.*) *Roxb.*, Fl. Ind., ed. 2, 2: 340 (1832); Lock in Kew Bull. 43: 341 (1988).
Cassia tora L. 1753: 538; Forssk. 1775: CXI No. 266; Schweinfurth 1896: 221; Schwartz 1939: 89.
Cassia sunsub Forssk. 1775: 86 (CXI No. 269; Cent. III. No. 61); Christensen 1922: 19.
VERNACULAR NAME. *Sunsub* (Arabic).
LOCALITY. Yemen: Taizz ("Frequens prope Taaes"), 1763.
MATERIAL. *Forsskål* no type specimen found.
NOTE. Schweinfurth considered *C. sunsub* to be a synonym of *C. tora*. The vernacular name *Aschrek* given after Forsskål's description (on p. 86) appears to be an editing error.

Sesbania grandiflora (*L.*) *Poir.*, Encycl. 7: 127 (1806); Deflers 1889: 131; Schweinfurth 1896: 248; Schwartz 1939: 108.
Aeschynomene grandiflora L., Sp. Pl. ed. 2, 1050 (1762); Vahl 1790: 54.
Dolichos arboreus Forssk. 1775: 134 (CXVII No. 444; Cent. V No. 14); Christensen 1922.
VERNACULAR NAME. *Sesebân* (Arabic).
LOCALITY. Yemen: Wadi Zabid ("Uadi Zebîd"), Apr. 1763.
MATERIAL. *Forsskål* 1230 (C — 1 sheet, type of *D. arboreus*, microf. 40: I. 3, 4).

Sesbania sesban (*L.*) *Merrill* in Philipp. Journ. Sci., Bot. 7: 235 (1912); Tackholm 1974: 258.
Sesbania aegyptiaca Pers. (1806) — Aschers. & Schweinf. 1887: 65.
Dolichos aeschynomene sesban Forssk. 1775: 134 (LXX No. 362; Cent. V No. 13); Christensen 1922: 24.
Aeschynomene sesban L. 1753: 714; Vahl 1790: 54.
VERNACULAR NAME. *Sejseban*; *Seisebân* (Arabic).
LOCALITY. Egypt: Cairo ("Ch = Kåhirae"), 1762.
MATERIAL. *Forsskål* 1231 (C — 1 sheet ex herb. Hornemann, microf. 39: III. 7, 8); 1232 (C — 1 sheet ex herb. Schumacher, microf. 40: I. 1, 2).
Forsskål s.n. (BM — 1 sheet ex herb. Banks "Cassia procumbens").
NOTE. Forsskål's name should not be hyphenated.

Sophora alopecuroides *L.* 1753: 373; Fl. Turk. 3: 11 (1972).
Hedysarum virginicum sensu Forssk. 1775: XXX No. 312.
Glycyrhiza astragaloides Vahl MS.
LOCALITY. Turkey: Istanbul, Belgrad ("Belgrad"), Aug. 1761.
MATERIAL. *Forsskål* 1325 (C — 1 sheet with field label "Co. 80. Hedysarum virginicum. Belgrad.", microf. 53: I. 7, 8).

Taverniera lappacea (*Forssk.*) *DC.*, Prodr. 2: 339 (1825); Schweinfurth 1896: 249; Schwartz 1939: 110.
Hedysarum lappaceum Forssk. 1775: 136 (CXVII No. 451; Cent. V No. 22); Christensen 1922: 24.
VERNACULAR NAME. *Höbb el adjais* (Arabic).
LOCALITY. Yemen: Al Luhayyah ('Lohajae'), Jan. 1763.
MATERIAL. *Forsskål* 1318 (C — 1 sheet, type of *H. lappaceum*, microf. 52: III. 3, 4); 1319 (C — 1 sheet ex herb. Schumacher, type of *H. lappaceum*, microf. 52: III. 5, 6).
Forsskål s.n. (BM — 1 sheet ex herb. Banks, type of *H. lappaceum*).
Forsskål s.n. (LD — 1 sheet ex herb. Retzius, type of *H. lappaceum*).

Teline monspessulana (*L.*) *C. Koch*, Dendrologie 1: 30 (1869); Fl. Europ. 2: 93 (1968).
LOCALITY. ? France.
MATERIAL. *Forsskål* 1781 (C — 1 sheet ex herb. Hornemann, microf. nil).

Tephrosia tomentosa (*Forssk.*) *Pers.*, Syn. Pl. 2: 329 (1807); Schwartz 1939: 108.
Lathyrus tomentosus Forssk. 1775: 135 (CXVII No. 446; Cent. V No. 19); Christensen 1922: 24.
Galega tomentosa (Forssk.) Vahl 1791: 84.
VERNACULAR NAME. *Sonaefa* (Arabic).
LOCALITY. Yemen: Wadi Mawr ('Môr'), 1763.
MATERIAL. *Forsskål* no type specimen found.

Teramnus repens (*Taub.*) *Bak.* subsp. **gracilis** (*Chiov.*) *Verdc.* in Kew Bull. 24: 275 (1970).
LOCALITY. ? Yemen: unspecified.
MATERIAL. *Forsskål* s.n. (LD — 1 sheet ex herb. Retzius).

Trifolium alexandrinum *L.*, Cent. Pl. 1: 25 (1755); Forssk. 1775: 139 (LXXI No. 382; Cent. V No. 34); Christensen 1922: 25; Fl. Palaest. 2: 185, pl. 265 (1972); Tackholm 1974: 242.
VERNACULAR NAME. *Bersim, Bersûm, Bersûn, Berzûm* (Arabic).
LOCALITY. Egypt: cult. 1761–2.
MATERIAL. *Forsskål* 1789 (C — 1 sheet ex herb. Hornemann, microf. 108: I. 3, 4); 1790 (4C — 1 sheet ex herb. Schumacher, microf. 108: I. 5, 6); 1791 (C — 1 sheet ex herb. Hornemann, microf. 108: I. 7, 8); 1792 (C — 1 sheet ex herb. Schumacher, microf. nil).
Forsskål s.n. (BM — 1 sheet ex herb. Banks MS 'Trifolium alpestre Forsk.').
Forsskål s.n. (LD — 1 sheet ex herb. Retzius).

Trifolium fragiferum *L.* 1753: 772; Forssk. 1775: IX No. 189, LXXI No. 383; Fl. Europ. 2: 165 (1968); Fl. Turk. 3: 409 (1970).
LOCALITY. France: Marseille ("Estac"), 9 May–3 June 1761. Egypt: Cairo, 1761–62.
MATERIAL. *Forsskål* s.n. (LD — 1 sheet ex herb. Retzius).
NOTE. Named and labelled by P. Lassen 1977.

Trifolium micranthum *Viv.*, Fl. Lib.: 45 (1824); Fl. Europ. 2: 166 (1968).
LOCALITY. Unspecified, prob. France.
MATERIAL. *Forsskål* 1796 (C — 1 sheet, microf. 132: III. 1, 2).

Trifolium repens sensu Forssk. 1775: 214 (CXVIII No. 461; Cent. VIII No. 70).
LOCALITY. Yemen: Al Hadiyah ("Hadîe").
MATERIAL. *Forsskål* no specimen found.
NOTE. Identity needs to be confirmed. It is probably *T. semipilosum* Fresen. in Flora 22: 52 (1839); F.T.E.A. Legum.-Papil.: 1027 (1971).

Trifolium resupinatum *L.* 1753: 771; Tackholm 1974: 241, pl. 76A.
T. bicorne Forssk. 1775: 139 (LXXI No. 381; Cent. V No. 35); Christensen 1922: 25.
VERNACULAR NAME. *Gurr, Djulbân* (Arabic).
LOCALITY. Egypt: Cairo ("Cd = Káhirae"), Dec. 1761.
MATERIAL. *Forsskål* 1793 (C — 1 sheet ex herb. Hornemann, type of *T. bicorne*, microf. 108: II. 1, 2).
Forsskål s.n. (BM — 1 sheet ex herb. Banks MS 'Trifolium spumasum Forsk.', type of *T. bicorne*).

Trifolium uniflorum *L.* 1953: 771; Fl. Europ. 2: 164 (1968); Fl. Turk. 3: 391 (1970).
LOCALITY. Turkey ? (see Note).
MATERIAL. *Forsskal* s.n. (LD — 1 sheet ex herb. Retzius).

NOTE. Although this sheet has 'Arabia' marked on it, the species occurs only in the Eastern Mediterranean region.

Trigonella hamosa *L.*, Syst. Nat. ed. 10, 1180 (1759); Forssk. 1775: 141 (LXXI No. 389; Cent. V No. 41); Christensen 1922: 25; Fl. Palaest. 2: 126 pl. 183 (1972); Tackholm 1974: 233 pl. 74C.
VERNACULAR NAME. *Daragrag, Adjelmaelek* (Arabic).
LOCALITY. Egypt: Cairo ("Cd" = Kåhirae), 1761–2.
MATERIAL. *Forsskal* 1797 (C — 1 sheet ex herb. Hornemann with field label "Trigonella hamosa. des. Cairi", microf. 108: II. 5, 6); 1798 (C — 1 sheet ex herb. Hornemann, microf. 108: II. 7, 8); 1799 (C — 1 sheet ex herb. Vahl, microf. 108: III. 1, 2).

Trigonella procumbens (*Besser*) *Reichenb.*, Pl. Crit. 4: 35 (1826); Fl. Turk. 3: 478 (1970).
LOCALITY. Turkey: Izmir ("ad Smyrnam" on sheet), 1761.
MATERIAL. *Forsskål* 1800 (C — 1 sheet ex herb. Hornemann, microf. nil).

Trigonella stellata *Forssk.* 1775: 140 (LXXI No. 390; Cent. V No. 40); Christensen 1922: 25; Fl. Palaest. 2: 127 pl. 186 (1972); Tackholm 1974: 233 pl. 74.
T. monspeliaca sensu Vahl 1791: 85, non L.
VERNACULAR NAME. *Gargas* (Arabic).
LOCALITY. Egypt: Cairo ("In desertis arenosis circa Kàhiram"), 1762.
MATERIAL. *Forsskål* 1801 (C — 1 sheet ex herb. Hornemann, microf. 108: III. 3, 4); 1802 (C — 1 sheet ex herb. Hornemann, microf. 108: III. 5, 6); 1803 (C — 1 sheet ex herb. Hornemann, microf. 108: III. 7, 8); 1804 (C — 1 sheet ex herb. Schumacher, microf. 109: I. 1, 2); 1805 (C — 1 sheet ex herb. Schumacher, microf. 109: I. 3, 4); 1806 (C — 1 sheet ex herb. Schumacher, microf. 109: I. 5, 6). (All types of *T. stellatum*). *Forsskål* s.n. (BM — 1 sheet ex herb. Banks, type of *T. stellatum*).

Ulex europaeus *L.* 1753: 741; Fl. Europ. 2: 102 (1968).
LOCALITY. Probably France: Marseille (not Alexandria, "ad Alexandriam" given on sheet).
MATERIAL. *Forsskål* 1810 (C — 1 sheet ex herb. Schumacher, microf. nil).

Vicia monantha *Retz.*, Obs. Bot. 3: 34 (1783); Fl. Europ. 2: 133 (1968); Fl. Turk. 3: 296 (1970).
LOCALITY. ? Egypt.
MATERIAL. *Forsskål* s.n. (LD — 1 sheet ex herb. Retzius).
NOTE. This is not likely to be Retzius' missing type (Burtt & Lewis in Kew Bull. 4: 497–515 (1949)) since its inflorescences are not 1-flowered.

Vicia peregrina *L.* 1753: 737; Forssk. 1775: IX No. 183; Fl. Europ. 2: 135 (1968).
LOCALITY. France: Marseille ("Estac"), 9 May–3 June 1761.
MATERIAL. *Forsskål* 1822 (C — 1 sheet ex herb. Hornemann, microf. 112: II. 1, 2). *Forsskål* s.n. (LD — 1 sheet ex herb. Retzius).

Vicia sativa *L.* var. **angustifolia** *L.*, Fl. Suec., ed. 2: 255 in obs. (1755); Fl. Europ. 2: 134 (1968) under subsp. *nigra*.
LOCALITY. Unspecified, Mediterranean region.
MATERIAL. *Forsskål* s.n. (LD — 1 sheet ex herb. Retzius).

Vigna aconitifolia (*Jacq.*) *Maréchal* in Bull. Jard. Bot. Belg. 39: 160 (1969).
Phaseolus aconitifolius Jacq., Obs. Bot. 3: 2, t. 52 (1768); Vahl 1790: 53; Schweinfurth 1896: 260; Schwartz 1939: 115.
Phaseolus palmatus Forssk. 1775: 214 (CXVII No. 435; Cent. VIII No. 68); Christensen 1922: 34.
VERNACULAR NAME. *Gotn* (Arabic).
LOCALITY. Yemen: Wadi Mawr ("Môr"), Feb. 1763.
MATERIAL. *Forsskål* 1568 (C — 1 sheet with field label "Dolichos * palmatus. Gotn. Mour", type of *P. palmatus*, microf. 79: I. 7, 8); 1569 (C — 1 sheet, type of *P. palmatus*, microf. 79: II. 1, 2).

Vigna luteola (*Jacq.*) *Benth.* in Mart., Fl. Bras. 15(1): 194, t. 50/2 (1859); Tackholm 1974: 281.
Dolichos sinensis sensu Forssk. 1775: 132 (LXX No. 361; Cent. V No. 8), non L.; Christensen 1922: 24.
VERNACULAR NAME. *Höllaech*; *Öllaeab* (Arabic).
LOCALITY. Egypt: Rashid ("Rs = Rosettae", "Ad fossas agrorum prope Nilum"), Oct. 1761.
MATERIAL. *Forsskål* 1234 (C — 1 sheet, microf. 40: I. 7, 8).
Forsskål s.n. (LD — 1 sheet ex herb. Retzius).
NOTE. Although this C sheet was annotated by Vahl "Cent. 5, pag. 133, No. 11" and placed in a cover labelled *Dolichos lubia* Cent. No. V. 11 it must be identified with *D. sinensis* sensu Forsskål, Cent. V. No. 8.

Vigna radiata (*L.*) *R. Wilczek* in Fl. Congo Belge 6: 386 (1954); F.T.E.A., Legum.-Papil.: 655 (1971).
Phaseolus radiatus L. 1753: 725; Forssk. 1775: 214 (CXVII No. 434; Cent. VIII No. 67); Christensen 1922: 34.
VERNACUALR NAME. *Koschâri* (Arabic).
LOCALITY. Yemen: Wadi Mawr ("Môr"), 1763.
MATERIAL. *Forsskål* 1570 (C — 1 sheet, microf. 79: II. 3, 4).
NOTE. This specimen had been named and labelled by B. Verdcourt 1987 as var. *setulosa* (Dalz.) Ohwi & Chashi in Journ. Jap. Bot. 44: 31 (1969).

Vigna unguiculata (*L.*) *Walp.*, Rep. 1: 779 (1842); F.T.E.A. Legum.-Papil.: 642 (1971); Tackholm 1974: 281.
Dolichos lubia Forssk. 1775: 133 (LXX No. 363; CXVII No. 440; Cent. V No. 11); Christensen 1922: 24.
VERNACULAR NAME. *Lubia baeledi*, *Lubia baelledi* (Egyptian Arabic); *Didjre* (Yemeni Arabic).
LOCALITY. Egypt: Cairo ("Ch" = Káhirae, "In agris Aegypti frequens cultus"), 1761–62. Yemen: Al Luhayyah ("Lhj" = Lohaja), 1763.
MATERIAL. *Forsskål* no specimen found.

Vigna unguiculata (*L.*) *Walp.* subsp. **sesquipedalis** (*L.*) *Verdc.* in Kew Bull. 24: 544 (1970); F.T.E.A. Legum.-Papil.: 643 (1971).
Dolichos didjre Forssk. 1775: 133 (Cent. V No. 12); Christensen 1922: 24.
VERNACULAR NAME. *Didjre* (Arabic).
LOCALITY. Yemen: Al Luhayyah ('Lohajae'), Jan. 1763.
MATERIAL. *Forsskål* no specimen found.

Vigna variegata Deflers 1889: 132.
Rhynchosia cf. *mensensis* Schweinfurth 1896: 256.
LOCALITY. Yemen: unspecified.
MATERIAL. *Forsskål* 1623 (C — 1 sheet, microf. 131: III. 7, 8); 1628 (C — 1 sheet, microf. nil).

LENTIBULARIACEAE

Utricularia inflexa *Forssk.* 1775: 9 (LIX No. 8; Cent. I No. 26); Vahl 1790: 6, 1805: 196; Christensen 1922: 11; P. Taylor in Kew Bull. 18: 185 (1964); Tackholm 1974: 509.
VERNACULAR NAME. *Hamûl* (Arabic).
LOCALITY. Egypt: Rashid ("Rosettae copiose in fossis inter agros Oryzae"), 2–6 Nov. 1761.
MATERIAL. *Forsskål* 355 (C — 1 sheet, type, microf. 111: III. 2, 3; 356 (C — 1 sheet, holotype of *U. inflexa* designated by P. Taylor, microf. 111: II. 8, III. 1); 1813 (C — 1 sheet with several specimens ex herb. Schumacher & Hornemann, isotypes of *U. inflexa*, microf. 111: II. 7).
Forsskål s.n. (P — 1 sheet, isotype of *U. inflexa*).

LINACEAE

Linum corymbulosum *Reichenb.*, Fl. Germ. Exc.: 834 (1832); Fl. Turk. 2: 435 (1967).
LOCALITY. Turkey ?
MATERIAL. *Forsskål* s.n. (LD — 1 sheet ex herb. Retzius).

Linum strictum *L.* 1753: 279; Tackholm 1974: 314.
LOCALITY. Egypt: Cairo, 1761–62.
MATERIAL. *Forsskål* 469 (C — 1 sheet with field label "Ca 149", microf. 129: III. 3, 4).
NOTE. The field label has the MS name in Forsskål's hand "Tetrastoma nov. gen. lanceolata". This name occurs nowhere in 'Flora Aegyptiaco-Arabica'.

Linum sp. indet.
LOCALITY. Unspecified.
MATERIAL. *Forsskål* s.n. (LD — 1 sheet ex herb. Retzius).
NOTE. A slender specimen with long aristate leaves and two capsules on long, slender stalks which has not yet been identified.

LORANTHACEAE

Plicosepalus acaciae (*Zucc.*) *Weins & Polhill* in Nord. J. Bot. 5(3): 221 (1985).
Loranthus acaciae Zucc., Abh. Akad. Wiss. 3: 249, t. 2 f.3 (1840); Schwartz 1939: 30; Fl. Palaest. 1: 46, t. 46 (1966).
LOCALITY. Yemen: Taizz, Apr. 1763.
MATERIAL. *Forsskål* 685 (C — 1 sheet with field label "Taaes", microf. 129: III. 5, 6).
NOTE. The field label also mentioned that it was a parasite of *Euphorbia*. Hepper and Wood also collected it on *E. amak* in 1975.

Viscum schimperi *Engl.*, Bot. Jahrb. 20: 132 (1894).
Viscum ? Forssk. 1775: CXXII No. 588.
LOCALITY. Yemen: ? M'harras ("M'harras"), near Chadra, 1763.
MATERIAL. *Forsskål* 686 (C — 1 sheet, microf. 112: II. 7, 8); 687 (C — 1 sheet, microf. 112: III. 1, 2).

LYTHRACEAE

Lawsonia inermis *L.* 1753: 349; Forssk. 1775: LXV no. 217; Schwartz 1939: 172; Bircher 1960: 550.
VERNACULAR NAME. *Tamrabenne* (Arabic).
LOCALITY. Egypt: Alexandria (on label), 1761–62.
MATERIAL. *Forsskål* 768 (C — 1 sheet with field label "Ac. 8", microf. 61: III. 1, 2).

L. spinosa L. 1753: 349; Forssk. 1775: LXV No. 216.
LOCALITY. Egypt: Cairo cult. ("Ch = Cairi plantae hortenses"), 1762.
MATERIAL. *Forsskål* 767 (C — 1 sheet, microf. 61: III. 3, 4); 748 (C — 1 sheet, microf. 61: III. 1, 2).
Forsskål s.n. (LD — 2 sheets ex herb. Retzius).

Lythrum hyssopifolium *L.* 1753: 447; Tackholm 1974: 377.
Pentaglossum linifolium Forssk. 1775: 11 (LIX No. 11; Cent. I No. 30); Christensen 1922: 11.
LOCALITY. Egypt: Alexandria and Cairo ("Alexandriae locis cultis, humilioribus, umbrosis prope Canalem. Kahirae ad Birket el hadj locis humidiuse-ulis"), 1761–2.
MATERIAL. *Forsskål* 1563 (C — 1 sheet with field label "L.I. Pentaglossum linophyllum", type, microf. 77: III. 7, 8); 1564 (C — 1 sheet, type of *P. linifolium*, microf. 78: I. 1, 2); 1565 (C — 1 sheet, type, microf. 78: I. 3, 4).

MALPIGHIACEAE

Caucanthus edulis *Forssk.* 1775: CXI No. 289; Schweinfurth 1899: 296; Christensen 1922: 19; Schwartz 1939: 130. = *Caucanthus* [*sp. without epithet*] Forssk. 1775: 91 (Cent. III No. 78).
C. arabicus Lam., Encycl. 1: 658 (1785).
C. forskahlei Raeusch., Nomencl. bot. ed. 3, 132 (1797).
Aspidopteris yemensis Deflers 1889: 117, t. 1.
VERNACULAR NAME. *Kaha, Kauka, Kouka* (Arabic).
LOCALITY. Yemen: Taizz ("In montibus ad *Taaes*" or "Haes" see CXI on field label "inter Oudae & Homarae."), 4 Apr. 1763. Oude is between Taizz and Hays.
MATERIAL. *Forsskål* 477 (C — 1 sheet with field label, bearing No. "10.1", type of *Caucanthus edulis*, microf. 22: III. 3, 4); 1068 (C — 1 sheet ex herb. Univ. Kiliensis, type, microf. nil).
NOTE. This is one of the cases where a new genus and species is described in the 'Centuriae', while an epithet is provided in the 'Florae'.

MALVACEAE

Abelmoschus esculentus (*L.*) *Moench*, Methodus 617 (1794); Schwartz 1939: 164.
Hibiscus esculentus L. 1753: 696; Forssk. 1775: 125 (LXX no. 343, 344; CXVII No. 426; Cent. IV No. 91); Deflers 1889: 114; Christensen 1922: 23.

VERNACULAR NAMES. *Bamia shâmi, stambûli, rumi, bamia uaki, baelledi* (Arabic).
LOCALITY. Egypt: Cairo ("Kâhirae"), 1761–62. Yemen: Tihama ("Cultae in planitae"), 1763.
MATERIAL. *Forsskål* no specimen found.

Abutilon bidentatum *Hochst. ex A. Rich.*, Tent. Fl. Abyss. 1: 68 (1847); Deflers 1889: 114; Schwartz 1939: 158.
Sida cordifolia Forssk. 1775: 124 (CXVI No. 410; Cent. IV No. 83), non L.
Abutilon indicum sensu Christensen 1922: 23, non (L.) Decne.
VERNACULAR NAME. *Rên* (Arabic).
LOCALITY. Yemen, Lohajae ('Lohajae'), Jan. 1763.
MATERIAL. *Forsskål* 596 (C — 1 sheet, type of *S. cordifolia*, microf. 98: III. 1, 2), 589 (C — 1 sheet, type of *S. cordifolia*, microf. 98: III. 3, 4); 1726 (C — 1 sheet, ? type, microf. 98: III. 5, 6); 1727 (C — 1 sheet ex herb. Schumacher, ? type, microf. 99: I. 1, 2).
NOTE. *Abutilon indicum* proper has not been recorded for the Arabian Peninsula.

Abutilon pannosum (*Forst. f.*) *Schlecht.*, Bot. Zeit. 9: 828 (1851); Tackholm 1974: 354.
LOCALITY. Unspecified, prob. Egypt, 1761–62.
MATERIAL. *Forsskål* 900 (C — 1 sheet ex herb. Vahl, microf. 130: II. 1, 2); 901 (C — sheet, microf. 98: III. 7, 8).

Althaea hirsuta *L.* 1753: 687; Fl. Europ. 2: 253 (1968).
LOCALITY. Unspecified (France to Turkey).
MATERIAL. *Forsskål* 927 (C — 1 sheet ex herb. Hofman Bang, microf. 130: I. 1, 2).

Althaea rosea (*L.*) *Cav.*, Diss. 2: 91 t. 29 f. 3 (1786); Bircher 1960: 494.
Alcea rosea L. 1753: 687; Tackholm 1974: 353.
Alcea ficifolia L. 1753: 687; Forssk. 1775: LXX No. 338.
VERNACULAR NAME. *Chatmiae* (Arabic).
LOCALITY. Egypt: Cairo garden, 1762.
MATERIAL. *Forsskål* 588 (C — 1 sheet, microf. 4: II. 3, 4).

Gossypium arboreum *L.* 1753: 693; Forssk. 1775: 125 (LXX No. 341, CXVI No. 417; Cent. IV No. 89); Christensen 1922: 23; Schwartz 1939: 166; Bircher 1960: 498.
VERNACULAR NAMES. *Cotn el sadjar* (Egypt Arabic); *Otb, Ödjaz* (Yemen Arabic).
LOCALITY. Egypt: Cairo garden, 1761–2; Yemen: Wadi Mawr ("Môr"), 1763.
MATERIAL. *Forsskål* no specimen found.

Gossypium herbaceum *L.* 1753: 693; Schwartz 1939: 166; Bircher 1960: 498.
G. rubrum Forrsk. 1775: 125 (LXX No. 340, CXVI No. 418; Cent. IV No. 88); Christensen 1922: 23.
VERNACULAR NAME. *Otb, Ödjâs* (Arabic).
LOCALITY. Egypt: Cairo garden, 1761–2; Yemen: Al Hadiyah ("Hadîee"), Mar. 1763.
MATERIAL. *Forsskål* 599 (C — 1 sheet, type of *G. rubrum*, microf. 52: I. 3, 4).

Hibiscus cannabinus *L.*, Syst. Nat. ed. 10: 1149 (1759); Deflers 1889: 114; Schwartz 1939: 164; Bircher 1960: 494.
LOCALITY. Unspecified, prob. Yemen, 1763.
MATERIAL. *Forsskål* s.n. (LD — 1 sheet ex herb. Retzius).

Hibiscus ficulneus *L.* 1753: 695; Forssk. 1775: 125 (LXX No. 342; Cent. IV No. 90); Christensen 1922: 23.
VERNACULAR NAME. *Bami* (Egypt Arabic).
LOCALITY. Egypt: Alexandria ('Alexandriae'), Cairo garden ("Ch"), 1761–62.
MATERIAL. *Forsskål* no specimen found.

Hibiscus deflersii *Schweinf. ex Cuf.*; Schwartz 1939: 163.
LOCALITY. Unspecified, prob. Yemen, 1763.
MATERIAL. *Forsskål* 1336 (C — 1 sheet ex herb. Schumacher, microf. 130: I. 7, 8). *Forsskål* s.n. (LIV — 1 sheet 8331).

Hibiscus ovalifolius (*Forssk.*) *Vahl* 1790: 50; Hepper & Wood in Kew Bull. 38: 84 (1983).
Urena ovalifolia Forssk. 1775: 124 (CXVI No. 419; Cent. IV No. 87); Christensen 1922: 23.
LOCALITY. Yemen: Taizz ("Montosa loca circa Taaes inhabitat"), 1763.
MATERIAL. *Forsskål* 578 (C — 1 sheet, microf. 54: II. 5, 6); 579 (C — 1 sheet, microf. 54: II. 3, 4); 578 (C — 1 sheet, microf. 54: II. 5, 6); 600 (C — 1 sheet, lectotype of *U. ovalifolia*, microf. 110: III. 5, 6); 1811 (C — 1 sheet, microf. 110: III. 7, 8).
NOTE. Christensen 1922: 23 and Schwartz 1939: 164 considered *U. ovalifolia* to be synonymous with *H. purpureus* Forssk.

Hibiscus palmatus *Forssk.* 1775: 126 (CXVII No. 423; Cent. IV No. 97); Christensen 1922: 23; Robson in Fl. Zamb. 2: 469, t. 89 (1961).
LOCALITY. Yemen: Al Mukham ("Prope Mochham"), May 1763.
MATERIAL. *Forsskål* 604 (C — 1 sheet, type of *H. palmatus*, microf. 54: II. 1, 2).

Hibiscus praecox *Forssk.* 1775: 125 (Cent. IV No. 92); Christensen 1922: 23.
VERNACULAR NAME. *Uaeki, Baeledi* (Arabic).
LOCALITY. Egypt: Cairo ("Káhirae"), 1761–62.
MATERIAL. *Forsskål* no type specimen found.
NOTE. In the absence of the type specimen this remains imperfectly known, although a validly published name. According to Forsskål's description it has the appearance of *Abelmoschus esculentus* (*Hibiscus esculentus*), with 5 (not 10) loculi in the hispid (not glabrous) fruit. The name should not be adopted if it upsets widely accepted nomenclature.

Hibiscus purpureus *Forssk.* 1775: 126 (CXVII No. 421; Cent. IV No. 95); Christensen 1922: 23; Schwartz 1939: 164; Hepper & Wood in Kew Bull. 38: 84 (1983).
VERNACULAR NAME. *Sech, Chobaes, Malât, Hotomtom* (Arabic).
LOCALITY. Yemen: Al Hadiya ("Hadie"), Mar. 1763.
MATERIAL. *Forsskål* 590 (C — 1 sheet with field label "Hibiscus hirtus. Sech. Hadie" holotype of *H. purpureum*, microf. 54: II. 7, 8).
NOTE. *Forsskål* 578 & 579, which were written up as "*H. purpureus*", are in fact *H. ovalifolius* (q.v.). The Linnaean epithet "*hirtus*" on No. 590 was not taken up in the Flora Aegyptiaco-Arabica.

Hibiscus syriacus *L.* 1753: 695; Forssk. 1775: LXX No. 345; Christensen 1922: 23.
LOCALITY. Egypt: Cairo garden, 1761–62.
MATERIAL. *Forsskål* no specimen found.

Hibiscus vitifolius *L.* 1753: 696; Schwartz 1939: 164; Hepper & Wood in Kew Bull. 38: 84 (1983).
H. tripartitus Forssk. 1775: 126 (CXVII No. 422; Cent. IV No. 96); Christensen 1922: 23.
LOCALITY. Yemen: Wasab ("Uahfad"), 1763.
MATERIAL. *Forsskål* s.n. (BM — 2 sheets ex herb. Banks).
NOTE. It is strange that no specimens of this species have been found at C.

Hibiscus sp.
LOCALITY. ? Yemen.
MATERIAL. *Forsskål* 602 (C — 1 sheet, microf. 130: I. 5, 6).
NOTE. This is still unidentified as it does not match any specimens in Kew Herbarium from Yemen or neighbouring countries. Somebody has written "Hibiscus villosis" on the reverse of the sheet.

Malva parviflora *L.*, Demonstr. Pl.: 18 (1753); Fl. Europ. 2: 251 (1968); Tackholm 1974: 349, pl. 119B.
M. rotundifolia ? sensu Forssk. 1775: LXX No. 339, non L.
VERNACULAR NAME. *Chobbeize* (Arabic).
LOCALITY. Egypt: Cairo ("Ch" = Cairi plantae hortenses, "Cd" = Cairi loca deserta), 1762.
MATERIAL. *Forsskål* 587 (C — 1 sheet, microf. 65 (2–2): II. 4, 5); 1448 (C —1 sheet, microf. 65(2–2): II. 6).
Forsskål s.n. (LD — 1 sheet ex herb. Retzius).

Malva sylvestris *L.* 1753: 689; Forssk. 1775: VIII No. 164, XIV No. 54; Fl. Europ. 2: 250 (1968).
LOCALITY. ? France: Marseille ("Estac"), or ? Malta, 1761.
MATERIAL. *Forsskål* 594 (C — 1 sheet, microf. 65(2–2): II. 2, 3); 595 (C — 1 sheet, microf. 65(2–2): II. 7, 8); 66: I. 1, 2).
Forsskål s.n. (LD — 1 sheet ex herb. Retzius "Fl. Estac" on reverse).

var. **eriocarpa** *Boiss.*, Fl. Orient. 1: 819 (1867).
M. tournef(ourtiana) sensu Forssk. 1775: XXIX No. 299, non L.
VERNACULAR NAME. *Mollocha* (Greek), *Aebedjumez* (Turkish).
LOCALITY. Turkey: Bozcaada I. ("Td" = Tenedos) & Dardanelles ("Dd"), July 1761.
MATERIAL. *Forsskål* 592 (C — 1 sheet, microf. 66: I. 3, 4); 593 (C — 1 sheet, microf. 66: I. 5, 6).
NOTE. Cullen in Fl. Turk. 2: 405 considered that Forsskål's record of *M. tournefortiana*, a Western Mediterranean species, was *M. moschata*, but this was based on an assumption as he was unaware of the existence of these specimens.

Malva verticillata *L.* 1753: 689; Fl. Europ. 2: 251 (1968); Schwartz 1939: 160; Tackholm 1974: 349.
M. nicaeensis sensu Vahl 1790: 50, non L.
M. montana Forssk. 1775: 124 (CXVI No. 415; Cent. IV No. 86); Christensen 1922: 23; Meikle 1977: 310 (note).
VERNACULAR NAME. *Hörod, Höbsen* (Arabic).
LOCALITY. Yemen: Al Hadiyah ("*Hadîe*"), Mar. 1763.
MATERIAL. *Forsskål* 591 (C — 1 sheet with field label "Malva * montana. Hörud Bolghose", type of *M. montana*, microf. 65(2–2): I. 8; II. 1).
Forsskål s.n. (LD — 1 sheet ex herb. Retzius, type of *M. montana*).

Pavonia flavo-ferruginea (*Forssk.*) *Hepper & Wood* in Kew Bull. 38: 85 (1983).
Hibiscus flavo-ferrugineus Forssk. 1775: CXVII No. 420.
H. flavus β Forssk. 1775: 126 (Cent. IV No. 94); Christensen 1922: 23.
LOCALITY. Yemen: Hays ("Circa Haes rarius", "Oude, Bulgose, Roboa"), 1763.
MATERIAL. *Forsskål* 603 (C — 1 sheet with field label "Hibiscus fl. flavo basi ferrugineo. Probe Robôa.", holotype of *H. flavo-ferrugineus*, microf. 54: I. 7, 8).

Pavonia hildebrandtii *Guerke & Ulbr.* in Engl. Bot. Jahrb. 48: 371 (1912); Hepper & Wood in Kew Bull. 38: 85 (1983).
Hibiscus microphyllus Vahl 1790: 50, non *P. microphylla* Casar (1842).
H. flavus sensu Forssk. 1775: 126 (CXVII No. 425; Cent. IV No. 94); Christensen 1922: 23, non L. (1753) nec *Pavonia flava* Spring ex Mart. (1837).
LOCALITY. Yemen: Al Luhayyah ("Lohojae") 1763.
MATERIAL. *Forsskål* 581 (C — 1 sheet, type of *H. microphyllus*, microf. 54: I. 5, 6); 601 (C —1 sheet, type of *H. microphyllus*, microf. 54: I. 3, 4).

Sida alba *L.* 1762: 960; Tackholm 1974: 353, pl. 120A.
Stewartia corchoroides Forssk. 1775: 126 (LXX No. 347; Cent. IV No. 98); Christensen 1922: 23.
Sida spinosa sensu Vahl 1791: 78, non L.
LOCALITY. Egypt: Cairo ("Cs" = Káhirae spontanae), 1762.
MATERIAL. *Forsskål* 575 (C — 1 sheet, type of *S. corchoroides*, microf. 104: III. 5, 6); 584 (C — 1 sheet, microf. 99: I. 5, 6); 586 (C — 1 sheet, type of *S. corchoroides*, microf. 104: II. 3, 4); 1763 (C — 1 sheet with field label "Ca. 54. Stewartia n. fruticosus ex herb. Vahl", microf. 104: II. 7, 8); 1764 (C — 1 sheet ex herb. Vahl, type of *S. corchoroides*, microf. 104: III. 1, 2).

Sida ciliata *Forssk.* 1775: CXVI No. 414; Christensen 1922: 37.
VERNACULAR NAME. *Vuzar, Tschaeba, Sockáa* (Arabic).
LOCALITY. Yemen: Wadi Surdud ("Srd"), 1763.
MATERIAL. *Forsskål* no type specimen found.
NOTE. An imperfectly known species, probably synonymous with another one, with a brief validating description: "Stipulis filiformibus, non ciliatis". The name should not be adopted if it upsets widely accepted nomenclature.

Sida ovata *Forssk.* 1775: 124 (CXVI No. 413; Cent. IV No. 84); Christensen 1922: 23; F.W.T.A. ed. 2, 1: 339 (1958).
LOCALITY. Yemen: Wadi Surdud ("Surdûd" on field label and in book), Feb. 1763.
MATERIAL. *Forsskål* 1728 (C — 1 sheet with field label "Sida * ovata. Surdud", lectotype proposed by O.A. Leistner 1967, microf. 99: I. 3, 4); 1729 (C — 1 sheet ex herb. Vahl with field label illegible, microf. 99: I. 7, 8); 1730 (C —1 sheet ex herb. Vahl, microf. 99: II. 1, 2).
Forsskål s.n. (BM — 1 sheet ex herb. Banks, type of *S. ovata*).

Sida urens *L.*, Syst. Nat. ed. 10: 1145 (1759); Schwartz 1939: 161.
LOCALITY. Yemen: Wadi Surdud ("Surdud" on field label), 1763.
MATERIAL. *Forsskål* 598 (C — 1 sheet with field label "Sida ch. fere ciliata. Surdud", microf. 130: I. 3, 4).

Thespesia populnea (*L.*) *Soland. ex Corr.* in Ann. Mus. Paris 9: 290 (1807); Schwartz 1939: 165.

LOCALITY. Unspecified, prob. Yemen.
MATERIAL. *Forsskål* s.n. (LD — 1 sheet ex herb. Retzius).

Wissadula amplissima (*L.*) *R.E. Fries* var. **rostrata** (*Schum. & Thonn.*) *R.E. Fries* in Kungl. Sv. Vet. Akad. Handl. 43, 4: 51, t. 6, 13–14 (1908); Schwartz 1939: 159.
Sida paniculata ? sensu Forssk. 1775: 124 (CXVI No. 412; Cent. IV No. 85), non L.; Christensen 1922: 23.
VERNACULAR NAME. *Rên*, *Ghobari* (Arabic).
LOCALITY. Yemen: Wadi Mawr ('Môr') & Wasab ('Uahfât'), 29 Mar. 1763.
MATERIAL. *Forsskål* 597 (C — 1 sheet, microf. 99: II. 3, 4).

MELIACEAE

Melia azedarach *L.* 1753: 384; Schwartz 1939: 129; Bircher 1960: 424.
LOCALITY. Egypt: Alexandria (on reverse of sheet), 1761–62.
MATERIAL. *Forsskål* 1457 (C — 1 sheet ex herb. Hornemann, microf. 133: I. 3, 4).

Trichilia emetica *Vahl* 1790: 31; Deflers 1889: 121; Schweinfurth 1896: 295; White in Bothalia 16: 157 (1986).
T. roka Forssk. ex Chiov., Fl. Somalia 2: 131 (1932).
Elcaja [*sp. without epithet*] Forssk. 1775: 127 (CXVI No. 409; Cent. IV No. 100); Christensen 1922: 24.
E. roka Forssk. 1775: XCV, non rite publ.
VERNACULAR NAME. *Roka* (Arabic).
LOCALITY. Yemen: Al Hadiya ("In montibus Yemen frequens", "Hadie"), Mar. 1763.
MATERIAL. *Forsskål* 478 (C — 1 sheet with original label "Hadîe", holotype of *T. roka* and *T. emetica*, microf. 41: II. 1, 2).
NOTE. The nomenclature of *Elcaja* and *Trichilia emetica* has been much debated. For a summary see Friis in Taxon 34: 663–664 (1984).

Turraea holstii *Guerke* in Engl., Bot. Jahrb. 19, Beibl. No. 47: 35 (1894).
LOCALITY. Yemen, 1763.
MATERIAL. *Forsskål* 464 (C — 1 sheet, microf. 138: I. 4, 5).

MENISPERMACEAE

Cocculus hirsutus (*L.*) *Theob.* in Mason, Burmah ed Theob. 2: 657 (1883); Schwartz 1939: 60; F.T.E.A. Menispermac.: 12 (1956).
Cocculus villosus (Lam.) DC., Syst. 1: 525 (1817).
Cebatha villosa (Lam.) Christensen 1922: 37.
Cebatha b) *foliis pubescentibus* Forssk. 1775: CXXII No. 586.
VERNACULAR NAME. *Kebath*, *Erdjadj* (Arabic).
LOCALITY. Yemen: Al Hadiyah ("Hadîe"), Mar. 1763.
MATERIAL. *Forsskål* 696 (C — 1 sheet, microf. 23: I. 3, 4).

Cocculus pendulus (*J.R. & G. Forst.*) *Diels* in Engl., Pflanzenr. IV, 94: 237 (1910); Schweinfurth 1912: 135; Schwartz 1939: 60; Fl. Palaest. 1: 216, pl. 317 (1966); Tackholm 1974: 144.

Cebatha a) *foliis glabris* Forssk. 1775: CXXII No. 585. = *Cebatha* [*sp. without epithet*] Forssk. 1775: 171 (Cent. VI No. 67); Christensen 1922: 29.
Cocculus cebatha DC., Syst. 1: 527 (1817).
Menispermum edule Vahl 1790: 80.
VERNACULAR NAME. *Kebath, Erdjadj* (Arabic).
LOCALITY. Yemen: Al Hadiyah ("Hadîe"), Mar. 1763.
MATERIAL. *Forsskål* 693 (C — 1 sheet with field label "* Cebatha edulis. Mour", type of *Cebatha* and *C. cebatha*, microf. 22: III. 5, 6); 694 (C — sheet, type of *Cebatha* and *C. cebatha*, microf. 23: I. 1, 2); 1462 (C — 1 sheet ex herb. Schumacher, type *Cebatha* and *C. cebatha*, microf. 22: III. 7, 8).

Leaeba [*sp. without epithet*] Forssk. 1775: 172 (LXXVII No. 539; Cent. VI No. 68); Christensen 1922: 29.
Menispermum leaeba Delile, Fl. Egypte: 284 (1812).
Cocculus leaeba (Delile) DC., Syst. 1: 529 (1817).
VERNACULAR NAME. *Lacbach el djebbel, Laebach el djaebbel* (Arabic).
LOCALITY. Egypt: Cairo ("Káhirae"), 1761–62.
MATERIAL. *Forsskål* 695 (C — 1 sheet with field label "Ca. 85", type of *Leaeba, M. leaeba, C. leaeba*, microf. 61: III. 7, 8); 1404 (C — 1 sheet, type of *Leaeba, M. leaeba, C. leaeba*, microf. 61: III. 5, 6); 1405 (C — 1 sheet ex herb. Hornemann, type of *Leaeba, M. leaeba, C. leaeba*, microf. 62: I. 1, 2).
NOTE. Neither *Cebatha* nor *Leaeba* was published with an epithet anywhere in '*Flora aegyptiaco-Arabica*'.

MENYANTHACEAE

Nymphoides cristata (*Griseb.*) *O. Kuntze*, Rev. Gen. 429 (1891).
LOCALITY. Unspecified, probably Yemen.
MATERIAL. ? *Forsskål* 1498 (1 sheet ex herb. Vahl, microf. 125: II. 1, 2); 1499 (1 sheet ex herb. Vahl, microf. 125: II. 3, 4).
NOTE. It is possible that these two specimens are Forster, not Forsskål, collections.

MOLLUGINACEAE

Mollugo cerviana (*L.*) *Ser.* var. **spathulifolia** *Fenzl* in Ann. Wien Mus. 1: 379 (1836); Schwartz 1939: 50; F.T.E.A. Aizoaceae: 17 (1961).
Pharnaceum umbellatum Forssk. 1775: 58 (CIX No. 217; Cent. II No. 94); Christensen 1922: 17.
LOCALITY. Yemen: Al Luhayyah ('Lohajae'), Jan. 1763.
MATERIAL. *Forsskål* 1567 (C — 1 sheet ex herb. Schumacher, type of *P. umbellatum*, microf. 79: I. 5, 6).
NOTE. Forsskål makes the interesting observation that it 'wakes' at 8 am and 'sleeps' at 3 pm — presumably referring to the opening and closing of the flowers.

Orygia villosa *Forssk.* 1775: CXIV No. 343; Christensen 1922: 37.
VERNACULAR NAME. *Horudj, Horudjrudj* (Arabic).
LOCALITY. Yemen: Bughah ("Boka"), 1763.
MATERIAL. *Forsskål* no type specimen found.
NOTE. The name *O. villosa* is validated by the description: "Flor. fulvis; foliis villoso-sericeis", but the species is unknown. The name should not be adopted if it upsets widely accepted nomenclature.

MORACEAE

Dorstenia foetida (*Forssk.*) *Schweinfurth* 1896: 120; Schwartz 1939: 24; Friis in Nordic J. Bot. 3: 536 (1983). Fig. 23.
Kosaria foetida Forssk. 1775: CXXI No. 532; Christensen 1922: 28. = *Kosaria* [*sp. without epithet*] Forssk. 1775: 164 (Cent. VI No. 34); Icones 6, t. 20 (1776); F.T.E.A., Morac.: 41 (1989).
VERNACULAR NAME. *Kosar* (Arabic).
LOCALITY. Yemen: Al Hadiyah ("Hadîe in Coffeae-cetis sub arboribus inter lapides"), Mar. 1763.
MATERIAL. *Forsskål* 327 (C — 1 sheet with field label "Ignota. Koser. [description]. Hadie et ante Bolghose", type of *K. foetida*, microf. 60: III. 7, 8).
NOTE. This is one of the cases where a description of a new genus and species occur in the 'Centuriae', while the epithet is provided in the 'Florae'.

Ficus carica *L.* 1753: 1059; Forssk. 1775: CXXIV No. 620 (see Note); Schwartz 1939: 25.
VERNACULAR NAME. *Tin* (? Arabic).
LOCALITY. Yemen: cult., 1763.
MATERIAL. *Forsskål* 778 (C — 1 sheet, microf. 44: III. 7).
NOTE. This large single leaf was determined as "Ficus carica L. var. fr. albo" by G. Schweinfurth 1895. It certainly is not the plant referred to from the Yemen as *F. carica*, which is also entered in the Flora lists for Turkey (XXXV No. 451) and Egypt (LXXVII No. 555). In the absence of further material, it cannot be excluded that the leaf represents an extreme form of *F. palmata*.

Ficus cordata *Thunb.* subsp. **salicifolia** (*Vahl*) *C.C. Berg* in Kew Bull. 43: 82 (1988); F.T.E.A., Morac.: 63 (1989).
F. salicifolia Vahl 1790: 82, t. 23; Enum. 1805: 195; Schweinfurth 1896: 133; Schwartz 1939: 26.
F. indica sensu Forssk. 1775: 179 (CXXIV No. 625; Cent. VI No. 97), non L.; Christensen 1922: 30.
VERNACULAR NAME. *Thaab* (Arabic).
LOCALITY. Yemen: Wasab and Wadi Zabid ("Ad Uahfât and Uadi Zebîd"), 29 Mar. 1763.
MATERIAL. *Forsskål* 780 (C — 1 sheet, type of *F. salicifolia*, microf. 44: III. 8); 1263 (C — 1 sheet ex herb. Vahl, type of *F. salicifolia*, microf. 45: I. 1, 2).

Ficus exasperata *Vahl* 1805: 197; Schwartz 1939: 26, F.T.E.A. Morac.: 52 (1989).
? *Ficus serrata* sensu Forssk. 1775: 179 (CXXIV No. 624; Cent. VI No. 96), non L.; Vahl 1790: 83 & 1805: 202 Schweinfurth 1896: 121; Christensen 1922: 30. (See note to this sp. and to *F. palmata*).
VERNACULAR NAME. *Haschref* (Arabic).
LOCALITY. Yemen: Bolghose ("Bolgosi, alibique"), Mar. 1763.
MATERIAL. *Forsskål* 785 (C — 1 sheet with original field label, type of *F. exasperata*, microf. 45: II. 3, 4).
NOTE. The type of *F. exasperata* is an Isert specimen from Ghana. No. 785 was identified by Vahl with Forsskål's *F. serrata*, a view accepted by Ascherson and J. Hutchinson, but perhaps not correct (see note under *F. palmata*).

Ficus palmata *Forssk.* 1775: 179 (CXXIV No. 623; Cent. VI No. 95); Schweinfurth 1896: 124; Schwartz 1939: 25.
VERNACULAR NAME. *Baeles* (Arabic).
LOCALITY. Yemen: Wasab ("Ad Uahfât"), Mar. 1763.

Fig. 23. **Dorstenia foetida** (Forssk.) Schweinf. — Icones, Tab. XX *Kosaria*.

MATERIAL. *Forsskål* 781 (C — 1 sheet with field label "inter Bolghose & Mockaja", type of *F. serrata* or *F. palmata*, microf. 45: I. 5, 6); 784 (C — 1 sheet, type of *F. palmata*, microf. 45: I. 7, 8).
NOTE. The original field label of No. 781 is inscribed "*Ficus serrata*" and the specimen has the partly ovate, partly palmately lobed leaves mentioned in Forsskål's description of *F. serrata*. It is likely that *F. serrata*, contrary to earlier statements, belongs here.

Ficus populifolia *Vahl* 1790: 82, t. 22; 1805: 81; Schweinfurth 1896: 129; Schwartz 1939: 26; F.T.E.A. Morac.: 68 (1989).
F. religiosa sensu Forssk. 1775: 180 (CXXIV No. 626; Cent. VI No. 98), non L.
VERNACULAR NAME. *Mudáh, Vudáh* (Arabic).
LOCALITY. Yemen: Wadi Zabid ("Uadi Zebîd"), 5–6 Apr. 1763.
MATERIAL. *Forsskål* 783 (C — 1 sheet, holotype of *F. populifolia*, microf. 45: II. 1, 2).

Ficus sur *Forssk.* 1775: 180 (CXXIV No. 619; Cent. VI No. 99); Vahl 1805: 199; Christensen 1922: 30; F.T.E.A. Morac.: 56, fig. 18 (1989).
VERNACULAR NAME. *Sur* (Arabic).
LOCALITY. Yemen: Jiblah ("Djöblae"), 30–31 Mar. 1763.
MATERIAL. *Forsskål* 782 (C — 1 sheet, holotype of *F. sur*, microf. 45: II. 5, 6).
F. capensis Thunb. (1786) — Schweinfurth 1896: 140 partly; Schwartz 1939: 26.

Ficus sycomorus *L.* 1753: 1059; Forssk. 1775: 180 (LXXVII No. 556, CXXIV No. 616; Cent. VI No. 100); Schweinfurth 1896: 142.
VERNACULAR NAME. *Djummeiz, Chanas, Öbre, Sokam* (Arabic).
LOCALITY. Egypt: Rashid ("Rosetta ad turrem Canopi"), Nov. 1761.
MATERIAL. *Forsskål* 779 (C — 1 sheet with field label "Ro. 19 ad terrem Canopi", microf. 45: II. 7, 8).

F. chanas Forssk. 1775: 219 (Cent. VIII No. 98); Muschler 1912: 247; Christensen 1922: 35.
VERNACULAR NAME. *Chanas, Öbre* (Arabic).
LOCALITY. Yemen: Wadi Surdud ("Surdûd in montosis"), 1763.
MATERIAL. *Forsskål* no type specimen found.
NOTE. In the absence of a type specimen for *F. chanas* Forssk. it has been decided to place it in the synonymy of *F. sycomorus* since Forsskål stated that it resembled that species.

Ficus taab *Forssk.* 1775: 219 (Cent. VIII No. 100); Schweinfurth 1912: 140; Wood in Kew Bull. 39: 134 (1984).
VERNACULAR NAME. *Táab* (Arabic).
LOCALITY. Yemen: Zabid ("Zebîd"), 1763.
MATERIAL. *Forsskål* no type specimen found.
NOTE. Wood (l.c.) considers that this is a *nomen ambiguum* and in the absence of a type specimen it should not be used for *F. ingens*, *F. salicifolia* or any other species.

Ficus vasta *Forssk.* 1775: 179 (CXXIV No. 621; Cent. VI No. 93); Schweinfurth 1896: 129; Christensen 1922: 30; F.T.E.A. Morac.: 64 (1989).
F. benghalensis sensu Vahl 1790: 82, non L.
VERNACULAR NAME. *Tålak, Túlak, Taluk, Delb* (Arabic).
LOCALITY. Yemen: Al Hadiyah and Taizz ("Hadîe", "Taaes", "In Yemen ubique"), 1763.

MATERIAL. *Forsskål* 776 (C — 1 sheet, type of *F. vasta*, microf. 45: III. 1, 2).
Forsskål s.n. (BM — 1 sheet ex herb. Banks, type of *F. vasta*).

Ficus sp. possibly young stage of *F. palmata* Forssk. — Schweinfurth 1896: 124.
F. morifolia Forssk. 1775: 179 (CXXIV No. 622; Cent. VI No. 94), non Lam. (1786); Christensen 1922: 30.
F. forskalaei Vahl 1805: 196.
VERNACULAR NAME. *Baeles* (Arabic).
LOCALITY. Yemen: Al Hadiyah ("Hadîe"), 29 Mar. 1763.
MATERIAL. *Forsskål* 777 (C — 1 sheet, microf. 45: I. 3, 4).
NOTE. J. Hutchinson 1913 labelled this "viz Ficus" while E.J.H. Corner 1954 marked it "Ficus palmata v. young?"
Forsskål s.n. (BM — 1 sheet ex herb. Banks, type of *F. morifolia*).

Morus sylvestris *Forssk.* 1775: XXXIII No. 405; Christensen 1922: 36.
LOCALITY. Turkey: ? Istanbul.
MATERIAL. *Forsskål* no specimen found.
NOTE. The description validates *M. sylvestris*: "Fructu parum succoso; sed seminibus clitiore; contra: *M. culta* fructu gaudet succulentiore, verum seminibus pauperiore". The name should not be taken up if it upsets widely accepted nomenclature.

MORINGACEAE

Moringa peregrina (*Forssk.*) *Fiori* in Agricolt. Colon. 5: 59 (1911); Fl. Palaest. 1: 340, pl. 495 (1966); Tackholm 1974: 211, pl. 65.
Hyperanthera peregrina Forssk. 1775: CVII No. 159; Christensen 1922: 17. = *Hesperanthera* [*sp. without epithet*] Forssk. 1775: 67 (Cent. III No. 10).
Gymnocladus arabica Lam., Encycl. 1: 733 (1785).
Hyperanthera semidecandra Vahl 1790: 30.
Moringa arabica (Lam.) Pers., Syn. 1: 460 (1805).
LOCALITY. Yemen: Bayt al Faqih ("Beit el fakíh"), Feb. 1763.
MATERIAL. *Forsskål* 1349 (C — 1 sheet ex herb. Horneman with field label "Arbor. 5: 1 falso d. seisbân. Beit el Faki exotica", type of *H. peregrina, G. arabica, H. semidecandra*, microf. 55: III. 5, 6); 1350 (C — 1 sheet ex herb. Hornemann, type of *H. peregrina, G. arabica, H. semidecandra*, microf. 56: I. 1, 2); 1351 (C — 1 sheet ex herb. Schumacher, type of *H. peregrina, G. arabica, H. semidecandra*, microf. nil).
NOTE. The statement on the field label that the plant is an exotic and the note in 'Flora Aegyptiaco-Arabica' explain the epithet *peregrina*. It has been suggested that this would indicate that this species might fit better with the widely cultivated *M. oleifera*, but B. Verdcourt has examined the material and confirms its identity with what is now known as *M. peregrina*. This is one of the cases where a new genus and species with epithet is described in the 'Centuriae', while the epithet is provided in the 'Florae'.

MYRSINACEAE

Maesa lanceolata *Forssk.* 1775: CVI No. 129; Christensen 1922: 17; Schwartz 1939: 179. = *Maesa* [*sp. without epithet*] Forssk. 1775: 66 (Cent. III No. 9).
Baeobothrys lanceolata (Forssk.) Vahl 1790: 19, t. 6.
VERNACULAR NAME. *Máas, Arar* (Arabic).
LOCALITY. Yemen: Al Udayn ("In montibus Yemen ad Öddein"), Feb.-Apr. 1763.

MATERIAL. *Forsskål* 361 (C — 1 sheet, type of *M. lanceolata*, microf. 65(2–2): I. 5, 6); 362 (C — 1 sheet, type of *M. lanceolata*, microf. 65(2–2): I. 3, 4); 363 (C — 1 sheet with field label 'M. Soudân', type of *M. lanceolata*, microf. 65(2–2): I. 1, 2); 1447 (C — 1 sheet ex herb. Schumacher, type of *M. lanceolata*, microf. 65(2–2): I. 7).
Forsskål s.n. (BM — 1 sheet, type of *M. lanceolata*, microf. nil).
NOTE. *Maesa lanceolata* is the type species of Forsskål's genus *Maesa*. This is one of the cases where a new genus and species without epithet is described in the 'Centuriae' while an epithet is provided in the 'Florae'.

Myrsine africana *L.* 1753: 196; Schwartz 1939: 179; F.T.E.A. Myrsinac.: 6, fig. 2 (1984).
Buxus dioica Forssk. 1775: 159 (CXXI No. 538; Cent. VI No. 15); Christensen 1922: 27.
Myrica montana Vahl 1791: 99, nom. illeg.
VERNACULAR NAME. *Katam* (Arabic).
LOCALITY. Yemen: Barad and Kusma ("In monte Barah", "Kurma"), Mar. 1763.
MATERIAL. *Forsskål* 729 (C — 1 sheet with field label "Kurma", type of *B. dioica* and *M. montana*, microf. 17: III. 7, 8); 730 (C — 1 sheet with field label "Bulghose", type of *B. dioica* and *M. montana*, microf. 18: I. 1, 2); 1024 (C — 1 sheet?, ex herb. Hornemann, type of *B. dioica* and *M. montana*, microf. 17: III. 5, 6); 1025 (C — 1 sheet, type of *B. dioica* and *M. montana*, microf. nil).

NEURADACEAE

Neurada procumbens *L.* 1753: 441; Forssk. 1775: 90 (LXVI No. 245; Cent. III No. 75); Christensen 1922: 19; Tackholm 1974: 219, pl. 68B.
VERNACULAR NAME. *Saadân* (Arabic).
LOCALITY. Egypt: Alexandria ("*Alexandriae*"), Nov. 1761.
MATERIAL. *Forsskål* 545 (C — 1 sheet, microf. 71: II. 1, 2); 546 (C — 1 sheet, microf. 71: I. 5, 6); 547 (C — 1 sheet, microf. 71: I. 7, 8); 1494 (C — 1 sheet ex herb. Hofman Bang, microf. 71: II. 3, 4); 1495 (C — 1 sheet ex herb. Schumacher, microf. 134: III. 1, 2).

NYCTAGINACEAE

Boerhavia plumbagineus Cav. var. **forskalei** Schweinfurth 1896: 167; Schwartz 1939: 48.
B. scandens sensu Forssk. 1775: 3 (CII No. 1; Cent. I No. 5), non L.; Deflers 1889: 192; Christensen 1922: 10.
VERNACULAR NAME. *Orkos* (Arabic).
LOCALITY. Yemen: Al Luhayyah and Wadi Mawr ("Lohajae and Môr", 1763.
MATERIAL. *Forsskål* no type specimen found.
NOTE. As the type is missing there is doubt about the status of this variety.

Boerhavia diandra sensu Forssk. 1775: 8 (CII No. 2; Cent. I No. 6), non L.; Christensen 1922: 10.
VERNACULAR NAME. *Vuddjef, Rokâma, Chaddir, Chadder* (Arabic).
LOCALITY. Yemen: Al Luhayyah, Wadi Mawr, Bayt al Faqih ("Lohajae, Môr, Beit el fakihn), 1763.
MATERIAL. *Forsskål* no specimen found.
NOTE. The status of this remains obscure.

Boerhavia repens *L.* var. **diffusa** (*L.*) *Boiss.*, Fl. Orient. 4: 1045 (1879); Schweinfurth 1896: 166; Schwartz 1939: 46.

B. repens L. var. *viscosa* Choisy — Deflers 1889: 192.
B. diffusa L. — Forssk. 1775: 3 (CII No. 3; Cent. I No. 7); Christensen 1922: 10.
LOCALITY. Yemen: ? ("Dahhi"), 1763.
MATERIAL. *Forsskål* 1006 (C — 1 sheet ? ex herb. Hornemann, microf. 15: II. 1, 2).

Commicarpus plumbagineus (*Cav.*) *Standl.* in Contrib. US. Nat. Herb. 18: 101 (1916).
Boerhavia dichotoma Vahl 1805: 290; Christensen 1922: 11.
B. plumbaginea var. *dichotoma* (Vahl) Aschers. & Schweinf. in Beitr. Fl. Aeth. 168 (1867); Schweinfurth 1896: 167; Schwartz 1939: 47.
Valeriana scandens sensu Forssk. 1775: 12 (CIII No. 34; Cent. I No. 31), non Loefl.
VERNACULAR NAMES. *Charad, Choddâra* (Arabic).
LOCALITY. Yemen: Wadi Surdud ('Surdûd'), Feb. 1763.
MATERIAL. *Forsskål* 533 (C — 1 sheet, type of *B. dichotoma*, microf. 111: III. 7, 8); 534 (C —1 sheet with field label "Valeriana * celtu. charad. Hadie, type of *B. dichotoma*, microf. 111: III. 5, 6); 1818 (C — 1 sheet ex herb. Schumacher, ? type of *B. dichotoma*, microf. 111: III. 4, 112: I. 1, 2).; 1819 (C — 1 sheet, ? type of *B. dichotoma*, microf. 133: I. 5, 6). *Forsskål* s.n. (LD — 1 sheet ex herb. Retzius, type of *B. dichotoma*).

NYMPHAEACEAE

Nymphaea lotus *L.* 1753: 511; Forssk. 1775: 100 (LXVII No. 279; Cent. IV No. 10); Christensen 1922: 20; Tackholm 1974: 144, pl. 39A.
VERNACULAR NAME. *Naufar* (Arabic).
MATERIAL. Egypt: Rashid ("Rosettae"), 2–6 Nov. 1961.
MATERIAL. *Forsskål* 613 (C — 1 sheet, microf. 71: II. 5, 6).

OCHNACEAE

Ochna inermis (*Forssk.*) *Schweinf.* apud Penzig in Atti Congr. Bot. Intern. Genova 1892: 335 (1893); Schweinfurth 1912: 148; Christensen 1922: 33; Robson in Fl. Zambes. 2: 237 (1963).
Euonymus inermis Forssk. 1775: 204 (CVII No. 157; Cent. VIII No. 15).
Ochna parvifolia Vahl 1790: 33; Deflers 1889: 120.
VERNACULAR NAME. *Öyun ennemr, el Benât* (Arabic).
LOCALITY. Yemen: Al Hadiyah ("Hadîe"), Mar. 1763.
MATERIAL. *Forsskål* 760 (C — 1 sheet with field label "inter Ersch & Alûdje", holotype of *E. inermis* and *O. parvifolia*, microf. 43: II. 7, 8).

OLEACEAE

Jasminum officinale *L.* 1753: 7; Forssk. 1775: XVIII No. 5, LIX No. 6; Bircher 1960: 626.
VASCULAR NAME. *Jasmin, Kajan* (Egyptian Arabic).
LOCALITY. Turkey: Istanbul 30 July–8 Sept. 1761, or Egypt: Cairo garden, 1761–62.
MATERIAL. *Forsskål* 1368 (C — 1 sheet, microf. 133: I. 7, 8).

Olea europaea *L.* 1753: 8; Forssk. 1775: 202 (XVIII No. 2; Cent. VIII No. 1); Christensen 1922: 32; Fl. Turk. 6: 155 (1978).
VERNACULAR NAME. *Elies* (Greek).

LOCALITY. Turkey: Aegaean Islands ("In *Archipelago* sylvestris"), July 1761.
MATERIAL. *Forsskål* no specimen found.

ONAGRACEAE

Epilobium angustifolium *L.* 1753: 347; Fl. Europ. 2: 309 (1968).
E. tetragonum sensu Forssk. 1775: XXV No. 185, non L.
LOCALITY. Turkey: Büyükdere ("Bj" = Buiuchtari), 1762.
MATERIAL. *Forsskål* 513 (C — 1 sheet, microf. 138: II. 8, III. 1).
NOTE. It seems reasonable to assume this identification, as Forsskål describes his 'E. tetragonum': "foliis non oppositis, sed alternis".

Ludwigia stolonifera (*Guill. & Perr.*) *Raven* in Reinwardtia 6: 390 (1963); Fl. Palaest. 2: 373 (1972).
Jussiaea diffusa Forssk. 1775: 210 (LXVI No. 235; Cent. VIII No. 45); Christensen 1922: 34.
J. repens L. - Tackholm 1974: 380 var. *diffusa* (Forssk.) Brenan in Kew Bull. 8: 171 (1953).
L. diffusa (Forssk.) Greene, Fl. Francisc. 1: 227 (1891), non Buch.-Ham. (1924).
VERNACULAR NAME. *Forgaa, Fraekal, Fôrgâa, Fraekahl* (Arabic).
LOCALITY. Egypt: Rashid ("Rosettae ad ripam Nili"), 31 Oct. 1761.
MATERIAL. *Forsskål* 511 (C — 1 sheet, type of *J. diffusa*, microf. 58: III. 1, 2); 514 (C — 1 sheet, type of *J. diffusa*, microf. 58: II. 7, 8).

OXALIDACEAE

Oxalis corniculata *L.* 1753: 435; Forssk. 1775: XIII No. 215; LXVI No. 244; CXII No. 93; Fl. Turk. 2: 490 (1967); Fl. Europ. 2: 192 (1968); Tackholm 1974: 293, pl. 94A; Schwartz 1939: 116.
LOCALITY. Malta, Turkey, Egypt or Yemen 1761–63.
MATERIAL. *Forsskål* 1540 (C — 1 sheet ex herb. Hornemann, microf. nil); 1541 (C — 1 sheet ex herb. Schumacher, microf. 75: II. 3, 4).
Forsskål s.n. (LD — 1 sheet ex herb. Retzius).

PAPAVERACEAE

Hypecoum aegyptiacum (*Forssk.*) *Aschers. & Schweinf.*, Ill. Fl. Egyp.: 37 (1887); Fl. Palaest. 1: 235, t. 346 (1966); Tackholm 1974: 162, pl. 46B.
Mnemosilla aegyptiaca Forssk. 1775: 122 (LXIX No. 314; Cent. IV No. 76); Christensen 1922: 23.
LOCALITY. Egypt: Alexandria ("Alexandriae in peninsula Ras ettên"), 1 Apr. 1762.
MATERIAL. *Forsskål* 702 (C — 1 sheet type of *M. aegyptiaca*, microf. 70: I. 3, 4); 703 (C — 1 sheet, type of *M. aegyptiaca*, microf. 70: I. 5, 6); 1491 (C — 1 sheet ex herb. Schumacher, type of *M. aegyptiaca*, microf. 70: I. 7, 8).
Forsskål s.n. (LD - 1 sheet ex herb. Retzius, type of *M. aegyptiaca*).

Roemeria hybrida (*L.*) *DC.*, Reg. Veg. Syst. Nat. 2: 92 (1821); Fl. Europ. 1: 251 (1964).
Chelidonium hybridum L. 1753: 506; Forssk. 1775: VII No. 120.
LOCALITY. France: Marseille ("Estac"), May 1761.
MATERIAL. *Forsskål* 704 (C — 1 sheet, microf. 26: I. 3, 4).

subsp. **dodecandra** (*Forssk.*) *Durand & Barratte*, Fl. Lib. Prodr.: 6 (1910); Tackholm 197: 154.
Chelidonium dodecandrum Forssk. 1775: 100 (LXVII No. 277; Cent. IV No. 8); Christensen 1922: 20.
Roemeria dodecandra (Forssk.) Stapf, Denkschr. Akad. Wien 51: 295 (1886).
VERNACULAR NAME. *Ridjlet el ghrâb* (Arabic).
LOCALITY. Egypt: Cairo ("In desertis Kahirinis"), 1762.
MATERIAL. *Forsskål* 705 (C — 1 sheet type of *C. dodecandra*, microf. 25. III. 5, 6); 706 (C —1 sheet, type of *C. dodecandra*, microf. 26: I. 1, 2); 1098 (C — 1 sheet ex herb. Schumacher, type of *C. dodecandra*, microf. 25: III. 7, 8).
Forsskål s.n. (LD — 1 sheet ex herb. Retzius, type of *C. dodecandra*).

PASSIFLORACEAE

Adenia venenata *Forssk.* 1775: 77 (CX No. 245; Cent. III No. 45); Deflers 1889: 139; Christensen 1922: 18; Schwartz 1939: 171.
VERNACULAR NAME. *Aden* (Arabic).
LOCALITY. Yemen: Al Hadiyah ("Hadîe"), Mar. 1763.
MATERIAL. *Forsskål* 655 (C — 1 sheet with field label "C.1", holotype of *A. venenata*, microf. 2: III. 3, 4).
NOTE. *Adenia venenata* is the type species of Forsskål's genus *Adenia*.

PEDALIACEAE

Sesamum indicum *L.* 1753: 634; Forssk. 1775: 113 (LXVIII No. 306; CXV No. 380; Cent. IV No. 46); Christensen 1922: 22; Schwartz 1939: 247; Fl. Turk. 6: 196 (1978).
VERNACULAR NAME. *Djyldjylân semsem* (seeds, Egypt), *Salît* (oil, Arabic).
LOCALITY. Egypt or Yemen ("Ubique in Arabia cultum"), 1763. (see Note).

S. orientale L. — Forssk. 1775: XXIX No. 284, LXVIII No. 305.
VERNACULAR NAME. *Sišami* (Greek).
LOCALITY. Turkey: Bozcaada I. ("Td" = Tenedos), July 1761.
LOCALITY. Egypt: Cairo garden ("Ch"), 1762.
MATERIAL. *Forsskål* 364 (C - 1 sheet, microf. 98: II. 5, 6); 366 (C — 1 sheet, microf. 98: II. 3, 4).
NOTE. It is not possible to localise the origin of the two specimens in Herbarium Forsskålii at C.

PHYTOLACCACEAE

Phytolacca americana *L.* 1753: 441; Forssk. 1775: LXVI No. 246; Meikle 1985: 1395.
LOCALITY. Egypt: Cairo ("Cs"), 1761–62.
MATERIAL. *Forsskål* 1579 (C — 1 sheet ex herb. Hornemann, microf. nil).

PLANTAGINACEAE

Plantago coronopus *L.* 1753: 115; Forssk. 1775: IV No. 44, XX No. 64; Fl. Europ. 4: 40 (1976); Fl. Turk. 7: 508 (1982).

LOCALITY. France: Marseille ("Estac"), 9 May-3 June 1761; Turkey: Dardanelles on label ("Imr" = Imros in book), July 1761.
MATERIAL. *Forsskål* 249 (C — 1 sheet partially — plant without long infl. —microf. 81: III. 1, 2); 252 (C - 1 sheet with field label 'Dardanelli', microf. 80: III. 1, 2); 1583 (C — 1 sheet, microf. 80: III. 3, 4).

Plantago crassifolia *Forssk.* 1775: 31 (LXII No. 99; Cent. II No. 6); Christensen 1922: 14; Tackholm 1974: 514; Fl. Palaest. 3: 223 (1978).
LOCALITY. Egypt: Alexandria ("Alexandriae"), 1761.
MATERIAL. *Forsskål* 261 (C — 1 sheet, type of *P. crassifolia* microf. 80: III. 5, 6).

Plantago cylindrica *Forssk.* 1755: 31 (LXII No. 97; Cent. II No. 4); Christensen 1922: 14; Tackholm 1974: 516; Fl. Palaest. 3: 226 (1978).
LOCALITY. Egypt: Cairo ("Cd"), 1761–62.
MATERIAL. *Forsskål* 255 (C — 1 sheet, type of *P. cylindrica*, microf. 80: III. 7, 8); 257 (C — 1 sheet, type of *P. cylindrica*, microf. 81: I. 3, 4); 260 (C —1 sheet, type of *C. cylindrica*, microf. 81: I, 1, 2); 1584 (C — 1 sheet ex herb. Hornemann, type of *P. cylindrica*, microf. nil).

Plantago lagopus *L.* 1753: 114; Fl. Europ. 4: 43 (1976); Fl. Turk. 7: 514 (1982).
LOCALITY. Unspecified.
MATERIAL. *Forsskål* 1585 (C — 1 sheet, microf. 138: I. 3).

Plantago lanceolata *L.* 1753: 113; Forssk. 1775: IV No. 43, XX No. 63, LXII No. 93, CV No. 91; Tackholm 1974: 511; Fl. Europ. 4: 42 (1976); Fl. Turk. 7: 513 (1982).
LOCALITY. France: Marseille ("Estac"), 9 May-3 June 1761; Turkey: Istanbul ("Bujuchtari") July 1761; Egypt: Cairo ("Cs"), 1761–62.
MATERIAL. *Forsskål* 263 (C — 1 sheet, microf. 81: II. 1, 2).

Plantago major *L.* 1753: 112; Forssk. 1775: XX No. 65; LXII No. 92; Tackholm 1974: 514; Fl. Europ. 4: 39 (1976); Fl. Turk. 7: 507 (1982).
VERNACULAR NAME. *Pen anevzon* (Greek), *Lissan el bamal* (Arabic).
LOCALITY. Turkey: Dardanelles ("Brg." = Borghas), July 1761; Egypt: Cairo ("Cs"), 1961–62.
MATERIAL. *Forsskål* 262 (C — 1 sheet, microf. 81: II. 3, 4).

Plantago ovata *Forssk.* 1775: 31 (LXII No. 98; Cent. II No. 5); Christensen 1922: 14; Tackholm 1974: 516; Fl. Palaest. 3: 226 (1978).
var. **ovata**.
VERNACULAR NAME. *Lokmet ennadji* (Arabic) — attributed to *P. decumbens* on p. 30.
LOCALITY. Egypt: Alexandria ("Alexandriae"), Apr. 1762.
MATERIAL. *Forsskål* 249 partly (C — 1 sheet inflorescence only, ? type, microf. 81: III. 1, 2); 250 (C — 1 sheet, type of *P. ovata*, microf. 81: II. 7, 8); 253 (C — 1 sheet, type of *P. ovata*, microf. 81: II. 5, 6).
NOTE. No. 249 is a mixed sheet with the leafy plant P. *coronopus*, named by K. Rahn 1969.

var. **decumbens** (*Forssk.*) *Zohary* in Palest. Journ. Bot. Jerusalem ser., 1: 227 (1938); Tackholm 1974: 516; Fl. Palaest. 3: 227 (1978).
P. decumbens Forssk. 1775: 30 (LXII No. 96; Cent. II No. 3); Christensen 1922: 13.

VERNACULAR NAME. *Senaemae, Lókmet en nági* (Arabic). On p. LXII the latter name is attributed to *P. ovata*.
LOCALITY. Egypt: Cairo. ("Kahirinis frequens"), 1761–62.
MATERIAL. *Forsskål* 254 (C — 1 sheet, type of *P. decumbens*, microf. 81: I. 7, 8); 259 (C — 1 sheet, type of *P. decumbens*, microf. 81: I. 5, 6).

Plantago scabra *Moench*, Meth.: 461 (1794); Fl. Turk. 2: 518 (1982).
P. psyllium L. nom. ambig., Forssk. 1775: IV No. 45, XX No. 62.
LOCALITY. France: Marseille ("Estac"), 9 May–3 June 1761; Turkey: Borghas ("Borghàs fons"), July 1761.
MATERIAL. *Forsskål* 256 (C — 1 sheet, microf. 81: III. 3, 4).
NOTE. Named and labelled by K. Rahn, 1964, as *P. indica* L. which is a nom. illegit. In Fl. Europ. 4: 43 (1976) it is under *P. arenaria* Waldst. & Kit.

Plantago serraria *L.*, Syst. Nat. ed. 10, 2: 896 (1759); Forssk. XIII No. 16; Fl. Europ. 4: 40 (1976).
LOCALITY. Malta, June 1761.
MATERIAL. *Forsskål* 251 (C — 1 sheet, microf. 81: III. 5, 6); 258 (C — 1 sheet, microf. 81: III. 7, 8).

PLATANACEAE

Platanus orientalis *L.* 1753: 999; Forssk. 1775: LXXV No. 487; Pl. Palaest. 2: 1, pl. 1 (1972).
VERNACULAR NAME. *Schinar* (Arabic).
LOCALITY. Egypt: Cairo garden (Ch = "Cairo plantae hortenses"), 1762.
MATERIAL. *Forsskål* 473 (C — 1 sheet, microf. 82: I. 1, 2).

PLUMBAGINACEAE

Goniolimon incanum (*L.*) *Hepper* in Fl. Turk. 10: 212 (1988).
Statice incana L. 1767: 59 ("Arabia"); Vahl 1790: 25 ('Egypt').
LOCALITY. Turkey (see below).
MATERIAL. *Forsskål* (cult. Uppsala, LINN 395: 7, microf. 210: I. 3).

S. speciosa sensu Forssk. 1775: XXIV No. 159 non L.
LOCALITY. Turkey: Gökçeada ("Imros, ad Dardanellos"), July 1761.
MATERIAL. *Forsskål* 523 (C — 1 sheet, microf. 104: I. 7, 8); 527 (C — 1 sheet, microf. 104: II. 1, 2); 528 (C — 1 sheet, microf. 104: I. 5, 6).
Forsskål s.n. (LD — 1 sheet ex herb. Retzius).
NOTE. This species has hitherto been known as *Goniolimon collinum* (Griseb.) Boiss. —Fl. Turk. 7: 477 (1982).

Limoniastrum monopetalum (*L.*) *Boiss.* in DC., Prodr. 12: 689 (1848); Tackholm 1974: 403.
Statice monopetala L. 1753: 276; Forssk. 1775: 59 (LXIV No. 190; Cent. II No. 97); Christensen 1922: 17.
VERNACULAR NAMES. *Saetj, Zaetja* (Arabic).
LOCALITY. Egypt: Alexandria ("*Alexandriae* in desertis circa catacombas"), 1 Apr. 1762.
MATERIAL. *Forsskål* 1762 (C — 1 sheet, microf. 103: III. 7, 8).

Limonium axillare (*Forssk.*) *Kuntze*, Rev. Gen. Pl.: 395 (1891); Tackholm 1974: 403.
Statice axillaris Forssk. 1775: 58 (CIX No. 225; Cent. II No. 96); Vahl 1790: 26 t. 9 f. 9; Schwartz 1939: 181.
LOCALITY. Yemen: Al Luhayyah ("*Lohojae*"), Jan. 1763.
MATERIAL. *Forsskål* 522 (C — 1 sheet with field label "Statice * axillaris. Lohajae. Lo. 13", type of *S. axillaris*, microf. 103: II. 7, 8).

Limonium cylindrifolium (*Forssk.*) *Kuntze*, Rev. Gen. Pl.: 395 (1891).
Statice cylindrifolia Forssk. 1775: 59 (CIX No. 226; Cent. II No. 98); Vahl 1790: 26 t. 10; Christensen 1922: 17; Schwartz 1939: 181.
LOCALITY. Yemen: ? Mukham ("Mochhae in littoral argillaceo."), Apr–May 1763.
MATERIAL. *Forsskål* 519 (C — 1 sheet, type of *S. cylindrifolia*, microf. 103: III. 1, 2); 520 (C — 1 sheet, type of *S. cylindrifolia*, microf. 103: III. 3, 4); 1761 (C — 1 sheet ex herb. Liebmann, type of *S. cylindrifolia*, microf. 103: III. 5, 6).
Forsskål s.n. (B.M. — 1 sheet ex herb. Banks, type of *S. cylindrifolia*, microf. nil.).

Limonium ferulaceum (*L.*) *O. Kuntze*, Rev. Gen. Pl. 2: 395 (1891); Fl. Europ. 3: 41 (1972).
LOCALITY. Unspecified, prob. France.
MATERIAL. *Forsskål* 1411 (C — 1 sheet ex herb. Hornemann, microf. 133: II. 1, 2).

Limonium gmelinii (*Willd*) *Kuntze*, Rev. Gen. Pl. 2: 395 (1891); Fl. Turk. 7: 95 (1982).
Statice speciosa sensu Forssk. 1775: XXIV No. 159, LXV No. 192, non L.
LOCALITY. Turkey: Gökçeada, Dardanelles, Izmir ("Imroz, Dardanelli, Smirna"), July 1761.
MATERIAL. *Forsskål* 525 (C — 1 sheet, microf. 104: I. 3, 4); 526 (C — 1 sheet, microf. 104: I. 1, 2).

Limonium pruinosum (*L.*) *Kuntze*, Rev. Gen. Pl.: 396 (1891); Tackholm 1974: 403.
Statice pruinosa L. 1767: 59; Vahl 1790: 26.
S. aphylla Forssk. 1775: 60 (LXV No. 191; Cent. II No. 99).
LOCALITY. Egypt: Alexandria ("Alexandriae ad Catacombas copiose"), 1 Apr. 1762.
MATERIAL. *Forsskål* 524 (C — 1 sheet, type of *S. aphylla*, microf. 103: II. 3, 4); 1760 (C — 1 sheet ex herb. Vahl, type of *S. aphylla*, microf. 103: II. 5, 6).
Forsskål s.n. (BM — 1 sheet ex herb. Banks, type of *S. aphylla*, microf. nil).

Limonium sp. indet. 1
LOCALITY. Unspecified.
MATERIAL. *Forsskål* 521 (C — 1 sheet, microf. 133: II. 3, 4).
NOTE. This is a sterile specimen.

Limonium sp. indet. 2
LOCALITY. Unspecified.
MATERIAL. *Forsskål* 1836 (C — 1 sheet ex herb. Vahl, microf. 133: II. 5, 6).
NOTE. Although this is good material the difficulty of identifying it is increased by the lack of provenance. The name on the reverse is "Statice cordata" which refers to a Rhodes endemic now known as *Limonium psilocladum* (Boiss.) O. Kuntze, which has spaced flowers on the rhachis, unlike this plant.

POLYGALACEAE

Polygala abyssinica *R. Br. ex Fresen.* in Mus. Senchenb. 2: 273 (1837); Schwartz 1939: 132.
P. paniculata sensu Forssk. 1775: CXVII No. 429, non L.
LOCALITY. Yemen: Bughah ("Boka"), Mar. 1763.
MATERIAL. *Forsskål* 611 (C — 1 sheet with field label "Polygala vulgaris ? panniculata? inter Mokhaja & Boka", microf. 83: I. 3, 4); 1592 (C — 1 sheet ex herb. Schumacher, microf. 83: I. 5, 6).
NOTE. On the back of No. 611 Vahl suggested the name "P. genistifolia".

Polygala tinctoria *Vahl* 1790: 50; Deflers 1889: 112; Schwartz 1939: 131.
P. bracteolata Forssk. 1775: 213 (CXVII No. 430; Cent. VIII No. 65); non L., Christensen 1922: 34.
VERNACULAR NAME. *Schadjaret el houer* (Arabic).
LOCALITY. Yemen: Al Hadiyah ("Hadîe"), Mar. 1763.
MATERIAL. *Forsskål* 497 (C — 1 sheet with field label "Polygala bracteolata. inter Ersch & Alûdja", type of *P. bracteolata*, microf. 82: III. 7, 8); 1593 (C — 1 sheet ex herb. Vahl, type of *P. bracteolata*, microf. 83: I. 1, 2).

POLYGONACEAE

Emex spinosus (*L.*) *Campd.*, Monogr. Rumex: 58, t. 1, f. 1 (1819); Schweinfurth 1896: 153; Tackholm 1974: 61, pl. 11A.
Rumex spinosus L. — Forssk. 1775: LXV No. 213.
R. glaber Forssk. 1775: 75 (Cent. III No. 40); Christensen 1922: 18.
VERNACULAR NAME. *Figl el djebbel, Ságarat el aguz, Raensah* (Arabic).
LOCALITY. Egypt: Cairo ("In desertis, Káhirinis"), 1761–62.
MATERIAL. *Forsskål* 1244 (C — 1 sheet, microf. 133: II. 7, 8).
NOTE. This specimen was not seen by Christensen.

Polygonum equisetiforme *Sibth. & Sm.*, Fl. Graecae Prodr. 1: 266 (1809); Tackholm 1974: 64, pl. 10B.
P. marit? sensu Forssk. 1775: LXV No. 219, non L.
VERNACULAR NAME. *Gaeddaba* (Arabic).
LOCALITY. Egypt: Alexandria, Cairo, Abu Qir, 1761–62.
MATERIAL. *Forsskål* 1597 (C — 1 sheet with field label "Polygonum caude fruticos. procumbente. maritimum? As 14", microf. 83: II. 6, 7); 1598 (C — 1 sheet with field label "D.3. Polygonum maritimum. Bükier", ex herb. Liebrmann, microf. 83: III. 2, 3); 1599 (C — 1 sheet with field label "Polygonum vere maritim. Alex. vorna Calesch", microf. 83: II. 8, III. 1).

Polygonum maritimum *L.* 1753: 361; Forssk. 1775: XXV No. 190; Fl. Turk. 2: 276 (1967).
LOCALITY. Turkey: Dardanelles ("Dd. Ecl."), July 1761.
MATERIAL. *Forsskål* 1595 (C — 1 sheet, microf. 83: II. 1, 2, 3); 1596 (C — 1 sheet, microf. 83: II. 4, 5).

Polygonum cf. **pulchellum** *Lois.* in Mém. Soc. Linn. Soc. Paris 6: 441 (1827); Fl. Turk. 2: 279 (1967).
P. aviculare sensu Forssk. 1775: XXV No. 191, non L.
LOCALITY. Turkey: Dardanelles ("Dd" = "Dardanellae"), July 1761.

MATERIAL. *Forsskål* 1600 (C — 1 sheet with field label "Polygonum aviculare stipulae setiferae. Bo. 28", microf. 83: I. 7, 8).

Polygonum salicifolium *Brouss. ex Willd.*, Enum. Hort. Berol. 1: 428 (1809); Fl. Turk. 2: 273 (1967); Tackholm 1974: 62, pl. 10A.
P. persicaria sensu Forssk. 1775: 81 (XXV No. 192, LXV No. 220; Cent. III. 52), non L.; Christensen 1922.
LOCALITY. Turkey: Istanbul, Büyükdere ("Bujuchtari"), July 1761; Egypt: Alexandria (or on field label "Rosette"), 1761–62.
MATERIAL. *Forsskål* 1594 (C — 1 sheet ex herb. Hofman Bang with field label No. "Polygonum stipulis. RO. 40", microf. 83: III. 4, 5).
NOTE. This specimen in Herb. Forssk. has been referred to *P. persica* but that species has ovate-lanceolate leaves as described there, while *P. salicifolium* has very narrow leaves. *Forsskål* s.n. (LD — 1 sheet ex herb. Retzius).

Rumex aegyptiacus *L.* 1753: 335; Tackholm 1974: 65.
R. aegypt. vel comosus Forssk. 1775: 76 (LXV No. 211; Cent. III. 42); Christensen 1922: 18.
LOCALITY. Egypt: Cairo ("Káhirae"), 1761–62.
MATERIAL. *Forsskål* 1642 (C — 1 sheet, type of *R. comosus*, microf. 88: III. 5, 6).

Rumex dentatus *L.* 1771: 226; Tackholm 1774: 65, pl. 11B.
R. obtusifolius sensu Forssk. 1775: LXV No. 212, non L.
VERNACULAR NAME. *Humaeid* (Arabic).
LOCALITY. Egypt: Cairo ("Cs"), 1761–62.
MATERIAL. *Forsskål* 1643 (C — 1 sheet, microf. 88: III. 7, 8).

Rumex lachanus *Forssk.* 1775: 209 (XXIV No. 180; Cent. VIII No. 42); Christensen 1922: 34.
VERNACULAR NAME. *Lachano* (Greek).
LOCALITY. Turkey: Istanbul, Belgrad ("Bg" = Belgrad, "Natoliae"), July 1761.
MATERIAL. *Forsskål* no type specimen found.
NOTE. This remains an obscure species not cited in Fl. Turkey (1967). The name should not be adopted if it upsets widely accepted nomenclature.

Rumex nervosus *Vahl* 1790: 27; Deflers 1889: 195; Schweinfurth 1896: 153; Schwartz 1939: 32.
R. persicarioides Forssk. 1775: 76 (CX No. 246; Cent. III No. 41); Christensen 1922: 18, non L. (1753).
VERNACULAR NAME. *Öthrob* (Arabic).
LOCALITY. Yemen: Al Hadiyah ("In montibus Hadiensibus" "Bolghose" on label), Mar. 1763.
MATERIAL. *Forsskål* 1639 (C — 1 sheet, type of *R. persicarioides*, microf. 89: I. 1, 2); 1640 (C — 1 sheet, type of *R. persicarioides*, microf. 89: I. 3, 4); 1641 (C — 1 sheet with field label "Rubus * persicarioides. Örthrob. Bolghose", type of *R. persicarioides*, microf. 89: I. 5, 6).

Rumex pictus *Forssk.* 1775: 77 (LXV No. 215; Cent. III No. 43); Christensen 1922: 18; Fl. Palaest. 1: 62, pl. 70 (1966); Tackholm 1974: 67.
VERNACULAR NAME. *Hemsis* (Arabic).
LOCALITY. Egypt: Rashid ("Rosettae"), 2–6 Nov. 1761.
MATERIAL. *Forsskål* 1644 (C — 1 sheet with field label "Ro. 41 Rumex fructus alatus venoso", type of *R. pictus*, microf. 89: I. 7, 8).

Rumex vesicarius *L.* 1753: 336; Forssk. 1775: LXV No. 214; Tackholm 1974: 67.
VERNACULAR NAME. *Humbaejt* (Arabic).
LOCALITY. Egypt: Cairo ("Cd"), 1761–62.
MATERIAL. *Forsskål* 1645 (C — 1 sheet ex herb. Hofman Bang, microf. 89: II. 3, 4); 1646 (C — 1 sheet with field label "Rumex vesciculosus. Caid Bey. Ca. 167", microf. 89: II. 1, 2).

PORTULACACEAE

Portulaca quadrifida *L.* 1767: 73; Deflers 1889: 412; Schweinfurth 1896: 171; Schwartz 1939: 52.
NOTE. Linnaeus described this species from plants raised from seeds from 'Egypt': probably Yemen by Forsskål; no type specimen is in Herb. LINN.
P. linifolia Forssk. 1775: 92 (CXII No. 299; Cent. III No. 79); Christensen 1922: 19.
VERNACULAR NAMES. *Mortah, Koraat errai* (Arabic).
LOCALITY. Yemen: Wadi Sordud ("Surdûd"), Feb. 1763.
MATERIAL. *Forsskål* 539 (C — 1 sheet, type of *P. linifolia*, microf. 84: II. 3, 4); 542 (C — 1 sheet with field label "Portulaca * linifolia. Circa Surdud", type of *P. linifolia*, microf. 84: II. 5, 6).

P. imbricata Forssk. 1775: 92 (CXII No. 300; Cent. III. No. 80); Christensen 1922: 19.
VERNACULAR NAMES. *Rozzi, Örnuba* (Arabic).
LOCALITY: Yemen: Bayt al Faqih ("Beit el Fakîh"), Feb. 1763.
MATERIAL. *Forsskål* 540 (C — 1 sheet with field label "Portulaca * imbricata. inter Kaxla (?) & Ödein", type of *P. imbricata*, microf. 84: II. 1, 2); 548 (C —1 sheet, type of *P. imbricata*, microf. 84: I. 7, 8).
NOTE. *P. hareschta* Forssk. 1775: CXII No. 301 nomen nudum, 92 (No. 81) generic name, is obscure and may be the same as the above, although Forsskål noted a different arabic name *hareschtam rai* at Wadi Mawr ("Môr") in Yemen.

Talinum portulacifolium (*Forssk.*) *Aschers. ex Schweinf.* 1896: 172; Fl. Zambes. 1: 372 (1961).
Orygia portulacifolia Forssk. 1775: 103 (CXIV No. 342; Cent. IV No. 19); Deflers 1889: 140; Christensen 1922: 20.
Portulaca cuneifolia Vahl 1790: 33.
Talinum cuneifolium (Vahl) Willd., Sp. Pl. 2: 864 (1800); Schwartz 1939: 52.
VERNACULAR NAME. *Hörudj, Hörudjrudj* (Arabic).
LOCALITY. Yemen: Wadi Surdud, Al Hadiyah ("Surdûd. Hadîe"), 1763.
MATERIAL. *Forsskål* no type specimen found.
NOTE. The specimen in Herbarium Forsskålii at Copenhagen labelled by Ascherson in 1881 as *O. portulacifolia* is in fact *O. decumbens* Forssk. of which Christensen indicates no material. Now the situation is reversed, there being no material of *O. portulacifolia*.

PRIMULACEAE

Anagallis arvensis *L.* 1753: 148; Tackholm 1974: 399, pl. 137A; Fl. Palaest. 3: 6 (1978) as var. *latifolia* (L.) Lange.
A. latifolia L. — Forssk. 1775: LXII No. 120.
LOCALITY. Egypt; Cairo ("Cs" = Cairo spontaneae), 1762.
MATERIAL. *Forsskål* 352 (C — 1 sheet, microf. 6: II. 1, 2); 938 (C — 1 sheet ex herb. Hornemann, microf. 6: I. 7, 8).
Forsskål s.n. (LD — 1 sheet ex herb. Retzius).

Lysimachia nummularia *L.* 1753: 148; Forssk. 1775: XXI No. 100; Fl. Turk. 6: 138 (1978).
LOCALITY. Turkey: Istanbul, Belgrad forest ("ad Belgrad." on sheet), 30 July–8 Sept. 1761.
MATERIAL. *Forsskål* 1446 (C — 1 sheet, microf. 135: III. 7, 8).

Primula verticillata *Forssk.* 1775: 42 (CVI No. 115; Cent. II No. 38); Vahl 1790: 15, t. 5; Deflers 1889: 161; Christensen 1922: 15; Schwartz 1939: 179.
LOCALITY. Yemen: Kusma ("In monte Kurma ad rivulos aquarum"), Mar. 1763.
MATERIAL. *Forsskål* 353 (C — 1 sheet, type of *P. verticillata*, microf. 84: III. 3, 4); 354 (C — 1 sheet, type of *P. verticillata*, microf. 84: III. 1, 2); 1603 (C — 1 sheet ex herb. Schumacher, type of *P. verticillata*, microf. 84: III. 5, 6).
Forsskål s.n. (BM — 1 sheet ex herb. Banks, type of *P. verticillata*, microf. nil).

RANUNCULACEAE

Clematis vitalba *L.* 1753: 544; Forssk. 1775: VII No. 128; Fl. Europ. 1: 221 (1964), 1: 267 (1993).
LOCALITY. France: Marseille ("Estac"), May 1761.
MATERIAL. *Forsskål* 700 (C — 1 sheet with field label "Em. 1", microf. 28: I. 1, 2).
Forsskål s.n. (LD — 1 sheet ex herb. Retzius).

Clematis simensis *Fresen.* in Mus. Senckenb. 2: 267 (1837); Schweinfurth 1896: 177 (as var.); Schwartz 1939: 59.
C. vitalba sensu Forssk 1775: 212 (CXIV No. 348; Cent. VIII No. 57), non L.; Christensen 1922: 34.
VERNACULAR NAME. *Scheradj* (Arabic).
LOCALITY. Yemen: Kusma ("Kurmae"), 1763.
MATERIAL. *Forsskål* no specimen found.
NOTE. Since *C. vitalba* does not occur in Yemen, it is likely that Forsskål actually recorded *C. simensis*.

Consolida aconiti (*L.*) *Lindley* in J. Roy. Hort. Soc. 6: 55 (1851); Fl. Turk. 1: 122 (1965).
Delphinium aconiti L. 1767: 77; Vahl 1790: 40.
LOCALITY. Turkey: Dardanelles ("Habitat in Dardanella"), July 1761.
MATERIAL. *Forsskål* s.n. (LINN — 1 sheet, presumably holotype grown from seed sent by Forsskål, microf. 357: I. 1).

Aconitum monogn. Forssk. 1775: XXVII No. 248.
LOCALITY. Turkey: Dardanelles ("Dd" = Dardanelli), July 1761.
MATERIAL. *Forsskål* 913 (C — 1 sheet with field label ex herb. Vahl, microf. 2: II. 3, 4); 914 (C — 1 sheet ex herb. Vahl, microf. 2: II. 5, 6).

Delphinium peregrinum *L.* 1753: 531; Fl. Turk. 1: 117 (1965).
D. grandiflorum sensu Forssk. 1775: 212 (XXVII No. 251; Cent. VIII No. 56), non L.; Christensen 1922: 34.
D. forskolii Reichenb., Illustr. Sp. Acon.: 5, t. 5 [68] (1827).
VERNACULAR NAME. *Agrio Zompolia* (Greek).
LOCALITY. Turkey: Dardanelles ("Ad Dardanellos"), July 1761.
MATERIAL. *Forsskål* 1218 (C — 1 sheet ex herb. Vahl, type of *D. forskalii*, microf. 38: I. 1, 2).
Forsskål s.n. (LD — 1 sheet ex herb. Retzins, type of *D. forskolii*).

NOTE. In DC., Syst. Veg. 1: 349 (1817) *D. grandiflorum* and Forsskål's specimen are tentatively and erroneously included with the Egyptian *D. nanum* DC. there described. Reichenbach, however, figures the specimens on the C sheet cited above as *D. forskolii* which seems to be no different from *D. peregrinum*.

Ranunculus chaerophyllos sensu Forssk. 1775: 102 (CXIV No. 349; Cent. IV No. 17), non L.; Christensen 1922: 20.
LOCALITY. Yemen: Kusma ("Kurmae"), 1763.
MATERIAL. *Forsskål* no specimen found.
NOTE. In the absence of a specimen this remains obscure.

Ranunculus multifidus *Forssk.* 1775: 102 (CXIV No. 350; Cent. IV No. 18); Schweinfurth 1896: 178; Christensen 1922: 20; Schwartz 1939: 59.
R. forskoehlii DC., Reg. Veg. Syst. Nat. 1: 303 (1817); Schwartz 1939: 59.
LOCALITY. Yemen: Taizz ("In fosea prope urbem *Taaes*"), Apr. 1763.
MATERIAL. *Forsskål* 698 (C — 1 sheet with field label, type of *R. multifidus*, microf. 86: I. 7, 8).
NOTE. The field label in pencil and the full text is difficult to read.

Ranunculus sceleratus *L.* 1753: 551; Forssk. 1775: XXVIII No. 256, LXVIII No. 288; Tackholm 1974: 139, pl. 35A; Fl. Turk. 1: 191 (1965).
VERNACULAR NAME. *Zagblil* (Arabic).
LOCALITY. Turkey: Izmir ("Smirna"), 30 June–10 July 1761. Egypt: Cairo ("Cs"), 1761–62.
MATERIAL. *Forsskål* 697 (C — 1 sheet, microf. 86: II. 1, 2).
Forsskål s.n. (LD — 1 sheet ex herb. Retzius).

Thalictrum foetidum *L.* 1753: 545; Fl. Europ. 1: 241 (1964), 1: 291 (1993).
LOCALITY. ? France.
MATERIAL. *Forsskål* 699 (C — 1 sheet, microf. 134: I. 7, 8).

RESEDACEAE

Caylusea hexagyna (*Forssk.*) *M.L. Green* in Nom. Prop. Brit. Bot.: 102 (1929); Taylor in Kew Bull. 13: 285 (1958); Fl. Palaest. 1: 339, pl. 494 (1966); Tackholm 1974: 211, pl. 64.
Reseda hexagyna Forssk. 1755: 92 (LXVII No. 253; Cent III No. 82); Christensen 1922: 19.
R. canescens L., Syst. Nat. ed. 12, 2: 330 (1767), non 1753; Vahl 1791: 52.
LOCALITY. Egypt, 1761–62.
MATERIAL. (LINN — 1 sheet 629.3, presumably plant grown from seed sent by Forsskål to Linnaeus, microf. 318: II. 7).
VERNACULAR NAME. *Dhenâba* (Arabic).
LOCALITY. Egypt: Cairo ("Inter rudera ad Caid Bey prope Káhiram"), 1762.
MATERIAL. *Forsskål* 1617 (C — 1 sheet ex herb. Vahl, designated lectotype by Abdallah, microf. 86: III. 7, 8); 1618 (C — 1 sheet ex herb. Schumacher, isotype, microf. 87: I. 1, 2); 1619 (C — 1 sheet ex herb. Hornemann, isotype?, microf. 134: II. 1, 2).

Reseda alba *L.* 1753: 449; Forssk. 1775: VI No. 103; XIII No. 35; XXVI No. 222; Fl. Europ. 1: 347 (1964); Fl. Turk. 1: 500 (1965); Tackholm 1974: 208.
VERNACULAR NAME. *Agricharthamo* (Greek).
LOCALITY. Marseille, ?Malta, ?Istanbul, 1761. Turkey: Bozcaada ("Tenedos"), July 1761.

MATERIAL. *Forsskål* 1612 (C — 1 sheet ex herb. Hornemann, microf. nil); 1613 (C — 1 sheet ex herb. Hornemann, "in ins. Tenedos", microf. 134: II. 7, 8).
Forsskål s.n. (BM — 1 sheet ex herb. Nolte — sterile specimen tentatively identified).

Reseda arabica *Boiss.*, Diagn. Pl. Or. Nov. 1(1): 6 (1843); Tackholm 1974: 210.
R. phyteuma sensu Forssk. 1775: LXVII No. 252, non L.
LOCALITY. Egypt: Cairo ("Cd"), 1761–62.
MATERIAL. *Forsskål* 609 (C — 1 sheet, microf. 87: II. 3, 4); 1621 (C — 1 sheet ex herb. Hornemann, microf. 87: II. 1, 2).

Reseda decursiva *Forssk.* 1775: LXVI No. 250; Christensen 1922: 36; Fl. Palaest. 1: 332, pl. 484 (1966); Tackholm 1974: 208.
R. tetragyna Forssk. 1775: 92 (LXVII No. 254; Cent. III No. 83); Christensen 1922: 19.
VERNACULAR NAME. *Romaejhh* (Arabic).
LOCALITY. Egypt: Alexandria ("Alexandriae"), Apr. 1762.
MATERIAL. *Forsskål* 606 (C — 1 sheet type of *R. decursiva* and *R. tetragyna*, microf. 86: III. 3, 4); 610 (C — 1 sheet, type of *R. decursiva*, microf. 86: III. 5, 6); 1616 (C — 1 sheet ex herb. Hornemann, type of *C. decursiva*, microf. nil); 1615 (C — 1 sheet, plant No. 1, ex herb. Hornemann, type of *R. tetragyna*, microf. 134: II. 3, 4); 1614 (C — 1 sheet, plant No. 2, ex herb. Hornemann, microf. 134: II. 5, 6).

Reseda lutea *L.* 1753: 449; Tackholm 1974: 210.
LOCALITY. Egypt: Cairo, 1761–62.
MATERIAL. *Forsskål* 1615 (C — 1 sheet, plant No. 2, ex herb. Hornemann, microf. 134: II. 3, 4).

Reseda luteola *L.* 1753: 448; Forssk. 1775: LXVI No. 251; Tackholm 1974: 210.
VERNACULAR NAME. ?*Uaeba* (Arabic).
LOCALITY. Egypt: Cairo ("Ch"), 1761–62.
MATERIAL. *Forsskål* 608 (C — 1 sheet, microf. 87: I. 3, 4); 612 (C — 1 sheet, microf. 87: I. 7, 8); 1620 (C — 1 sheet, microf. 87: I. 5, 6).

Reseda undata *L.*, Nat. Syst. ed. 10: 1046 (1758); Fl. Europ. 1: 347 (1964).
LOCALITY. Uncertain.
MATERIAL. *Forsskål* 1614 (C — 1 sheet, plants NO. 1 & 3, ex herb. Hornemann, microf. 134: II. 5, 6).
NOTE. This is normally considered to be a Spanish species.

RHAMNACEAE

Ziziphus spina-christi (*L.*) *Desf.*, Fl. Atlant. 1: 201 (1798); Schwartz 1939: 151; Tackholm 1974: 345, pl. 117 C.
Rhamnus nabeca sensu Forssk. 1775: 204 (LXIII No. 139, CVI No. 142, Cent. III No. 14), non L.; Christensen 1922: 33.
VERNACULAR NAMES. *Nabk, Sidr, Ghasl, Aelb, Ardj, Örredj* (Arabic).
LOCALITY. Egypt: Cairo ("Ch"); Yemen (see below).

Rhamnus divaricatus Forssk. 1775: 204 (CVI No. 143; Cent. III. No 14a); Christensen 1922: 33.
LOCALITY. Yemen: Wadi Mawr ('Môr'), Apr. 1763.
MATERIAL. *Forsskål* 499 (C — 1 sheet, type of *R. divaricatus*, microf. 87: II. 5, 6).

Rhamnus rectus Forssk. 1775: 204 (CVI No. 142; Cent. 8: 14b); Christensen 1922: 33.
LOCALITY. Yemen 1763.
MATERIAL. *Forsskål* 500 (C — 1 sheet, type of *R. rectus*, microf. 87: III. 1, 2); 501 (C — 1 sheet, type of *R. rectus*, microf. 87: II. 7, 8).
Forsskål s.n. (LD — 1 sheet ex herb. Retzius).
NOTE. While *R. divaricatus* is the normal spiny plant, *R. rectus* appears to be the unarmed form sometimes known as var. *inermis* Boiss.

ROSACEAE

Alchemilla sp.
LOCALITY. Unspecified.
MATERIAL. *Forsskål* s.n. (LD — 1 sheet ex herb. Retzius).

Crataegus monogyna *Jacq.*, Fl. Austr. 3: 50 (1775); Fl. Europ. 2: 75 (1968).
Crataegus oxyac(antha) sensu Forssk. 1775: VII No. 113, non L. s.s.
LOCALITY. France: Marseille ("Estac"), May 1761.
MATERIAL. *Forsskål* 761 (C — 1 sheet, microf. 33: II. 5, 6); 762 (C — 1 sheet with original field label "M4", microf. 33: II. 3, 4).

Potentilla dentata *Forssk.* 1775: 98 (CXIII No. 329; Cent. IV No. 3); Christensen 1922: 20.
LOCALITY. Yemen: Wasab ("Uahfad"), 29 Mar. 1763.
MATERIAL. *Forsskål* 1602 (C — 1 sheet ex herb. Hornemann, type of *P. dentata*, microf. 84: II. 7, 8).
NOTE. Forsskål (p. 98) notes that the specimen was found by Niebuhr.
P. hispanica Zimm. — Schwartz 1939: 81.

Potentilla supina *L.* 1753: 497; Forssk. 1775: LXVII No. 275; Tackholm 1974: 217 (as var. *aegyptiaca* Vis).
LOCALITY. Egypt: Cairo ("Cs" = "Cairi spontanae"), 1761–62.
MATERIAL. *Forsskål* 1601 (C — 1 sheet ex herb. Hornemann, microf. nil).

Prunus spinosa *L.* 1753: 475.
LOCALITY. Unspecified.
MATERIAL. *Forsskål* 1604 (C — 1 sheet ex Schumacher, microf. 134: III. 3, 4).
NOTE. A sterile specimen identified on the reverse as 'Pyrus spinosa' but annotated by K.I. Christensen, 1987, as *Prunus spinosa* L. for Flora Hellenica.

Pyrus amygdaliformis *Vill.*, Cat. Meth. Jard. Strasb.: 323 (1807); Fl. Turk. 4: 163 (1972).
P. spinosa sensu Forssk. 1775: 211 (XXVII No. 234; Cent. VIII No. 53), non L.; Christensen 1922: 34.
VERNACULAR NAME. *Angoriza* (Greek).
LOCALITY. Turkey: Gokçeada I. and coast ("Imros and in littore Natoliae"), July 1761.
MATERIAL. *Forsskål* 765 (C — 1 sheet, microf. 85: III. 5, 6).

Rosa andegavensis *Bast.*, Essai Fl. Maine Loire: 189 (1809); Fl. Europ. 2: 10 (1968).
R. canina sensu Forssk. 1775: VII No. 115, non L.
LOCALITY. France: Marseille ("Estac"), May 1761.

MATERIAL. *Forsskål* 1631 (C — 1 sheet ex herb. Hornemann with field label "M.1", microf. 88: I. 5, 6); 1632 (C — 1 sheet, microf. 88: I. 7, 8).
NOTE. Named and labelled by F. Crépin as "Rosa canina L. in group R. andegavensis Bast." and this determination has been accepted here.

Rosa ephemera *Forssk.* 1775: XXVII No. 236; Christensen 1922: 36.
VERNACULAR NAME. *Symbadjul* (Turkish & Greek).
LOCALITY. Turkey: Istanbul ("Cph" = Constaniopolitanae hortenses).
MATERIAL. *Forsskål* no specimens found.
NOTE. A garden plant. The epithet is validated by the description: "Foliis obovatis, obtusis. Semel quovis anno floret: flos exphicatus ante meridem ruber; p. m. pallidus; altero die albus".

Rosa moschata *J. Herrm.*, Diss.: 15 (1762).
LOCALITY. Unspecified.
MATERIAL. *Forsskål* s.n. (LD — 2 sheets ex herb. Retzius).
NOTE. Named and labelled by F. Crépin, 1893.

Rubus arabicus (*Deflers*) *Schweinfurth* 1896: 204; Schwartz 1939: 81.
R. fruticosus sensu Forssk. 1775: CXIII No. 328, non L.
VERNACULAR NAMES. *Naefaes*, *Hömmaes* (Arabic).
LOCALITY. Yemen: Bolghose ("Blg"), 1763.
MATERIAL. *Forsskål* no specimen found.

Rubus sanctus *Schreber*, Icon. Descr. Pl.: 15, t. 18 (1766); Fl. Turk. 4: 33 (1972).
LOCALITY. Unspecified.
MATERIAL. *Forsskål* 1636 (C — 1 sheet ex herb. Hornemann with field label "U.2" (or "M2"?) microf. nil).
NOTE. The field label does not agree with No. 1636 being an Arabian plant.

Sorbus domestica *L.* 1753: 477; Forssk. 1775: XXVII No. 231; Fl. Turk. 4: 148 (1972).
LOCALITY. Turkey: Istanbul, Büyükdere ("Cph. Bj." = "Constantinopoli hortensis Bujuchtari"), July–Sept. 1761.
MATERIAL. *Forsskål* 763 (C — 1 sheet, microf. 102: III. 5, 6).

RUBIACEAE

Anthospermum herbaceum *L.f.*, Suppl. Pl.: 440 (1781); F.T.E.A. Rubiac.: 325 (1976).
LOCALITY. Yemen: between Mukhajah and Bughah (see label), Mar. 1763.
MATERIAL. *Forsskål* 952 (C — 1 sheet ? ex herb. Vahl, with field label "inter Mokhaja and Boka", microf. 134: III. 5, 6); 953 (C — 1 sheet, microf. 134: III. 7, 8); 954 (C — 1 sheet, microf. 135: I. 7, 8).
A. muriculatum A. Rich. (1848) — Schwartz 1939: 263.

Asperula arvensis *L.* 1753: 103; Forssk. 1775: IV No. 37; Fl. Europ. 4: 13 (1976).
LOCALITY. France: Marseille ("Estac"), 1761.
MATERIAL. *Forsskål* 1842 (C — 1 sheet, microf. nil).

Breonadia salicina (*Vahl*) *Hepper & Wood* in Kew Bull. 36: 860 (1982).
Nerium salicinum Vahl 1791: 45, partly (leaves).
Nerium foliis ternatis Forssk. 1775: CVII No. 174; Christensen 1922: 33, partly (leaves only).
= *Nerium* [*sp. without locality*] Forssk. 1775: 205 (Cent. VIII No. 18).
VERNACULAR NAME. *Daerah* (Arabic).
LOCALITY. Yemen: Al Hadiyah ("Hadîe"), 1763.
MATERIAL. *Forsskål* 236 (C — 1 sheet, type of *N. salicinum* as to leaves, microf. 70: III. 3, 4); 237 (C — 1 sheet, type of *N. salicinum* as to leaves, microf. 70: III. 5, 6).
NOTE. These are mixed sheets with flowers of *Nerium oleander* and leaves of the tree hitherto known as *Adina microcephala* (Del.) Hiern — Schwartz 1939: 262.

Coffea arabica *L.* 1753: 172; Forssk. 1775: CVI No. 128; Schwartz 1939: 262.
VERNACULAR NAME. *Bunn* (Arabic).
LOCALITY. Yemen: Al Hadiyah ("In montibus Hadiensibus" according to Christensen on the folder), Mar. 1763.
MATERIAL. *Forsskål* 1126 (C — 1 sheet, microf. 28. III. 5, 6).

Crucianella aegyptiaca *L.* 1767: 38.
LOCALITY. Egypt, 1761–62.
MATERIAL. *Forsskål* (no type specimen in herb. LINN, although described from plants raised in Uppsala).

C. herbacea Forssk. 1775: 30 (LXII No. 91; Cent. II No. 2); Tackholm 1974: 424; Fl. Palaest. 3: 238 (1978).
LOCALITY. Egypt: Alexandria ("Alexandriae"), Apr. 1762.
MATERIAL. *Forsskål* 1170 (C — 1 sheet, type of *C. herbacea*, microf. 35: I. 5, 6); 1169 (C — 1 sheet ex herb. Schumacher, type of *C. herbacea*, microf. 35: I. 3, 4).

Crucianella angustifolia *L.* 1753: 108; Forssk. 1775: IV No. 41; Fl. Europ. 4: 4 (1976).
LOCALITY. France: Marseille ("Estac"), May 1761.
MATERIAL. *Forsskål* 1168 (C — 1 sheet, microf. nil).

Crucianella latifolia *L.* 1753: 109; Fl. Europ. 4: 4 (1976).
LOCALITY. Probably Turkey: Gokçeada I. ("Imros", "Im" on field label), July 1761.
MATERIAL. *Forsskål* 1171 (C — 1 sheet with field label "Im 3", microf. nil); 1172 (C — 1 sheet, microf. nil); 1173 (C — 1 sheet, microf. 35: I. 1, 2).

Crucianella maritima *L.* 1753: 109; Forssk. 1775: LXI No. 90; Tackholm 1974: 424, pl. 147A.
LOCALITY. Egypt: Alexandria ("As") 1761.
MATERIAL. *Forsskål* 1174 (C — 1 sheet, microf. 35: I. 7, 8); 1175 (C — 1 sheet ex herb. Hornemann, microf. 135: I. 1, 2).
NOTE. There are three specimens on the sheet, two being numbered 1 "legi in marit. Gallia occ." (presumably not Forsskål's), and the small specimen numbered 2 "e collect. Forsk.").

Galium album *Miller* subsp. **pycnotrichum** (*H. Braun*) *Krendl* in Öst. Bot. Zeitschr. 114: 539 (1967); Fl. Turk. 7: 791 (1982).
G. album sensu Forssk. 1775: XX No. 69, partly.
LOCALITY. Turkey: Izmir ("Smirna"), July 1761.
MATERIAL. *Forsskål* 1288 (C — 1 sheet with field label "D. 7", microf. 49: I. 7, 8).
NOTE. The other material of Forsskål's *G. album* is *G. heldreichii*.

Galium cf. **aparine** *L.* 1753: 108; Fl. Europ. 4: 35 (1976).
LOCALITY. Unspecified, probably France.
MATERIAL. *Forsskål* 1290 (C — 1 sheet ex herb. Schumacher, microf. 135: I. 5, 6).

Galium aparinoides *Forssk.* 1775: 30 (CV No. 87; Cent. II No. 1); Vahl 1791: 30; Schweinfurth 1912: 114; Christensen 1922: 13; Schwartz 1939: 264; Verdcourt in Kew Bull. 28: 60 (1973) & F.T.E.A. Rubiac. 1: 389 (1976).
VERNACULAR NAME. *Schebette, Schöbodh bodha, Meschaerreba* (Arabic).
LOCALITY. Yemen: Bolghose, Al Hadiya ("Bulghose, Hadia, in umbrosis"); Mukhajah ('Mokhaja' on label), 1763.
MATERIAL. *Forsskål* 1291 (C — 1 sheet with field label "Galium * aparinoides. Mokhaja", lectotype selected by Verdcourt (l.c.), microf. 49: II. 5, 6); 1292 (C — 1 sheet, isolectotype, microf. 49: I. 3, 4).
Forsskål s.n. (BM — 1 sheet with field label 'Bolghose').
G. hamatum Hochst. ex A. Rich. (1847).

Galium glaucum *L.* 1753: 107; Fl. Europ. 4: 27 (1976).
LOCALITY. Unspecified, prob. France.
MATERIAL. *Forsskål* 1293 (C — 1 sheet ex herb. Vahl; microf. nil).

Galium heldreichii *Halácsy* in Öst. Bot. Zeitschr. 47: 94 (1897); Fl. Turk. 7: 793 (1982).
G. album Forssk 1775: XX No. 69 nomum nudum, partly, non Miller (1768).
LOCALITY. Turkey: Izmir ("Sm"), July 1761.
MATERIAL. *Forsskål* 1289 (C — 1 sheet, microf. 49: II. 1, 2).
NOTE. Named and labelled by F. Ehrendorfer 1981.

Galium paschale *Forssk.* 1775: 203 (XX No. 71; Cent. VIII No. 7); Vahl 1792: 29; Fl. Turk. 7: 800 (1982).
VERNACULAR NAME. *Rizari* (Greek).
LOCALITY. Turkey: Instanbul ("Graeci Constantinop."), Aug. 1761.
MATERIAL. *Forsskål* 1294 (C — 1 sheet with field label "CO. 68", type of *G. paschale*, microf. 49: III. 1, 2).
Forsskål s.n. (LD — 1 sheet ex herb. Retzius, type of *G. paschale*).

Galium rubioides *L.* 1753: 105; Fl. Europ. 4: 19 (1976).
LOCALITY. Unspecified, prob. France.
MATERIAL. *Forsskål* 1295 (C — 1 sheet ex herb. Vahl, microf. nil); 1296 (C — 1 sheet, microf. nil).

Galium tricornutum *Dandy* in Watsonia 4: 47 (1957); Fl. Europ. 4: 35 (1976).
LOCALITY. Unspecified, prob. France.
MATERIAL. *Forsskål* 1297 (C — 1 sheet ex herb. Schumacher, microf. nil).

Galium (or **Asperula**) sp.
LOCALITY. France: Marseille ("Estac" on field label), May-June 1761.
MATERIAL. *Forsskål* 1298 (C — 1 sheet, microf. 135: I. 3, 4).
NOTE. The specimen is without flowers.

Kohautia caespitosa *Schnizl.* in Flora 25, Beibl. 1: 145 (1842); Bremekamp in Verh. K. Nederl. Akad. Wet., Afd. Natuurk., ser. 2. 48(2): 104 (1952).
Hedyotis herbacea ? sensu Forssk. 1775: CV No. 88, non L.

LOCALITY. Yemen: Bayt al Faqih, 1763.
MATERIAL. *Forsskål* 1314 (C — 1 sheet with field label "Hedyotis cfr. herbacea. prope Beikelfaki", microf. 52: II. 3, 4); 1315 (C — 1 sheet, microf. 52: II. 5, 6); 1316 (C — 1 sheet ex herb. Schumacher, microf. 52: II. 7, 8).
NOTE. The plants, which all lack basal parts, have been identified as var. *schimperi* (Presl) Bremek. by D. Hillcoat, 1973.

Oldenlandia capensis *L.f.*, Suppl. Pl.: 127 (1781); Tackholm 1974: 418.
LOCALITY. ? Egypt.
MATERIAL. *Forsskål* s.n. (LD — 1 sheet ex herb. Retzius).

Pavetta longiflora *Vahl* 1794: 12; Deflers 1889: 142; Christensen 1922: 39.
Ixora occidentalis ? sensu Forssk. CV No. 90, non L.
VERNACULAR NAMES. *Schunf, Ghoraejeb* (Arabic).
LOCALITY. Yemen: Taizz ("Taaes"), Mar. 1763.
MATERIAL. *Forsskål* 1559 (C — 1 sheet, type of *P. longiflora*, microf. 58: I. 7, 8).
Forsskål s.n. (H — 1 sheet ex herb. Steven).

Pavetta villosa *Vahl* 1794: 12; Christensen 1922: 39; Bridson in Kew Bull. 41: 312 (1986).
LOCALITY. Yemen, 1763.
MATERIAL. *Forsskål* 1560 (C — 1 sheet, type of *P. villosa*, microf. 135: II. 1, 2); 1561 (C — 1 sheet, type of *P. villosa*, microf. 135: II. 3, 4).

Pentas lanceolata (*Forssk.*) *Deflers* 1889: 142; Verdcourt in Kew Bull. 6: 377 (1951) and in F.T.E.A. Rubiac.: 208 (1976).
Ophiorrhiza lanceolata Forssk. 1775: 42 (CVI No. 117; Cent. II No. 39); Christensen 1922: 15.
Manettia lanceolata (Forssk.) Vahl 1790: 12.
VERNACULAR NAME. *Laaeja* (Arabic).
LOCALITY. Yemen: Al Hadiyah ("In montibus altioribus Hadîe, alibique"), Mar. 1763.
MATERIAL. *Forsskål* 1525 (C — 1 sheet, type of *O. lanceolata*, microf. 74: I. 5, 6); 1526 (C —1 sheet ex herb. Liebmann, type of *O. lanceolata*, microf. 74: I. 7, 8); 1527 (C — 1 sheet ex herb. Hornemann, type of *O. lanceolata*, microf. nil).
Forsskål s.n. (BM — 1 sheet ex herb. Banks, type of *O. lanceolata*, microf. nil).

Rubia peregrina *L.* 1753: 109; Fl. Europ. 2: 38 (1976); Fl. Turk. 7: 860 (1982).
R. tinctor[*ium*] sensu Forssk. 1775: XX No. 68, non L.
LOCALITY. Turkey: Dardanelles ("Dd"), July 1761.
MATERIAL. *Forsskål* 1635 (C — 1 sheet, with field label "D.9", microf. 88: II. 1, 2).

Valantia hispida *L.*, Syst. Nat. ed. 10, 2: 1307 (1759); Tackholm 1974: 420, pl. 148C.
Valantia [*sp. without epithet*] Forssk. 1775: XIV No. 85.
V. aspera Forssk. 1775: LXXVII No. 546; Christensen 1922: 37.
LOCALITY. Egypt: Alexandria, 1 Apr. 1762.
MATERIAL. *Forsskål* 1816 (C — 1 sheet, microf. 111: II. 3, 4).
NOTE. Named and labelled by F. Ehrendorfer, 1981.
V. aspera is validated by the brief description: "Seminibus asperis, 4-fariam imbricatis".

Valantia muralis *L.* 1753: 1051; Fl. Europ. 4: 38 (1976).
LOCALITY. Uncertain.
MATERIAL. *Forsskål* 1817 (C — 1 sheet ex herb. Schumacher, microf. 111: II. 5, 6).

RUTACEAE

Citrus aurantium *L.* 1753: 782.
VERNACULAR NAMES. *Narendj haelu, n. malech* (Arabic). Forssk. 1775: 142 (LXXI No. 392; Cent. V No. 43).
LOCALITY. Egypt cult., 1761–62.
MATERIAL. *Forsskål* no specimens found.
VERNACULAR NAME. *Narendj Bortughal* (Arabic). Forssk. 1775: 142 (LXXI No. 393; Cent. V No. 44).
LOCALITY. Egypt cult., 1761–62.
MATERIAL. *Forsskål* no specimens found.
VERNACULAR NAME. *Chommaesch, Turundj* (Arabic). Forssk. 1775: CXVIII No. 467, 468.
LOCALITY. Yemen cult., 1763.
MATERIAL. *Forsskål* no specimen found.

Citrus medica *L.* 1753: 782; Forssk. 1775: 142 (LXXII No. 394, Cent. V No. 45).
VERNACULAR NAME. *Limun malech* (Arabic). Forssk. 1775: 142 (LXXII No. 395; Cent. V No. 46).
VERNACULAR NAME. *Limun haelu* (Arabic). Forssk. 1775: 142 (LXXII No. 396; Cent. V No. 47).
VERNACULAR NAME. *Idalia haelu* (Arabic). Forssk. 1775: 142 (LXXII No. 397; Cent. V No. 48).
VERNACULAR NAME. *Idalia malech* (Arabic). Forssk. 1775: 142 (LXXII No. 398; Cent. V No. 49).
VERNACULAR NAME. *Limun sjaeiri* (Arabic). Forssk. 1775: 142 (LXXII No. 399, Cent. V No. 50).
VERNACULAR NAME. *Kabbad* (Arabic). Forssk. 1775: 142 (LXXII No. 400; Cent. V No. 51).
VERNACULAR NAME. *Naeffasch* (Arabic). Forssk. 1775: 142 (LXXII No. 401; Cent. V No. 52).
VERNACULAR NAME. *Turundj baeledi* (Arabic). Forssk. 1775: 142(LXXII No. 402; Cent. V No. 53).
VERNACULAR NAME. *Turundj m'sabba* (Arabic).
LOCALITY. Egypt, cult., 1761–62. (all above).
MATERIAL. *Forsskål* no specimens found.
VERNACULAR NAMES. *Lîm., Limûn* (Arabic). Forssk. 1775: CXVIII No. 466.
LOCALITY. Yemen, wild, 1763.
MATERIAL. *Forsskål* no specimens found.

Haplophyllum buxbaumii (*Poir.*) *G. Don*, Gen. Syst. 1: 780 (1831); Fl. Turk. 2: 503 (1967).
Ruta linifolia sensu Forssk. 1775: XXV No. 196.
LOCALITY. Turkey: Dardanelles ("Dd., Brg."), July 1761.
MATERIAL. *Forsskål* 1648 (C — 1 sheet ex herb. Hornemann, microf. 89: II. 5, 6). *Forsskål* s.n. (LD — 1 sheet ex herb. Retzius).

Haplophyllum tuberculatum (*Forssk.*) *A. Juss.*, Mém. Mus. Hist. Nat. Paris 12: 528, t. 17 No. 10 (1825); Townsend in Kew Bull. 20: 99 (1966) and Hook. Ic. Pl. (1986); Fl. Palaest. 2: 292, pl. 432, 432a (1972); Tackholm 1974; 334, pl. 112.
Ruta tuberculata Forssk. 1775: 86 (LXVI No. 226; Cent. III No. 64); Christensen 1922: 19.
VERNACULAR NAMES. *Meddjenninae, Maeddjenninae* (Arabic).
LOCALITY. Egypt: Cairo ("In desertis Káhirinis"), 1762; Yemen, 1763.

MATERIAL. *Forsskål* 755 (C — 1 sheet, type of *R. tuberculata*, microf. 89: II. 7, 8); 756 (C —1 sheet, type of *R. tuberculata*, microf. 89: III. 1, 2); 1649 (C — 1 sheet ex herb. Hofman Bang, type of *R. tuberculata*, microf. 89: III. 3, 4); 1650 (C — 1 sheet ex herb. Hornemann, type of *R. tuberculata*, microf. 117: I. 7, 8); 1651 (C — 1 sheet, type of *R. tuberculata*, microf. 135: III. 1, 2).
Forsskål s.n. (LD — 1 sheet ex herb. Retzius, type of *R. tuberculata*).

Ruta chalepensis *L.* 1767: 69; Schweinfurth 1899: 279; Fl. Palaest. 2: 289, pl. 427 (1972).
R. graveolens sensu Forssk. 1775: CXI No. 275.
VERNACULAR NAME. *Schedâb* (Arabic).
LOCALITY. Yemen cult. ("Cultae in planitie"), 1763.
MATERIAL. *Forsskål* 1253 (C — 1 sheet ex herb. Hornemann, partly with *Fagonia arabica*, microf. 135: II. 5, 6); 1647 (C — 1 sheet ex herb. Hornemann, microf. 135: II. 7, 8).
NOTE. Although Forsskål called this *R. graveolens* the description of the petals as "ciliatis" clearly identifies it as *R. chalepensis*.

SALICACEAE

Salix acmophylla *Boiss.*, Diagn. ser. 1, 7: 98 (1846); Fl. Palaest. 1: 26 (1966).
LOCALITY. Unspecified.
MATERIAL. *Forsskål* s.n. (LD — 1 sheet ex herb. Retzius).

Salix glabra *Forssk.* 1775: XXXV No. 431; Christensen 1922: 36.
LOCALITY. Turkey: Dardanelles ("Dd"), July 1761.
MATERIAL. *Forsskål* no specimen found.
NOTE. The brief description validates *S. glabra*: "Foliis serrato-lanceolatis, glabris". It is not included in the Flora of Turkey. The name should not be taken up if it upsets accepted nomenclature.

Salix subserrata *Willd.*, Sp. Pl. ed. 4, 4: 671 (1805); Tackholm 1974: 51, pl. 5B.
S. safsaf baelledi Forssk. 1775: LXXVI No. 527.
LOCALITY. Egypt: Cairo garden ("Ch"?), 1762
MATERIAL. *Forsskål* 461 (C — 1 sheet 'fasciculus secundus', microf. 135: III, 3, 4); 462 (C — 1 sheet 'fasciculus secundus', microf. 135: III. 5, 6).
NOTE. These sterile specimens have larger leaves than is typical of *S. subserrata*; if the identification is correct, then the specimens may have been gathered late in the season. It is not certain that Forsskål saw his *safsaf baelledi* in a garden, as he states "indigena Salix Aegyptiaca".

S. aegyptiaca sensu Forssk. 1775: 170 (LXXVI No. 523; Cent. VI No. 6), non L.; Christensen 1922: 29.
VERNACULAR NAME. *Bân*, *Chalâf* (Arabic).
LOCALITY. Egypt; Cairo ("Káhirae"), 1761–62.
MATERIAL. *Forsskål* 459 (C — 1 sheet, microf. 91: III. 7, 8).

Salix cf **tetrasperma** *Roxb.*, Pl. Corom. 1: 66, pl. 97 (1795); Tackholm 1974: 54, pl. 5A.
LOCALITY. Egypt: ? Cairo, 1762.
MATERIAL. *Forsskål* 460 (C — 1 sheet, microf. 92: I. 1, 2).
NOTE. The mature leaves, glabrous stems and buds are a good match for this species but it is sterile. If the identification is correct, then *S. tetrasperma*, a native of SE Asia, was an early introduction to Egypt. It is difficult to know which it is of Forsskål's species listed on p. LXXVI.

SALVADORACEAE

Dobera glabra (*Forssk.*) *Juss. ex Poir.* in Lam., Encycl., Suppl. 2: 493 (1812); Deflers 1889: 162; Schwartz 1939: 183.
Tomex glabra Forssk. 1775: 32 (CV No. 97; Cent. II No. 9); Christensen 1922: 14.
VERNACULAR NAME. *Dober* (Arabic).
LOCALITY. Yemen: Wadi Surdud ("Surdûd, frequens"), Feb. 1763.
MATERIAL. *Forsskål* 470 (C — 1 sheet with field label "Tomex. Dobaer. Surdud", type of *T. glabra*, microf. 107: III. 1, 2).

Salvadora persica *L.* 1753: 122; Vahl 1790: 12; Deflers 1889: 162; Schwartz 1939: 184.
Cissus arborea Forssk. 1775: 32 (CV No. 95; Cent. II No. 8); Christensen 1922: 14.
VERNACULAR NAMES. *Redif, Rak, Örk*, fruit *Kebath* (Arabic).
LOCALITY. Yemen: Wadi Surdud, Dahhi, Wasab ("Surdûd, Uahfad"), 1763.
MATERIAL. *Forsskål* 518 (C — 1 sheet, type of *C. arborea*, microf. 27: II. 5, 6); 683 (C — 1 sheet, type of *C. arborea*, microf. 27: II. 1, 2); 1117 (C — 1 sheet ex herb. Liebmann, type of *C. arborea*, microf. 27: II. 3, 4).

SANTALACEAE

Osyris alba *L.* 1753: 1022; Forssk. 1775: XXXV No. 432; Fl. Turk.
LOCALITY. Turkey: Dardanelles ("Dd"), July 1761.
MATERIAL. *Forsskål* 536 (C — 1 sheet, microf. 75: II. 1, 2); 537 (C — 1 sheet, with field label, microf. 75: I. 7, 8).
Forsskål s.n. (LD — 1 sheet ex herb. Retzius).

Thesium linophyllon *L.* 1753: 207; Forssk. 1775: V No. 65; Fl. Europ. 1: 71 (1964), 1: 84 (1993).
LOCALITY. France: Marseille ("Estac"), May 1761.
MATERIAL. *Forsskål* 538 (C — 1 sheet, with field label "Thesium linophyllum", microf. 107: I. 1, 2).

SAPOTACEAE

Mimusops laurifolia (*Forssk.*) *Friis* in Kew Bull. 35: 787, fig. 1 (1981); Friis, Hepper & Gasson in Journ. Egypt. Arch. 72: 201 (1986).
Binectaria laurifolia Forssk. 1775: CX No. 252; Christensen 1922: 19. = *Binectaria* [*sp. without epithet*] Forssk. 1775: 82 (Cent. III No. 54).
Mimusops kauki sensu Vahl 1790: 27, non L.
LOCALITY. Yemen: Bayt al Faqih ("Beit el fakih"), 1763.
MATERIAL. *Forsskål* 359 (C — 1 sheet, type of *B. laurifolia*, microf. 15: I. 7, 8); 360 (C — 1 sheet, type of *B. laurifolia*, microf. 15. I: 5, 6).
NOTE. This species was hitherto known as *M. schimperi* Hochst. The sheets are labelled by D. Hillcoat 1973 and I. Friis 1980. It is one of these cases when the generic name and the description of the new genus and species appear in the 'Centuriae' and the epithet is supplied in the 'Florae'.

SCROPHULARIACEAE

Alectra parasitica *A. Rich.*, Tent. Fl. Abyss. 2: 117 (1847); Deflers 1896: 327; Schwartz 1939: 246.
LOCALITY. Yemen: 1763.
MATERIAL. *Forsskål* s.n. (BM — 1 sheet ex herb. Banks).

Anarrhinum forskaolii (*J.F. Gmel.*) *Cufod.* in Bull. Jard. Bot. Brux. 33 (4 Suppl.): 891 (1963); Sutton, Rev. Tribe Antirr.: 257 (1988).
Simbuleta [*sp. without epithet*] Forssk. 1775: 115 (CXV No. 389; Cent. IV No. 54); Vahl 1791; 67; Christensen 1922: 22.
S. forskahlii J.F. Gmel., Syst. 2: 242 (1791).
S. arabica Poir., Encycl. Méth. 7: 194 (1806).
Anarrhinum arabicum (Poir.) Jaub. & Spach, Ill. Pl. Or. 5: 50, t. 446 (1855).
VERNACULAR NAMES. *Symbulet ennesem, Susal* (Arabic).
LOCALITY. Yemen: Kusma ("Kurma"), Mar. 1763.
MATERIAL. *Forsskål* 1736 (C — 1 sheet ex herb. Schumacher, type of *Simbuleta*, microf. 100: II. 1, 2); 1737 (C — 1 sheet ex herb. Vahl, with field label "inter Bosca & Kurma", type of *Simbuleta*, microf. 100: II. 3, 4).
Forsskål s.n. (BM — 1 sheet ex herb. Banks, type of *Simbuleta*).

A. orientale Benth. - Schwartz 1939: 243.
NOTE. Nowhere in the 'Flora Aegyptiaco-Arabica' is an epithet provided for the new species in Forsskål's new genus *Simbuleta*.

Anticharis linearis (*Benth.*) *Hochst. ex Ascherson* in Monatsber. Akad. Wiss. Berl. 1866: 882 (1866); Tackholm 1974: 494, pl. 173B.
Antirrhinum linaria sensu Forssk. LXVIII No. 300, non L.
VERNACULAR NAME. *Aeisj el maelik* (Arabic).
LOCALITY. Egypt: Cairo garden, ("Ch"), 1762.
MATERIAL. *Forsskål* 950 (C — 1 sheet ex herb. Vahl, with original label "Ca. 117", microf. 8: I. 7, 8); 951 (C — 1 sheet ex herb. Schumacher, microf. 8: II. 1, 2).

Antirrhinum supinum sensu Forssk. 1775: 213 (XXIX No. 282; Cent VIII No. 63), non L.
LOCALITY. Turkey: Istanbul ("Circa Constantinop."), 1761.
MATERIAL. *Forsskål* no specimen found.

Bacopa monnieri (*L.*) *Pennell* in Proc. Acad. Nat. Sci. Philad. 98: 94 (1946); Philcox in Kew Bull. 33: 679 (1979).
Herpestis monniera (L.) H.B. & K. - Schwartz 1939: 244.
Limosella calycina Forssk. 1775: 112 (CXV No. 375; Cent. IV No. 43); Christensen 1922: 22.
LOCALITY. Yemen: Wadi Zabid ("Uadi Zebîd"), 1763.
MATERIAL. *Forsskål* no specimen found.

Celsia ramosa *Forssk.* 1775: CXV No. 373; Christensen 1922: 37.
LOCALITY. Yemen: Barad ("Barah"), 1763.
MATERIAL. *Forsskål* no type specimen found.
NOTE. Presumably a *Verbascum* but the brief validating description is insufficiently diagnostic: "Flor. flavis, racemosis".

Cistanche phelypaea (*L.*) *Coutinho*, Fl. Port.: 571 (1913). Tackholm 1974: 509.
Orobanche tinctoria Forssk. 1775: 112 (CXV No. 376; Cent. IV No. 44); Vahl 1791: 70; Christensen 1922: 22.
Phelypaea tinctoria (Forssk.) Walp., Repert. 3: 462 (1844).
VERNACULAR NAMES. *Hödar, Zybb alkaa, Zybbelka* (Arabic).
LOCALITY. Yemen: Wadi Mawr ("Môr"), Feb. 1763.
MATERIAL. *Forsskål* 1537 (C — 1 sheet ex herb. Vahl, type of *O. tinctoria*, microf. 75: I. 3, 4).

Lathraea quinquefida Forssk. 1775: 111 (LXVIII No. 299; Cent. IV No. 39), Christensen 1922: 22.
VERNACULAR NAME. *Haluk, Halue* (Arabic).
LOCALITY. Egypt: Alexandria ("Alexandriae frequens ad omnes ripas arenosis"), 1761–62.
MATERIAL. *Forsskål* no type material found.
NOTE. *L. quinquefida* may be the plant that is usually known as *C. tubulosa* (Schenk) Wight (*C. lutea* Hoffm. & Link) — Schwartz 1939: 2, with yellow flowers, which appears to be a colour form of *C. phelypaea* — Schwartz 1939: 247.

Cycniopsis humifusa (*Forssk.*) *Engl.* in Engl., Bot. Jahrb. 36: 233 (1905); Schwartz 1939: 246.
Buchnera humifusa (Forssk.) Vahl 1794: 81.
Cycnium humifusum (Forssk.) Benth. & Hook.f., Gen. Pl. 2: 969 (1876); Deflers 1889: 180.
Striga humifusa (Forssk.) Benth. in Hook., Comp. Bot. Mag. 1: 362 (1835).
Browallia humifusa Forssk. 1775: 112 (CXV No. 374; Cent. IV No. 42); Christensen 1922: 22.
LOCALITY. Yemen: Al Hadiyah ("Hadîe"), Mar. 1763.
MATERIAL. *Forsskål* 401 (C — 1 sheet, lectotype of *B. humifusa*, microf. 17: I. 3, 4); 404 (C — 1 sheet, isolectotype of *B. humifusa*, microf. 17: I. 1, 2).

Kickxia aegyptiaca (*L.*) *Nábelek*, Publ. Fac. Sci. Univ. Masaryk 70: 31; (1926); Tackholm 1974: 490, pl. 171A; Sutton, Rev. Tribe Antirr.: 185 (1988).
Linaria aegyptiaca (L.) Dum. — Cours. (1802).
Antirrhinum aegyptiacum L. 1753: 613; Forssk. 1775: 112 (LXVIII No. 301; Cent. IV No. 41); Christensen 1922: 22.
VERNACULAR NAMES. *Asjib ed dib, Aeschib ed dib, Doraejse, Doraeise* (Arabic).
LOCALITY. Egypt: Cairo ("Káhirae"), 1761–62.
MATERIAL. *Forsskål* 398 (C — 1 sheet, microf. 7: III. 3, 4); 946 (C — 1 sheet ex herb. Schumacher, microf. 7: III. 5, 6); 947 (C — 1 sheet ex herb. Vahl, microf. 7: III. 7, 8).

Linaria haelava (*Forssk.*) *F. G. Dietr.*, Nachtr. Vollst. Lexic.: 400 (1818); Tackholm 1974: 488, pl. 170B; Fl. Palaest. 3: 190, pl. 318 (1978); Sutton, Rev. Tribe Antirr.: 348 (1988).
Antirrhinum haelava Forssk. 1775: 111 (LXVIII No. 302; Cent. IV No. 40); Christensen 1922: 22.
VERNACULAR NAME. *Haelava* (Arabic).
LOCALITY. Egypt: Cairo ("In desertis Káhirinis"), 1762.
MATERIAL. *Forsskål* 397 (C — 1 sheet, type of *A. haelava*, microf. 8: I. 5, 6); 948 (C — 1 sheet ex herb. Schumacher, type of *A. haelava*, microf. 8: I. 1, 2); 949 (C — 1 sheet ex herb. Vahl, type of *A. haelava*, microf. 8: I. 3, 4).
Forsskål s.n. (BM — 1 sheet ex herb. Banks, type of *A. haelava*).
Forsskål s.n. (LD — 1 sheet ex herb. Retzius, type of *A. haelava*).

Lindernia crustacea (*L.*) *Muell.*, Syst. Census Austral. Pl. 1: 97 (1882); Philcox in Kew Bull. 22: 17 (1968).
LOCALITY. Unspecified.
MATERIAL. *Forsskål* s.n. (LD — 1 sheet ex herb. Retzius).

Orobanche aegyptiaca *Pers.*, Syn. Pl. 2: 181 (1806); Tackholm 1974: 508.
Lathraea phelypaea sensu Forssk. 1775: 111 (LXVIII No. 298; Cent. IV No. 38), non L.
Orobanche ramosa sensu Vahl 1791: 71, non L.
VERNACULAR NAME. *Haluk rihi* (Arabic).
LOCALITY. Egypt: Alexandria ("Alexandriae"), Mar. 1762.
MATERIAL. *Forsskål* 357 (C — 1 sheet, microf. 61: II. 5, 6); 1395 (C — 1 sheet ex herb. Vahl with field label "Av 120", microf. 61: II. 7, 8).

Orobanche crenata *Forssk.* 1775: 113 (LXVIII No. 304; Cent. IV No. 45); Christensen 1922: 22; Tackholm 1974: 506, pl. 177A.
VERNACULAR NAME. *Haluk metabi* (Arabic).
LOCALITY. Egypt: Cairo ("Káhirae"), Alexandria, 1761-62.
MATERIAL. *Forsskål* 358 (C — 1 sheet, type of *O. crenata*, microf. 74: III. 7, 8); 1533 (C — 1 sheet ex herb. Vahl, type of *O. crenata*, microf. nil); 1534 (C — 1 sheet ex herb. Vahl, "Alex. vern. ad Calish" on field label, type of *O. crenata*, microf. nil); 1535 (C — 1 sheet ex herb. Vahl, type of *O. crenata*, microf. nil).
NOTE. No. 1535 has been annotated "O. forskalei" by Vahl. This name has not been validated.

Orobanche ramosa *L.* 1753: 633; Tackholm 1974: 506, pl. 177c.
LOCALITY. Unspecified.
MATERIAL. *Forsskål* 1536 (C — 1 sheet ex herb. Schumacher, microf. 75: I. 1, 2).

Scoparia dulcis *L.* 1753: 116; Schwartz 1939: 245.
S. ternata Forssk. 1775: 31 (CV No. 93; Cent. II No. 6); Vahl 1790: 12; Christensen 1922: 14.
S. dubia? Forssk. 1775: CV No. 94; Christensen 1922: 37 (see note).
VERNACULAR NAME. *Dfar* (Arabic).
LOCALITY. Saudi Arabia: Al Qunfudhah ("Ghomfude, Gomfodam"), Oct. 1762; Yemen: Wadi Mawr ("Môr"), Jan–Feb. 1763.
MATERIAL. *Forsskål* 1708 (C — 1 sheet ex herb. Rottböll, "ad Mour" on back, type of *S. ternata*, microf. 97: I. 3, 4); 1709 (C — 1 sheet with field label "Scoparia an dulcis. prope. Gomfodam", type of *S. ternata*, microf. 97: I. 5, 6).
NOTE. *S. dubia*? has no standing according to the International Code of Botanical Nomenclature (Art. 23.6) in spite of the brief description.

Verbascum pinnatifidum *Vahl* 1791: 39; Fl. Turk. 6: 546 (1978).
V. sinuatum sensu Forssk. 1775: XXI No. 96, non L.
LOCALITY. Turkey: Dardanelles ("Dd"), July 1761.
MATERIAL. *Forsskål* 402 (C — 1 sheet, type of *V. pinnatifidum*, microf. 112: I. 7, 8); 1820 (C — 1 sheet, ex herb. Schumacher, type of *V. pinnatifidum*, microf. 112: I. 5, 6).

Verbascum sinuatum *L.* 1753: 178; Fl. Turk. 6: 538.
LOCALITY. Unspecified, probably Turkey.
MATERIAL. *Forsskål* 403 (C — 1 sheet not now in Herb. Forsk., microf. 112: I. 3, 4). *Forsskål* s.n. (LD — 1 sheet ex herb. Retzius).

SOLANACEAE

Datura metel *L.* 1753: 179; Forssk. 1775: LXIII No. 127, CVI No. 131; Tackholm 1974: 479.
VERNACULAR NAMES. *Beudj, Mandj* (Arabic Yemen).

LOCALITY. Egypt: Cairo garden ("Ch"), 1761–62; Yemen: Wadi Mawr ("Môr"), Bayt al Faqih (Btf.), Jiblah ("ad Djöblam"), 1763.
MATERIAL. *Forsskål* 1216 (C — 1 sheet, ex herb. Hornemann, microf. 136: I. 1, 2); 1217 (C — 1 sheet ex herb. Hornemann, microf. 136: I. 3, 4).
Forsskål s.n. (LD — 1 sheet ex herb. Retzius).

Hyoscyamus muticus *L.* 1767: 45; Tackholm 1974: 483, pl. 166. Fig. 24A.
MATERIAL. Egypt: Cairo?, *Forsskål* (plant grown from seeds sent to Linnaeus, no type specimen in Herb. LINN.).

H. datora Forssk. 1775: 45 (LXIII No. 128; Cent. II No. 47), & Icones t. 5A (1776); Christensen 1922: 15.
VERNACULAR NAMES. *Datôra, Saecarân* (Arabic).
LOCALITY. Egypt: Cairo ("In desertis Káhirinis arensis, apricis, copiose"), 1761–62.
MATERIAL. *Forsskål* 423 (C — 1 sheet, type of *H. datora*, microf. 55: III. 3, 4); 1347 (C — 1 sheet ex herb. Hofman Bang, type of *H. datora*, microf. 55: III. 1, 2); 1837 (C — 1 sheet ex herb. Hornemann, type of *H. datora*, microf. 136: I. 5, 6).
Forsskål s.n. (BM — 1 sheet ex herb. Banks, type of *H. datora*, microf. nil).

Lycium europaeum *L.* 1753: 192; Forssk. 1775: V No. 62; Fl. Europ. 3: 194 (1972).
LOCALITY. France: Marseille ("Estac"), May 1761.
MATERIAL. *Forsskål* 408 (C — 1 sheet, microf. 65 (1–2): III. 1, 2); 1441 (C — 1 sheet ex herb. Hornemann, microf. 65 (1–2): III. 3, 4).

Nicotiana tabacum *L.* 175: 180; Forssk. 1775: LXIII No. 129; Bircher 1960: 689; Meikle 1985: 1195.
VERNACULAR NAME. *Doccban* (Arabic).
LOCALITY. Egypt: Cairo garden ("Ch"), 1761–62.
MATERIAL. *Forsskål* 1496 (C — 1 sheet ex herb. Hornemann, microf. 137: I. 3, 4).

Physalis alkekengi *L.* 1753: 183; Bircher 1960: 678; Fl. Turk. 6: 444 (1978).
LOCALITY. ? Turkey: ("ad Aqueduct ad Bagin" on sheet).
MATERIAL. *Forsskål* 1576 (C — 1 sheet ex herb. Hornemann, microf. 136: I. 7, 8).
NOTE. Of the two specimens on the sheet No. 2 is Forsskål's.

Physalis angulata *L.* 1753: 183; F.W.T.A. ed. 2, 2: 329 (1963).
P. ramosa Forssk. 1775: 204 (CVI No. 134; Cent. VIII No. 12); Christensen 1922: 33.
LOCALITY. Yemen: Wassb ("Ad Uahfâd"), 29 Mar. 1763.
MATERIAL. *Forsskål* 417 (C — 1 sheet, with field label "Physalis * subgemina. Uahfêt", type of *P. ramosa*, microf. 80: I. 7, 8).

Solanum armatum *Forssk.* 1775: 47 (CVII No. 150; Cent. II No. 54); Christensen 1922: 15.
S. arabicum Dunal, Hist. Solan.: 240 (1813).
VERNACULAR NAMES. *Bockeme, Bokaeme, Bonkom* (Arabic).
LOCALITY. Yemen: Hays, Zabid ("Haes. Zebîd") 1763.
MATERIAL. *Forsskål* no type specimen found.
NOTE. Schwartz 1939: 237 considered *S. armatum* to be synonymous with *S. xanthocarpum* Schrad. & Wendl. (1795) i.e. *S. virginianum* L. (1753) — Hepper & Jaeger in Kew Bull. 41: 43 (1986).

Solanum carense *Dunal* in DC., Prodr. 13(1): 90 (1852).
LOCALITY. Yemen, 1763.
MATERIAL. *Forsskål* 412 (C — 1 sheet, microf. 102: I. 1, 2).

FIG. 24. (A) **Hyoscyamus muticus** L. — Icones, Tab. V *Hyoscyamus datora*.

(B) **Swertia polynectaria** (Forssk.) Gilg — Icones, Tab. V *Parnassia polynectaria*.

Solanum coagulans *Forssk.* 1775: 47 (CVII No. 149; Cent. II No. 55); Vahl 1791: 41; Christensen 1922: 15.
VERNACULAR NAMES. *Bejkaman, Soraej saban* (Arabic).
LOCALITY. Yemen: Wadi Mawr ("Frequens in Yemen"), 1763.
MATERIAL. *Forsskål* 409 (C — 1 sheet with field label "Maur", type of *S. coagulans*, microf. 101: III. 5, 6); 410 (C — 1 sheet, type of *S. coagulans*, microf. 101: III. 3, 4); 1744 (C — 1 sheet ex herb. Schumacher, type of *S. coagulans*, microf. 101: III. 1, 2).
NOTE. This species was long known as *S. dubium* Fres. which is now a straight synonym. Unfortunately the name *S. coagulans* was considered to be a synonym of *S. incanum* L. — Schwartz 1939: 237. Its true identity was first recognised by D. Hillcoat who labelled the sheets in 1973.

Solanum cordatum *Forssk.* 1775: 47 (CVII No. 154; Cent. II No. 56); Christensen 1922: 15; Wood in Kew Bull. 39: 134 (1984).
VERNACULAR NAME. *Hadak* (Arabic).
LOCALITY. Yemen: Suq Ar Rubu' ("Roboa"), 1763.
MATERIAL. *Forsskål* no type specimens found: see note.
NOTE. *Wood* 2327 has been designated the neotype (Wood 1984 q.v.).
S. gracilipes Decne. in Jacquemont Voy. 4: 113 (1844). Type: India *Jacquemont* (P); Schwartz 1939: 237.
S. hadaq Deflers in Bull. Soc. Bot. Fr. 43: 122 (1896). Type: S. Yemen, *Deflers* 377 (K, P).

Solanum dulcamara *L.* 1753: 185; Forssk. 1775: XXI No. 103; Fl. Turk. 6: 441 (1978).
LOCALITY. Turkey: Istanbul, Izmir ("Cph, Sm" = Constantinopolini hortensi, Smirna), July–Sept 1761.
MATERIAL. *Forsskål* 418 (C — 1 sheet with field label "Co.O", microf. 102: I. 3, 4).

Solanum forskalii *Dunal* 181: 237; Hepper & Wood in Kew Bull. 38: 85 (1983).
S. villosum Forssk. 1775: 47 (CVII No. 151; Cent. II No. 57), non Miller (1768); Christensen 1922: 15.
VERNACULAR NAME. *Bockaeme* (Arabic).
LOCALITY. Yemen: Wadi Surdud ("Surdud"), Feb. 1763.
MATERIAL. *Forsskål* 414 (C — 1 sheet, type of *S. villosum*, microf. 102: II. 7, 8).
S. albicaule Kotschy ex Dunal — Schwartz 1939: 236.

Solanum glabratum *Dunal*, Hist. Solan.: 240 (1813) and in DC., Prodr. 13, 1: 283 (1852).
S. sepicula Dunal in DC., Prodr. 13(1): 283 (1852); Deflers 1889: 176; Schwartz 1939: 236.
S. bahamense ? sensu Forssk. 1775: 46 (CVII No. 152; Cent. II No. 53), non L.; Christensen 1922: 15.
VERNACULAR NAMES. *Melihaemi, Homaesch, Habak* (Arabic).
LOCALITY. Yemen: Wadi Surdud ("Surdud"), Uahfad, Feb. 1763.
MATERIAL. *Forsskål* 411 (C — 1 sheet, syntype of *S. glabratum*, microf. 101: II. 5, 6); 429 (C — 1 sheet, syntype of *S. glabratum*, microf. 101: II. 7, 8).

Solanum incanum *L.* 1753: 188; Forssk. 1775: 46 (CVII No. 148; Cent. II No. 52); Christensen 1922: 15; Schwartz 1939: 237; Hepper & Jaeger in Kew Bull. 40: 387 (1985).
VERNACULAR NAMES. *Ennaema, Aejn el bagar, Ersan* (Arabic).
LOCALITY. Yemen: Jiblah ("Djöblae"), 30–31 Mar. 1763.
MATERIAL. *Forsskål* 415 (C — 1 sheet, microf. 102: II. 1, 2); 1745 (C — 1 sheet ex herb. Schumacher, microf. 102: I. 5, 6); 1746 (C — 1 sheet ex herb. Schumacher microf. 136: III. 7, 8); 1747 (C — 1 sheet ex herb. Schumacher with field label "Mour", microf. 107: I. 7, 8).

Forsskål s.n. (BM - 1 sheet ex herb. Banks, with old label "Akêb No. 52"); s.n. (BM — 1 sheet ex herb. Banks, with field label "in Mour").

Solanum microcarpum *Vahl* 1791: 40; Christensen 1922: 39.
S. diphyllum sensu Forssk. 1775: LXIII No. 134, non L.
LOCALITY. Egypt: Rashid (Rosette), 1761.
MATERIAL. *Forsskal* 1751 (C — 1 sheet ex herb. Hofman Bang, with field label "Solanum diphyllum Ro.8", type of *S. microcarpum*, microf. 101: III. 7, 8).

Solanum nigrum *L.* 1753: 186; Forssk. 1775: 46 (V No. 61; XXI No. 104; LXIII No. 132; CVII No. 145; Cent. II No. 49); Christensen 1922: 15; Tackholm 1974: 473, pl. 163; Fl. Turk. 6: 439 (1978).
VERNACULAR NAMES. *Skilosaphilo* (Greek); *Enabeddib* (Egyptian Arabic); *Mesaeleha* (Yemen Arabic).
LOCALITY. France, Turkey, Egypt, Yemen, 1761–63.
MATERIAL. *Forsskål* 1748 (C — 1 sheet ex herb. Hornemann, with locality "ad Alexandriam" on back, microf. 136: III. 3, 4); 1749 (C — 1 sheet ex herb. Hornemann, microf. 136: II. 3, 4).
NOTE. There may be some confusion between true *S. nigrum* and the black-fruited variety b of *S. aegyptiacum* (see *S. villosum*).

Solanum palmetorum *Dunal* in DC., Prodr. 13(1): 282 (1852).
LOCALITY. Yemen: 1763.
MATERIAL. *Forsskål* 1750 (C — 1 sheet ex herb. Schumacher, microf. 136: II. 1, 2).
NOTE. Named and labelled *S. palmetorum* Dun. by Bitter, the Solanum authority. This species is glabrous, whereas the closely related *S. glabratum* (*S. sepicula*) is stellate-pubescent.

Solanum platacanthum *Dunal* in DC., Prodr. 13, 1: 285 (1852); Schwartz 1939: 238.
LOCALITY. Yemen: between ? Mukhajah and Bughah ("Habak inter Mokaia et Boka" on reverse of sheet).
MATERIAL. *Forsskål* 413 (C — 1 sheet, microf. 137: I. 1, 2).

Solanum terminale *Forssk.* 1775: 45 (CVII No. 146; Cent. II No. 48); Vahl 1791: 40; Christensen 1922: 15; Schwartz 1939: 235.
LOCALITY. Yemen: ? Mukham ("ad Mochham in umbrosis"), Mar. 1763 (see note).
MATERIAL. *Forsskål* 406 (C — 1 sheet, type of *S. terminale*, microf. 102: II. 5, 6); 419 (C — 1 sheet with field label (see note), type, of *S. terminale*, microf. 102: II. 3, 4).
NOTE. Although "Mochham" is cited in the Flora, the locality on the original ticket is given as "Mokhaja" i.e. Mukhajah, Jebel Barad, and a general distribution is given as "Yemen in montis altioribus".

Solanum villosum *Miller*, Gard. Dict. ed. 8, art. Solanum No. 2 (1768); Schwartz 1939: 234; Tackholm 1974: 473 (as var. of *S. nigrum*).
S. aegyptiacum a. Forssk. 1775: 46 (Cent. II No. 50); Christensen 1922: 15.
S. nigrum var. *villosum* Miller—Vahl 1791: 40.
VERNACULAR NAME. *Enab eddîb* (Arabic).
LOCALITY. Egypt: 1761–62.
MATERIAL. *Forsskål* 407 (C — 1 sheet, microf. 101: I. 5, 6); 420 (C — 1 sheet, microf. 101: I. 3, 4); 1752 (C — 1 sheet ex herb. Schumacher, microf. 136: II. 5, 6); 1753 (C — 1 sheet ex herb. Vahl, microf. 136: II. 7, 8).

S. [*aegyptiacum*] b. Forssk. 1775: 46 (Cent. II. No. 51); Christensen 1922: 15.
MATERIAL. *Forsskål* 405 (C — 1 sheet, microf. 101: II. 3, 4); 421 (C — 1 sheet, microf. 101: I. 7, 8); 422 (C — 1 sheet, microf. 101: II. 1, 2).
NOTE. a. is orange-fruited and b. is black-fruited. See also *S. nigrum* which is black-fruited.

[**Solanum** sp. indet. 1.
NOTE. This sheet was erroneously attributed to Forsskål as No. 425 in the Herbarium Forsskålii at Copenhagen and appears in the microfiche 136: III. 1, 2].

[**Solanum** sp. indet. 2.
NOTE. This sheet was erroneously attributed to Forsskål as No. 427 in the Herbarium Forsskålii at Copenhagen and appears in the microfiche 136: III. 5, 6].

Withania somnifera (*L.*) *Dunal* in DC., Prodr. 13(1): 453 (1852); Schwartz 1939: 233; Tackholm 1974: 474, pl. 164.
Physalis somnifera L. — Forssk. 1775: XXII No.- 107; LXIII No. 130, CVI No. 133.
P. curassiv[*ica*] sensu Forssk. 1775: LXIII No. 131, non L.
VERNACULAR NAMES. *Saekarân* (Egypt Arabic). *Barde, Öbab, Uárak esschefa* (Yemen Arabic).
LOCALITY. Rhodes, 21 Sept. 1761; Egypt: Cairo ("Cs") 1761–62; Yemen: Wadi Mawr ("Môr"), Al Hadiyah ("Hadie"), Bayt al Faqih ("Btf") 1763.
MATERIAL. *Forsskål* 416 (C — 1 sheet with field label "Ni. 21", microf. 80: II. 3, 4); 1577 (C — 1 sheet ex herb. Schumacher, with field label "Physalis curassavica via Cairum Ni. 21", microf. 80: II. 1, 2); 1578 (C — 1 sheet, microf. nil).
Forsskål s.n. (LD — 1 sheet ex herb. Retzius).

STERCULIACEAE

Melhania velutina *Forssk.* 1775: 64 (CVII No. 158; Cent. III No. 6); Deflers 1895: 421; Christensen 1922: 17.
Pentapetes velutina (Forssk.) Vahl 1790: 49.
LOCALITY. Yemen: Milhan ("Melhan"), 1763.
MATERIAL. *Forsskål* 510 (C — 1 sheet, type of *M. velutina*, microf. 67: I. 5, 6).
Forsskål 1909. LBG. 9245 (32) (LIV — 1 sheet, type of *M. velutina*).
Forsskål s.n. (BM — 1 sheet ex herb. Banks, type of *M. velutina*, microf. nil).

M. abyssinica sensu Deflers 1889: 115, non A. Rich.
M. ferruginea A. Rich. — Schwartz 1939: 167.
NOTE. This species is the type of the genus *Melhania* Forssk.

Sterculia africana (*Lour.*) *Fiori* in Agric. Colon., Ital. 5 suppl.: 37 (1912).
Culhamia [*sp. without epithet*] Forssk. 1775: 96 (CXII No. 295; Cent. III. 100); Christensen 1922: 20.
Culhamia hadiensis J.F. Gmel., Syst. Nat. 2: 754 (1791).
Sterculia planifolia Vahl 1790: 80; Schwartz 1939: 168.
S. arabica T. Anders. in J. Linn. Soc. 5, suppl. 1: 9 (1860).
VERNACULAR NAME. *Kulham* (Arabic).
LOCALITY. Yemen: Al Hadiyah ("in montibus Hadiensibus"), Mar. 1763.
MATERIAL. *Forsskål* 605 (C — 1 sheet, holotype of *C. hadiensis* & *S. planifolia*, microf. 35: III. 3, 4).
NOTE. The new genus *Culhamia* is described with a new species in the 'Centuriae', but nowhere in the 'Flora Aegyptiaco-Arabica' is this new species provided with an epithet.

TAMARICACEAE

Reaumurea hirtella *Jaub. & Spach*, Ill. Pl. Or. 3: 54, 55, t. 244 (1848); Tackholm 1974: 369, pl. 126B.
R. vermiculata sensu Forssk. 1775: 101 (LXVIII No. 284; Cent. IV No. 14), non L.; Christensen 1922: 20.
VERNACULAR NAMES. *Mullaeh, Adhbe* (Arabic).
LOCALITY. Egypt: Alexandria ("Alexandriae"), Oct. 1761 or Apr. 1762.
MATERIAL. *Forsskål* 482 (C — 1 sheet type of *R. vermiculata*, microf. 86: II. 7, 8); 1610 (C —1 sheet ex herb. Schumacher, type of *R. vermiculata*, microf. 86: III. 1, 2).
Forsskål s.n. (LD — 1 sheet ex herb. Retzius).

Tamarix aphylla (*L.*) *Karsten*, Deutsch. Fl.: 641 (1882); Tackholm 1974: 365, pl. 125; Baum in Gen. Tamarix: 81, pl. 20 (1978).
T. orientalis Forssk. 1775: 206 (LXIV No. 182, CIX No. 215; Cent. VIII No. 29); Christensen 1922: 33.
T. articulata Vahl 1791: 48, t. 32.
VERNACULAR NAME. *Atl, Atle* (Arabic).
LOCALITY. Egypt: Alexandria ("Alexandria Al.3" on field label, "Cairi plantae hortenses" in book), 1761 or 1762.
MATERIAL. *Forsskål* 1776 (C — 1 sheet ex herb. Hofman Bang, with field label "Tamarix. atle. Alexandriae Al.1", type of *T. orientalis* and *T. articulata*, microf. 106: I. 5, 6).
LOCALITY. Saudi Arabia: Al Qunfudhah ("Ghomfuda"), Oct. 1762. Yemen: Wadi Mawr ("Môr"), 1763.
MATERIAL. *Forsskål* 480 (C — 1 sheet, type of *T. orientalis* and *T. articulata*, microf. 106: I. 1, 2); 481 (C — 1 sheet, type of *T. orientalis* and *T. articulata*, microf. 106: I. 7, 8); 1775 (C — 1 sheet ex herb. Hofman Bang, type of *T. orientalis* and *T. articulata*, microf. 106: I: 3, 4); 1777 (C — 1 sheet ex herb. Schumacher, type of *T. orientalis* and *T. articulata*, microf. 106: II. 1, 2).

Tamarix tetragyna *Ehrenb.* in Linnaea 2: 247–258 (1827); Tackholm 1974: 367; Baum, Gen. Tamarix: 128, pl. 37 (1978).
T. gallica sensu Forssk. 1775: XXIV No. 157, LXIV No. 181, non L.
VERNACULAR NAMES. *Tarfa*; *Hattab achmar* (Arabic).
LOCALITY. Egypt: Alexandria ("Alex. ad Calisch" on label, Cairo in book), 1761–62.
MATERIAL. *Forsskål* 1774 (C — 1 sheet ex herb. Hornemann, with field label "Tamarix gallica, sed floribus ... 92", microf. 105: III. 7, 8).

THYMELAEACEAE

Thymelaea hirsuta (*L.*) *Endl.*, Gen. Suppl. 4, 2: 65 (1848); Vahl 1790: 29; Tackholm 1974: 360.
Passerina metnan Forssk. 1775: 81 (LXV No. 218; Cent. III No. 51); Christensen 1922: 19.
VERNACULAR NAME. *Metnân* (Arabic).
LOCALITY. Egypt: Alexandria ("Alexandriae"), 1 April 1962.
MATERIAL. *Forsskål* 532 (C — 1 sheet, type of *P. metnan*, microf. 77: II. 5, 6); 1558 (C — 1 sheet, type of *P. metnan*, microf. 77: II. 7, 8).

TILIACEAE

Corchorus depressus (*L.*) *Christensen* 1922: 34; Chatterjee in Kew Bull 3: 372 (1949).
Antichorus depressus L. 1767: 64; Vahl 1790: 27.
MATERIAL. ? Yemen ("Arabia"), 1763, (LINN 487.1 holotype of *A. depressus* cult. Uppsala from *Forsskål*'s seeds, microf. 250: III. 5).

Corchorus antichorus Raeusch. (1797) — Deflers 1889: 116.
Jussiaea edulis Forssk. 1775: 210 (CXI No. 283; Cent. VIII No. 44).
VERNACULAR NAME. *Uaeki* (Arabic).
LOCALITY. Yemen: Al Luhayyah ("Lohajae"), Jan. 1763.
MATERIAL. *Forsskål* 512 (C — 1 sheet, type of *J. edulis*, microf. 58: III. 3, 4); 515 (C — 1 sheet, type of *J. edulis*, microf. 58: III. 5, 6); 1375 (C — 1 sheet ex herb. Schumacher, type of *J. edulis*, microf. 58: III. 7, 8).

Corchorus olitorius *L.* 1753: 529; Forssk. 1775: 101 (LXVIII No. 281, CXIV No. 345; Cent. IV No. 12); Christensen 1922: 20; Tackholm 1974: 346, pl. 118A.
VERNACULAR NAME. *Melochia* (Arabic).
LOCALITY. Egypt: gardens, 1761–62. Yemen: Wadi Surdud ("Srd"), 1763.
MATERIAL. *Forsskål* s.n. (BM — 1 sheet ex herb. Banks).
Forsskål s.n. (LD — 1 sheet ex herb. Retzius).
Forsskål s.n. (LE — 1 sheet ex herb. Pallas).

Corchorus trilocularis *L.*, Syst. Nat. ed. 12, 2: 369 (1767) and 1767: 77; Schwartz 1939: 155; Meikle 1977: 316.
MATERIAL. "Arabia. *Forskahl*" (LINN 691.2, 3 type specimens grown at Uppsala from Forsskål's seeds, microf. 356: I. 1, 2).

C. aestuans Forssk. 1775: 101 (CXIV No. 346; Cent. IV No. 13); Vahl 1791: 63; Christensen 1922: 20.
VERNACULAR NAME. *Melochia* (Arabic).
LOCALITY. Yemen: Wadi Mawr ('Môr'), 1763.
MATERIAL. *Forsskål* no type specimen found (but see *C. trilocularis* LINN types).

Grewia arborea (*Forssk.*) *Lam.*, Encycl. 3: 45 (1789); Schwartz 1939: 157; Wood in Kew Bull. 39: 137 (1984).
Chadara arborea Forssk. 1775: 105 (CXIV No. 339; Cent. IV No. 24); Christensen 1922: 21.
Grewia excelsa Vahl 1790: 35.
VERNACULAR NAME. *Saerak* (Arabic).
LOCALITY. Yemen: Wasab ("Uahfad"), 1763.
MATERIAL. *Forsskål* 490 (C — 1 sheet, type of *C. arborea*, microf. 24: I. 7, 8).
Forsskål s.n. (BM — 1 sheet ex herb. Banks, type of *C. arborea*).

Grewia tenax (*Forssk.*) *Fiori* in Agric. Colon., Ital. 5 suppl.: 23 (1912).
Chadara tenax Forssk. 1775: CXIV No. 338; Christensen 1922: 21. = *Chadara* [*sp. without epithet*] Forssk. 1775: 105 (Cent. IV No. 23).
Grewia populifolia Vahl 1790: 33; Schwartz 1939: 156.
VERNACULAR NAMES. *Chadâr*, *Nabbâ* (Arabic).
LOCALITY. Yemen: Surd[u]d ("*Surdûd* alibique frequens"); Taizz, Apr. 1763.
MATERIAL. *Forsskål* 1085 (C — 1 sheet ex herb. Schumacher, type of *C. tenax*, microf. 24: II. 3, 4); 491 (C — 1 sheet with field label 'Taes', type of *C. tenax*, microf. 24: II. 1, 2).
Forsskål s.n. (BM — 1 sheet ex herb. Banks with field label 'Surdud', type of *C. tenax*).

NOTE. This is one of the examples of a new species in a new genus described without an epithet in 'Centuriae'; one is provided in the 'Florae'. See discussion of the validity of the genus *Chadara* in Taxon 33: 665 (1984).

Grewia velutina (*Forssk.*) *Vahl* 1790: 35; Deflers 1889: 116; Schwartz 1939: 157; Wood in Kew Bull. 39: 138 (1984).
Chadara velutina Forssk. 1775: 106 (CXIV No. 340; Cent. IV No. 25); Christensen 1922: 21.
VERNACULAR NAME. *Nescham, Neschamm* (Arabic).
LOCALITY. Yemen: Al Hadiyah ("Hadîe"), Apr. 1763.
MATERIAL. *Forsskål* 489 (C — 1 sheet, holotype of *C. velutina*, microf. 24: II. 5, 6).

Grewia villosa *Willd.*, Ges. Nat. Fr. Neue Schr. 4: 205 (1803); Deflers 1889: 116; Schwartz 1939: 157.
LOCALITY. Yemen: Al Hadiyah ("Hadie" on label), Mar. 1763.
MATERIAL. *Forsskål* 467 (C — 1 sheet with field label "Ignota fol. subrotundo ... Hadie", microf. 137: I. 5, 6).

Grewia sp. indet.
LOCALITY. Yemen: without locality, 1763.
MATERIAL. *Forsskål* 1311 (C — 1 sheet, microf. 138: I. 1, 2).
NOTE. According to B. Mathew this and *Schweinfurth* 288 belong to *Grewia* sect. *Axillares*, nearest to *G. bicolor* Juss. sens. lat. It might be *G. canescens* Hochst. ex A. Rich., but these specimens have not been compared with the type.

Triumfetta rhomboidea *Jacq.*, Enum. Syst. pl: 22 (1760); Schwartz 1939: 157.
T. glandulosa Forssk. 1775: CXII No. 297, with validating description; Vahl 1794: 62.
LOCALITY. Yemen: ? Al Hadiyah ("Mm"), Mar. 1763.
MATERIAL. *Forsskål* 483 (C — 1 sheet, type of *T. glandulosa*, microf. 109: II. 3, 4); 1808 (C — 1 sheet, type of *T. glandulosa*, microf. 109: II. 5, 6); 1809 (C — 1 sheet ex herb. Schumacher, type of *T. glandulosa*, microf. 137: I. 7, 8).

UMBELLIFERAE

Ammi majus *L.* 1753: 243; Forssk. 1775: 54 (XXIII No. 144, LXIV No. 171; Cent. II No. 82); Christensen 1922: 16; Fl. Turk. 4: 427 (1972) Tackholm 1974: 390.
VERNACULAR NAMES. *Asperokephalos* (Greek), *Chaelle* (Arabic).
LOCALITY. Turkey: Bozcaada I. ("Td."), July 1761.
MATERIAL. *Forsskål* 676 (C — 1 sheet with field label "Bo. 30", microf. 5: III. 5, 6).
LOCALITY. Egypt: Rashid (" Rosettae"), 31 Oct. 1761.
MATERIAL. *Forsskål* 675 (C — 1 sheet, microf. 5: III. 4, 5).

Ammi visnaga (*L.*) *Lam.*, Fl. Fr. 3: 462 (1778); Fl. Turk. 4: 426 (1972).
Daucus gingid[*ium*] sensu Forssk. 1775: XXIII No. 142, non L.
Daucus b.) [*sp. without epithet*] Forssk. 1775: XXIII No. 141.
LOCALITY. Turkey: Izmir ("Smirna"), July 1761.
MATERIAL. *Forsskål* 673 (C — 1 sheet, microf. 37: III. 7, 8).
LOCALITY. Turkey: Istanbul, Büyükdere ("Bujuchtari"), 1761; Egypt: between Alexandria and Cairo (on sheet), 1761–62.
MATERIAL. *Forsskål* 672 (C — 1 sheet, microf. 37: III. 5, 6); 673 (C — 1 sheet, microf. 37: III. 7, 8); 674 (C — 1 sheet, microf. 137: II. 7, 8); 679 (C — 1 sheet with field label "an itinerare terrestri inter Alex. et Cairo", microf. 137: II. 3, 4).

NOTE. No. 672 is inscribed "ded. Schumacher" and it probably came to Herbarium Forsskålii before the entire Schumacher herbarium was acquired. Most specimens from Schumacher's herbarium have modern numbers above 900.

Bupleurum flavum *Forssk.* 1775: 205 (XXIII No. 146; Cent. VIII No. 23); Christensen 1922: 33; Fl. Turk. 4: 402 (1972).
LOCALITY. Turkey: Dardanelles ("Ad Dardanellos in arvis"), July–Aust. 1761.
MATERIAL. *Forsskål* 681 (C — 1 sheet, type of *B. flavum*, microf. 17: III. 1, 2); 1023 (C — 1 sheet ex herb. Hornemann, type of *B. flavum*, microf. 17: III. 3, 4).
Forsskål s.n. (BM — 1 sheet ex herb. Banks, type of *B. flavum*).

? **Daucus broteri** *Ten.*, Fl. Nap. 4, Syll. App. 3: 4 (1830); Fl. Turk. 4: 533 (1972).
Artedia muricata sensu *Forssk.* 1775: 206 (XXIII No. 139; Cent. VIII No. 27), non L.; Christensen 1922: 33.
VERNACULAR NAME. *Aziggano* (Greek).
LOCALITY. Bozcaada I. and Dardanelles ("Tenedos & ad Dardanellos"), July 1761.
MATERIAL. *Forsskål* no specimen found.
NOTE. If Forsskål's plant was correctly identified as *Artedia muricata* L. ie. *Daucus muricata* (L.) L. then it would not occur in Turkey, since the species native there is *D. broteri.*

Daucus glaber (*Forssk.*) *Thell.* in Mem. Soc. Nat. Cherbourg 38: 407 (1912); Meikle 1977: 713.
Caucalis glabra Forssk. 1775: 206 (LXIV No. 169; Cent. VIII No. 25); Christensen 1922: 33.
Daucus littoralis Sibth. & Sm. var. *forskahlei* Boiss., Fl. Orient. 2: 1074 (1872); Tackholm 1974: 396.
LOCALITY. Egypt: Alexandria, ("As" = Alexandriae), 1761–62.
MATERIAL. *Forsskål* (LD — 1 sheet ex herb. Retzius, type of *C. glabra*).

Echinophora tenuifolia *L.* 1753: 239; subsp. **sibthorpiana** (*Guss.*) *Tutin* in Feddes Rep. 74: 31 (1767); Fl. Turk. 4: 309 (1972).
LOCALITY. Turkey, 1761.
MATERIAL. *Forsskål* 669 (C — 1 sheet, microf. nil); 1237 (C — 1 sheet ex herb. Schumacher, microf. nil).

Eryngium bourgatii *Gouan*, Ill. Obs. Bot. 7: t. 3 (1773); Vahl 1791: 47; Fl. Turk. 4: 300 (1972), see note below.
E. foetidum sensu Forssk. 1775: XXIII No. 135, non L.
LOCALITY. Turkey: Bozcaada I., Gökçeada I. ("Tenedos, Imros"), July 1761.
MATERIAL. *Forsskål* 670 (C — 1 sheet, microf. 42: I. 3, 4).
NOTE. Although this sheet was labelled by Ascherson as *E. amethystinum* — Forssk. 1775: XXIII No. 137, the specimen is not that species; Vahl's annotation in the reverse suggests that it is *E. foetidum* (Forssk. 1775: XXIII No. 135). However, *E. bourgatii*, is not recorded from Tenedos in the Flora of Turkey.

Eryngium maritimum *L.* 1753: 233; Forssk. 1775: XXIII No. 136; Fl. Turk. 4: 294 (1972).
LOCALITY. Turkey: Burgaz, Dardanelles ("Borghàs"), July 1961.
MATERIAL. *Forsskål* 680 (C — 1 sheet, microf. 42: I. 5, 6).

Pimpinella menachensis *Schweinf. ex Wolff* in Fedde Rep. 16: 238 (1919); Schwartz 1939: 177.
LOCALITY. Yemen: without locality, 1763.
MATERIAL. *Forsskål* 1582 (C — 1 sheet ex herb. Viborg, microf. nil).

Pimpinella anisum *L.* 1753: 264; Fl. Europ. 2: 331 (1968).
LOCALITY. Unspecified.
MATERIAL. *Forsskål* 678 (C — 1 sheet, microf. 137: II. 5, 6).

Pseudorlaya pumila (*L.*) *Grande* in Nuov. Giorn. Bot. Ital., n.s. 32: 86 (1925); Fl. Palaest. 2: 400, pl. 577 (1972); Tackholm 1974: 394; Meikle 1977: 708.
Caucalis pumila L., Syst. Nat. ed. 10: 955 (1759); Forssk. 1775: 206 (LXIV No. 168; Cent. VIII No. 24).
LOCALITY. Egypt: Alexandria ("Alexandriae"), 1761–62.
MATERIAL. *Forsskål* 1067 (C — 1 sheet ex herb. Hornemann, microf. nil).
NOTE. In 1871 J. Lange labelled this *Orlaya maritima* (Gou.) Koch which is a synonym of the above.

Sanicula elata *D. Don*, Prodr. Fl. Nepal.: 183 (1825); F.T.E.A. Umbellif.: 17, fig. 4 (1989).
S. europaea sensu Schwartz 1939: 175, non L.
LOCALITY. Yemen: Mukhajah, J. Barad ("Mokhaja, Barah" on label), Mar 1763.
MATERIAL. *Forsskål* 671 (C — 1 sheet with field label, microf. 137: II. 1, 2).

Seseli cf. **tortuosum** *L.* 1753: 260; Fl. Europ. 2: 337 (1968).
LOCALITY. France: Marseille ('Estac'), May 1761.
MATERIAL. *Forsskål* 677 (C — 1 sheet, microf. 98: II. 7, 8).

Torilis arvensis (*Huds.*) *Link*, Enum. Hort. Berol. Alt. 1: 265 (1821); Schwartz 1939: 175; Fl. Palaest. 2: 395, pl. 571 (1972); Tackholm 1974: 397.
Caucalis angustifolia Forssk. 1775: 206 (CIX No. 208; Cent. VIII No. 26); Christensen 1922: 33.
LOCALITY. Yemen: Al Hadiyah ("Hadîe"), Mar. 1763.
MATERIAL. *Forsskål* 668 (C — 1 sheet, type of *C. angustifolia*, microf. 22: III. 1, 2).

Scandix infesta L., Syst. Nat. ed. 12: 732 (1767); Forssk. 1775: 58 (LXIV No. 175; Cent. II No. 93).
Torilis infesta (L.) Hoffm. (1824) — Muschler 1912: 714.
VERNACULAR NAMES. *Cellae, Gazar malaiki, Gazar sjaeitani* (Arabic).
LOCALITY. Egypt: Alexandria ("Alexandriae, copiosissime"), 1761–62.
MATERIAL. LINN — (1 sheet 364.10 cult. hort. Uppsala, from Forsskål's seeds, type of *S. infesta*, microf. 198: I.5).
NOTE. Forsskål proposed the name *Scandix infesta* which was not published until 1775, after his death. In the meantime Linnaeus must have received seeds from Forsskål's collection via D. Zoega at Copenhagen; presumably the name "Scandix infesta" was with the seeds and it was taken up and published by Linnaeus in 1767.

Umbelliferae gen. & sp. indet.
LOCALITY. Unspecified.
MATERIAL. *Forsskål* s.n. (LD — 1 sheet ex herb. Retzius).
NOTE. This flowering specimen is without fruits or basal leaves.

ULMACEAE

Celtis toka (*Forssk.*) *Hepper & Wood* in Kew Bull. 38: 86 (1983).
Ficus toka Forssk. 1775: 219 (CXXIV No. 618; Cent. VIII No. 99); Christensen 1922: 35.
VERNACULAR NAME. *Toka* (Arabic).

LOCALITY. Yemen: Milhan ("Melhân"), 1763.
MATERIAL. *Forsskål* no type specimen found.
NOTE. This tree has hitherto been known as *Celtis integrifolia* Lam.

URTICACEAE

Debregesia saeneb (*Forssk.*) *Hepper & Wood* in Kew Bull. 38: 86 (1983).
Rhus saeneb Forssk. 1775: 206 (CIX No. 216; Cent. VIII No. 28); Christensen 1922: 33.
VERNACULAR NAMES. *Saeneb, Bajad, Baejad* (Arabic).
LOCALITY. Yemen: Al Hadiyah ("in montibus Hadiensibus"), 1763.
MATERIAL. *Forsskål* no type specimen found.
NOTE. This plant has hitherto been known as *Debregaesia bicolor* (Roxb.) Wedd. — Schwartz 1939: 28.

Droguetia iners (*Forssk.*) *Schweinf.* 1896: 146; Schwartz 1939: 29; F.T.E.A. Urticac.: 56 (1989).
Urtica iners Forssk. 1775: 160 (CXXI No. 540; Cent. VI No. 17); Christensen 1922: 27.
U. urens L. γ *iners* (Forssk.) Wedd. in DC., Prodr. 16(1): 40 (1869).
U. verticillata Vahl 1790: 76.
VERNACULAR NAME. *Hamsched* (Arabic).
LOCALITY. Yemen: Al Hadiyah ("In montibus Hadîensibus"), 1763.
MATERIAL. *Forsskål* 774 (C — 1 sheet with field label "Urtica * in ... Hamscher. Bolghose", holotype of *U. iners*, microf. 111: I. 3, 4).

Forsskåolea (Forskohlea) tenacissima *L.*, Amoen. Acad 7: 73 (1764), & Opobals. Decl.: 18 (1764), & Mant. Pl.: 11 (1767); Deflers 1889: 409; Schwartz 1939: 28; Fl. Palaest. 1: 43, pl. 42 (1966); Tackholm 1974: 57, pl. 7A.
LOCALITY. Egypt: Cairo, 1762.
MATERIAL. LINN (2 sheets Nos. 605, 1, 2 grown from Forsskål's seeds, types of *F. tenacissima*, microf. 314. I. 6, 7).

Caidbeja adhaerens Forssk. 1775: 82 (LXV No. 221; Cent. III No. 55); Christensen 1922: 19.
VERNACULAR NAMES. *Lussaq, Hamsched* (Arabic).
LOCALITY. Egypt: Cairo ("in desertis Kâhirinis orientalibus ad Caid Bey"), 1761–62.
MATERIAL. *Forsskål* 775 (C — 1 sheet, type of *C. adhaerens*, microf. 19: III. 5, 6); 1039 (C — 1 sheet ex herb. Vahl, type of *C. adhaerens*, microf. 19: III. 7, 8).
NOTE. The new genus *Caidbeja* and the new species *C. adhaerens* were described in the 'Centuriae' and provided with an epithet there.

Girardinia diversifolia (*Link*) *Friis* in Kew Bull. 36: 145 (1981); F.T.E.A. Urticac.: 13 (1989).
Urtica heterophylla Vahl 1790: 76, non Decne.
Girardinia heterophylla (Vahl) Decne. in Jacquemont, Voy. Inde 4, Bot.: 151, t. 153 (1844).
Urtica palmata Forssk. 1775: 159 (CXXI No. 539; Cent. VI No. 16), non *Giradinia palmata* Bl.; Christensen 1922: 27.
VERNACULAR NAMES. *Schadjâret el mehâbbe*(*r*), *Horokrok* (Arabic).
LOCALITY. Yemen: Bolghose ("Blg" = Bulgose, "Yemen in montibus"), 1763.
MATERIAL. *Forsskål* 771 (C — 1 sheet, type of *U. palmata* and *U. heterophylla*, microf. 111: I. 5, 6).
G. condensata (Hochst. ex Steud.) Wedd. — Schwartz 1939: 28.

Laportea aestuans (*L.*) *Chew* in Gardens Bull. Singapore 21: 200 (1965); F.T.E.A. Urticac.: 23 (1989).

Fleurya aestuans (L.) Gaud. in Freycinet, Voy. Uran.: 497 (1830); Schwartz 1939: 27.
Urtica divaricata sensu Forssk. 1775: 160 (CXXI No. 542; Cent. VI No. 19), non L.; Christensen 1922: 27.
U. hirsuta Vahl 1790: 77.
VERNACULAR NAME. *Mehaerreka* (Arabic).
LOCALITY. Yemen: Al Hadiyah ("Hadie"), Mar. 1763.
MATERIAL. *Forsskål* 772 (C — 1 sheet, with field label "Urtica divaricata vel canadensis. Mehaerreka. Hadîe", type of *U. hirsuta*, microf. 111: I. 1, 2).

Parietaria alsinifolia *Delile*, Descr. Fl. Egypte: 137 (1813); Tackholm 1974: 57, pl. 7D.
P. judaica sensu Forssk. LXXVII No. 548, non L.
VERNACULAR NAME. *Roqrek* (Arabic).
LOCALITY. Egypt: Cairo ("Cd., Cs."), 1761–62.
MATERIAL. *Forsskål* 769 (C — 1 sheet, microf. 77: I. 7. 8).

Pouzolzia parasitica (*Forssk.*) *Schweinf.* 1896: 145; Schwartz 1939: 28; Friis & Jellis in Kew Bull. 39: 593 (1984); F.T.E.A. Urticac.: 51 (1989).
Urtica parasitica Forssk. 1775: 160 (CXXI No. 541; Cent. VI No. 18); Christensen 1922: 27.
U. muralis Vahl 1790: 77.
VERNACULAR NAMES. *Naedjaa, Nedjára* (Arabic).
LOCALITY. Yemen: Al Hadiyah ("Hadîe"), Mar. 1763.
MATERIAL. *Forsskål* 770 (C — 1 sheet with field label "Naedja. Urtica * ... Bolghosi", type of *U. parasitica* and *U. muralis*, microf. 111: I. 7, 8).

Urtica caudata *Vahl* 1791: 96, non Burm.f.
Urtica dubia? Forssk. 1775: CXXI No. 544; Christensen 1922: 37.
LOCALITY. Yemen: J. Barad ("Barah") 1763.
MATERIAL. *Forsskål* no type specimens found.
NOTE. The name *U. dubia* is cited in the International Code of Botanical Nomenclature (Art. 23.6) as having no standing, being a dubious epithet.

Urtica urens *L.* 1753: 984; Forssk. 1775: CXXI No. 543; Schwartz 1939: 27.
VERNACULAR NAME. *Kolaebleba* (Arabic).
LOCALITY. Yemen: Bolghose ("Bulghose"), Mar. 1763.
MATERIAL. *Forsskål* 773 (C — 1 sheet with field label "Urtica ? urentis. inter Bölghose & Mokhaja", microf. 111: II. 1, 2).

VALERIANACEAE

Centranthus calcitrapa (*L.*) *Dufr.*, Hist. Nat. Méd. Fam. Valér.: 39 (1811); Fl. Europ. 4: 56 (1976); Fl. Turk. 4: 559 (1972); Meikle 1985: 838.
LOCALITY. Unspecified.
MATERIAL. *Forsskål* 248 (C — 1 sheet, microf. nil).

VERBENACEAE

Lantana viburnoides (*Forssk.*) *Vahl* 1790: 45; Deflers 1889: 184; Schwartz 1939: 215; F.T.E.A. Verbenac.: 40 (1992).
Charachera viburnoides Forssk. 1775: 116 (CXV No. 379; Cent. IV No. 56); Christensen 1922: 22.

LOCALITY. Yemen: J. Barad ("Barah"), 1763.
MATERIAL. *Forsskål* 323 (C — 1 sheet with field label "inter Bolghose et Moxhaja", type of *C. viburnoides*, microf. 24: III. 1, 2); 1086 (C — 1 sheet ex herb. Horneman, type of *C. viburnoides*, microf. 24: III. 3, 4); 1087 (C — 1 sheet ex herb. Schumacher, type of *C. viburnoides*, microf. 24: III. 5, 6).
Forsskål s.n. (LD — 1 sheet ex herb. Retzius, type of *C. viburnoides*).

Charachera tetragona Forssk. 1775: CXV No. 378; Christensen 1922: 22. = *Charachera* [*sp. without epithet*] Forssk. 1775: 115 (Cent. IV No. 55).
Lantana tetragona (Forssk.) Schweinf. 1912: 145, nomen. illeg.
VERNACULAR NAMES. *Frefrân, Mekatkata* (Arabic).
LOCALITY. Yemen: Al Hadiyah ("In montibus *Hadiensibus*"), Mar. 1763.
MATERIAL. *Forsskål* 315 (C — 1 sheet, holotype of *C. tetragona*, microf. 24: II. 7, 8).
NOTE. The genus *Charachera* is an example of a new genus with two new species, all described in the 'Centuriae', but *C. tetragona* is only provided with an epithet in the 'Florae'.

Phyla nodiflora (*L.*) *Greene* in Brittonia 4: 46 (1899); F.T.E.A. Verbenac.: 25 (1992).
Lippia nodiflora (L.) A. Rich. — Deflers 1889: 185; Schwartz 1939: 216; Tackholm 1974: 452.
Verbena capitata Forssk. 1775: 10 (LXI No. 10, CIII No. 32; Cent. I No. 28); Christensen 1922: 11.
LOCALITY. Egypt: Rashid ("Rosettae ad littora Nili"), 1762. Yemen: mountain region, 1763.
MATERIAL. *Forsskål* s.n. (LD — 1 sheet ex herb. Retzius).

Priva adhaerens (*Forssk.*) *Chiov.* in Bull. Soc. Ital. 1923: 115 (1923); Schwartz 1939: 216; Verdcourt in Kew Bull. 43: 671 (1988); F.T.E.A. Verbenac.: 22 (1922).
Ruellia adhaerens Forssk. 1775: 114 (CXV No. 388; Cent. IV No. 51); Christensen 1922: 22.
Phryma? Forssk. 1775: CXV No. 372.
Verbena Forskalaei Vahl 1794: 6.
Priva dentata Juss., Ann. Mus. Nat. Hist. Paris. 7: 70 (1806).
P. Forskalii Jaub. & Spach, Illustr. 5: 59, t. 455 (1855).
P. Forskaolii (Vahl) E. Mey., Comm. Pl. Afr. Austr.: 275 (1836).
Zapania arabica Poir., Encyc. Méth. Bot. 8: 844 (1808).
VERNACULAR NAME. *Hamsched* (Arabic).
LOCALITY. Yemen: Al Hadiyah ("Hadie"), Mar. 1763.
MATERIAL. *Forsskål* 1814 (C — 1 sheet ex herb. Vahl, type of *R. adhaerens* & *V. forskalaei* microf. 79: III. 5, 6); 1815 (C — 1 sheet ex herb. Hofmann Bang, type of *R. adhaerens* & *V. forskalaei*, microf. 79: III. 7, 8).

Verbena supina *L.* 1753: 21; Tackholm 1974: 454.
V. procumbens Forssk. 1775: 10 (LIX No. 9; Cent. I No. 27); Christensen 1922: 11.
LOCALITY. Egypt: Rashid ("Rs"), 1761–62.
MATERIAL. *Forsskål* 1821 (C — 1 sheet ex herb. Liebmann, type of *V. procumbens*, microf. nil).
NOTE. This type specimen has been overlooked and was not seen by Christensen (1922); it is now possible to confirm the synonymy.

Vitex agnus-castus *L.* 1753: 638; Forssk. 1775: 213 (XXIX No. 283, LXVIII No. 307; Cent. VIII No. 64); Fl. Europ. 3: 122 (1972); Fl. Turk. 7: 34 (1982).
VERNACULAR NAME. *Kaf marjam* (Arabic); *Ligaria* (Greek). Egypt: Alexandria ("Alexandriae"); Turkey: islands and mainland ("In Archipelago and Natolia ubique copiose"), 1761.
MATERIAL. *Forsskål* no specimens found.

VIOLACEAE

Hybanthus enneaspermus (*L.*) *F. Muell.*, Fragm. Phytogr. Austral. 10: 81 (1876); Schwartz 1939: 170.
Viola arborea Forssk. 1775: CXX No. 515, non L.
VERNACULAR NAMES. *Sidr, Rábba* (Arabic).
LOCALITY. N. Yemen: Wasab ("Uahfud"), Mar. 1763.
MATERIAL. *Forsskål* 621 (C — 1 sheet, microf. 112: II. 3, 4); 622 (C — 1 sheet, microf. 112: II. 5, 6); 1823 (C — 1 sheet ex herb. Schumacher, microf. nil).

Viola repens *Forssk.* 1775: CXX No. 516; Christensen 1922: 37.
LOCALITY. Yemen: "montium regio media", 1763.
MATERIAL. *Forsskål* no type specimen found.
NOTE. The brief description: "Foliis parvis, reniform" validates the name. Its identify remains obscure. The name should not be taken up if it upsets widely accepted nomenclature.

VITACEAE

Cissus glandulosa *J.F. Gmel.*, Syst. Nat. 2: 256 (1791).
Saelanthus glandulosus Forssk. 1775: 34 (CV No. 99; Cent. II No. 12); Christensen 1922: 14, non rite publ.
VERNACULAR NAME. *Mimiae* (roots) (Arabic).
LOCALITY. Yemen: Kudmiyah ("Kudmîe"), 1763.
MATERIAL. *Forsskål* no type specimen found.

Cissus quadrangularis *L.* 1767: 39; Schwartz 1939: 154. Fig. 25.
Vitis quadrangularis (L.) Wall. — Deflers 1889: 125.
Saelanthus quadrangonus Forssk. 1775: 33 (CV No. 98; Cent. II No. 11) & Icones t. 2 (1776); Christensen 1922: 14.
VERNACULAR NAMES. *Saela* (general name); *Dakari* (angles hispid, considered to be male); *Entai* (angles glabrous, considered to be female) (Arabic).
LOCALITY. Yemen: Wadi Sordud ("Uadi Surdud", "In sylvis Arabiae felicis vulgaris"), 1763.
MATERIAL. *Forsskål* no type specimen found.
NOTE. Four new species of the new genus *Saelanthus* are described in the 'Centuriae', all with epithets. However, there is no separate generic diagnosis or description, and all names are therefore invalid.

Cissus rotundifolia *Vahl* 1794: 19; Schwartz 1939: 153. Fig. 26.
Saelanthus rotundifolius Forssk. 1775: 35 (CV No. 100; Cent. II No. 14) & Icones t. 4 (1776); Christensen 1922: 14, non rite publ.
Vitis rotundifolius (Forssk.) Deflers 1889: 125.
VERNACULAR NAMES. *Haelaes, Halka* (Arabic).
LOCALITY. Yemen: Wadi Sordud ("Surdûd frequens"), Feb. 1763.
MATERIAL. *Forsskål* 682 (C — 1 sheet, type of *S. rotundifolius*, microf. 90: II. 8, III. 1); 684 (C — 1 sheet, type of *S. rotundifolius*, microf. 90: III. 2, 3).
NOTE. See note under *C. quadrangularis*.

Cissus sp. indet.
LOCALITY. Unspecified (see note below).

FIG. 25. **Cissus quadrangularis** L. — Icones, Tab. II *Saelanthus quadragonus*.

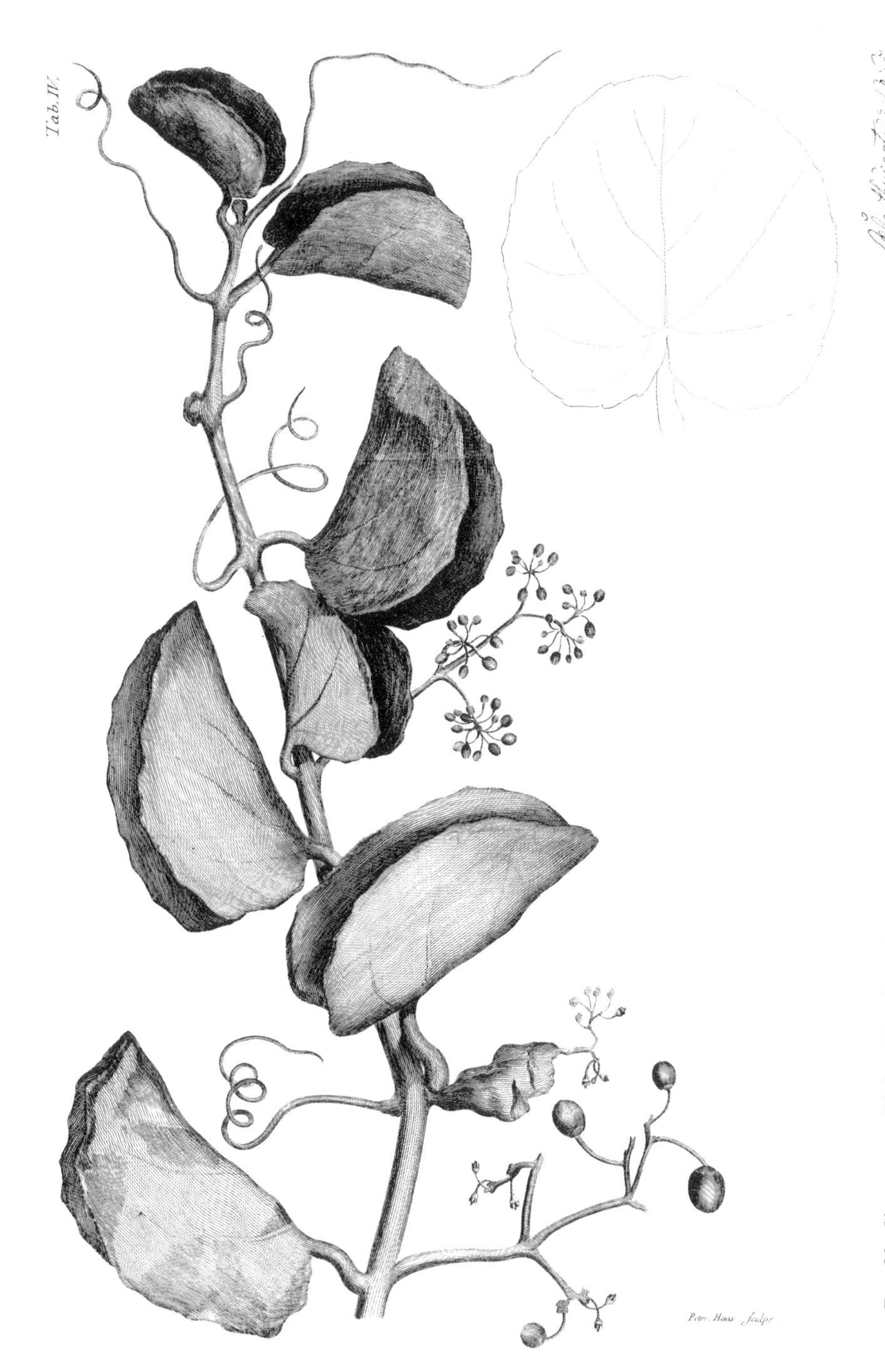

FIG. 26. **Cissus rotundifolius** (Forssk.) Vahl — Icones, Tab. IV *Saelanthus rotundifolius.*

FIG. 27. **Cyphostemma digitatum** (Lam.) Descoings — Icones, Tab. III ***Saelanthus digitatus.***

MATERIAL. *Forsskål* s.n. (LD — 1 sheet ex herb. Retzius).
NOTE. This large-leaved, fruiting specimen has not been matched with species of *Cissus* or allied genera. It is probably from Yemen.

Cyphostemma digitatum (*Lam*) *Descoings* in Nat. Monspel., Sér. Bot., Fasc. 18: 220 (1967). Fig. 27.
Cissus digitatus Lam., Tabl. Encycl. 1: 332 (1793); Schwartz 1939: 154.
Vitis digitata (Lam.) Deflers 1889: 125.
Saelanthus digitatus Forssk. 1775: 35 (CV No. 102; Cent. II No. 13), & Icones t. 3 (1776); Christensen 1922: 13, non rite publ.
VERNACULAR NAME. *Haluaek, Haelvaek* (Arabic).
LOCALITY. Yemen: Al Hadiyah ("Hadîe"), 1763.
MATERIAL. *Forsskål* no type specimen found.
NOTE. See note under *Cissus quadrangularis*.

Cyphostemma ternatum (*J.F. Gmel.*) *Descoings* in Nat. Monspel., Sér. Bot., Fasc. 18: 229 (1967).
Cissus ternata J.F. Gmel., Syst. Nat. 2: 256 (1791); Schwartz 1939: 154.
Saelanthus ternatus Forssk. 1775: 35 (CV No. 101; Cent. II No. 15); Christensen 1922: 14, non rite publ.
VERNACULAR NAMES. *Hanka, Hankaja* (Arabic).
LOCALITY. Yemen: Hays ("Haes"), 1763.
MATERIAL. *Forsskål* no type specimen found.
NOTE. See note under *Cissus quadrangularis*.

Vitis vinifera *L.* 1753: 202; Fl. Turk. 2: 522 (1967).
LOCALITY. Turkey: Istanbul, Belgrad ("ad Belgrad" on reverse of sheet), July–Sept. 1761.
MATERIAL. *Forsskål* 1824 (C — 1 sheet ex herb. Hornemann, microf. nil).
NOTE. "Fructu ovato" is written on the reverse of the sheet, together with the name and above locality.

ZYGOPHYLLACEAE

Fagonia arabica *L.* 1753: 386; Forssk. 1775: 88 (Cent. III No. 68); Deflers 1889: 118; Christensen 1922: 19; Schwartz 1939: 119.
F. cretica sensu Forssk. 1775: CXI No. 280, non L. (1753).
VERNACULAR NAMES. *Schoaeka, Schouki, Schoki* (Arabic).
LOCALITY. Yemen: Wadi Mawr ("Ad Môr"), Jan. 1763.
MATERIAL. *Forsskål* 1251 (C — 1 sheet ex herb. Schumacher, microf. 43: III. 1, 2); 1252 (C — 1 sheet ex herb. Hornemann and Hofman Bang, microf. 43: III. 3, 4); 1253 (C — 1 sheet partly with *Ruta chalepensis*, ex herb. Hornemann, microf. 135: II. 5, 6).
NOTE. It is strange that this species has not otherwise been recorded from Yemen.

Fagonia cretica *L.* 1753: 386; Tackholm 1974: 307, pl. 97C.
F. scabra Forssk. 1775: 88 (LXVI No. 231; Cent. III No. 69); Christensen 1922: 19.
VERNACULAR NAMES. *Djaemde, Djamdae* (Arabic).
LOCALITY. Egypt: Cairo ("*Kahirae* in desertis"), 1762.
MATERIAL. *Forsskål* 740 (C — 1 sheet, type of *F. scabra*, microf. 43: III. 7, 8); 741 (C — 1 sheet, type of *F. scabra*, microf. 44: I. 1, 2); 742 (C — 1 sheet, type of *F. scabra*, microf. 44: I. 3, 4 & plate in Bot. Tidssk. 72 (1): 30); 1254 (C — 1 sheet ex herb. Schumacher, type of *F. scabra*, microf. 43: III. 5, 6).

Nitraria retusa (*Forssk.*) *Aschers.* in Verh. Bot. Ver. Prov. Brandenb. 18: 94 (1876); Fl. Palaest. 2: 257, pl. 371 (1972); Tackholm 1974: 313, pl. 101A.
Peganum retusum Forssk. 1775: 211 (LXVI No. 248; Cent. VII No. 51).
VERNACULAR NAME. *Gharghed, Gharghaeàd* (Arabic).
LOCALITY. Egypt: Alexandria ("Alexandriae"), 1 Apr. 1761.
MATERIAL. *Forsskål* 463 (C — 1 sheet, type of *P. retusum*, microf. 77: III. 5, 6); 758 (C — 1 sheet, type of *P. retusum*, microf. 77: III. 1, 2); 759 (C — 1 sheet, type of *P. retusum*, microf. 77: III. 3, 4); 1562 (C — 1 sheet ex herb. Schumacher, type of *P. retusum*, microf. nil).

Tribulus pentandrus *Forssk.* 1775: 88 (LXVI No. 232; Cent. III No. 70); Hadidi in Taeckholmia 9: 61 (1978) and in F.T.E.A. Zygophyllac.: 3 (1985).
VERNACULAR NAMES. *Gatba, Eddraejsi,* (var. b:) *Kótaba* (Arabic).
LOCALITY. Egypt: Cairo ("Kahirae"), 1761–62. Yemen: Al Luhayyah ("Lohajae"), Jan. 1763.
MATERIAL. *Forsskål* 743 (C — 1 sheet, type of *T. pentandrus*, microf. 107: III. 7, 8); 1788 (C — 1 sheet, type of *T. pentandrus*, microf. 108: I. 1, 2).
NOTE. This is *T. alatus* Del. nom. nud. and *T. longipetalus* Viv. — Tackholm 1974: 311.

Tribulus terrestris *L.* 1753: 387; Tackholm 1974: 313; Hadidi in Taeckholmia 9: 60 (1978), in F.T.E.A. Zygophyllac.: 7 (1985).
T. lanuginosus L. 1753: 387; Forssk. 1775: LXVI No. 234.
LOCALITY. Egypt, 1761–62.
MATERIAL. *Forsskål* 744 (C — 1 sheet, microf. 107: III. 5, 6).

Zygophyllum album *L.f.*, Dec. 1, 11, t. 6 (1762) and L. 1762: 551; Tackholm 1974: 309, pl. 99C. Fig. 28A.
Z. proliferum Forssk. 1775: 87 (LXVI No. 228, CXI No. 278; Cent. III. No. 65), & Icones t. 12A (1776); Christensen 1922: 19.
VERNACULAR NAMES. *Chraesi, Hmada, Hamd* (Arabic).
LOCALITY. Egypt: Alexandria ("Alexandriae"), 1 Apr. 1762.
MATERIAL. *Forsskål* 1829 (C — 1 sheet, type of *Z. proliferum*, microf. 113: III. 4, 5); 1830 (C — 1 sheet ex herb. Hornemann, type of *Z. proliferum*, microf. 113: III. 6, 7).
Forsskål s.n. (BM — 1 sheet ex herb. Banks, type of *Z. proliferum*).

Zygophyllum coccineum *L.* 1753: 386; Tackholm 1974: 309, pl. 99B. Fig. 29.
Z. desertorum Forssk. 1775: 87 (LXVI No. 229; Cent. III No. 66), & Icones t. 11 (1776); Christensen 1922: 19.
VERNACULAR NAMES. *Kamôn karamânr, Rotraejt, Rottraejt* (Arabic).
LOCALITY. Egypt: between Cairo and Suez ("inter Káhiram & Suês"), Aug. 1762.
MATERIAL. *Forsskål* 753 (C — 1 sheet, type of *Z. desertorum*, microf. 113: II. 2, 3).
Forsskål s.n. (BM — 1 sheet ex herb. Banks, type of *Z. desertorum*); s.n. (BM —1 sheet ex herb. Banks, with field label 'Ca. 86', type of *Z. desertorum*); s.n. (BM — 1 sheet ex herb. Banks, with old label, type of *Z. desertorum*).
Forsskål s.n. (LD — 1 sheet ex herb. Retzius, type of *Z. desertorum*).

Zygophyllum simplex *L.* 1767: 68; Schwartz 1939: 122; Fl. Palaest. 2: 254, pl. 367 (1972); Tackholm 1974: 309, pl. 99D. Fig. 27B.
MATERIAL. 'Arabia' (more likely Egypt than Yemen) description based on plants raised from seeds collected by Forsskål and sent to Linnaeus; no type specimen in LINN 544.

Z. portulacoides Forssk. 1775: 88 (LXVI No. 230, CXI No. 279; Cent. III No. 67), & Icones t. 12B (1776); Christensen 1922: 19.

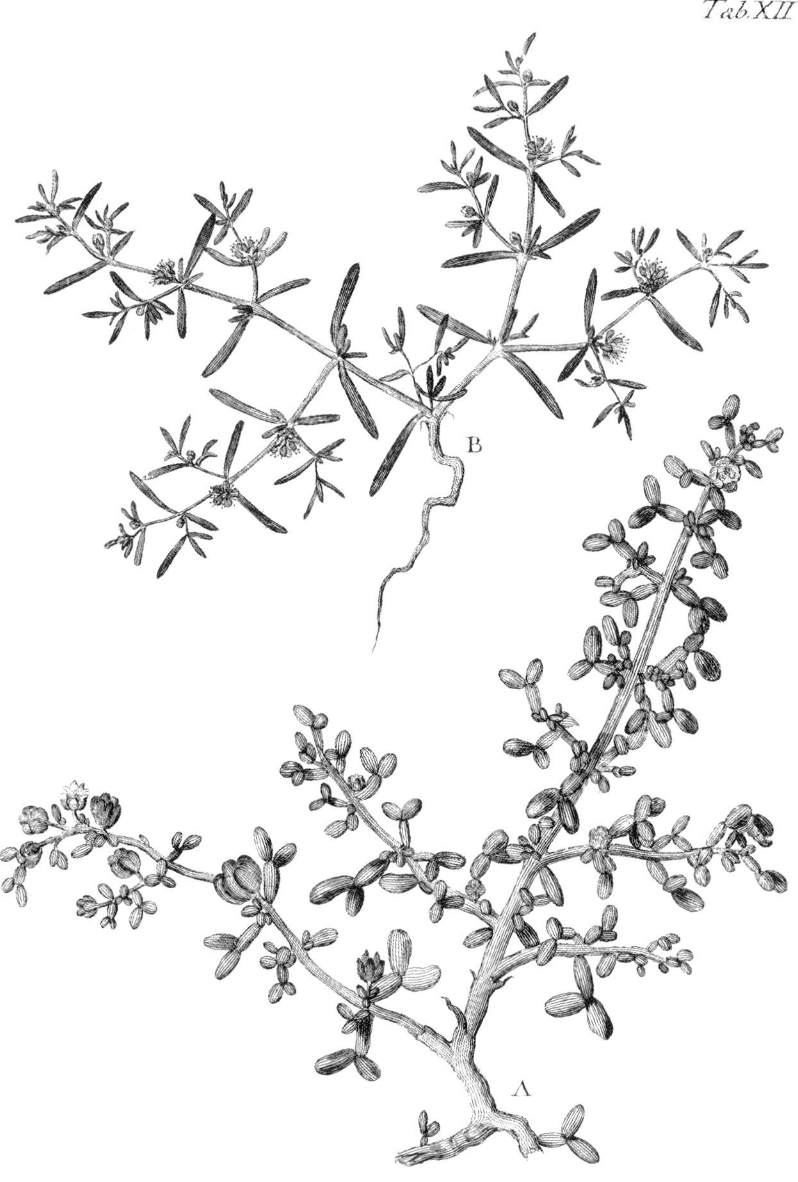

FIG. 28. (A) **Zygophyllum album** L. — Icones, Tab. XII *Zygophyllum proliferum*.
(B) **Zygophyllum simplex** L. — Icones, Tab. XII *Zygophyllum portulacoides*.

Fig. 29. **Zygophyllum coccineum** L. — Icones, Tab. XI *Zygophyllum desertorum*.

VERNACULAR NAMES. *Djarmal*, *Garmal* (Egyptian Arabic); *Kermel*, *Djirmel* (Yemeni Arabic).
LOCALITY. Egypt: Cairo or Yemen: Wadi Mawr, Ghurab ("Môr, Ghorab"), 1762–3.
MATERIAL. *Forsskål* 474 (C — 1 sheet, type of *Z. portulacoides*, microf. 113: II. 8, III. 1); 751 (C — 1 sheet, type of *Z. portulacoides*, microf. 113: II. 4, 5); 752 (C — 1 sheet, type of *Z. portulacoides*, microf. 113: II. 6, 7); 1828 (C — 1 sheet ex herb. Schumacher, type of *Z. portulacoides*, microf. 113: III. 2, 3).
Forsskål s.n. (LD — 1 sheet ex herb. Retzius, type of *Z. portulacoides*).

UN-NAMED DICOTYLEDONS

Dicot. fam., gen. & sp. indet. 1
LOCALITY. Unspecified.
MATERIAL. *Forsskål* 471 (C — 1 sheet, microf. 138: I. 8, II. 1).
NOTE. A suffrutex with brown-lanate lower stem, silvery-pubescent slender stems and undersurface of the opposite leaves.

Dicot. fam., gen. & sp. indet. 2
LOCALITY. Unspecified.
MATERIAL. *Forsskål* s.n. (LD — 1 sheet ex herb. Retzius).
NOTE. This sterile shoot has stellate-pubescent leaves.

Dicot. fam., gen. & sp. indet. 3
LOCALITY. Unspecified.
MATERIAL. *Forsskål* 468 (C — 1 sheet, microf. 138: I. 6, 7).
NOTE. Sterile twig with large leathery entire leaves having a pulvinus at each end of the long petioles. Attributed to Forsskål but unlikely to be one of his collection.

Dicot. fam., gen. & sp. indet. 4
LOCALITY. Unspecified.
MATERIAL. *Forsskål* 667 (C — 1 sheet, microf. 123: I. 3, 4).
NOTE. A sterile climber with lobed and simple leaves considered by C. Jeffrey not to be a cucurbit.

Dicot. fam., gen. & sp. indet. 5
LOCALITY. Unspecified.
MATERIAL. *Forsskål* 1227 (C — 1 sheet, microf. nil).
NOTE. A sterile twig with a terminal tuft of leaves.

Dicot. fam., gen. & sp. indet. 6
LOCALITY. Unspecified.
MATERIAL. *Forsskål* 472 (C — 1 sheet, microf. 138: II. 2, 3).
NOTE. A climber (?) with rather large, long-petioled, 2- or 3-foliate leaves; possibly Capparaceae.

Dicot. fam., gen. & sp. indet. 7
LOCALITY. Unspecified.
MATERIAL. *Forsskål* s.n. (LD — 1 sheet ex herb. Retzius).

Dicot. fam., gen. & sp. indet. 8
LOCALITY. Unspecified.
MATERIAL. *Forsskål* 471 (C — 1 sheet, microf. 138: I. 8–II. 1).
NOTE. A sterile subshrub with opposite leaves, arachnoid beneath.

Mesua glabra *Forssk.* 1775: CXIV No. 341; Christensen 1922: 20.
VERNACULAR NAME. *Chadar* (Arabic).
LOCALITY. Yemen ("Ph = Planities argillacea humida, montibus propior"), 1763.
MATERIAL. *Forsskål* no type specimen found.
NOTE. This is an undescribed species still unidentified. The Arabic name indicates that it is a species of *Grewia.*

"**Cinna arundinacea**" sensu Forssk. 1775: 3 (CII No. 8; Cent. I No. 8), non L.; Christensen 1922: 10.
LOCALITY. Yemen: J. Barad ("in monte Barah"), 1763.
MATERIAL. *Forsskål* no specimen found.
NOTE. *Cinna arundinacea* is a North American species not known in Arabia. In the absence of a specimen it is impossible to identify it from the brief description alone.

ADDITIONAL SPECIES

Lepidium sativum L. 1753: 644; Forssk. 1775: CXVI No. 394; Schwartz 1939: 72; Fl. Europ. 1: 331 (1964), 1: 400 (1993).
VERNACULAR NAME. *Half* (Arabic).
LOCALITY. Yemen: Bayt al Faqih ("Btf."), 1763.
MATERIAL. *Forsskål* s.n. (C — 1 sheet ex herb. Vahl).
NOTE. This specimen was recovered from Vahl's herbarium while this book was in proof. On the reverse is written in Vahl's hand 'Lepidium forskalei'. It seems to be an unusual form of this species.

MONOCOTYLEDONS

AGAVACEAE

Dracaena sp. indet.
LOCALITY. Unspecified.
MATERIAL. *Forsskål* 1235 (C — 1 sheet ex herb. Hofman Bang, microf. 129: II. 3, 4).
NOTE. I have been unable to make a positive determination. "Dracaena graminifolia" is written on the reverse of the sheet.

ALISMATACEAE

Alisma plantago-aquatica *L.* 1753: 342; Forssk. 1775: XXIV No. 182 as *A. plantago*; Fl. Turk. 8: 7 (1984).
LOCALITY. Turkey: Izmir ("Smyrna" on sheet), Istanbul, Belgrad ("Bg" in book), 1761.
MATERIAL. *Forsskål* 923 (C — 1 sheet ? ex herb. Hornemann, microf. 4: II. 5, 6).

AMARYLLIDACEAE

Allium sphaerocephalon *L.* 1753: 297; Fl. Turk. 8: 177 (1984).
A. marit[imum] Forssk. 1775: XXIV No. 163; Christensen 1922: 36.
LOCALITY. Turkey: Dardanelles ("Dd"), July 1761.
MATERIAL. *Forsskål* 6 (C — 1 sheet with field label, microf. 4: III. 3, 4); 7 (C — 1 sheet, microf. 4: III. 5, 6).

Allium curtum *Boiss. & Gaill.* in Boiss., Diagn. ser. 2(4): 116 (1859); Tackholm 1974: 650.
LOCALITY. Egypt: Alexandria, Ras etin, 1 April 1762.
MATERIAL. *Forsskål* 18 (C — 1 sheet with original field label "Av. 51" and "Ras ettim", microf. 129: I. 7, 8).
NOTE. The specimen was labelled and named with some doubt by F. Kollman, 1983.

Allium desertorum *Forssk.* 1775: 72 (LXV No. 200; Cent. III No. 25); Christensen 1922: 18; Tackholm 1974: 652, pl. 238B.
VERNACULAR NAME. *Zaaetemân, Zaeitaeman* (Arabic).
LOCALITY. Egypt: Cairo ("In desertis Káhirinis"), 1762.
MATERIAL. *Forsskål* 19 (C — 1 sheet, type of *A. desertorum*, microf. 4: II. 7, 8); 922 (C — 1 sheet ex herb. Hofman Bang, type of *A. desertorum*, microf. 4: III. 1, 2).
Forsskål s.n. (LD — 1 sheet ex herb. Retzius, type of *A. desertorum*).

Crinum album (*Forssk.*) *Herb.*, Amaryll.: 272 (1837).
Amaryllis alba Forssk. 1775: 209 (CIX No. 232; Cent. VIII No. 37); Christensen 1922: 34.
VERNACULAR NAME. *Soraf* (Arabic).
LOCALITY. Yemen: Kusma ("Kurmae"), 1763.
MATERIAL. *Forsskål* 21 (C — 1 sheet, type of *A. alba*, microf. 117: I. 1–2).
NOTE. Neither Herbert nor Christensen saw the type specimen of *A. alba*. This species was hitherto known as *C. yemense* Deflers 1889: 209 (Yemen: Schibâm *Deflers* 335, s.n., isotypes P); Schwartz 1939: 354.

Pancratium maritimum *L.* 1753: 291; Tackholm 1974: 657, pl. 241.
P. illyricum sensu Forssk. 1775: 209 (LXV No. 196; Cent. VIII No. 36); Christensen 1922: 34.
VERNACULAR NAME. *Susann* (Arabic).
LOCALITY. Egypt: Alexandria garden ("Alexandriae, hortensis"), 1761–62.
MATERIAL. *Forsskål* 1542 (C — 1 sheet ex herb. Hornemann, microf. nil); 1543 (C — 1 sheet ex herb. Hofman Bang, microf. 116: III. 6, 7).

Pancratium maximum *Forssk.* 1775: 72 (CIX No. 231; Cent. III No. 24); Christensen 1922: 18.
LOCALITY. Yemen: Taizz ("Taaes"), 1763.
MATERIAL. *Niebuhr* in *Herb. Forsskål* 23 (C — 1 sheet, type of *P. maximum*, microf. 75: II. 7, 8).
NOTE. This is one of the few plants in the collection attributed to Niebuhr.

Scadoxus multiflorus (*Martyn*) *Raf.*, Fl. Tellur. 4: 19 (1838); Friis & Nordal in Norw. Journ. Bot. 23: 64 (1976).
Amaryllis coccineus sensu Forssk. 1775: 75 (CX No. 244; Cent. III No. 36), non L.; Christensen 1922: 18.
VERNACULAR NAMES. *Hömhömet el hanasch, Voket et hannasch* (Arabic).
LOCALITY. Yemen: Al Hadiyah ("Hadie"), 1763.
MATERIAL. *Forsskål* no specimen found.
NOTE. It is likely that the record of a specimen of this species in Herbarium Forsskålii by Christensen is due to the confusion with a specimen of *Chlorophytum tetraphyllum* (L.f.) Bak. (q.v.). Such a specimen was found in the cover marked 'Haemanthus coccineus' by Christensen. *H. coccineus* is a South African species, while the somewhat similar *S. multiflorus* occurs in the Yemen.

ARACEAE

Arisaema bottae *Schott*, Prodr. Aroid.: 42 (1860); Schwartz 1939: 345; Hepper & Wood in Kew Bull. 38: 83 (1983).
Arum pentaphyllum Forssk. 1775: 157 (CXX No. 526; Cent. VI. No. 6), non L.; Christensen 1922: 27.
VERNACULAR NAME. *Dochaf* (Arabic).
LOCALITY. Yemen: Jebel Barad ("in Monte Barah"), 1763.
MATERIAL. *Forsskål* 4 (C — 1 sheet, type of *A. pentaphyllum*, microf. 10: I. 2–4).
NOTE. According to a note pinned to the sheet Schott doubted it was *A. bottae* and he identified it as *A. schimperianum*, but we are convinced that it is *A. bottae*. The separate leaf is almost certainly of *Sauromatum venosum* (Ait.) Kunth.

Arisaema flavum (*Forssk.*) *Schott*, Prodr. Syst. Aroid.: 40 (1860); Schwartz 1939: 345.
Arum flavum Forssk. 1775: 157 (CXX No. 525, Cent. VI No. 5); Christensen 1922: 27.
VERNACULAR NAME. *Dochaf* (Arabic).
LOCALITY. Yemen: Taizz ("Ad Taaes"), June 1763.
MATERIAL. *Forsskål* 2 (C — 1 sheet, type of *A. flavum*, microf. 9: III. 6, 7); 3 (C — 1 sheet, type of *A. flavum*, microf. 10: I. 1, 2).

Dracunculus vulgaris *Schott & Endl.*, Melet. Bot.: 17 (1832); Fl. Turk. 8: 63 (1984).
LOCALITY. Unspecified, field label not clear, probably Turkey.
MATERIAL. *Forsskål* 5 (C — 1 sheet with field label, microf. 117: I. 3, 4).
NOTE. Named by Engler who considered it to be more like *D. canariensis* than *D. vulgaris*.

COMMELINACEAE

Aneilema forskalei *Kunth*, Enum. Pl. 4: 71 (1843); Schweinfurth 1894: 58; Schwartz 1939: 346.
Commelina tuberosa sensu Forssk. 1775: 12 (CIII No. 39; Cent. I. 33), non L. (1763); Christensen 1922: 11.
C. paniculata Vahl 1805: 179, non Hill (1773).
VERNACULAR NAME. *Vaalan* (Arabic).
LOCALITY. Yemen: Al Hadiyah ("Hadîe", "Bolghose" on field label), Mar. 1763.
MATERIAL. *Forsskål* 31 (C — 1 sheet, type of *C. paniculata* & *A. forskalei,* microf. 122: I. 3, 4); 32 (C — 1 sheet with field label, type of *C. paniculata* & *A. forskalei*, microf. 29: I. 7, 8); 33 (C — 1 sheet, type of *C. paniculata* & *A. forskalei*, microf. 29: II. 1, 2).
NOTE. Annotated by R.B. Faden, 1978.

Commelina africana *L.* 1753: 41.
C. divaricata Vahl 1805: 169; Schwartz 1939: 348; Christensen 1922: 11.
C. benghalensis sensu Forssk. 1775: 12 (CIII: 38; Cent. I No. 34), non L.
LOCALITY. Yemen: Al Hadiyah ("Hadîe"), Mar. 1763.
MATERIAL. *Forsskål* 37 (C — 1 sheet, microf. 29: I. 5, 6).
NOTE. Annotated by R.B. Faden, 1978.

Commelina benghalensis *L.* 1753: 41; Schwartz 1939: 347.
C. canescens Vahl 1805: 173; Christensen 1922: 38.
LOCALITY. Yemen: unspecified, 1763.
MATERIAL. *Forsskål* 36 (C — 1 sheet, type of *C. canescens*, microf. 121: III. 7, 8).
NOTE. Annotated by R.B. Faden, 1978.

Commelina commelinoides *Forssk.* 1775: 12 (CIII No. 40; Cent. I No. 35); Deflers 1896: 330; Christensen 1922: 11; Schwartz 1939: 348.
VERNACULAR NAME. *Kunan* (Arabic).
LOCALITY. Yemen: unspecified ("Mm" = Montium regio media), 1763.
MATERIAL. *Forsskål,* no type specimen found.
NOTE. An obscure species. The name should not be adopted if it upsets widely accepted nomenclature.

Commelina forskalei *Vahl* 1805: 172; Deflers 1889: 214; Schweinfurth 1896: 56; Schwartz 1939: 347.
LOCALITY. Yemen: unspecified, 1763.
MATERIAL. *Forsskål* 35 (C — 1 sheet, type of *C. forskalei*, microf. 122: I. 1, 2); 1129 (C — 1 sheet, type of *C. forskalei*, microf. nil).

Cyanotis sp.
Tillandsia (err. *Tradescantia*) *decumbens* Forssk. 1775: 72 (CIX No. 229; Cent. III No. 23); Vahl 1790: 27.
LOCALITY. Yemen: Taizz ("Taaes"), 1763.
MATERIAL. *Forsskål* 34 (C — 1 sheet, microf. 107: II. 7, 8); 1785 (C — 1 sheet ex herb. Schumacher, microf. 107: II. 5, 6).
Forsskål s.n. (LD — 1 sheet ex herb. Retzius).
NOTE. The LD specimen was formerly named on reverse "*Tillandsia decumbens*" and "*Tradescantia papilionacea*", also named and labelled "Possibly *Cyanotis tuberosa* (Roxb.) Schult.f. a stunted deformed specimen" by Rao and Kammathag (?) 1963, but that is an Indian species.

CYPERACEAE

Bolboschoenus maritimus (*L.*) *Palla* in Allg. Bot. Zeitschr. 57. Beil 3 (1911).
Scirpus maritimus L. 1753: 51; Vahl 1805: 258.
S. tuberosus Desf. — Tackholm & Drar 1950: 27; Tackholm 1974: 778, pl. 287.
S. corymbosus Forssk. 1775: 14 (LX No. 27; Cent. 1: 44); Christensen 1922: 12.
VERNACULAR NAME. *Depsjae* (Arabic).
LOCALITY. Egypt: Alexandria & Cairo ("Káhirae in locis inundatis ad agros"), Dec. 1761.
MATERIAL. *Forsskål* 1702 (C — 1 sheet with field label "A.124. Alex. vern. ad Caliss.", ? type of *S. corymbosus*, microf. 96: II. 7, 8).
NOTE. While there is no doubt that this specimen is *B. maritimus*, there is reason to be uncertain whether it is the type of *Scirpus corymbosus* since the description does not quite fit and Forsskål's label indicates Alexandria not Cairo.

Cladium mariscus (*L.*) *Pohl*, Tentamen Fl. Bohem. 1: 32 (1809); Tackholm & Drar 1950: 10; Tackholm 1974: 773.
LOCALITY. Egypt: without locality, 1761–2.
MATERIAL. *Forsskål* 1118 (C — 1 sheet, microf. nil).
NOTE. This sheet was labelled by Böckeler as *C. mariscus*, and by C.B. Clarke 1892 as *C. jamaicense* Crantz.

Cyperus alopecuroides *Rottb.*, Descr. Ic. Nov. Pl.: 38, t. 8 f. 2 (1773); Vahl 1805: 368; Schweinfurth 1894: 49; Schwartz 1939: 337; Tackholm & Drar 1950: 89; Tackholm 1974: 789, pl. 292.
C. fastigiatus Forssk. 1775: 14 (LIX No. 18; Cent. I No. 41); Rottb., Descr. Pl. Rar. Ic. Progr.: 18 (1772), Descr. Ic. Nov. Pl.: 321, t. 7, f. 2 (1773); Vahl 1805: 367; Christensen 1922: 11.
VERNACULAR NAMES. *Samâr dabbûs, Samsûr dubbus* (Arabic).
LOCALITY. Egypt: Rashid ("Rosettae"), 1761–62.
MATERIAL. *Forsskål* no type specimen found.
NOTE. Tackholm & Drar, Fl. Egypt 2: 86 (1950) entered *C. fastigiatus* with a ? under *C. dives* Delile (1813).

Cyperus articulatus *L.* 1753: 44; Rottb., Descr. Ic. Nov. Pl.: 26 (1773); Vahl 1790: 7; Deflers 1889: 216; Schwartz 1939: 337; Tackholm & Drar 1950: 82, Tackholm 1974: 789, pl. 290C.
C. niloticus Forssk. 1775: 13 (LX No. 22; Cent. I No. 37); Vahl 1805: 302; Christensen 1922: 11.
LOCALITY. Egypt: Rashid ("Ad ripam Nili occidentalem limnosam Rosettae"), Nov. 1761.
MATERIAL. *Forsskål* 1206 (C — 1 sheet ex herb. Vahl, type of *C. niloticus*, microf. 37: II. 3, 4).

Cyperus capitatus *Vandelli*, Fasc. Pl. 5 (1771); Tackholm & Drar 1950: 52; Hooper in Israel Journ. Bot. 26: 98 (1977).
Schoenus mucronatus L. 1753: 42; Vahl 1805: 213.
Scirpus kalli Forssk. 1775: 15 (LX No. 28; Cent. I No. 48), as *Scirpus kalli 3 Alpini* (see note below); Christensen 1922: 12.
VERNACULAR NAME. '*Sae aed*' (Arabic).
LOCALITY. Egypt: Alexandria ("Alexandriae") 1761–2.
MATERIAL. *Forsskål* 1705 (C — 1 sheet ex herb. Hornemann, with field label "34. Scirpus kalli novus. Alex. ad salinas", type of *S. kalli*, microf. 96: III. 5–6); 1706 (C — 1 sheet ex herb. Vahl, type of *S. kalli*, microf. 96: III. 7, 8); 1707 (C — 1 sheet ex herb. Hornemann,

with Hornemann's label "257. Scirp. mucron. L. Scirp. kalli Forsk. ad salinas Alexandria", type, microf. 97: I. 1–2).
NOTE. S.S. Hooper in Israel Journ. Bot. 26: 98–99 (1977) besides discussing the nomenclature, explains the name *S. kalli 3 Alpina* as being a reference to *P. Alpinus*, Hist. Aegypti Nat. (1735).

Cyperus conglomeratus *Rottb.*, Descr. Pl. Rar. Progr.: 16 (1772), & Descr. Ic. Rar. Nov. Pl.: 21, t. 15, f. 7 (1773); Tackholm & Drar 1950: 49; Tackholm 1974: 784.
C. complanatus Forssk. 1775: 14 (LX No. 19; Cent. I No. 42); Christensen 1922: 11.
VERNACULAR NAME. *Saeaed* (Arabic).
LOCALITY. Egypt: Rashid ("Rosettae, in agris oryzae humidis; copiosae"), Oct. 1761.
MATERIAL. *Forsskål* 1196 (C — 1 sheet ex herb. Vahl, type of *C. complanatus*, microf. 123: III. 3, 4).

C. jeminicus Rottb., Descr. Pl. Rar. Progr.: 24, t. 15 f. 7 (1772), & Descr. Ic. Nov. Pl.: 25, t. 8 (1773).
C. conglomeratus var. *jeminicus* (Rottb.) Kük. (1936).
LOCALITY. Saudi Arabia: Al Qunfudhah ("Gomfoda"), Nov. 1762.
MATERIAL. *Forsskål* 1197 (C — 1 sheet ex herb. Vahl, with field label "Gomfoda", type of *C. jeminicus*, microf. 123: III. 5, 6).
NOTE. Rottböll evidently thought that 'Gomfoda' was in Yemen, hence the epithet *jeminicus*.Tackholm & Drar (Fl. Egypt 2: 54 (1950)) placed *C. complanatus* in the synonymy of *C. difformis*. It is not certain that No. 1196 came from Egypt, rather than Saudi Arabia.

Cyperus cruentus *Rottb.*, Descr. Pl. Rar. Progr.: 17 (1772), & Descr. Ic. Nov. Pl.: 21, t. 5, f. 1 (1773); Vahl 1790: 8; 1805: 351.
C. conglomeratus Rottb. var. *multi-glumis* (Boek.) Kük. (1936) — Schwartz 1939: 339.
C. globosus Forssk. 1775: 13 (CIII No. 46; Cent. I No. 40); Christensen 1922: 11.
VERNACULAR NAME. *Zaráa* (Arabic).
LOCALITY. Yemen: Al Hadiyah ("Hadie"), 1763.
MATERIAL. *Forsskål* 1188 (C — 1 sheet with field label "Bolghose", type of *C. cruentus* & *C. globosus*, microf. nil).
NOTE. This specimen was not seen by Christensen.

Cyperus effusus *Rottb.*, Descr. Pl. Rar. Progr.: 16 (1772) & Descr. Ic. Nov. Pl.: 22, t. 12, fig. 3 (1773).
C. conglomeratus var. *effusus* (Rottb.) Boiss. — Schwartz 1939: 340.
LOCALITY. Yemen: Al Mukha ("ad Mokkam") May 1763.
MATERIAL. *Forsskål* 1189 (C — 1 sheet ex herb. Vahl, microf. 123, III. 7, 8); 1190 (C — 1 sheet, microf. 124: I. 1, 2).
NOTE. This species was not included in Forsskål's book. The locality is written on the sheet. Determined as *C. conglomeratus* Rottb. var. *effusus* (Rottb.) Kük. by C. Kükenthal (1930).

Cyperus fuscus *L.* 1753: 46; Rottb., Descr. Pl. Rar. Prog.: 21 (1772); Vahl 1805: 336; Tackholm & Drar 1950: 55; Tackholm 1974: 785.
C. ferrugineus Forssk. 1775: 14 (LX No. 20; Cent. I: 43); Christensen 1922: 11.
C. forskolei Dietr., Sp. Pl. 2: 251 (1833).
VERNACULAR NAMES. *Sööd, Saeaed, N'ghil* (Arabic).
LOCALITY. Egypt: Cairo and Rashid ("Gramen omnium copiosissimum & fere unicum pratorum humidorum juxta Nilum"), 1761–62.

MATERIAL. *Forsskål* 1191 (C — 1 sheet ex herb. Vahl, type of *C. ferrugineus*, microf. 37: I. 1, 2); 1192 (C — 1 sheet, type of *C. ferrugineus*, microf. 37: I. 3, 4); 1193 (C — 1 sheet ex herb. Hofman Bang, type of *C. ferrugineus*, microf. 37: I.5, 6); 1194 (C — 1 sheet, label "229 ad Rosette", mixed with "292" *Frimbristylis bisumbellata* & "18" *C. laevigatus*, microf. 123: III. 1, 2).

NOTE. Of *Forsskål* "229" S. Blake in 1948 noted "Is this the type of *Cyperus lateralis* Forssk., of which I have no description, Kükenthal cites *Forsskål* 38 from Rosetta under *Cyperus laevigatus*". Sheet No. 1194 seems to come from Hornemann and is stated to have been in the herbarium of Hofman Bang.

Cyperus fuscus L. forma **virescens** (*Hoffm.*) *Vahl* (1806) — Tackholm & Drar 1950: 56: Tackholm 1974: 785.

LOCALITY. Egypt: Rashid (Rosetta), Nov. 1761.

MATERIAL. *Forsskål* 1195 (C — 1 sheet, microf. nil).

NOTE. Although not indicated, No. 1195 seems to have been in the Liebmann herbarium.

Cyperus involucratus *Rottb.*, Descr. Pl. Rar.: 22 (1772).

C. flabelliformis Rottb., Descr. Ic. Nov. Pl.: 42, t. 12, f. 2 (1773); Vahl 1805: 322; Deflers 1889: 216; Schweinfurth 1894: 48.

C. alternifolius L. ssp. *flabelliformis* (Rottb.) Kük. (1936) — Schwartz 1939: 339.

C. gradatus Forssk. 1775: 13 (CIII No. 44; Cent. I No. 39); Christensen 1922: 11.

LOCALITY. Yemen: Wadi Zabid ("Uadi Zebîd"), 1763.

MATERIAL. *Forsskål* no type specimen found.

Cyperus laevigatus *L.* 1771: 179; Tackholm & Drar 1950: 43; Tackholm 1974: 784, pl. 289B.

C. mucronatus Rottb., Descr. Pl. Rar. Progr.: 17 (1772); Descr. Ic. Nov. Pl.: 19, t. 8, f. 4 (1773); Vahl 1790: 7.

C. lateralis Forssk. 1775: 13 (LX No. 23; Cent. I No. 38); Christensen 1922: 11.

LOCALITY. Egypt: Rashid ("Rs"), Nov. 1761.

MATERIAL. *Forsskål* 1204 (C — 1 sheet ex herb. Vahl, type of *C. mucronatus* and *C. lateralis*, microf. 37: I. 7, 8); 1205 (C — 1 sheet ex herb. Vahl, type of *C. mucronatus* and *C. lateralis*, microf. 37: II. 1, 2); 1194 (C — 1 sheet with field label "Ro. 18" see note under *C. fuscus*, microf. 123: III. 1, 2).

Cyperus longus *L.* 1753: 45; Tackholm 1974: 786, pl. 291B; Fl. Turk. 9: 35 (1985).

LOCALITY. Turkey: Izmir ("Smirna"), 30 June–10 July 1761.

MATERIAL. *Forsskål* 1203 (C — 1 sheet with locality "ad Smyrnam" written on it, microf. nil).

LOCALITY. ? Egypt, 1761–62.

MATERIAL. *Forsskål* 1201 (C — 1 sheet ex herb. Vahl, microf. 124: I. 3, 4); 1202 (C — 1 sheet ex herb. Vahl, microf. 124: I. 5, 6).

NOTE. These specimens were not referred to by either Forsskål or Christensen and the species was not included in their works. There is no indication on the sheets that Nos. 1201 & 1202 came from Egypt, but Vahl stated with certainty (Vahl 1806: 346) that the species occurs in Egypt.

Cyperus minimus sensu Forssk. 1775: 202 (XVIII No. 18; Cent. VIII No. 3), non L; Christensen 1922: 33.

LOCALITY. Turkey: Dardanelles ("Dardanelli", "Ad littora Maris Marmora").

MATERIAL. *Forsskål* no specimen found.

NOTE. The identity of this plant is uncertain from the description alone.

Cyperus rotundus *L.* 1753: 45; Tackholm & Drar 1950: 69; Tackholm 1974: 786, pl. 291A.
C. hexastachyos Rottb., Descr. Pl. Rar. Progr: 25, t. 8 (1772), & Descr. Ic. Nov. Pl.: 28, t. 14, f. 2 (1773).
LOCALITY. Egypt: Cairo, 1761–62.
MATERIAL. *Forsskål* 1199 (C — 1 sheet ex herb. Vahl, type of *C. hexastachyos*, microf. 123: II. 5, 6); 1200 (C — 1 sheet ex herb. Rottboell, type of *C. hexastachyos*, microf. 123: II. 7, 8); 1198 (C — 1 sheet ex herb. Vahl, with field label "Ca. 3", see note, microf. 36, III. 7, 8).
NOTE. The name *C. ferrugineus* is written on the back of sheet "Ca. 3" and it is kept in a folder with other sheets of that species, but clearly it does not belong there.

Eleocharis geniculata (*L.*) *Roem. & Schult.*, Syst. Veg. 2: 150 (1817).
LOCALITY. Unspecified, probably Yemen as it is a tropical species.
MATERIAL. *Forsskål* 1243 (C — 1 sheet ex herb. Hofmann Bang, microf. 124: I. 7, 8).

Fimbristylis bis-umbellata (*Forssk.*) *Bubani*, Dodecanthea: 30 (1850); Tackholm & Drar 1950: 13; Tackholm 1974: 774, pl. 285C.
F. dichotoma Vahl 1805: 287, as to descr. and citations only.
Scirpus bis-umbellatus Forssk. 1775: 15 (LX No. 25, Cent. I No. 46); Christensen 1922: 12.
LOCALITY. Egypt: Rashid and Cairo ("habitat in locis littorelis inundatis insularum Niloticarum"), 1761.
MATERIAL. *Forsskål* 1194 (C — 1 sheet with label by Hornemann "292", mixed with *C. fuscus* & *laevigatus* with label 'Rosette', microf. 123: III. 1, 2); C — 1 sheet, type of *S. bis-umbellatum*, microf. 96: II. 5, 6).
Forsskål s.n. (BM — 1 sheet, type of *S. bis-umbellatus*).

Kyllinga monocephala *Rottb.*, Descr. Ic. Nov. Pl. 13, t. 4, f. 4 (1973).
LOCALITY. Unspecified, probably Yemen.
MATERIAL. *Forsskål* s.n. (LD — 1 sheet ex herb. Retzius).

Schoenoplectus articulatus (*L.*) *Palla* in Engl. Jahrb. 10: 299 (1889).
Scirpus articulatus L. 175: 47; Vahl 1790: 8, 1805: 258.
S. fistulosus Forssk. 1775: 14 (LX No. 26; Cent. 1: 45); Christensen 1922: 12; Tackholm & Drar 1950: 21; Tackholm 1974: 777.
LOCALITY. Egypt: Rashid ("Rosettae in agro humido Trifoliis consito"), Nov. 1761.
MATERIAL. *Forsskål* 1703 (C — 1 sheet, type of *S. fistulosus*, microf. 96: III. 1–2); 1704 (C — 1 sheet, type of *S. fistulosus*, microf. 96: III. 3, 4).

Schoenus incanus *Forssk.* 1775: 12 (CIII No. 42; Cent. I No. 36); Christensen 1922: 11.
VERNACULAR NAME. *Aejn el bagar* (Arabic).
LOCALITY. Yemen: Wadi Mawr ("*Môr*"), 1763.
MATERIAL. *Forsskål* no type specimen found.
NOTE. This name has been ignored by Schweinfurth, Deflers and Schwartz presumably because no specimen is extant, but it is validly published and it may well be the earliest for the species, whatever that may prove to be.

Scirpoides holoschoenus (*L.*) *Soják* in Cas. Nár. Muz (Prague): 140 (3–4): 127 (1972).
Scirpus holoschoenus L. — Schwartz 1939: 342; Tackholm 1974: 778, pl. 288D.
LOCALITY. Unspecified.
MATERIAL. *Forsskål* 1699 (C — 1 sheet ex herb. Hornemann, microf. nil); 1700 (C — 1 sheet ex herb. Hornemann, microf. nil).

Scirpus lateralis *Forssk.* 1775: 15 (CIII No. 47; Cent. I. No. 47); Vahl 1805: 280; Christensen 1922: 12.
VERNACULAR NAME. *Hallâl* (Arabic).
LOCALITY. Yemen: Wadi Zabid ("Uadi Zebîd"), 1763.
MATERIAL. *Forsskål* no type specimen found.
NOTE. As the type is missing the identity of this species is uncertain. Christensen (1922: 12) thought it was probably *Isolepis uninodis* Del. (*Scirpus supinus* L. var. *uninodis* (Delile) Asch. & Schweinf.) now *Scirpus uninodis* (Delile) Boiss.

GRAMINEAE

Aegilops bicornis (*Forssk.*) *Jaub. & Spach*, Illustr. Pl. Or. 4: 10, t. 309 (1850); Tackholm & Drar 1941: 269; Tackholm 1974: 702; Fl. Palaest. 4: 170, pl. 208 (1986).
Triticum bicorne Forssk. 1775: 26 (LXI No. 86; Cent. I No. 91); Christensen 1922: 13.
LOCALITY. Egypt: Alexandria ("*Alexandriae*"), April 1762.
MATERIAL. *Forsskål* no type specimen found.

Aegilops geniculata *Roth*, Bot. Abh.: 45 (1787); Fl. Europ. 5: 201 (1980).
A. ovata L. 1753: 1050 pro parte; Forssk. 1775: XII No. 256.
LOCALITY. France: Marseille ("Estac"), May 1761.
MATERIAL. *Forsskål* 1250 (C — 1 sheet ex herb. Hornemann, microf. 3: I. 3, 4).

Aegilops kotschyi *Boiss.*, Diagn. ser. 1(7): 129 (1846); Fl. Turk. 9: 241 (1985).
LOCALITY. Without locality, ? Turkey.
MATERIAL. *Forsskål* s.n. (BM — 1 sheet ex herb. Banks).
Forsskål s.n. (LD — 1 sheet ex herb. Retzius).

Aeluropus lagopoides (*L.*) *Trin. ex Thwaites*, Enum. Pl. Zeyl.: 3: 374 (1864); Tackholm & Drar 1941: 196; Tackholm 1974: 692, pl. 256B.
LOCALITY. Egypt: Alexandria, 1761–62.
MATERIAL. *Forsskål* 42 (C — 1 sheet, microf. 125: III. 3, 4); 67 (C — 1 sheet with field label "As. 2"; microf. 125: III. 1, 2); 140 (C — 1 sheet, microf. 125: III. 7, 8); 1209 (C — 1 sheet ex herb. Hofman Bang, microf. 37: II. 5, 6); 1210 (C — 1 sheet ex herb. Hofman Bang, microf. 37: II. 7, 8); 1211 (C — 1 sheet ex herb. Hofman Bang, microf. 37: III. 1, 2); 1212 (C — 1 sheet ex herb. Hofman Bang, microf. 37: III. 3, 4).

Cynosurus lima sensu Forssk. 1775: CIV No. 73, non L.
LOCALITY. Saudi Arabia: Al Qunfudhah, 1762.
MATERIAL. *Forsskål* 57 (C — 1 sheet, "Gumfuda" on field label, microf. 36: III. 5, 6).
LOCALITY. Yemen: Ghurab & Al Mukha ("Ghorâb, Mochha"), 1763.
MATERIAL. *Forsskål* 1185 (C — 1 sheet ex herb. Liebmann given by Hofman Bang, microf. 36: III. 3, 4).

Ammophila arenaria (*L.*) *Link*, Hort. Berol. 1: 105 (1827); Fl. Europ. 5: 236 (1980).
LOCALITY. Unspecified.
MATERIAL. *Forsskål* 933 (C — 1 sheet ex herb. Hornemann, microf. 126: I. 1, 2).

[**Andropogon bicornis** L.]
LOCALITY. A South American species.
MATERIAL. Attributed to *Forsskål* "572" (C — 1 sheet ex herb. Schumacher, microf. 125: III. 5, 6).
NOTE. See also under *Cymbopogon caesius* (Hook. & Arn.) Stapf.

Andropogon ramosum *Forssk.* 1775: 173 (CXXIII No. 592; Cent. VI No. 71); Christensen 1922: 29.
VERNACULAR NAME. *Auvid* (Arabic).
LOCALITY. Yemen: Al Hadiyah ("Hadîe"), 1763.
MATERIAL. *Forsskål* no type specimen found.
NOTE. In the absence of a type this remains unknown.

Aristida adscensionis *L.* 1753: 82; Forssk. 1775: CIV No. 80; Vahl 1791: 25; Schwartz 1939: 322.
A. paniculata Forssk. 1775: 25 (CIV No. 81; Cent. I. No. 88); Christensen 1922: 13.
VERNACULAR NAME. *Daku esschaeha, Dhenneb et tôr, Höbb el adjais* (Arabic).
LOCALITY. Yemen: Hays to Zabid ("Inter Haes et Zebid" according to label on No. 124), 5 Apr. 1763.
MATERIAL. *Forsskål* 119 (C — 1 sheet, type of *A. paniculata*, microf. 9: II. 5, 6); 123 (C — 1 sheet, type of *A. paniculata*, microf. 8: III. 3, 4); 124 (C — 1 sheet with field label, type of *A. paniculata*, microf. 8: III. 1, 2).

Arundo donax *L.* 1753: 81; Forssk. 1775: 23 (LXI No. 73, CIV No. 78; Cent. I No. 85) pro parte, name but not description which applies to *Phragmites australis;* Christensen 1922: 13; Schwartz 1939: 332; Tackholm & Drar 1941: 203.
VERNACULAR NAME. *Kasab* (Arabic).
LOCALITY. Yemen: unlocalised ("Ad rivos Yemenis frequens"), 1763.
MATERIAL. *Forsskål* no specimen found.
LOCALITY. Egypt: Alexandria ("Alex. in hortis" on own label on reverse of sheet), 1761–2.
MATERIAL. *Forsskål* 965 (C — 1 sheet ex herb. Hornemann, microf. 126: I. 3, 4).

Avena elata *Forssk.* 1775: XIX No. 40; Christensen 1922: 35.
LOCALITY. Turkey: Istanbul, Büyükdere ("Bj = Bujuchtari"), Sept. 1761.
MATERIAL. *Forsskål* no type specimen found.
NOTE. The name *Avena elata* is validated by the brief description: "Parva, subspicata; caule & foliis villosis". It was not taken up in the Flora of Turkey. Baum, 'Oats: wild and cultivated' suggests that it may be *Helictotrichon bromoides* (Gouan) C.E. Hubbard. The name should not be adopted if it upsets widely accepted nomenclature.

Bothriochloa ischaemum (*L.*) *Keng* in Contr. Biol. Lab. Sci. Soc. China, Bot. Ser. 10: 201 (1936); Fl. Turk. 9: 612 (1985).
Andropogon ischaemum L. 1753: 1047; Forssk. 1775: 219 (XXXV No. 443; Cent. VIII No. 97); Christensen 1922: 35.
LOCALITY. Turkey: Istanbul, Belgrad ("Belgrad prope Constantinop.").
MATERIAL. *Forsskål* no specimen found.

Brachiaria mutica (*Forssk.*) *Stapf* in F.T.A. 9: 526 (1919); Tackholm & Drar 1941: 430; F.W.T.A. ed. 2, 3: 443 (1972); Tackholm 1974: 743; Fl. Palaest. 4: 304, pl. 406 (1986).
Panicum muticum Forssk. 1775: 20 (LX No. 45; Cent. I No. 66); Christensen 1922: 12.
Urochloa mutica (Forssk.) Nguyen in Nov. Sist. Vys. Rast. 1966: 13 (1966).
LOCALITY. Egypt: Rashid ("Rosettae"), 2–6 Nov. 1761.
MATERIAL. *Forsskål* 86 (C — 1 sheet, type of *P. muticum*, microf. 76: II. 7, 8).

Panicum appressum Forssk. 1775: 20 (LX No. 44 CIV No. 53; Cent. I No. 65); Christensen 1922: 12.
VERNACULAR NAME. *Faelek, Faelaek, Eflik* (Arabic).

LOCALITY. Egypt: Alexandria & Rashid, 1761; Yemen: Al Luhayyah & Wadi Mawr ('Lohajae, Môr'), Jan. 1763.
MATERIAL. *Forsskål* 87 (C — 1 sheet, type of *P. appressum*, microf. 76: II. 5, 6). *Forsskål* s.n. (BM — 2 sheets).
NOTE. J.R.I. Wood (pers. com.) doubts whether the Yemen record is correct as this species is not otherwise recorded from there.

Brachypodium phoenicioides (*L.*) *Roem. & Schult.*, Syst. Veg. 2: 740 (1817); Fl. Europ. 5: 190 (1980).
LOCALITY. Unspecified, prob. France.
MATERIAL. *Forsskål* 1009 (C — 1 sheet ex herb. Hornemann, microf. nil); 1010 (C — 1 sheet ex herb. Hofman Bang, microf. 16: I. 5, 6).
NOTE. A species of the Mediterranean area which does not occur in Egypt or Yemen. Sheet No. 1010 carries the inscription "dedit Dom. Hornemann" on the back.

Brachypodium retusum (*Pers.*) *P. Beauv.*, Ess. Agrost: 101, 155 (1812); Fl. Europ. 5: 190 (1980).
Bromus pinnatus sensu Forssk. 1775: III No. 20, non L.
LOCALITY. France: Marseille ("Estac"), May 1761.
MATERIAL. *Forsskål* 99 (C — 1 sheet, microf. 16: I. 3, 4); 100 (C — 1 sheet, microf. 16: I. 1, 2).

Bromus hordeaceus *L.* 1753: 77; Fl. Europ. 5: 187 (1980); Fl. Turk. 9: 278 (1985).
LOCALITY. Unspecified.
MATERIAL. *Forsskål* 1013 (C — 1 sheet ex herb. Vahl, microf. 126: I. 5, 6).

Bromus intermedius *Guss.*, Fl. Sic. Prodr. 1: 114 (1827); Fl. Turk. 9: 280, fig. 11/2 (1985).
B. rubens sensu Forssk. 1775: XIX No. 36, non L.
LOCALITY. Turkey: Gokçeada I. ("Imros"), July 1761.
MATERIAL. *Forsskål* 92 (C — 1 sheet with field label "Im. 5", microf. 16: II. 1, 2).

Bromus japonicus *Thunb.*, Fl. Jap.: 52, t. 11 (1784); Tackholm & Drar 1941: 152; Tackholm 1974: 678.
LOCALITY. Egypt: Cairo, 1761–62.
MATERIAL. *Forsskål* 1014 (C — 1 sheet ? ex herb. Hornemann with field label "Cairi 246", microf. 126: I. 7, 8).

Bromus madritensis *L.*, Cent. Pl. 1: 5 (1755); Vahl 1791: 23; Tackholm & Drar 1941: 160; Tackholm 1974: 681.
Bromus villosus Forssk. 1775: 23 (LXI No. 65; Cent. I No. 79); Christensen 1922: 13.
LOCALITY. Egypt: Alexandria ("*Alexandriae*"), Apr. 1762.
MATERIAL. *Forsskål* 89 (C — 1 sheet, type of *B. villosus*, microf. 16: III. 1, 2); 90 (C — 1 sheet, type of *B. villosus*, microf. 16: II. 7, 8); 91 (C — 1 sheet, type of *B. villosus*, microf. 16: III. 3, 4); 93 (C — 1 sheet, type of *B. villosus*, microf. 16: III. 7, 8).

Bromus rigidus *Roth* in Bot. Mag. (Roemer & Usteri) 4 (10): 2 (1790); Tackholm 1974: 681.
LOCALITY. Egypt: Alexandria, Ras etin ("Ras Atir"), 1761.
MATERIAL. *Forsskål* 1831 (C — 1 sheet ex herb. Hornemann, with field label "58 Bromus nov. crinitus Alex Ras Atir", microf. 16: III. 5, 6).

Bromus rubens *L.*, Cent. Pl. 1: 5 (1755); Tackholm & Drar 1941: 157; Tackholm 1974: 681, pl. 251B.
LOCALITY. Egypt: Alexandria, Ras etin ("Alex. Ras. Atir" on field label), 1761.
MATERIAL. *Forsskål* 1017 (C — 1 sheet ex herb. Hornemann, with field label "57", microf. 126: II. 1, 2); 1018 (C — 1 sheet ex herb. Liebmann, microf. 126: II, 3, 4).
NOTE. Sheet No. 1017 is correctly annotated "Bromus rubens" on the original label.

Bromus squarrosus *L.* 1753: 76; Forssk. 1775: 203 (III No. 23; Cent. VIII No. 5); Christensen 1922: 33; Fl. Europ. 5: 188 (1980).
LOCALITY. France: Marseille ("Estac"), May 1761.
MATERIAL. *Forsskål* 95 (C — 1 sheet, microf. 16: II. 3, 4); 96 (C — 1 sheet, microf. 16: II. 5, 6).

Bromus sp.
Bromus arvens[is] sensu Forssk. 1775: CIV No. 64, non L.
LOCALITY. Yemen: Bughah ("Baka" in book, "inter Boka & Kurma" on field label), 1763.
MATERIAL. *Forsskål* 1012 (C — 1 sheet ex herb. Hornemann, with field label, microf. nil).

[? **Calamagrostis purpurascens** R.Br.]
Arundo calamagrostis sensu Forssk. 1775: 23 (LXI No. 78; Cent. I No. 84), non L.; Christensen 1922: 23.
LOCALITY: Egypt: near Suez ("In locis palustribus, Ghobeibe prope Suês"), 10 Oct. 1762.
MATERIAL. *Forsskål* 83 (C — 1 sheet, microf. 10: I. 5, 6).
NOTE. There is evidently a confusion of the material with a North American specimen attributed to Forsskål's collection. T.A. Cope considers it to be "a reasonable match with" the American *C. purpurascens*.

Catapodium rigidum *(L.) C.E. Hubbard ex Dony*, Fl. Bedfordshire: 437 (1953).
Dezmazeria rigida (L.) Tutin (1952) — Fl. Europ. 5: 158 (1980).
LOCALITY. Unspecified.
MATERIAL. *Forsskål* 60 (C — 1 sheet, microf. 128: II. 3, 4); 1065 (C — 1 sheet, microf. 128: II. 1, 2).

Cenchrus biflorus *Roxb.*, Fl. Indica 1: 238 (1820).
Elymus caput-medusae sensu Forssk. 1775: 25 (CIV No. 82; Cent. I No. 89), non L.; Christensen 1922: 13.
VERNACULAR NAME. *Höbb el adjais* (Arabic).
LOCALITY. Yemen: Munayrah ("Menejrae"), Jan. 1763.
MATERIAL. *Forsskål* 72 (C — 1 sheet, microf. 41: II. 3, 4).
Forsskål s.n. (BM — 1 sheet, mixed with *C. echinatus*).

Cenchrus ciliaris *L.*, 1771: 302.
Pennisetum ciliare (L.) Link — Schwartz 1939: 318.
Panicum glaucum β Forssk. 1775: 20 (? CIV No. 54, Cent. I No. 68); Christensen 1922: 12.
VERNACULAR NAME. *Aebaed* (Arabic).
LOCALITY. Presumably Yemen (see last).
MATERIAL. *Forrskal* no specimen found at C.
Forsskål s.n. (BM — 1 sheet ex herb. Banks, type of *P. glaucum β*).

Cenchrus echinatus *L.* 1753: 1050; Fl. Palaest. 4: 313, pl. 418 (1986).
LOCALITY. Unspecified.
MATERIAL. *Forsskål* 1069 (C — 1 sheet, microf. nil).
Forsskål s.n. (BM — 1 sheet, mixed with *C. biflorus*).

Cenchrus setigerus *Vahl* 1806: 395; Schweinfurth 1894: 24; Schwartz 1939: 317; F.T.E.A. Gramin.: 694 (1982).
Panicum glaucum α Forssk. 1775: 20 (CIV No. 54; Cent. I No. 68); Christensen 1922: 12.
VERNACULAR NAME. *Aebaed* (Arabic).
LOCALITY. Saudi Arabia: Al Qunfudhah ("Gumfoda", see Note). Yemen: Surdud & Mawr ("Surdûd, Môr"), 1763.
MATERIAL. *Forsskål* 116 (C — 1 sheet with field label "Gumfuda", microf. 76: I. 5, 6).
NOTE. Sheet No. 116 has been identified by Vahl, Ascherson and Christensen with Forsskål's *Panicum glaucum* α, said to come from Surdud and Mawr in Yemen; however, the field label carries the note "Panicum setiger. simile glaucum. Gumfuda." indicating that the plant was collected at the port of Al Qunfudhah, Saudi Arabia, before arrival in Yemen proper. It also shows that Forsskål had intended to call the plant *Panicum setigerum*. This casts doubt on the identification with *Panicum glaucum* α in Flora Aegyptiaco-Arabica.

Centropodia forskalei (*Vahl*) *Cope* in Kew Bull. 37: 658 (1983).
Avena forskalei Vahl 1791: 25.
Danthonia forskalei (Vahl) R. Br. in Denham & Clapperton, Narr. Trav. North & Centr. Africa, App.: 244 (1826).
Asthenatherum forskalei (Vahl) Nevski in Act. Univ. As. Med., Ser. 8b, Bot. fasc. 17: 8 (1834); Tackholm & Drar 1941: 331; Fl. Iraq 9: 382 (1968); Tackholm 1974: 713.
Avena penssylvannica sensu Forssk. 1775: 23 (LXI No. 69; Cent. I No. 81), non L.; Christensen 1922: 17.
LOCALITY. Egypt: Cairo ("In desertis Káhirinis"), 1761–62.
MATERIAL. *Forsskål* 40 (C — 1 sheet, type of *A. forskalei*, microf. 14: III. 1, 2); 43 (C — 1 sheet, type of *A. forskalei*, microf. 14: II. 7, 8); 980 (C — 1 sheet ex herb. Schumacher, type, microf. 14: III. 3, 4); 981 (C — 1 sheet ex herb. Liebmann, type of *A. forskalei*, microf. 14: III. 5, 6).

Chloris barbata *Swartz*, Fl. Ind. Occ. 1: 200 (1797).
LOCALITY. Unspecified.
MATERIAL. *Forsskål* 1099 (C — 1 sheet mixed with *Eustachys paspaloides*, microf. 127: I. 3, 4).
C. virgata Swartz — Schwartz 1939: 327; Tackholm 1974: 732.

Crypsis schoenoides (*L.*) *Lam.*, Tab. Encycl. 1: 166, t. 42/1 (1791); Tackholm 1974: 719, pl. 263A; F.T.E.A. Gramin.: 353 (1974).
C. aculeata (L.) Ait. (1789) — Tackholm & Drar 1941: 349.
Phalaris vaginiflora Forssk. 1775: 18 (LX No. 37; Cent. I No. 57); Christensen 1922: 12.
Crypsis vaginiflora (Forssk.) Opiz, Naturalientausch 8: 83 (1824).
LOCALITY. Egypt: Alexandria ("Alexandriae"), 1761–62.
MATERIAL. *Forsskål* 52 (C — 1 sheet, type of *P. vaginiflora*, microf. 78: III. 5, 6); 53 (C — 1 sheet, type of *P. vaginiflora*, microf. 78: III. 3, 4).

Cutandia dichotoma (*Forssk.*) *Trabut* in Batt. & Trab., Fl. Alg. Monocot.: 237 (1895); Tackholm & Drar 1941: 170; Tackholm 1974: 686, pl. 253B; Fl. Palaest. 4: 243, pl. 314 (1986).

Festuca dichotoma Forssk. 1775: 22 (LXI No. 61; Cent. I No. 75); Christensen 1922: 13.
Scleropoa dichotoma (Forssk.) Parl., Fl. Ital. 1: 471 (1850).
Triticum maritimum sensu Vahl 1790: 12, 1791, 26 partly, non L.
LOCALITY. Egypt: Alexandria ("*Alexandriae*"), Rashid ("Rosette" field label), Apr. 1762.
MATERIAL. *Forsskål* 47 (C — 1 sheet, microf. 44: II. 3, 4); 1257 (C — 1 sheet ex herb. Schumacher, microf. 44: II. 5, 6); 1258 (C — 1 sheet one specimen with field label "Rosette", lectotype of *F. dichotoma* selected here by C. Stace, mixed with *Cutandia memphitica*, microf. 44: II. 7, 8).
NOTE. There are two original field labels No. 1258, presumably one referring to the plant which is *C. dichotoma*, the other to the plant which is *C. memphitica*.

Cutandia memphitica (*Spreng.*) *Benth.* in Journ. Linn. Soc. Bot. 19: 118 (1881); Tackholm & Drar 1941: 169; Tackholm 1974: 686, pl. 253A.
LOCALITY. Egypt: Rashid (see Note), Apr. 1762.
MATERIAL. *Forsskål* 1259 (C — 1 sheet, microf. 44: II. 1, 2); 1258 (C — 1 sheet one specimen with field label "Rosette", mixed with *C. dichotoma*, microf. 44: II. 7, 8).
NOTE. See note to *C. dichotoma*.

Cutandia maritima (*L.*) *Barbey*, Fl. Sard. Comp.: 72 (1885); Tackholm & Drar, 1941: 169; Tackholm 1974: 684, as (L.) Benth.
Scleropoa maritima (L.) Parl., Fl. Ital. 1: 468 (1850).
Triticum maritimum L. 1762: 127; Vahl 1791: 26 partly.
Festuca lanceolata Forssk. 1775: 22 (LXI No. 60; Cent. I No. 76); Christensen 1922: 13.
LOCALITY. Egypt: Alexandria, Ras etin ("*Alexandriae*"), Apr. 1762.
MATERIAL. *Forsskål* 49 (C — 1 sheet, type of *F. lanceolata*, microf. 44: III. 3, 4); 1261 (C — 1 sheet ex herb. Hornemann, with field label "31 Ras ettin", type of *F. lanceolata*, microf. 127: I. 7, 8).

Cymbopogon caesius (*Nees ex Hook. & Arn.*) *Stapf* in Kew Bull. 1906: 360 (1906); Schwartz 1939: 309.
Andropogon bicorne sensu Forssk. 1775: 173 (CXXIII No. 593; Cent. VI No. 72), non L.; Christensen 1922: 29.
VERNACULAR NAME. *M'hâh* (Arabic).
LOCALITY. Yemen: Al Hadiyah ("Hadîe"), Mar. 1763.
MATERIAL. *Forsskål* 126 (C — 1 sheet, microf. 7: II. 3, 4); 941 (C — 1 sheet ex herb. Hofman Bang, microf. 7: II. 1, 2); 942 (C — 1 sheet ex herb. Schumacher, microf. 7: II. 5, 6).

Cynodon dactylon (*L.*) *Pers.*, Syn. Pl. 1: 85 (1805); Tackholm & Drar 1941: 378; Tackholm 1974: 732, pl. 268B.
Panicum dactylon L. — Forssk. 1775: III. No. 10; XIII No. 6; XIX No. 24; LX No. 43; CIV No. 61.
VERNACULAR NAME. *Agria* or *Agriada* (Greek), *Nisjil, Nedjil* (Arabic, Egypt) *Sabak, Ubal* (Arabic, Yemen).
LOCALITY. Egypt: Cairo, 1761–62.
MATERIAL. *Forsskål* 1544 (C — 1 sheet ex herb. Hofman Bang with field label "Ca. 253", microf. 75: III. 3, 4); 1545 (C — 1 sheet ex herb. Hornemann, microf. nil); 1546, 1547 (C — 2 sheets ex herb. Liebmann, microf. 126: II. 5, 6 and 7, 8).

Cynosurus echinatus *L.* 1753: 72; Forssk. 1775: III. No. 18; Fl. Europ. 5: 171 (1980).
LOCALITY. France: Marseille ("Estac"), May 1761.
MATERIAL. *Forsskål* 1183 (C — 1 sheet, microf. 36: II. 5, 6).

Cynosurus effusus *Link* in Schrader, J. Bot. 1799 (2): 315 (1800); Fl. Turk. 9: 514 (1985).
C. echinatus sensu Forssk. 1775: XX No. 56, non L.
LOCALITY. Turkey: Istanbul, Belgrad ("Belgrad"), Aug. 1761.
MATERIAL. *Forsskål* 56 (C — 1 sheet, microf. 36: II. 3, 4).

Cynosurus ternatus *Forssk.* 1775: 21 (CIV No. 72; Cent. I No. 72); Christensen 1922: 13.
VERNACULAR NAME. *Saher* (Arabic).
LOCALITY. Saudi Arabia: Al Qunfudhah ("Ghomfude"), Nov. 1762.
MATERIAL. *Forsskål* no specimen found.
LOCALITY. Yemen: Wadi Mawr ("Mor"), 1763.
MATERIAL. *Forsskål* 1186 (C — 1 sheet ex herb. Hornemann, with field label "Cynosurus * ternatus, Sâher, Mour", microf. 126: III. 5); 1187 (C — 1 sheet ex herb. Hornemann, microf. nil).
NOTE. The application of this name is still in doubt.

Dactylis glomerata *L.* 1753: 71; Forssk. 1775: III. No. 17; Fl. Europ. 5: 171 (1980); Fl. Turk. 9: 510 (1985).
LOCALITY. France: Marseille ("Estac"), May 1761; Turkey: Gökçeada ("Imros"), July 1761.
MATERIAL. *Forsskål* 1208 (C — 1 sheet ex herb. Hornemann & Hofman Bang, microf. 126: III. 1, 2).

Dactyloctenium aegyptium (*L.*) *Willd.*, Enum. Hort. Berol.: 1029 (1809); Tackholm & Drar 1941: 392; Tackholm 1974: 734, pl. 269A.
Cynosurus aegptius L. 1753: 72; Forssk. 1775: LXI No. 56.
LOCALITY. Egypt: Rashid ("Rosettae"), 31 Oct. 1761.
MATERIAL. *Forsskål* 1213 (C — 1 sheet ex herb. Schumacher, microf. 126: III. 3); 1214 (C — 1 sheet, microf. 126: III. 4).

Dactyloctenium aristatum *Link*, Hort. Reg. Bot. Berol. 1: 59 (1827).
Cynosurus aegyptiacus Forssk. 1775: CIV No. 75.
LOCALITY. Saudi Arabia: Al Qunfudhah ("Ghomfude"), Nov. 1762.
MATERIAL. *Forsskål* 1179 (C — 1 sheet ex herb. Hofman Bang, with field label "Ghomfude", microf. 36: I. 3, 4).

Dasypyrum villosum (*L.*) *Cand.* in Arch. Biol. Vég. Athénes 1: 35, 65 (1901); Fl. Turk. 9: 255 (1985).
Secale cretica sensu Forssk. 1775: XIX No. 45, non L.
LOCALITY. Turkey: Istanbul, Büyükdere ('Bujuchtari'), July–Sept. 1761.
MATERIAL. *Forsskål* 1713 (C — 1 sheet ex herb. Schumacher, microf. 97: II. 5, 6).

Desmostachya bipinnata (*L.*) *Stapf* in Dyer, Fl. Cap. 7: 632 (1900); Tackholm & Drar 1941: 177; Tackholm 1974: 690, pl. 255A.
Cynosurus durus sensu Forssk. 1775: 21 (LX No. 55 CIV No. 74; Cent. I No. 71), non L.; Christensen 1922: 12.
VERNACULAR NAME. *Chalfi, Hȟalfe* (Arabic).
LOCALITY. Egypt: Cairo ("In hortis Káhirinis spontaneus"), 1761–62.
MATERIAL. *Forsskål* 1180 (C — 1 sheet ex herb. Liebmann, microf. 36: I. 5, 6); 1181 (C —1 sheet, with field label "CA 50", microf. 36: I. 7, 8); 1182 (C — 1 sheet, ex herb. Hofman Bang, microf. 36: II. 1, 2).

Dichanthium annulatum (*Forssk.*) *Stapf* in Fl. Trop. Afr. 9: 178 (1917); Tackholm & Drar 1941: 515; Tackholm 1974: 760, pl. 280C; F.T.E.A. Gramin.: 725 (1982); Fl. Palaest. 4: 323, pl. 428 (1986).
Andropogon annulatum Forssk. 1775: 173 (LXXVII No. 541; Cent. VI No. 70); Vahl 1791: 102; Christensen 1922: 29.
LOCALITY: Egypt: Rashid ("Rs = Rosettae", "Passim ad ripas Nili"), June 1762.
MATERIAL. *Forsskål* 127 (C — 1 sheet with original label "Ca. 236", type of *A. annulatum*, microf. 7: I. 7, 8).
Forsskål s.n. (LD — 1 sheet ex herb. Retzius, type of *A. annulatum*).

Digitaria sanguinalis (*L.*) *Scop.*, Fl. Carn. ed. 2, 1: 52 (1771); Fl. Turk. 9: 594 (1985).
Panicum sangu[inale L.] — Forssk. 1775: XVIII No. 20.
LOCALITY. Turkey: Izmir ("Sm."), 30 June–10 July 1761.
MATERIAL. *Forsskål* 114 (C — 1 sheet with field label "Sm. 37", microf. 76: III. 1, 2).

Digitaria velutina (*Forssk.*) *P. Beauv.*, Ess. Agrost.: 51 (1812); Tackholm & Drar 1941: 424; F.T.E.A. Gramin.: 652 (1982).
Phalaris velutina Forssk. 1775: 17 (Cent. I No. 55).
Panicum sanguinale Vahl 1790: 8, non L.
Panicum forskalii C. Christensen 1922: 12.
LOCALITY. Yemen: Bolghose ("Bulghose, montibus Hadiensibus" on specimen, unlocalised in book), Mar. 1763.
MATERIAL. *Forsskål* 112 (C — 1 sheet with field label "Phalaris * velutina. Bolghose", type of all above names, microf. 79: I. 1, 2); 115 (C — 1 sheet, type of all above names, microf. 78: III. 7, 8).

Echinochloa colona (*L.*) *Link*, Hort. Berol. 2: 209 (1833); Tackholm & Drar 1941: 446; Tackholm 1974: 749, pl. 276A.
LOCALITY. Unspecified, probably Egypt.
MATERIAL. *Forsskål* s.n. (1 sheet ex herb. Retzius).

Echinochloa crus-galli (*L.*) *P. Beauv.*, Ess. Agrost.: 53 (1812); Tackholm & Drar 1941: 449; Tackholm 1974: 749, pl. 276B; Fl. Turk. 9: 591 (1985).
Panicum crusgalli L. 1753: 56; Forssk. 1775: 19 (LX No. 41; Cent. I No. 63); Christensen 1922: 12.
VERNACULAR NAME. *Kechzi* (Greek).
LOCALITY. Turkey: Istanbul & Izmir ("Belgrad", Smirna"), 1761.
MATERIAL. *Forsskål* 1236 (C — 1 sheet ex herb. Hornemann, with field label "Belgrad. Co. O" and marked "2", microf. nil).
LOCALITY. Egypt: Rashid ("*Rosettae* ad littora Nili"), Nov. 1761.

MATERIAL. *Forsskål* no specimen found.
NOTE. As the above specimen No. 1236 has Forsskål's original label with the locality Belgrad it cannot be the one from Rosetta (Cent. I No. 63) which is missing. This specimen bears the number 2 presumably relating to Hornemann's herbarium. It was seen by C. Mez in 1906–07 and named *Panicum crus galli* L. subsp. *microstachys* var. *vulgare* f. *submuticum*.

Eleusine floccifolia (*Forssk.*) *Spreng.*, Syst. Veg. 1: 350 (1824); Deflers 1889: 219; Schweinfurth 1894: 35; Schwartz 1939: 329; F.T.E.A. Gramin.: 267 (1974).
Cynosurus floccifolius Forssk. 1775: 21 (CIV No. 76; Cent. I No. 73); Vahl 1790: 10; Christensen 1922: 13.
LOCALITY. Yemen: Taizz ("In pratis humidis circa *Taaes* frequens"), 1763.
MATERIAL. *Forsskål* 44 (C — 1 sheet, type of *C. floccifolius*, microf. 36: III. 1, 2); 45 (C — 1 sheet, type of *C. floccifolius*, microf. 36: II. 7, 8); 1184 (C — 1 sheet ex herb. Hornemann, with field label "inter Taaes & Kaada", type of *C. floccifolius*, microf. 126: III. 6, 7).

Elymus subulatus *Forssk.* 1775: 26 (LXI No. 82; Cent. I No. 90); Christensen 1922: 13.
LOCALITY. Egypt: Alexandria ("*Alexandriae* ad margines agrorum"), Apr. 1762.
MATERIAL. *Forsskål* no specimen found.
NOTE. The application of this name is still in doubt. Tackholm & Drar, Fl. Egypt 1: 273 (1941) cited it as a tentative synonym of *E. delileana* Schult. The name should not be adopted if it upsets widely accepted nomenclature.

Elymus sp. indet.
Triticum junceum sensu Forssk. 1775: XX No. 55, non (L.) L.
LOCALITY. Turkey: Dardanelles, Izmir, Burgaz, ("Dd = Dardanelles, Sm = Smryna, Brg = Borghàs"), Aug. 1761.
MATERIAL. *Forsskål* 1586 (C — 1 sheet, with partly cut-off field label, microf. 109: II. 1, 2).
NOTE. The specimen is inadequate for full identification.

Eragrostis aegyptiaca (*Willd.*) *Delile*, Descr. Fl. Egypte: 157, t. 4, f. 2 (1814); Tackholm & Drar 1941: 191; Tackholm 1974: 691.
Poa amabilis ? Forssk. 1775: LXI No. 57, non L.
LOCALITY. Egypt: Cairo ("Cs"), 1761–62.
MATERIAL. *Forsskål* 54 (C — 1 sheet, microf. 82: I. 5, 6); 61 (C — 1 sheet, microf. 82: I. 3, 4).
Forsskål s.n. (BM — 1 sheet ex herb. Banks).

Eragrostis cilianensis (*All.*) *Vignolo & Janchen* in Mitt. Naturwiss. Vereins Univ. Wein 5(9): 110 (1907); F.T.E.A. Gramin.: 232, fig. 65 (1974), 859 (1982); Tackholm 1974: 691, pl. 255B.
LOCALITY. Yemen?
MATERIAL. *Forsskål* s.n. (BM — 1 sheet ex herb. Banks).

Eragrostis ciliaris (*L.*) *R. Br.* in Tuckey, Narr. Exp. Congo App.: 478 (1818); Schwartz 1939: 332; Tackholm & Drar 1941: 186; Tackholm 1974: 691.
LOCALITY. Egypt, 1761–62.
MATERIAL. *Forsskål* 1247 (C — 1 sheet ex herb. Schumacher, inscribed "ex Aegypto" on the back, microf. 126: III. 8).

Eragrostis kiwuensis *Jedwabn.* in Bot. Archiv. 5: 206 (1924); F.T.E.A. Gramin.: 229 (1974).
Poa multiflora Forssk. 1775: 21 (CIV No. 69; Cent. I No. 70); Christensen 1922: 12.
Eragrostis multiflora (Forssk.) Aschers., Fl. Prov. Brand. 1: 841 (1864), non Trin. (1830); Schwartz 1939: 333.
LOCALITY. Yemen: Al Hadiyah ("Hadîe"), 1763.
MATERIAL. *Forsskål* 59 (C — 1 sheet, with field label "Poa * 13 flora? inter Boka & Kurma", type of *P. multiflora*, microf. 82: I. 7, 8).

Eragrostis pilosa (*L.*) *P. Beauv.*, Ess. Agrost.: 162, 175 (1812); Tackholm & Drar 1941: 193; Tackholm 1974: 692.
LOCALITY. Egypt: without locality.
MATERIAL. *Forsskål* 38 (C — 1 sheet with original field label "Ni. 1", microf. 127: I. 1, 2); 1248 (C — 1 sheet ex herb. Hornemann & Hofman Bang, microf. 82: II. 1, 2).

Eustachys paspaloides (*Vahl*) *Lanza & Mattei* in Boll. Ort. Bot. Palermo 9: 56 (1910).
LOCALITY. Yemen, 1763.
MATERIAL. *Forsskål* 1099 (C — 1 sheet mixed with *Chloris* sp., microf. 127: I. 3, 4).
NOTE. The type of Vahl's *Cynosurus paspaloides* is a Thunberg specimen from the Cape of Good Hope in herb. Vahl.

Festuca ovina *L.* 1753: 73; Fl. Europ. 5: 145 (1980).
LOCALITY. Unspecified.
MATERIAL. *Forsskål* 98 (C — 1 sheet, microf. 127: II. 1, 2).

Helictotrichon bromoides (*Gouan*) *C.E. Hubbard* in Kew Bull. 1939: 101 (1939).
Avenula bromoides (Gouan) H. Scholz (1974) — Fl. Europ. 5: 215 (1980).
LOCALITY. Unspecified, prob. France.
MATERIAL. *Forsskål* 1327 (C — 1 sheet, microf. nil).

Helictotrichon sp. indet.
LOCALITY. Unspecified.
MATERIAL. *Forsskål* 71 (C — 1 sheet microf. 14: II. 3, 4); 1328 (C — 1 sheet ex herb. Schumacher, microf. 14: II. 5, 6).

Heteropogon contortus (*L.*) *P. Beauv. ex Roem. & Schultz.*, Syst. Veg. 2: 836 (1817); Schwartz 1939: 309.
Andropogon contortum L. 1753: 1045; Forssk. 1775: 173 (CXXIII No. 594; Cent. VI No. 73); Christensen 1922: 29.
LOCALITY. Yemen: Al Luhayyah ("Lohajae in horto"), Jan. 1763.
MATERIAL. *Forsskål* 943 (C — 1 sheet ex herb. Vahl, microf. 7: II. 7, 8).

Hordeum marinum *Hudson*, Fl. Angl. ed. 2, 1: 57 (1778) subsp. **marinum** — Tackholm & Drar 1941: 276; Tackholm 1974: 705.
LOCALITY. Egypt: Cairo, 1761–62.
MATERIAL. *Forsskål* 1344 (C — 1 sheet with field label "C. 23", microf. 127: II. 5, 6).

Hordeum marinum *Hudson* subsp. **gussoneanum** (*Parl.*) *Thell.* in Vjschr. naturf. Ges. Zürich 52: 441 (1908); Tackholm & Drar 1941: 277.
H. geniculatum All. (1785) — Tackholm 1974: 705.

LOCALITY. Unspecified, probably Egypt.
MATERIAL. *Forsskål* 51 (C — 1 sheet, microf. 127: II. 3, 4); 1345 (C — 1 sheet ex herb. Hofman Bang, microf. 127: II. 7, 8).

Hordeum murinum *L.* 1753: 85; Tackholm & Drar 1941: 274; Forssk. 1775: XIX No. 50; LXI No. 84 subsp. **glaucum** (*Steud.*) *Tzveler* in Nov. Sist. Vysshikh Rast. 8: 67 (1971); Fl. Turk. 9: 265 (1985).
H. glaucum Steud. — Tackholm 1974: 705.
VERNACULAR NAME. *Abu stirs* (Arabic).
LOCALITY. Egypt: Cairo (on sheet), Alexandria ("As" in book), 1761–62.
MATERIAL. *Forsskål* 139 (C — 1 sheet, microf. 55: II. 3, 4).

H. imrinum Forssk. 1775: XIX No. 52.
LOCALITY. Turkey: Gökçeada ("Imros" on field label), 1761.
MATERIAL. *Forsskål* 1346 (C — 1 sheet with field label "Hordeum ... Imros", microf. 127: III. 1, 2).
NOTE. The short description of *H. imrinum* in Flora Aegyptiaco-Arabica agrees very well with the descriptive phrases on the field label of No. 1346; this is enough to validate the name.

Hordeum peruersum *Forssk.* 1775: XIX No. 51 with descr.; Christensen 1922: 36.
VERNACULAR NAME. *Kophochorto* (Greek).
LOCALITY. Turkey: Burgaz ("Borghàs"), July 1761.
MATERIAL. *Forsskål* no specimen found.
NOTE. The identity is not known. The name should not be taken up if it upsets widely accepted nomenclature.

Hordeum vulgare *L.* 1753: 84; Tackholm & Drar 1941: 278.
H. hexastichon L. 1753: 85; Forssk. 1775: LXI No. 83, CIV No. 83.
VERNACULAR NAME. *Sjaeir* (Egypt, Arabic), *schaeir* (Yemen, Arabic).
LOCALITY. Egypt: Cairo cult. ("Ch"), 1761–62; Yemen, 1763.
MATERIAL. *Forsskål* 66 (C — 1 sheet, microf. 55: II. 1, 2).

Imperata cylindrica (*L.*) *Raeuschel*, Nom. Bot. ed. 3: 10 (1797); Tackholm & Drar 1941: 482; Tackholm 1974: 757, pl. 279.
Arundo epigejos sensu Forssk. 1775: 23 (XIX No. 43, LXI No. 76; Cent. I No. 82), non L.; Christensen 1922: 13.
VERNACULAR NAME. *Halfe* (Arabic).
LOCALITY. Turkey: Gökçeada ("Im" on field label), Tekirdag ("Ecl."). Egypt: Alexandria ("*Alexandriae*"), 1761–62.
MATERIAL. *Forsskål* 73 (C — 1 sheet, microf. 10: II. 5, 6); 966 (C — 1 sheet, ex herb. Schumacher, microf. 10: II. 7, 8); 967 (C — 1 sheet ex herb. Hornemann, with field label "Im. 20", microf. nil).
Forsskål s.n. (BM — 1 sheet ex herb. Banks).

[**Iseilema prostrata** (*L.*) *Anderss.*]
MATERIAL. *Unknown collector* (C — 1 sheet ex herb. Hofman Bang, microf. 125: III. 7, 8).
NOTE. Although this sheet is attributed to Forsskål it is unlikely to be his since it is an Indian species.

Lagurus ovatus *L.* 1753: 81; Forssk. 1775: XIX No. 34; LXI No. 72; Tackholm & Drar 1941: 344; Tackholm 1974: 716, pl. 248/2; Fl. Turk. 9: 357 (1985).
LOCALITY. Turkey: Dardanelles ("Dd"), July 1761.
MATERIAL. *Forsskål* 77 (C — 1 sheet with field label "C. 19", microf. 61: I. 5, 6).
LOCALITY. Egypt: Alexandria ("As" = Alexandriae), 1 Apr. 1762.
MATERIAL. *Forsskål* 78 (C —1 sheet with field label "Al. 30", microf. 61: I. 7, 8); 1390 (C —1 sheet ex herb. Hornemann, with field label "49 Lagurus ovatus Alex. ad salinas", microf. nil).

Lamarckia aurea (*L.*) *Moench.*, Meth.: 201 (1794); Tackholm & Drar 1941: 202; Tackholm 1974: 696, pl. 257C.
LOCALITY. Egypt: Alexandria, Ras etin, 1761–62.
MATERIAL. *Forsskål* 1391 (C — 1 sheet ex herb. Liebmann, microf. 127: III. 3); 1392 (C —1 sheet ex herb. Hornemann, with field label "Ras ettin", microf. 127: III. 8).

Lasiurus scindicus *Henrard* in Blumea 4: 514 (1941); Tackholm 1974: 756, pl. 280 (as *L. hirsuta*); Cope in Kew Bull. 35: 451 (1981).
Saccharum hirsutum Forssk. 1775: 16 (LX No. 31; Cent. I No. 51); Christensen 1922: 12, non *Rottboellia hirsuta* Vahl (1790) nec *Lasiurus hirsutus* (Forssk.) Boiss. (1859) — Tackholm & Drar 1941: 480.
LOCALITY. Egypt: Cairo ("In desertis Kahirinis"), 1761–62.
MATERIAL. *Forsskål* 48 (C — 1 sheet, type, microf. 90: I. 1, 2); 1652 (C — 1 sheet with field label "Ca", type, microf. 90: I. 3, 4); 1653 (C — 1 sheet ex herb. Hofman Bang, type, microf. 90: I. 5, 6); 1654 (C — 1 sheet ex herb. Liebmann, type, microf. 90: I. 7); 1653 (C — 1 sheet ex herb. Liebmann, type, microf. 90: I. 8, 90: II. 1); 1656 (C — 1 sheet ex herb. Hofman Bang, type, microf. 90: II. 2, 3); 1657 (C — 1 sheet ex herb. Hofman Bang, type, microf. 90: II. 4, 5); 1658 (C — 1 sheet, type, microf. 90: II. 6, 7); 1659 (C — 1 sheet, microf. nil); 1660 (C — 1 sheet ex herb. Hornemann, microf. nil).
Forsskål s.n. (BM — 1 sheet ex herb. Banks).
Forsskål s.n. (LD — 1 sheet ex herb. Retzius).
All types of *S. hirsutum.*
NOTE. For a discussion and clarification of the nomenclatural confusion see Cope l.c. (1981). See also synonymy of *Triticum aegilopoides* below. For *Lasiurus hirsutus* (Vahl) Boiss. hitherto considered to have been based on *Saccharum hirsutum*. No. 1652 has on the field label in Forsskål's hand "Panicum * ... (epithet illegible, perhaps "lanigerum"); no similar name has been traced in the printed work.

Leptochloa fusca (*L.*) *Kunth*, Rev. Gram. 1: 91 (1829).
Bromus polystachios Forssk. 1775: 23 (LXI No. 68; Cent. I No. 78); Christensen 1922: 13.
LOCALITY. Egypt: Alexandria ("Alexandriae; in humidis ad fossas"), 1 Apr. 1762.
MATERIAL. *Forsskål* 1016 (C — 1 sheet, type of *B. polystachios*, microf. 16: I. 8).

Festuca fusca L. — Forssk. 1775: LXI No. 59.
Diplachne fusca (L.) P. Beauv. — Tackholm 1974: 730.
LOCALITY. Egypt: Cairo ("Cs."). 1761–62.
MATERIAL. *Forsskål* 1215 (C — 1 sheet ex herb. Hofman Bang, with two specimens, each with "Festuca fusca. Ca. 11, Ni.18" respectively, one with vernacular name "Aburugbi", microf. ex herb. Hofman Bang, microf. 44: III. 1, 2).

? **Lolium multiflorum** *Lam.*, Fl. Fr. 3: 621 (1779); Tackholm & Drar 1941: 303; Tackholm 1974: 707, pl. 260c.
L. perenne sensu Forssk. 1775: LXI No. 80, non L.
VERNACULAR NAME. *Haschîsch el farras* (Arabic).
LOCALITY. Egypt: Cairo ("Cs" = Cairi spontaneae), 1761–62.
MATERIAL. *Forsskål* 1421 (C — 1 sheet specimen 2 with field label "Ca. 257", "Lolium superv. ... an novum", microf. 63: III. 7, 8).

Lolium perenne *L.* 1753: 83; Forssk. 1775: XIX No. 44; Fl. Turk. 9: 446 (1985).
LOCALITY. Turkey: Izmir, July 1761.
MATERIAL. *Forsskål* 1421 (C — 1 sheet, specimen 1 with field label "Smirna ad hortus", microf. 63: III. 7, 8).

Mibora minima (*L.*) *Desv.*, Obs. Pl. Angers: 45 (1818); Fl. Europ. 5: 172 (1980).
LOCALITY. Unspecified, prob. France.
MATERIAL. *Forsskål* 1469 (C — 1 sheet ex herb. Hofman Bang, with old label "74. Agrostis co. minima Gouan", microf. 127: III. 6, 7).

Ochthochloa compressa (*Forssk.*) *Hilu* in Kew Bull. 36: 560 (1981).
Eleusine compressa (Forssk.) Asch. & Schweinf. ex Christensen 1922: 12; Tackholm & Drar 1941: 389; Tackholm 1974: 734.
E. flagellifera Nees (1842) — Schwartz 1939: 328.
Panicum compressum Forssk. 1775: 18 (CIII No. 52; Cent. I No. 58); Christensen 1922: 12.
LOCALITY. Yemen: Al Hadiyah ("In montibus Hadiensibus"), Mar. 1763.
MATERIAL. *Forsskål* 46 (C — 1 sheet, type of *P. compressum*, microf. 75: III. 1, 2).
NOTE. Wood considers this to be a *Paniceae*, not the desert genus *Ochthochloa* which could not occur inland on the mountains of Al Hadiyah. Forsskål's description seem to be of a 2-flowered *Panicum*.

Odyssea mucronata (*Forssk.*) *Stapf* in Hook. Ic. Pl.: t. 3100 (1922); Schwartz 1939: 335.
Festuca mucronata Forssk. 1775: 22 (CIV No. 77; Cent. I No. 74); Christensen 1922: 13.
Aeluropus mucronatus (Forssk.) Asch. & Schweinf., Beitr.: 297 (1867), non *Eragrostis mucronata* (L.) Roem. & Schult. — Deflers, 1889: 220.
Festuca pungens Vahl 1790: 10, nom. illeg.
Eragrostis pungens (Vahl) Benth. — Schweinfurth 1894: 43.
VERNACULAR NAME. *Schocham, Schoncham* (Arabic).
LOCALITY. Yemen: Beit al Fakih (also Al Luhayyah & Al Mukha) ("Circa Beit el fakih copiose in collibus arenosis"), Apr. 1763.
MATERIAL. *Forsskål* 94 (C — 1 sheet ex herb. Schumacher, type of *F. mucronata*, microf. 44: III. 5, 6); 1262 (C — 1 sheet ex herb. Vahl, type of *F. mucronata*, microf. nil). *Forsskål* s.n. (BM — 1 sheet, type of *F. mucronata*).
NOTE. The marking of a specimen with both the old series of Forsskål numbers (No. 94) and "Herb. Schum[acher]" is unusual; possibly a stamp "Hb. Hort. Bot. Hafn." is an attempt to correct the markings. This is the normal stamp for the old numbers below 900.

Panicum coloratum *L.* 1767: 30; Tackholm & Drar 1941: 439; Tackholm 1974: 746, pl. 275A.
P. miliaceum sensu Forssk. 1775: LX No. 51, non L.
LOCALITY. Egypt: Cairo ("Cs"), 1761–62.
MATERIAL. *Forsskål* 65 (C — 1 sheet, microf. 76: II. 3, 4); 1553 (C — 1 sheet ex herb. Hofman Bang with field label "Ca. 237", microf. 76: II. 1, 2).

Panicum miliaceum *L.* 1753: 58; Tackholm & Drar 1941: 437.
LOCALITY. Egypt: 1761–62.
MATERIAL. *Forsskål* 1552 (C — 1 sheet, microf. nil).

Panicum muricatum *Forssk.* 1775: XIX No. 26; Christensen 1922: 35.
LOCALITY. Turkey: Tekirdag ("Ecl."), July 1761.
MATERIAL. *Forsskål* no type specimen found.
NOTE. The brief description "Totum hispidum" may validate the epithet. However, its identity is not known, and it should not be taken up.

Panicum polygamum *Forssk.* 1775: 27 (CIV No. 63; Cent. I No. 95); Christensen 1922: 13.
LOCALITY. Yemen: Taizz ("In pratis versus Taaes parcius"), 1763.
MATERIAL. *Forsskål* no specimen found.
NOTE. The application of this name is still in doubt and the name should not be taken up. Forsskål suggested it might represent a new genus.

Panicum repens *L.* 1762: 87; Tackholm & Drar 1941: 438; Tackholm 1974: 746.
P. grossarium ? sensu Forssk. 1775: 19 (LX No. 50; Cent. I No. 61), non L.; Christensen 1922: 12.
VERNACULAR NAME. *N'gîl, Nesi* (Arabic).
LOCALITY. Egypt: Rashid ("Rosettae in ripa Nili"), 2–6 Nov. 1761.
MATERIAL. *Forsskål* 1551 (C — 1 sheet, ex herb. Hofman Bang, with field label "Panicum ? grossarium. novum. Rosette. Ro.42", microf. 76: I. 7, 8).
Forsskål s.n. (BM — 1 sheet ex herb. Banks).

Panicum tetrastichon *Forssk.* 1775: 19 (LX No. 46; Cent. I No. 62); Christensen 1922: 12.
LOCALITY. Egypt: Rashid ("In pratis Nili littoreis"), 1761.
MATERIAL. *Forsskål* no specimen found.
NOTE. In the absence of a type specimen this is an unresolved species, and the name should not be taken up.

Panicum turgidum *Forssk.* 1775: 18 (LX No. 47; Cent. I No. 60); Christensen 1922: 12; Tackholm & Drar 1941: 435; F.W.T.A. ed. 2, 3: 433 (1972); Tackholm 1974: 746, pl. 275B; Fl. Palaest. 4: 302, pl. 403 (1986).
VERNACULAR NAME. *Bochar* (Arabic).
LOCALITY. Egypt: Cairo ("In desertis Káhirinis"), 1761–62.
MATERIAL. *Forsskål* 85 (C — 1 sheet, type of *P. turgidum*, microf. 76: III. 3, 4); 138 (C — 1 sheet ex herb. Hornemann, type of *P. turgidum*, microf. 76: III. 7, 8); 1554 (C — 1 sheet ex herb. Hornemann, with field label "Ca. 201", type of *P. turgidum*, microf. 76: III. 5, 6).

? **Parapholis strigosa** (*Dumort.*) *C.E. Hubbard* in Blumea, Suppl. 3: 14 (1946); Fl. Europ. 5: 243 (1980).
LOCALITY. Unspecified, prob. France.
MATERIAL. *Forsskål* 1555 (C — 1 sheet ex herb. Hofman Bang, microf. 127: III. 4, 5).

Paspalidium geminatum (*Forssk.*) *Stapf* in Fl. Trop. Afr. 9: 583 (1920); Tackholm & Drar 1941: 442; Tackholm 1974: 746, pl. 274A; F.T.E.A. Gramin.: 552 (1982); Fl. Palaest. 4: 307, pl. 410 (1986).
P. desertorum sensu Schwartz 1939: 313, non (A. Rich.) Stapf
Panicum geminatum Forssk. 1775: 18 (LX No. 49; Cent. I No. 59); Christensen 1922: 12.

P. fluitans Retz., Obs. 5: 18 (1783); Vahl 1790: 8.
Echinochloa geminata (Forssk.) Roberty in Bull. Inst. Franc. Afr. Noire Sér. A, 17: 66 (1955).
LOCALITY. Egypt: Rashid ("Rosettae in pratis ad littora Nili"), 2–6 Nov. 1761.
MATERIAL. *Forsskål* 113 (C — 1 sheet, type of *P. geminatum*, *P. fluctans*, microf. 76: I. 3, 4).

Pennisetum divisum (*J.F. Gmel.*) *Henr.* in Blumea 2: 162 (1938); F.W.T.A. ed. 2, 3: 463 (1972); Tackholm 1974: 754, pl. 275C; Fl. Palaest. 4: 311, pl. 416 (1986).
Panicum divisum J.F. Gmel., Syst. Nat. ed. 13: 2: 156 (1791).
Panicum dichotomum Forssk. 1775: 20 (LX No. 48; Cent. I No. 64), non L. (1758); Christensen 1922: 12.
Pennisetum dichotomum (Forssk.) Delile, Fl. d'Egypte: 159 (1813); Tackholm & Drar 1941: 465.
VERNACULAR NAME. *Tummâm* (Arabic).
LOCALITY. Egypt: Cairo, 1761–62.
MATERIAL. *Forsskål* 1548 (C — 1 sheet ex herb. Schumacher, type of *P. dichotomum*, microf. 75: III. 5, 6); 1549 (C — 1 sheet ex herb. Hofman Bang, type of *P. dichotomum*, microf. 75: III. 7, 8); 1550 (C — 1 sheet, ex herb. Hofman Bang, type of *P. dichotomum*, microf. 76: I. 1, 2).
Forsskål s.n. (BM — 1 sheet ex herb. Banks).
Forsskål s.n. (LD — 1 sheet ex herb. Retzius).
NOTE. Although the specimens cited above are credited to Egypt some or all might have been collected in Yemen where the species also occurs.

Pennisetum glaucum (*L.*) *R. Br.*, Prodr. Fl. Nov. Holl. 1: 195 (1810); F.T.E.A. Gramin.: 672 (1982).
LOCALITY. Yemen: Wadi Mawr ('Mour' on label), 1763.
MATERIAL. *Forsskål* 106 (C — 1 sheet, with field label "Panicum ital ? Dochm. Dochna sativa. Mour", microf. 54: III. 5, 6).

Pennisetum setaceum (*Forssk.*) *Chiov.* in Bull. Soc. Bot. Ital. 1923: 113 (1923); Tackholm & Drar 1941: 467; Tackholm 1974: 755; F.T.E.A. Gramin.: 675 (1982).
Phalaris setacea Forssk. 1775: 17 (LX No. 36; Cent. I No. 56); Christensen 1922: 12.
VERNACULAR NAME. *Raetam* (Arabic).
LOCALITY. Egypt: Cairo ("In desertis Káhirinis), 1761–62.
MATERIAL. *Forsskål* no type specimen found.
LOCALITY. Yemen: Mukhajah, J. Barad ("Mokhaja, Mt. Barak" on field label), Mar. 1763.
MATERIAL. *Forsskål* 117 (C — 1 sheet with field label, microf. 78: III. 1, 2).
Forsskål s.n. (LD — 1 sheet ex herb. Retzius).

Pennisetum typhoides (*Burm.*) *Stapf & C.E. Hubbard* in Fl. Trop. Afr. 9: 1050 (1934); Tackholm & Drar 1941: 469.
Holcus racemosus Forssk. 1775: 175 (CXXIII No. 596; Cent. VI No. 77).
LOCALITY. Yemen ("Ubique in Yemen"), 1763.
MATERIAL. *Forsskål* no type specimen found.

Phleum paniculatum *Hudson*, Fl. Angl.: 23 (1762); Fl. Europ. 5: 240 (1980).
LOCALITY. Unspecified, probably France.
MATERIAL. *Forsskål* s.n. (LD – 1 sheet ex herb. Retzius).

Phragmites australis (*Cav.*) *Trin. ex Steud.*, Nom. Bot. ed. 2, 2: 324 (1841); F.T.E.A. Gramin.: 117 (1970); Tackholm 1974: 697.

P. mauritianus Kunth (1830) — Schwartz 1939: 332.
P. communis (L.) Trin. (1820) — Tackholm & Drar 1941: 209.
Arundo phragmites L. 1753: 81; Forssk. 1775: 23 (LXI No. 77; Cent. I No. 83; Christensen 1922: 13.
LOCALITY. Egypt: Suez ("In locis palustribus Ghobeibe prope Suês"), Sept. 1762.
MATERIAL. *Forsskål* 967 (C — 1 sheet, microf. 10: III. 1, 2); 968 (C — 1 sheet with field label Ghobeibe, microf. 10: III. 3, 4); 969 (C — 1 sheet, with original label incl. words "Bûz" and "Ghobeibe", microf. 10: III. 5, 6); 970 (C — 1 sheet, ex herb. Hofman Bang, microf. 11: I. 1, 2); 55 (C — 1 sheet, microf. 10: III. 7, 8).

Phragmites australis (*Cav.*) *Trin. ex Steud.* subsp. **altissimus** (*Benth.*) *W.D. Clayton* in Taxon 17: 169 (1968) and F.T.E.A. Gram.: 118 (1970); Tackholm 1974: 697.
Arundo donax sensu Forssk. 1775: 23 (LXI No. 74; Cent. I No. 85) pro parte descr., non L.; Christensen 1922: 13.
VERNACULAR NAME. *Buz Haggni* (Egyptian Arabic).
LOCALITY. Egypt: Rashid ("*Rosettae* in fossis"), 2–6 Nov. 1761.
MATERIAL. *Forsskål* 963 (C — 1 sheet with field label "Ro.36", microf. 10: I. 7, 8); 964 (C — 1 sheet, ex herb. Hofman Bang, microf. 10: II. 3, 4); 58 (C —1 sheet, microf. 10: II. 1, 2).

Arundo maxima [donax maxima] Forssk. 1775: 24 (LXI No. 75; Cent. I No. 86), see note.
A. isiaca Del., Descr. Egypt. Hist. Nat. 2: 52 (1813), see note below.
LOCALITY. Egypt: Suez region and Nile.
MATERIAL. *Forsskål* no specimen found.
NOTE. The name *Arundo maxima* is cited as a synonym by Clayton (l.c.) who states that it is a nomen dubium, although it might in fact be *A. donax*. He also cites *A. isiaca* as a nomen superfluum since it is based on *A. maxima*, and many derivations of these names.

This species appears in many works (e.g. Tackholm & Drar, Fl. Egypt 1: 209 (1941)) as *P. communis* (L.) Trin.

Poa schimperiana *A. Rich.*, Tent. Fl. Abyss. 2: 423 (1851).
Poa spiculis 3-floris Forssk. 1775: CIV No. 70.
LOCALITY. Yemen: Jebel Barad ("Barah"), 1763.
MATERIAL. *Forsskål* 39 (C — 1 sheet, microf. 82: II. 3, 4).

Polypogon monspeliensis (*L.*) *Desf.*, Fl. Alt. 1: 67 (1768); Tackholm & Drar 1941: 339; Tackholm 1974: 715; Fl. Turk. 9: 356 (1985).
Phalaris crinita Forssk. 1775: XIX No. 27; Christensen 1922: 35.
LOCALITY. Turkey: Izmir ("Smirna"), 6 June to 1 July 1761.
MATERIAL. *Forsskål* 80 (C — 1 sheet, microf. 128: I. 5, 6); 75 (C — 1 sheet with field label "Phalaris * crinita", microf. 78: I. 5, 6).

Phalaris cristata [aristata?] Forssk. 1775: 17 (LX No. 35; Cent. I No. 54); Christensen 1922: 12.
VERNACULAR NAME. *Dêl el fâr* (Arabic).
LOCALITY. Egypt: Cairo ("Circa Káhiran frequens"), 1762.
MATERIAL. *Forsskål* 68 (C — 1 sheet, type?, microf. 78: I. 7, 8); 69 (C — 1 sheet, type?, microf. 78: II. 1, 2).

Polypogon viridis (*Gouan*) *Breistr.* in Bull. Soc. Bot. Fr. (Sess. Extr.): 110 (1966).
Agrostis viridis Gouan, Hort. Monsp.: 546 (1762).
Phalaris semiverticillata Forssk. 1775: 17 (LX No. 33; Cent, I. No. 52); Christensen 1922: 12.

Agrostis semiverticillata (Forssk.) C. Christensen 1922: 12; Tackholm & Drar 1941: 335.
Polypogon semiverticillatus (Forssk.) Hyl., Nomencl. und Syst. Nord. Gefässpl. in Uppsala Univ. Årsskr. n. 7; 74 (1945); Bor in Fl. Iraq 9: 318, pl. 116 (1968); Tackholm 1974: 715.
VERNACULAR NAME. *Naaejm* (Arabic).
LOCALITY. Egypt: Rosetta & Cairo ("Rosettae & Káhirae"), Apr. 1762.
MATERIAL. *Forsskål* 63 (C — 1 sheet, type of *A. semiverticillata*, microf. 78: II. 5, 6); 1566 (C — 1 sheet ex herb. Liebmann, type of *A. semiverticillata*, microf. 78: II. 7, 8).

Rhynchelytrum repens (*Willd.*) *C.E. Hubbard* in Kew Bull. 1934: 110 (1934).
LOCALITY. Yemen: Taizz ("Iter ad Taaes" on original label), Apr. 1763.
MATERIAL. *Forsskål* 121 (C — 1 sheet with field label "Holchus * racemosus, Iter ad Taaes", microf. 128: I. 7, 8).
NOTE. There seems to be a confusion over the field label as the description of "*Holchus racemosus*" (Cent. VI No. 77) does not fit *Rhynchelytrum* species.

Rostraria cristata (*L.*) *Tzvelev* in Novit. Syst. Pl. Vasc. (Leningrad) 7: 47 (1971); Fl. Turk. 9: 328 (1985).
Bromus poi-formis Forssk. 1775: 23 (LXI No. 68; Cent. I No. 80); Christensen 1922: 13.
VERNACULAR NAME. *Samme* (Arabic).
LOCALITY. Egypt: Alexandria ("Alexandria ad salinas" on field label), 1762.
MATERIAL. *Forsskål* 1633 (C — 1 sheet, microf. 128: I. 3, 4); 1634 (C — 1 sheet ex herb. Schumacher, microf. 128: I. 1, 2); 1015 (C — 1 sheet ex herb. Hornemann, with field label, type of *Bromus poi-formis*, microf. nil).
NOTE. The original label on sheet 1015 is inscribed "Bromus var. poiformis Alex. ad salinas". The specimen has been identified by A. Hansen and T. Cope.

Saccharum officinarum *L.* 1753: 54; Forssk. 1775: 15 (CIII No. 50; Cent. I No. 49); Christensen 1922: 12; Schwartz 1939: 304.
VERNACULAR NAME. *Qvasab* (Arabic); *Muddardjend* (? Indian).
LOCALITY. Yemen: without locality ("montium regio media"), 1763.
MATERIAL. *Forsskål* no specimen found.

Saccharum spontaneum *L.* subsp. **aegyptiacum** (*Willd.*) *Hack.* in DC., Monogr. Phan. 6: 115 (1889); Tackholm & Drar 1941: 488; Tackholm 1974: 757.
S. biflorum Forssk. 1775: 50 (LX No. 36; Cent. I No. 50); Christensen 1922: 12.
VERNACULAR NAME. *Ganisch, Buz farsi* (Arabic).
LOCALITY. Egypt: Rashid ("Rosettae in locis campestris"), 1761–62.
MATERIAL. *Forsskål* 64 (C — 1 sheet, type of *S. biflorum*, microf. 89: III. 5, 6); 74 (C — 1 sheet, type, microf. 89: III. 7, 8).

Schismus arabicus *Nees*, Fl. Afr. Austr.: 422 (1841); Bor in Fl. Iraq 9: 378 (1968).
S. barbatus sensu Tackholm & Drar 1941: 314; Tackholm 1974: 708, non (L.) Thell.
Festuca calycina Forssk. 1775: LXI No. 63, nomen nudum.
LOCALITY. Egypt: Alexandria & Cairo ("Alexandriae spontaneae, Cairi vel Káhirae loca deserta"), 1761.
MATERIAL. *Forsskål* 84 (C — 1 sheet, microf. 44: I. 5, 6); 1255 (C — 1 sheet mixed with *S. barbatus* with field label "Ca. 162" microf. 127: I. 5, 6); 1256 (C — 1 sheet ex herb. Liebmann, microf. 44: I. 7, 8).
Forsskål s.n. (BM — 1 sheet ex herb. Banks).

Schismus barbatus (*L.*) *Thell.* in Bull. Herb. Boiss. ser. 2, 7: 391 (1907).
LOCALITY. Egypt: Alexandria, Ras etin ("Ras ettin" on label), 1761.

MATERIAL. *Forsskål* 1255 (C — 1 sheet with field label "62 Ras ettin", mixed with *Schismus arabicus*, microf. 127: I. 5, 6).

Sehima ischaemoides *Forssk.* 1775: 178 (CXXIII No. 599; Cent. VI No. 91).
VERNACULAR NAME. *Saehim, Sehîm* (Arabic).
LOCALITY. Yemen: Al Hadiyah ("Yemen in montibus ad Hadîe"), 1763.
MATERIAL. *Forsskål* no type specimen found.
NOTE. An obscure species.

Setaria verticillata (*L.*) *P. Beauv.*, Ess. Agrost: 51, 178 (1812); Schweinfurth 1894: 24; Schwartz 1939: 316.
Panicum adhaerens Forssk. 1775: 20 (CIV No. 55; Cent. I No. 67); Christensen 1922: 12.
Setaria adhaerens (Forssk.) Chiov. in Nuov. Giorn. Bot. Ital. n.s. 26: 77 (1919).
VERNACULAR NAME. *Sara erra, Saera erra* (Arabic).
LOCALITY. Yemen: Al Hadiyah ("Hadîe"), 1763.
MATERIAL. *Forsskål* s.n. (LD — 1 sheet ex herb. Retzius).

Sorghum bicolor (*L.*) *Moench*, Meth.: 207 (1794); Tackholm & Drar 1941: 529.
Holcus durra Forssk. 1775: 174 (LXXVII No. 543, CXXIII No. 595; Cent. VI No. 76); Christensen 1922: 29.
Sorghum durra (Forssk.) Stapf var. *eois* (Burkill) J.D. Snowden, Cult. races of Sorghum: 198 (1936).
H. sorghum L. — Vahl 1790: 80.
VERNACULAR NAME. *Táam* (Arabic).
LOCALITY. Yemen: Wadi Mawr ("Mour" on label"), 1763.
MATERIAL. *Forsskål* 111 (C — 1 sheet with field label, type of *H. durra*, microf. 54: III. 7, 8).

Holcus dochna Forssk. 1775: 174 (LXXVII No. 544; Cent. VI No. 74); Christensen 1922: 29.
Sorghum dochna (Forssk.) J.D. Snowden var. *obovatum* (Hack.) J.D. Snowden, Cult. races of Sorghum: 100 (1936) (see *Forsskål* 108).
S. membranaceum Chiov. var. *ehrenbergianum* (Koern.) J.D. Snowden: l.c. 77 (1936) (see *Forsskål* 107).
VERNACULAR NAME. *Dachn, Dochn* (Arabic).
LOCALITY. Egypt: Rashid ("Rosettae"), 1761 & Yemen ("ubiqui"), 1763.
MATERIAL. *Forsskål* 107 (C — 1 sheet, type of *H. dochna*, microf. 54: III. 1, 2); 108 (C — 1 sheet, type of *H. dochna*, microf. 54: III. 3, 4).

Sorghum halepense (*L.*) *Pers.*, Syn. 1: 101 (1805); Tackholm & Drar 1941: 521; Tackholm 1974: 762.
Holcus halepensis L. 1753: 1047; Forssk. 1775: XXXV No. 441; Vahl 1790: 81.
VERNACULAR NAME. *Kalamágra* (Greek).
LOCALITY. Turkey: Izmir ("Smirna" on field label), June–July 1761.
MATERIAL. *Forsskål* 1343 (C — 1 sheet ex herb. Hornemann, with field label "Holcus halepens.? Smirna ad hortas", microf. 55: I. 7, 8).

Holcus exiguus Forssk. 1775: 174 (LXXVII No. 545; Cent. VI No. 75); Christensen 1922: 29.
LOCALITY. Egypt: Rashid ("Rs = Rosettae", "Ad ripam Nili"), Nov. 1761.
MATERIAL. *Forsskål* 50 (C — 1 sheet, type of *H. exiguus*, microf. 55: I. 1, 2); 109 (C — 1 sheet, type of *H. exiguus*, microf. 55: I. 3, 4); 110 (C — 1 sheet, type of *H. exiguus*, microf. 55: I. 5, 6).

Sphenopus divaricatus (*Gouan*) *Reichb.*, Fl. Germ. Excurs.: 45 (1830); Tackholm & Drar 1941: 174; Tackholm 1974: 688.
LOCALITY. Egypt: between Cairo and Alexandria ("Inter Cairo et Alexandriam" on reverse of sheet), 1 Apr. 1762.
MATERIAL. *Forsskål* 70 (C — 1 sheet, microf. 128: II. 7, 8); 1756 (C — 1 sheet, microf. 128: II. 6); 1757 (C — 1 sheet ex herb. Hornemann with field label "Ave 109. Agrostis nondum examinata", microf. nil).

Sphenopus sp. aff. **divaricatus** Reichb., Fl. Germ. Excurs.: 45 (1830).
LOCALITY. Egypt: Alexandria to Caleseh, 1761–62.
MATERIAL. *Forsskål* 1758 (C — 1 sheet ex herb. Liebmann, with field label "Av. 108 Aira nond. examinata. Alex. vernal.", microf. 128: II. 5).
NOTE. T.A. Cope 1987 could not improve on Hubbard's above determination and states that the "lemma is awned — I've never seen another specimen like it".

Sporobolus arenarius (*Gouan*) *Duval-Jouve* in Bull. Soc. Bot. Fr. 16: 294 (1869); Tackholm & Drar 1941: 346.
S. virginicus sensu Tackholm 1974: 719, non (L.) Kunth (1829).
Phalaris disticha Forssk. 1775: 17 (LX No. 34; Cent. I No. 53); Christensen 1922: 12.
Agrostis disticha (Forssk.) Schweig. ex Steud., Nom. ed. 2, 1: 39, 40 (1840).
LOCALITY. Egypt: Alexandria ("*Alexandriae* copiose in arena littorea ad catacombas"), Oct. 1761.
MATERIAL. *Forsskål* 132 (C — 1 sheet, type of *P. disticha*, 78: II. 3, 4).

Sporobolus pyramidalis *P. Beauv.*, Fl. Owar. Benin 2: 36, t. 80 (1816); F.T.E.A. Gram.: 373 (1974).
Agrostis indica ? sensu Forssk. 1775: CIV No. 66, non L.
VERNACULAR NAME. *Sorak* (Arabic).
LOCALITY. Yemen: Al Hadiyah, Bolghose ("Hadíe, Bulgose"), 1763.
MATERIAL. *Forsskål* 79 (C — 1 sheet, microf. 3: III. 7, 8).
Forsskål s.n. (BM — 1 sheet ex herb. Banks).

Sporobolus spicatus (*Vahl*) *Kunth*, Rev. Gram. 1: 67 (1829); Deflers 1889: 219; Schweinfurth 1894: 28; Schwartz 1939: 323; Tackholm & Drar 1941: 344; F.T.E.A. Gramin.: 369 (1974); Tackholm 1974: 716 pl. 262B.
Agrostis spicata Vahl 1790: 9.
Agrostis virginica sensu Forssk. 1775: 20 (CIV No. 68; Cent. I No. 69), non L.; Christensen 1922: 12.
VERNACULAR NAME. *Samma* (Arabic).
LOCALITY. Egypt: Cairo ("Cd" = Cairi deserta), Suez ("In deserto arenoso versus Sués"), 28–30 Aug. 1762.
MATERIAL. *Forsskål* 129 (C — 1 sheet, type of *A. spicata*, microf. 4: I. 5, 6); 142 (C — 1 sheet, type of *A. spicata*, microf. 4: I. 3, 4); 920 (C — 1 sheet ex herb. Hofman Bang, with field label "Ca 204", type of *A. spicata*, microf. 4: I. 7, 8).
LOCALITY. Yemen: Al Luhayyah ("Lohaja"), 31 Dec. 1762.
MATERIAL. *Forsskål* 130 (C — 1 sheet, type of *A. spicata*, microf. 4: I. 1, 2); 921 (C — 1 sheet ex herb. Hofman Bang, type of *A. spicata*, microf. 4: II. 1, 2).
NOTE. The Yemen specimens were distinguished by Forsskål as a variety of the species but it is not formally recognised taxonomically. There are three original labels in Forsskål's writing and one now missing that appears in the microf. 4: II. 1, 2.

Stenotaphrum dimidiatum (*L.*) *Brongn.* in Duperrey, Bot. Voy. Coquille: 127 (1832).
S. secundatum (Walt.) Kuntze (1891) — Tackholm & Drar 1941: 428.
LOCALITY. Unlocalised. Egypt or Yemen.
MATERIAL. Attributed to *Forsskål* s.n. (BM — 1 sheet ex herb. Banks).
NOTE. This is the only record of this species from Western Asia at BM so it may not have been collected by Forsskål. On the other hand it is an old specimen and presumably was part of Banks's herbarium.

Stipa capensis *Thunb.*, Prodr. Fl. Cap.: 19 (1794); Tackholm & Drar 1941: 354; Tackholm 1974: 722, pl. 264.
LOCALITY. Unspecified.
MATERIAL. *Forsskål* 128 (C — 1 sheet, microf. 128: III. 1, 2).

Stipagrostis lanata (*Forssk.*) *De Winter* in Kirkia 3: 135 (1963); Tackholm 1974: 725; Fl. Palaest. 4: 278, pl. 364 (1986).
Aristida lanata Forssk. 1775: 25 (LXI No. 79, CIV No. 79; Cent. I No. 87); Christensen 1922: 13; Tackholm & Drar 1941: 360.
VERNACULAR NAMES. *Dhraejrae, Dhraeirae, Sjaefsjuf* (Arabic).
LOCALITY. Egypt: Rashid & Cairo ("Rosettae in arena mobili circa turrim canope. In desertis quoque Káhirinis habitat"), Mar. 1762.
MATERIAL. *Forsskål* no type specimen found (see note and note to *Stipagrostis plumosa*).
NOTE. T.A. Cope writes: The description provided by Forsskål is very clear on the point that the lateral awns are plumose. The specimens, however, all have glabrous lateral awns and agree with *A. plumosa* L. (= *Stipagrostis plumosa* (L.) Munro ex T. Anderson).

Tausch (Flora, Jena 19: 506, 1836) has assumed that *A. lanata* was described from a mixture of material which is separated in the following way:

Species A: In his diagnosis of *A. lanata* Forsskål stated: "corollae arista media lanata". Tausch took this to mean "therefore lateral awns glabrous" and equates *A. lanata* with *A. plumosa* L.

Species B: In his description of *A. lanata*, Forsskål stated "Aristae laterales in medio villosae". Tausch used this as the basis of his *A. forskolii*.

Henrard (Critical Revision of the Genus Aristida, p. 186) has taken the view that Forsskål's diagnosis is intended to distinguish *A. lanata* from the next species (*A. paniculata*) the latter having a glabrous central awn. To consider "central awn plumose" as meaning "lateral awns glabrous" is, in my view, reading more into Forsskål's statement than Forsskål himself intended.

(The situation is further complicated by Tausch's use of "A. tomentosa Forsk. descr. 25 (partim)" as the basis of his new name when presumably he meant *A. lanata*).

In conclusion, all the material of "*A. lanata*" sent to Kew by Copenhagen has been determined as *Stipagrostis plumosa*. No specimen which could be considered as a type of *A. lanata* is extant in this material. It is reasonable to suppose that Forsskål's description was accurate, so a specimen with 3 plumose awns collected by him in Egypt must be looked for.

Stipagrostis plumosa (*L.*) *Munro ex T. Anders.* in Journ. Linn. Soc. Bot. 5, Suppl. 1: 40 (1860); Tackholm 1974: 726.
Aristida lanata sensu Christensen 1922: 13, non Forssk. (see note under *Stipagrostis lanata*; Tackholm & Drar 1941: 360.
LOCALITY. Egypt: Sinai, Rashid, Cairo (on field labels), 1761–62.
MATERIAL. *Forsskål* 62 (C — 1 sheet, microf. 8: III. 7, 8); 125 (C — 1 sheet, microf. 9: I. 1, 2); 1765 (C — 1 sheet ex herb. Hornemann with field label "Aristida nova lanata. Rosette. Ro. 31", microf. 8: III. 5, 6); 1706 (C — 1 sheet, ex herb. Liebmann, microf. 9: I. 3, 4); 1767 (C — 1 sheet ex herb. Hofman Bang, with field label "Aristida lanata. M.

Horeb", microf. 9: I. 5, 6); 1768 (C — 1 sheet ex herb. Hofman Bang, microf. 9: I. 7, 8); 1769 (C — 1 sheet with field label "Aristida * lanata. deserta Cairi", microf. 9: II. 1, 2); 1770 (C — 1 sheet ex herb. Hofman Bang, microf. 9: II. 3, 4).
Forsskål s.n. (LD — 1 sheet ex herb. Retzius).
NOTE. Note that several of the original field labels clearly indicate that they were intended to represent Forsskål's new species *Aristida lanata* (*Stipagrostis lanata*). The specimen from Sinai (Mt. Horeb) must have been collected by Niebuhr or von Haven.

Themeda triandra *Forssk.* 1775: 178 (CXXIII No. 598; Cent. VI No. 92); Christensen 1922: 30; Schwartz 1939: 310; Tackholm & Drar 1941: 513; Tackholm 1974: 759; F.T.E.A. Gramin.: 830 (1982), for synonymy.
Themeda polygama J.F. Gmel., Syst. Nat. 2: 149 (1791).
Anthistiria forskalii Kunth, Rév. Gram. 1: 162 (1829).
Themeda forskalii Hack. in DC., Monogr. Phan. 6: 659 (1889).
VERNACULAR NAME. *Thaemed, Themed, Alaf* (Arabic).
LOCALITY. Yemen: Al Hadiyah ("Hadîe"), 1763.
MATERIAL. *Forsskål* no type specimen found.

Trachynia distachya (*L.*) *Link*, Hort. Bot. Berol. 1: 43 (1827); Tackholm & Drar 1941: 161; Tackholm 1974: 683, pl. 252A.
Bromus distachyos L., Amoen. Acad. 4: 304 (1759); Forssk. 1775: LXI No. 66.
LOCALITY. Egypt: Alexandria ("Alexandriae"), 1 Apr. 1762.
MATERIAL. *Forsskål* 104 (C — 1 sheet, microf. 15: III. 7, 8).
Forsskål s.n. (LD — 1 sheet ex herb. Retzius).
NOTE. Some botanists consider the name of this species to be *Brachypodium distachyum* (L.) P. Beauv.

Tragus racemosus (*L.*) *All.*, Fl. Pedem. 2: 24 (1785); Fl. Turk. 9: 587 (1985).
Phalaris muricata Forssk. 1775: 202 (XIX No. 30; Cent. VIII No. 4); Christensen 1922: 33.
Tragus muricatus (Forssk.) Moench, Meth.: 53 (1794).
LOCALITY. Turkey: Dardanelles ("Dardanellei", "ad littora Maris Marmorae"), July 1761.
MATERIAL. *Forsskål* no specimen found.

Trisetaria linearis *Forssk.* 1775: LX No. 52; Christensen 1922: 13; Tackholm 1974: 709; Fl. Palaest. 4: 208, pl. 261 (1986). = *Trisetaria* [*sp. without epithet*] Forssk. 1775: 27 (Cent. I No. 99).
T. forskahlei J.F. Gmel., Syst. Nat. 2: 177 (1791).
Trisetum lineare (Forssk.) Boiss., Diagn. Ser. 1, 13; 49 (1853); Tackholm & Drar 1941: 317.
LOCALITY. Egypt: Alexandria, Ras etin ("Alexandiae in peninsula Râs-ettin"), April 1762.
MATERIAL. *Forsskål* 118 (C — 1 sheet with field label "60. Trisetaria linearis. Alex. Ras ettin", type of *T. linearis*, microf. 109: I. 7, 8).
NOTE. This is one of the cases where the generic name and description occurs in the 'Centuriae', while the epithet is provided in the 'Florae'.

Triticum aegilopoides *Forssk.* 1775: 26 (LXI No. 87; Cent. I No. 94); Christensen 1922: 13.
VERNACULAR NAME. *Qamh staejri* (Arabic).
LOCALITY. Egypt: Alexandria, 1761–62.
MATERIAL. *Forsskål* no type specimen found.
NOTE. The application of this name is still in doubt. Christensen gives the synonymy *Rottboellia hirsuta* Vahl = *Elionurus hirsutus* (Forssk.) Munro.

Triticum durum *Desf.*, Fl. Atlan. 1: 114 (1798); Tackholm & Drar 1941: 228.
LOCALITY. Egypt: Cairo, 1762.
MATERIAL. *Forsskål* 1807 (C — 1 sheet with field label "Ca. 245").
NOTE. T.A. Cope 1987 named and labelled this specimen on the assumption that the axes are tough. If fragile it could be *T. dicoccum.*

Triticum spelta *L.* 1753: 86; *α* **villosum** *Forssk.* 1775: 26 (LXI No. 85; Cent. I No. 92); Christensen 1922: 13.
VERNACULAR NAME. *Qamh, Hunta* (Arabic)
var. a) Qamh nac aejghe
var. b) Qamh m'ghaejir
LOCALITY. Egypt: Alexandria ("*Alexandriae* fine Apr. maturum"), 1761–62.
MATERIAL. *Forsskål* no specimen of a) or b).

Triticum spelta L. *β* **glabrum** *Forssk.* 1775: 26 (Cent. I No. 93); Christensen 1922: 13.
VERNACULAR NAME. *Qamh staejri* (Arabic).
LOCALITY. Egypt.
MATERIAL. *Forsskål* no specimen found.

Urochondra setulosa (*Trin.*) *C.E. Hubbard* in Hook., Ic. Pl.: 35 t. 3457 (1947).
LOCALITY. Unspecified.
MATERIAL. *Forsskål* 1812 (C — 1 sheet ex herb. Hornemann, microf. nil).
Forsskål s.n. (BM — 1 sheet ex herb. Banks).

Vulpia fasciculata (*Forssk.*) *Fritsch*, K. Erkursionflora ed. 2, 74 & 692 (1909); Sampaio, Herb. Portug. 24 (1913); Fl. Palaest. 4: 236, pl. 301 (1986).
Festuca fasciculata Forssk. 1775: 22 (LXI No. 62; Cent. I No. 77); Christensen 1922: 13.
LOCALITY. Egypt: Alexandria ("*Alexandriae*"), Apr. 1762.
MATERIAL. *Forsskål* 1260 (C — 1 sheet, ex herb. Hornemann, with field label "R[?] 45", type of *F. fasciculata*, microf. nil).
NOTE. Tackholm & Drar Fl. Egypt 1: 165 (1941) include this in *Vulpia membranacea* (L.) Link (1827); but according to Stace they are distinct species.

Andropogonoides [sp. without epithet, preliminary designation] Forssk. 1775: 27 (Cent. I No. 96); Christensen 1922: 13.
LOCALITY. Yemen: Al Hadiyah ("*Hadîe*"), 1763.
MATERIAL. *Forsskål* no specimen found.
NOTE. Any of the additional specimens in the Forsskål collection could be these undetermined plants but without further evidence it is unlikely to be resolved, which applies to Cent. I No. 96, 97 & 98 respectively.

"**Gramen**" [gen. & sp. indet.] Forssk. 1775: 27 (Cent. I No. 97); Christensen 1922: 13.
LOCALITY. Yemen: Al Hadiyah ("Ad *Hadîe*"), 1763.
MATERIAL. *Forsskål* no specimen found.
NOTE. See note after Cent. I No. 96 *Andropogonoides.*

"**Gramen**" [gen. & sp. indet.] Forssk. 1775: 27 (LXI No. 88; Cent. I No. 98); Christensen 1922: 13.
LOCALITY. Egypt: Alexandria ("*Alexandriae*"), 1762.
MATERIAL. *Forsskål* no specimen found.
NOTE. See note under Cent. I No. 96 *Andropogonoides.*

HYDROCHARITACEAE

Ottelia alismoides (*L.*) *Pers.*, Syn. 1: 400 (1805); Tackholm 1974: 613, pl. 223.
Stratiotes alismoides L. 1753: 535; Forssk. 1775: 101 (LXVIII No. 285; Cent. IV No. 15); Christensen 1922: 20.
LOCALITY. Egypt: Farseh I., Rashid ("In fossis agrorum insulae Farseh prope Rosettam"), 2–6 Nov. 1761.
MATERIAL. *Forsskål* 20 (C — 1 sheet with field label "Stratiotes fol. cordata. Ro. 28. Rosette. alismoides", microf. 104: III. 3, 4).

HYPOXIDACEAE

Hypoxis [sp. without epithet] *Forsskål.* 1775: 74 (CX No. 243; Cent. III No. 35); Christensen 1922: 18.
LOCALITY. Yemen: Taizz ("Ad Taaes lecta"), 1763.
MATERIAL. *Forsskål* no specimen found.
NOTE. Forsskål gave only the generic name, followed by two words that are not part of the name. Although there is a full description it has not been possible to identify the plant, even to confirming it as a *Hypoxis* which has yellow, not violet, flowers.

IRIDACEAE

Gynandriris sisyrinchium (*L.*) *Parl.*, Nuovi Gen. e sp. Monocot: 52 (1854).
Iris sisyrinchium L. 1753: 40; Forssk. 1775: 12 (LIX NO. 16; cent. I No. 32); Christensen 1922: 11 Tackholm & Drar 1954: 463; Tackholm 1974: 659.
VERNACULAR NAME. *Zambac, Zambak* (Arabic).
LOCALITY. Egypt: Giza ("Circa Pyramides copiose in campis"), 1762.
MATERIAL. *Forsskål* 24 (C — 1 sheet, microf. 57: III. 1, 2); 1362 (C — 1 sheet, microf. 129: I. 1, 2).

JUNCACEAE

Juncus acutus *L.* 1753: 325; Tackholm & Drar 1950: 460; Tackholm 1974: 664, pl. 245B; Fl. Palaest. 4: 141, pl. 188 (1986).
J. spinosus Forssk. 1775: 75 (XXIV No. 173 LXV No. 207; Cent. III No. 38); Christensen 1922: 18.
VERNACULAR NAME. *Brolo* (Greek), *Sammâr* (Arabic).
LOCALITY. Egypt: Alexandria ("As" = Alexandriae), 1761–62.
MATERIAL. *Forsskål* 29 (C — 1 sheet, type of *J. spinosus*, microf. 58: II. 1, 2).
NOTE. Forsskål records this from both Turkey (XXIV) and Egypt (LXV) so the type specimen may come from either country as there is no indication on the sheet, but the description on p. 75 includes the Egyptian name.

Juncus bufonius *L.* 1753: 328; Tackholm & Drar 1950: 456; Tackholm 1974: 662, pl. 245A.
LOCALITY. Egypt: Cairo, 1761–62.
MATERIAL. *Forsskål* 28 (C — 1 sheet, microf. 129: I. 3, 4); 1372 (C — 1 sheet ex herb. Hornemann, microf. nil).

Juncus effusus *L.* 1753: 326; Forssk. 1775: XXIV No. 175; Fl. Turk. 9: 10 (1985).
VERNACULAR NAME. *Borla* (Greek).
LOCALITY. Turkey: Istanbul, Belgrad ("Bg"), July–Sept. 1761.
MATERIAL. *Forsskål* 1373 (C — 1 sheet ex herb Hornemann, microf. nil).
NOTE. No. 1373 has been named *J. inflexus* L. by S. Snogerup, 1988.

Juncus striatus *Schousb. ex E.H.F. Meyer*, Syn. Junc.: 27 (1822); Fl. Europ. 5: 110 (1980); Fl. Turk. 9: 20 (1985).
LOCALITY. Unspecified, prob. Turkey.
MATERIAL. *Forsskål* 1374 (C — 1 sheet ex herb. Hofman Bang, microf. 129: I. 5, 6).

Juncus subulatus *Forssk.* 1775: 75 (LXV No. 208; Cent. III. No. 37); Christensen 1922: 18; Tackholm & Drar 1950: 455; Tackholm 1974: 662, pl. 245 C; Fl. Turk. 9: 9 (1985); Meikle 1985: 1654; Fl. Palaest. 4: 144, pl. 190 (1986).
VERNACULAR NAME. *Hallaen* (Arabic).
LOCALITY. Egypt: Alexandria ("Alexandriae"), Apr. 1762.
MATERIAL. *Forsskål* 30 (C — 1 sheet with field label "Juncus novus subulatus. Alex. Av. 111", type of *J. subulatus*, microf. 58: II. 3, 4).

LILIACEAE*

Aloe arborea *Forssk.* 1775: CX No. 241; Christensen 1922: 37; Wood in Kew Bull. 38: 21 (1983).
VERNACULAR NAME. *Kobab* (Arabic).
LOCALITY. Yemen: Jiblah ("Djöbla"), 1762.
MATERIAL. *Forsskål* no type specimen found.
NOTE. The name *A. arborea* is validated by the description: "Pedunculo racemoso. Venenum e succo paratur". Wood thought that *A. arborea* could be *A. rivierei* Lavranos & Newton (1977) — if so the name would have to be replaced by *A. arborea* — "the only *Aloe* that could grow at the type locality (Jiblah) and be described as *arborea* is [*rivieri*]. I have not seen it at Jiblah and *A. arborea* Forssk. is best treated as a nomen ambiguum".

Aloe inermis *Forssk.* 1775: 74 (CX No. 240; Cent. III No. 33); Christensen 1422: 18; Schwartz 1939: 351; Wood in Kew Bull. 38: 16 (1983).
VERNACULAR NAME. *Aebliae* (Arabic).
LOCALITY. Yemen: Taizz ("Ad urbem Taaes"), 1763.
MATERIAL. *Forsskål* no type specimen found.

Aloe pendens *Forssk.* 1775: 74 (Cent. III No. 32); Schweinfurth 1894: 72; Schwartz 1939: 351; Wood in Kew Bull. 38: 19 (1983).
VERNACULAR NAME. *Besesil, Fyll asfar* (Arabic).
LOCALITY. Yemen: Al Hadiyah ("Hadîe"), 1763.
MATERIAL. *Forsskål* 16 (C — 1 sheet, lectotype designated by Wood for *A. pendens*, microf. 4: III. 7, 8).

A. variegata sensu Forssk. 1775: 74 (CX No. 239; Cent. III No. 31); Christensen 1922: 18.
VERNACULAR NAME. *Besesil, Beselil, Bselil, Fyll asfar* (Arabic).
LOCALITY. Yemen: Al Hadiyah ("Hadie"), 1763.

* Liliaceae is here maintained in the traditional sense

MATERIAL. *Forsskål* (see note).
NOTE. Wood (l.c.) argues that *Forsskål* 16 matches the description of *A. pendens* better than *A. variegata* which seems to have been described from a very young specimen without flowers.

Aloe vera (*L.*) *Burm.f.* var. **officinalis** (*Forssk.*) *Bak.* in Journ. Linn. Soc. 18: 176 (1880); Schweinfurth 1894: 59; Schwartz 1939: 350; Wood in Kew Bull. 38: 22 (1983).
A. officinalis Forssk. 1775: 73 (CX No. 238; Cent. III No. 30); Deflers 1889: 211; Christensen 1922: 18.
VERNACULAR NAME. *Sabr* (Arabic).
LOCALITY. Yemen: Wadi Mawr ("Môr"), 1763.
MATERIAL. *Forsskål* no type specimen found.

A. maculata Forssk. 1775: 73 (Cent. III No. 29); Christensen 1922: 18.
A. variegata sensu Forssk. 1775: LXV No. 206.
VERNACULAR NAME. *Sabbâre* (Arabic).
LOCALITY. Yemen: Al Luhayyah ("Lohajae"), Jan. 1763.
MATERIAL. *Forsskål* no type specimen found.
NOTE. Wood (q.v.) has no doubt that these are conspecific — *A. maculata* appears to be the yellow-flowered plant that yields the aloe of commerce and medicine found cultivated in Egypt and elsewhere in the Mediterranean area, of which *A. officinalis* is the red-flowered form found growing wild in Yemen.

Aloe vacillans *Forssk.* 1775: 74 (CX No. 242; Cent. III No. 34); Christensen 1922: 18; Schweinfurth 1894: 62; Schwartz 1939: 350; Wood in Kew Bull. 38: 23 (1983).
VERNACULAR NAME. *Charchara* (Arabic).
LOCALITY. Yemen: Kusma ("Kurmae"), 1763.
MATERIAL. *Forsskål* no type specimen found.
NOTE. Wood (q.v.) cites the synonyms *A. dhalensis* Lavranos and *A. audhalica* Lavrananos & Hardy under this name.

Anthericum asphodelum *Forssk.* 1775: 209 (XXIV No. 171; Cent. VIII No. 40); Christensen 1922: 34.
LOCALITY. Rhodes ("In insula Rhodo"), 21–22 Sept. 1761.
MATERIAL. *Forsskål* no type specimen found.
NOTE. This validly published name has not been applied to any species in recent literature. It should not be adopted if it upsets widely accepted nomenclature.

Asparagus acutifolius *L.* 1753: 314; Forssk. 1775: XXIV No. 181; Fl. Turk. 8: 76 (1984).
VERNACULAR NAME. *Sparaggria* (Greek).
LOCALITY. Turkey: Gokceada I & Dardanelles ("Imros, Borghàs fons"), July 1761.
MATERIAL. *Forsskål* 13 (C — 1 sheet, microf. 12: I. 5, 6); 14 (C — 1 sheet with field label, microf. 12: I. 7, 8); 983 (C — 1 sheet ex herb. Hornemann, microf. 12: II. 1, 2).
Forsskål s.n. (LD — 1 sheet ex herb. Retzius).
NOTE. Sheet 983 (12. II. 1, 2) has a different aspect from the other two and it probably represents a distinct gathering.

Asparagus africanus *Lam.*, Encycl. Méth. Bot. 1: 295 (1783); Schweinfurth 1894: 77; Tackholm 1974: 645.
A. scaberulus sensu Deflers 1889: 210, non A. Rich.
A. retrofractus sensu Forssk. 1775: 73 (CIX No. 235; Cent. III No. 27), non L.; Christensen 1922: 18.

VERNACULAR NAMES. *Schadjaret ennemr, Hömumer, Heniet ennemr* (Arabic).
LOCALITY. Yemen: Al Hadiyah ("inter Ersch & Alûdje" on label), Mar. 1763.
MATERIAL. *Forsskål* 12 (C — 1 sheet with field label, microf. 12: II. 3, 4).
NOTE. Christensen l.c. and Schwartz 1939: 353 considered this to be *A. asiaticus* L.

Asparagus agul *Forssk.* 1775: LXV No. 203; Christensen 1922: 36.
LOCALITY. Egypt: ?Alexandria, 1761–62.
MATERIAL. *Forsskål* no type specimen found (see Note).
NOTE. This is an obscure plant and even the epithet is of doubtful meaning, yet it is validated by the brief description: "Sine stipulis ad exitum ramor". It should not be adopted if it upsets widely accepted nomenclature.

Asparagus falcatus *L.* 1753: 313; Schwartz 1939: 354.
LOCALITY. Unspecified, prob. Yemen.
MATERIAL. *Forsskål* s.n. (LD — 1 sheet ex herb. Retzius).

Asparagus officinalis *L.* 1753: 313; Schweinfurth 1894: 77; Schwartz 1939: 353.
LOCALITY. Unspecified, prob. Yemen.
MATERIAL. *Forsskål* 982 (C — 1 sheet, microf. 129: II. 1, 2).

Asparagus stipularis *Forssk.* 1775: 72 (LXV No. 202; Cent. III No. 26); Christensen 1922: 18; Tackholm 1974: 645, pl. 237A.
LOCALITY. Egypt: Alexandria ("*Alexandriae*"), Oct. 1761.
MATERIAL. *Forsskål* 15 (C — 1 sheet, type of *A. stipularis*, microf. 12: II. 5, 6); 984 (C — 1 sheet ex herb. Hornemann, type of *A. stipularis*, microf. 12: II. 7, 8); 985 (C — 1 sheet with field label "Al. 18", type of *A. stipularis*, microf. 12: III. 1, 2).
Forsskål s.n. (LD — 1 sheet ex herb. Retzius, type of *A. stipularis* with field label "Al. 18").

Asphodelus fistulosus *L.* var. **tenuifolius** (*Cav.*) *Bak.* in Journ. Linn. Soc. 15: 272 (1876).
A. tenuifolius Cav. — Schwartz 1939: 349.
Ornithogalum flavum Forssk. 1775: 209 (CIX No. 234; Cent. VIII No. 38); Christensen 1922: 34.
VERNACULAR NAME. *Vassal er robah, Chosar errobah, Bassal eddjinn, Bassal errobah* (Arabic).
LOCALITY. Yemen; Surdud ("Surdûd"), Feb. 1763.
MATERIAL. *Forsskål* 8 (C — 1 sheet, type of *O. flavum*, microf. 74: III. 3–4).

Chlorophytum tetraphyllum (*L.f.*) *Bak.* in Journ. Linn. Soc. 15: 328 (1876); Fl. Trop. Afr. 7: 500 (1898).
Scilla tetraphylla L.f., Suppl. 200 (1781).
LOCALITY. Unspecified (see note).
MATERIAL. *Forsskål* (LINN — 1 sheet 429.14 "Fabr. 27", type of *S. tetraphylla*, microf. 230: I. 7).
Forsskål s.n. (BM — 1 sheet ex herb. Banks, type of *S. tetraphylla*.
Forsskål 1697 (C — 1 sheet, type of *S. tetraphylla*, microf. nil); 1698 (C — 1 sheet ex herb. Vahl annotated "Haemanthus coccineus Cent. 3. No. 36" on back by Vahl, microf. 52: I. 7, 8).
NOTE. Linnaeus fil. provides the information on provenance of the Linnaean specimen as "Habitat in Africa. D.D. Fabricius, Prof. Havn." The specimen (No. 1698) annotated "Haemanthus coccineus" is probably the basis of a report of material of that species by

Christensen, 1922. J.C. Fabricius was Professor of Zoology in Copenhagen; he never went to Africa but was associated with the royal institutions where Forsskål's plants at first, were kept, and where Fabricius and Zoëga had access to them.

Dipcadi erythraeum *Webb & Berth.*, Phyt. Canar. 3: 341 (1848); Tackholm 1974: 647, pl. 234.
Hyacinthus serotinus? Forssk. 1775: 209 (LXV No. 205; Cent. VIII No. 41); Christensen 1922: 34.
VERNACULAR NAME. *Borraejt, Zaaeteman* (Arabic).
LOCALITY. Egypt: Cairo (Káhirae in desertis"), 1761–62.
MATERIAL. *Forsskål* 17 (C — 1 sheet, microf. 55: II. 5, 6, III. 7, 8); 692 (C — sheet, microf. 55: II. 7, 8).

Merendera abyssinica *A. Rich.*, Tent. Fl. Abyss. 2: 337 (1850); Deflers 1889: 213; Schwartz 1939: 349.
Colchicum montanum Forssk. 1775: 77 (CX No. 248; Cent. III No. 44), non L.; Christensen 1922: 18.
LOCALITY. Yemen: Kusma ("Kurmae"), Mar. 1763.
MATERIAL. *Forsskål* 27 (C — 1 sheet, type of *C. montanum*, microf. 28: III. 7–8).
NOTE. Christensen noted this as "Dubia. A single flower only is present". However, there is no doubt now as to its identity as the narrow tepals are clearly free, making it *Merendera* rather than *Colchicum* or *Romulea.*

Ornithogalum narbonense *L.* Cent. Pl. 2: 15 (1756); Fl. Europ. 5: 37 (1980).
LOCALITY. Unspecified.
MATERIAL. *Forsskål* s.n. (LD — 1 sheet ex herb. Retzius).

Sansevieria forskaliana (*Schult.f.*) *Hepper & Wood* in Kew Bull. 38: 83 (1983).
Smilacina forskaliana Schult.f., Syst. Veg. 7: 304 (1829). Type as below.
Convallaria racemosa Forssk. 1775: 73 (CX No. 236; Cent. III No. 28), non L.; Christensen 1922: 21.
VERNACULAR NAME. *Daenag* (Arabic).
LOCALITY. Yemen: Al Hadiyah ("Hadie", "Aludje' on field label), 1763.
MATERIAL. *Forsskål* 9 (C — 1 sheet with field label, type of *C. racemosa*, microf. 30: II. 8, III. 1).
Sansevieria guineensis sensu Swartz 1939: 352.

Scilla hycacinthina (*Roth*) *J.F. Macbride* in Contrib. Gray Herb. n.s. 56: 14 (1918).
LOCALITY. Unspecified, prob. Yemen.
MATERIAL. *Forsskål* 11 (C — 1 sheet, microf. 129: II. 7, 8); 1696 (C — 1 sheet ex herb. Vahl, microf. 129: II. 5, 6).

Urginea maritima (*L.*) *Baker* in Journ. Linn. Soc. 13: 221 (1872); Fl. Turk. 8: 213 (1984); Meikle 1985: 1634.
Anthericum aphyllum Forssk. 1775: 209 (XXIV No. 170; Cent. VIII No. 39); Christensen 1922: 34.
LOCALITY. Rhodes ("In insula Rhodo"), 21–22 Sept. 1761.
MATERIAL. *Forsskål* 10 (C — 1 sheet, type of *A. aphyllum*, microf. 129: III. 1, 2); 945 (C — 1 sheet, ex herb. Hornemann, microf. nil).

ORCHIDACEAE

Eulophia streptopetala *Lindl.* var. **rueppelii** (*Rchb.f.*) *Cribb* in Kew Bull. 33. 676 (1979).
Orchis flava Forssk. 1775: 156 (CXX No. 518; Cent. VI No. 1), non L.; Christensen 1922: 27.
Lissochilus arabicus Lindl., Gen. & Sp. Orch.: 192 (1833); Schwartz 1939: 357.
VERNACULAR NAME. *Djissâb* (Arabic).
LOCALITY. Yemen: Barad ("In monte Barah"), Mar. 1763.
MATERIAL. *Forsskål* 1528 (C — 1 sheet ex herb. Vahl with field label "Mokhajam", type of *O. flava* and *L. arabicus*, microf. 74: II. 1, 2).

Habenaria aphylla (*Forssk.*) *R. Br.*, Prodr.: 312, in adnot (1810).
Holothrix aphylla (Forssk.) Rchb.f., Otia Bot. Hamburg.: 119 (1881); Schwartz 1939: 356.
Orchis aphylla Forssk. 1775: 156 (CXX No. 519; Cent. VI No. 1); Christensen 1922: 27.
LOCALITY. Yemen: Wasab & Kusma ("Ad Uahfât & Kurmâ"), 1763.
MATERIAL. *Forsskål* no type specimen found.

Orchis viridis *Forssk.* 1775: 156 (CXX no. 520; Cent VI No. 3); Christensen 1922: 27.
LOCALITY. Yemen: Taizz ("Taaes", "Ad M'hàrras"), 1763.
MATERIAL. *Forsskål* no type specimen found.
NOTE. An obscure species. The name should not be taken up.

PALMAE

Chamaerops humilis *L.* 1753: 1187; Forssk. 1775: XII No. 265; Fl. Europ. 5: 267 (1980).
LOCALITY. France: Monpellier Botanical Garden, May 1761.
MATERIAL. *Forsskål* s.n. (LD — 1 sheet, ♂ inflorescence only, with "C.C.1").

Phoenix dactylifera *L.* 1753: 1188; Forssk. 1775: LXXVIII No. 575; CXXVI No. 692; Tackholm 1974: 763.
VERNACULAR NAME. *Nacbl* (Arabic).
LOCALITY. Egypt: Cairo ("Ch"), 1761–62.
MATERIAL. *Forsskål* 1 (C — 1 sheet with field label, microf. 75: II. 5, 6).

PANDANACEAE

Pandanus odoriferus (*Forssk.*) *Chiov.* in Bull. Soc. Bot. Ital. 1923: 116.
Keura odorifera Forssk. 1775: 172 (CXXII No. 590; Cent. VI No. 69); Christensen 1922: 29.
Pandanus odoratissimus L.f., Suppl. Pl.: 424 (1781); Deflers 1889: 214; Schweinfurth 1894: 7.
P. tectorius Sol. — Schwartz 1939: 300.
VERNACULAR NAME. *Kadi, Kabua Kadi* (Arabic), *Keura* (Indian = "Banjani").
LOCALITY. Yemen; Al Hadiyah ("Hadîe"), 1763.
MATERIAL. *Forsskål* no type specimen found.
NOTE. This is a new genus and species described and provided with an epithet in the 'Centuriae'. The same epithet is provided in the 'Florae'.

RUPPIACEAE

Ruppia maritima *L.* 1753: 127; Fl. Europ. 5: 11 (1980).
Zannichellia [*sp. without epithet*] Forssk. 1775: XIV No. 83.

LOCALITY. Malta, 12 June 1761.
MATERIAL. *Forsskål* 25 (C — 1 sheet with field label "Zannichellia. Malta in fonte dulci ad salinas", microf. 112: III. 7, 8).

ZOSTERACEAE

Halophila stipulacea (*Forssk.*) *Aschers.* in Sitz.-Ber. Ges. Naturf. Fr. Berlin 1867: 3 (1867); Schwartz 1939: 302; den Hartog in Verh. Konink Nederl. Akad. Wetensch., Afd. Nat., 59 (1): 258, fig. 63 (1970); Tackholm 1974: 613, pl. 222A.
Zostera stipulacea Forssk. 1775: 158 (CXX No. 529; Cent. VI No. 9); Christensen 1922: 27.
LOCALITY. Yemen: Al Mukha ("Omnes hae Zosterae ad littora Mochhae, gramina referentes submarina"), 1763.
MATERIAL. *Forsskål* no type specimen found.

Halodule uninervis (*Forssk.*) *Aschers.* in Boiss., Fl. Orient. 5: 24 (1882).
Diplanthera uninervis (Forssk.) Aschers. in Engl. & Prantl, Pflanzenfam. Nachtr. 1: 37 (1897); Schwartz 1939: 302.
Zostera uninervis Forssk. 1775: 157 (CXX No. 527; Cent. VI No. 7); Vahl 1804: 14; Christensen 1922: 27.
VERNACULAR NAME. *Djezavi* (Arabic?).
LOCALITY. Yemen: Al Mukha ("Mochha") 1763.
MATERIAL. *Forsskål* no type specimen found.

Thalassodendron ciliatum (*Forssk.*) *den Hartog* in Verh. Konink Nederl. Akad. Wetensch, Afd. Nat., 59(1): 188 fig. 52 (1970).
Cymodocea ciliata (Forssk.) Ehrh. ex Aschers. in Sitzb. Ges. Naturf. Fr. Berlin: 3 (1867); Schwartz 1939: 301.
Zostera ciliata Forssk. 1775: 157 (CXX No. 528; Cent. VI No. 8); Christensen 1922: 27.
Thalassia ciliata (Forssk.) Koenig in Koenig & Sims, Ann. Bot. 2: 97 (1805).
VERNACULAR NAMES. *Koschar, Kanaf* (Arabic).
LOCALITY. Yemen: Al Mukha ("Moccha"), May 1763.
MATERIAL. *Forsskål* 26 (C — 1 sheet, type of *Z. ciliata*, microf. 113: I. 7, 8, II. 1). *Forsskål* s.n. (BM — 1 sheet ex herb. Banks, type of *Z. ciliata*).

Zostera "dubia?" Forssk. 1775: CXX No. 530; den Hartog in Verh. Konink Nederl. Akad. Wetensch., Afd. Nat. 59(1): 97 (1970).
VERNACULAR NAME. *Ölefi* (Arabic?).
LOCALITY. Yemen: Jibla ("Djöbla"), 1763.
MATERIAL. *Forsskål* no specimen found.
NOTE. This cannot be *Zostera* from mountainous "Djöbla", nor is the name validly published since the Code specifically excludes Forsskål's "dubia" epithets. It was assigned to the "Excluded species" by the monographer den Hartog l.c.

FERNS AND FERN-ALLIES*

Acrostichum filare *Forssk.* 1775: 184 (CXXIV No. 628; Cent. VII No. 1); Christensen 1922: 30.
VERNACULAR NAME. *Meschat, Meschat elghorab* (Arabic).
LOCALITY. Yemen: without precise locality ("Yemen in sylvis montosis"), 1763.
MATERIAL. *Forsskål* no type specimen found.
NOTE. See note under *Asplenium aethiopicum.*

Actiniopteris radiata (*Swartz*) *Link*, Fil. sp.: 80 (1841).
A. australis (L.f.) Link, Fil. sp.: 80 (1841); Schwartz 1939: 20.
Acrostichum australe L.f., Suppl. Plant Syst. Veg.: 444 (1781); Vahl 1790: 84, t. 25.
Acrostichum dichotomum sensu Forssk. 1775: 184 (CXXIV No. 627; Cent. VII No. 2), non L.; Christensen 1922: 30.
VERNACULAR NAME. *Mejabese* (Arabic).
LOCALITY. Yemen: Al Hadiyah ("Hadîe"), Mar. 1763.
MATERIAL. *Forsskål* 815 (C — 1 sheet, microf. 2: II. 7, 8); 820 (C — 1 sheet, microf. 2: III. 1, 2).
Forsskål s.n. (LD — 1 sheet ex herb. Retzius).

Adiantum capillus-veneris *L.* 1753: 1096; Forssk. 1775: CXXV No. 649; Schwartz 1939: 19.
VERNACULAR NAME. *Schecb mabdjar* (Arabic).
LOCALITY. Yemen: Wasab; Al Hadiyah, 1763.
MATERIAL. *Forsskål* 804 (C — 1 sheet with field label "ad Uahfad", microf. 2: III. 7, 8); 805 (C — 1 sheet with field label "Hadîe", microf. 2: III. 5, 6).

Adiantum flabellulatum *L.* 1753: 1095.
LOCALITY. Unspecified.
MATERIAL. *Forsskål* 811 (C — 1 sheet, microf. 134: I. 3, 4); 812 (C — 1 sheet, microf. 134: I. 5, 6).

Adiantum incisum *Forssk.* 1775: 187 (CXXV No. 648; Cent. VII No. 19); Christensen 1922: 31; F.Z. Pterid.: 108 (1970).
VERNACULAR NAME. *Meschât el ghorâb* (Arabic).
LOCALITY. Yemen: Al Hadiyah ("Hadîe"), 1763.
MATERIAL. *Forsskål* 813 (C — 1 sheet, type of *A. incisum*, microf. 3: I. 1, 2); 1843 (C — 1 sheet with original field label, type of *A. incisum*, microf. nil).
A. caudatum sensu Schwartz 1939: 19.

Adiantum poiretii *Wikstr.*, Vet. Akad. Hal. 1825: 443 (1826).
LOCALITY. Unspecified.
MATERIAL. *Forsskål* 806 (C — 1 sheet, microf. 133: III. 7, 8); 823 (C — 1 sheet, microf. 134: I. 1, 2).

Arthropteris orientalis (*J.F. Gmel.*) *Posthumus* in Rec. Trav. Bot. Néerl. 21: 218 (1924); Schwartz 1939: 17; F.W.T.A. Ferns: 52, fig. 12 (1959); F.Z. Pterid.: 163 (1970).
Polypodium orientale J.F. Gmel., Syst. Nat. ed. 13, 2: 1312 (1791).

* Genera in alphabetical order

Dryopteris orientalis (J.F. Gmel.) C. Chr., Index Filic.: 281 (1905).
Polypodium pectinatum sensu Forssk. 1775: 185 (CXXV No. 645; Cent. VII No. 5), non L.; Christensen 1922: 30.
LOCALITY. Yemen: Bolghose ("Bulghose"), 1763.
MATERIAL. *Forsskål* 807 (C — 1 sheet, type of *P. pectinatum* & *P. orientale*, microf. 84: I. 1, 2); 822 (C — 1 sheet, type of *P. pectinatum* & *P. orientale*, microf. 84: I. 3, 4).

Asplenium adiantum-nigrum *L.* 1753: 1081; Schwartz 1939: 17.
A. ruta muraria sensu Forssk. 1775: CXXV No. 643, non L.
LOCALITY. Yemen: Mukhajah, J. Barad ("Mochaja Barad" on field label, "Kurma" in book), 1763.
MATERIAL. *Forsskål* 816 (C — 1 sheet with original field label, microf. 12: III. 7, 8); 821 (C — 1 sheet, microf. 133: III. 3, 4).
Forsskål s.n. (LD — 1 sheet ex herb. Retzius).

Asplenium aethiopicum (*Burm. f.*) *Becherer* in Candollea 6: 23 (1935); F.Z. Pterid.: 181 (1970).
A. praemorsum Sw. (1788) — Schwartz 1939: 17.
A. lanceolatum Forssk. 1775: 185 (CXXV No. 644; Cent. VII No. 4), non Huds.; Deflers 1889: 221; Christensen 1922: 30.
LOCALITY. Yemen: Kusma ("Kurmae", "Kurma" on field label), 1763.
MATERIAL. *Forsskål* 817 (C — 1 sheet with field label, type of *A. lanceolatum*, microf. 12: III. 3, 4); 986 (C — 1 sheet ? ex herb. Hornemann, type of *A. lanceolatum*, microf. 12: III. 5, 6).
Forsskål s.n. (1 sheet ex herb. Retzius, type of *A. lanceolatum*).
NOTE. Christensen (1922: 30) considered *Acrostichum filare* Forssk. 1775: 184 (Cent. VII No. 1) to belong here in the absence of any material.

Asplenium rutifolium (*Berg.*) *Kunze* var. **bipinnatum** (*Forssk.*) *Schelpe* in Journ. S. Afr. Bot. 30: 194 (1964), & F.Z. Pterid.: 185 (1970).
A. bipinnatum (Forssk.) C. Chr. ex Hieron. in Wiss. Ergebn. Deutsch. Zent.-Afr. Exped. 2: 11 (1910), non Roxb. (1844).
Lonchitis bipinnata Forssk. 1775: 184 (CXXIV No. 640; Cent. VII No. 3); Christensen 1922: 30.
Asplenium achilleifolium (Lam.) C. Chr. var. *bipinnatum* (Forssk.) C. Christensen 1922: 30; Schwartz 1939: 17.
LOCALITY. Yemen: Bolghose ("Blg" = Bolghose), 1763.
MATERIAL. *Forsskål* 810 (C — 1 sheet with field label "inter Bolghose et Mokhaja", type of *L. bipinnata*, microf. 64: I. 1, 2).
NOTE. On the field label in Forsskål's hand "Lonchitis vel nov. gen. * bipinnata".

Cheilanthes farinosa (*Forssk.*) *Kaulf.*, Enum. Fil.: 212 (1824); Schwartz 1939: 18; F.Z. Pterid.: 122 (1970).
Pteris decursiva Forssk. 1775: 186 (CXXIV No. 633; Cent. VII No. 13); Christensen 1922: 31.
Pteris farinosa Forssk. 1775: 187 (CXXIV No. 638; Cent. VII No. 18); Christensen 1922: 31.
LOCALITY. Yemen: Al Hadiyah ("montium Hadiensium media inhabitant"), 1763.
MATERIAL. *Forsskål* no type specimen of either sp. at C.
Forsskål s.n. (LD — 1 sheet ex herb. Retzius, type of *P. farinosa*/*P. decursiva*).

Christella dentata (*Forssk.*) *Brownsey & Jermy* in Brit. Fern. Gaz. 10: 338 (1978).
Thelypteris dentata (Forssk.) *E. St. John* in Amer. Fern Journ. 26: 44 (1936); F.Z. Pterid. 197 (1970).

Polypodium dentatum Forssk. 1775: 185 (CXXV No. 647; Cent. VII No. 7); Christensen 1922: 30.
Dryopteris dentata (Forssk.) C. Chr. in Kongel. Dansk Vid. Selsk., Afd. 8, 6: 24 (1920); Schwartz 1939: 16.
Cyclosorus dentatus (Forssk.) Ching in Bull. Fan. Mem. Inst. Biol. Bot. 8: 206 (1938); F.W.T.A. Ferns: 62 (1959).
VERNACULAR NAME. *Màas* (Arabic).
LOCALITY. Yemen: Bolghose ("sum priore in montibus Yemen ad Bolghose"), 1763.
MATERIAL. *Forsskål* 809 (C — 1 sheet with field label "Polypodium * dentatum. Maas. Hadie", type of *P. dentatum*, microf. 83: III. 6, 7).

Doryopteris concolor (*Langsd. & Fisch.*) *Kuhn* in Von Deck., Reisen, Bot. 3, 3: 19 (1879); F.Z. Pterid.: 121 (1970).
LOCALITY. Unspecified, prob. Yemen.
MATERIAL. *Forsskål* s.n. (LD — 1 sheet ex herb. Retzius).

Equisetum ramosissimum *Desf.*, Fl. Atl. 2: 398 (1800); Fl. Turk. 1: 32 (1965).
E. arvense sensu Forssk. 1775: XXXVI No. 452.
LOCALITY. Turkey: Burgaz, Dardanelles ("Borghàs"), July 1761.
MATERIAL. *Forsskål* 797 (C — 1 sheet with original field label "Bo. 0", microf. 41: III. 3, 4).

Hypodematium crenatum (*Forssk.*) *Kuhn* in Von Deck., Reisen, Bot., 3, 3: 37 (1879); F.Z. Pterid.: 230, t. 66 (1970).
Polypodium crenatum Forssk. 1775: 185 (CXXV No. 646, Cent. VII No. 6); Christensen 1922: 30.
Aspidium crenatum (Forssk.) Kuhn, Fil. Afr.: 129 (1868).
Lastrea crenata (Forssk.) Bedd., Ferns Brit. Ind., Suppl.: 18 (1876).
Nephrodium crenatum (Forssk.) Bak., Fl. Maurit.: 49 (1877).
Dryopteris crenata (Forssk.) Kuntze, Rev. Gen. Pl. 2: 811 (1891); Schwartz 1939: 16.
VERNACULAR NAME. *Maschôt* (Arabic).
LOCALITY. Yemen: Bolghose ("cum priore [i.e. *Arthropteris orientalis*] in montibus Yemen ad Bolghose"), 1763.
MATERIAL. *Forsskål* no type specimen found.
Polypodium speluncae sensu Vahl 791: 105, non L.

Lycopodiella cernua (*L.*) *Pic. Ser.* in Webbia 23: 166 (1968).
Lycopodium cernuum L. 1753: 1103.
Lepidotis cernua (L.) Beauv., Magas. Encycl. 5: 479 (1804); Fl. Europ. 1: 3 (1964).
LOCALITY. Unspecified.
MATERIAL. *Forsskål* 1442 (C — 1 sheet ex herb. Hofman Bang, microf. 129, III. 7, 8).

Nephrolepis biserrata (*Sw.*) *Schott*, Gen. Fil.: sub. t. 3 (1834).
LOCALITY. Unspecified.
MATERIAL. *Forsskål* s.n. (LD — 1 sheet ex herb. Retzius).

Pellaea quadripinnata (*Forssk.*) *Prantl* in Engl., Bot. Jahrb. 3: 420 (1882); Schwartz 1939: 18; F.Z. Pterid.: 133 (1970).
Pteris quadripinnata Forssk. 1775: 186 (CXXIV No. 631; Cent. VII No. 11); Christensen 1922: 31.
LOCALITY. Yemen: Al Hadiyah ("montium Hadiensium"), 1763.
MATERIAL. *Forsskål* no type material found.

Pellaea viridis (*Forssk.*) *Prantl* in Engl., Bot. Jahrb. 3: 420 (1882); F.Z. Pterid.: 133 (1970).
Adiantum viride (Forssk.) Vahl 1791: 104.
Pteris viridis Forssk. 1775: 186 (CXXIV No. 635; Cent. VII No. 15).
LOCALITY. Yemen: Al Hadiyah ("montium Hadiensium"), 1763.
MATERIAL. *Forsskål* no type specimen found.

Phyllitis scolopendrium (*L.*) *Newman*, Hist. Brit. Ferns, ed. 2, 10 (1844); Fl. Europ. 1: 17 (1964); Fl. Turk. 1: 52 (1965).
Asplenium scolopendrium L. 1753: 1079; Forssk. 1775: XXXVI No. 454.
LOCALITY. Turkey: Istanbul ("Cp" = Constantinopolitanae), Aug. 1761.
MATERIAL. *Forsskål* 801 (C — 1 sheet, microf. 13: I. 1, 2).

Polystichum setiferum (*Forssk.*) *Woynar* in Mitt. Naturm. Ver. Steierm. 49: 181 (1913); Fl. Turk. 1: 57 (1965).
Polypodium setiferum Forssk. 1775: 185 (XXXVI No. 453; Cent. VII No. 8); Christensen 1922: 31.
P. aculeatum Vahl 1791: 105, non (L.) Roth (1799).
LOCALITY. Turkey: Dardanelles ("Ad Dardanellos"), July 1761.
MATERIAL. *Forsskål* 814 (C — 1 sheet with field label "Polypod. ort ... foliolis serratose ...", type of *P. setiferum*, microf. 84: I. 5, 6).

Pteridium aquilinum (*L.*) *Kuhn* in Von Deck, Reisen, Bot. 3, 3: 11 (1879), s.l.; Fl. Europ. 1: 12 (1964), 1: 16 (1993).
LOCALITY. France: Montpellier, May 1761.
MATERIAL. *Forsskål* 798 (C — 1 sheet with field label "Pteris aquilina. Monsp.", microf. 133: III. 5, 6).

Pteris cretica *L.* 1767: 130; F.Z. Pterid.: 116 (1970).
P. semiserrata Forssk. 1775: 186 (CXXIV No. 634; Cent. VII No. 14); Christensen 1922: 31.
LOCALITY. Yemen: Al Hadiyah ("montium Hadiensum"), 1763.
MATERIAL. *Forsskål* 799 (C — 1 sheet with field label "Pteris * semiserrata", type of *P. semiserrata*, microf. 85: II. 7, 8); 800 (C — 1 sheet, type of *P. semiserrata*, microf. 85: III. 1, 2).

Pteris dentata *Forssk.* 1775: 186 (CXXIV No. 636; Cent. VII No. 16); Christensen 1922: 31; Schwartz 1939: 20; F.Z. Pterid.: 117 (1970).
LOCALITY: Yemen: Al Hadiyah ("montium Hadiensium"), 1763.
MATERIAL. *Forsskål* no type specimen found.

Pteris regularis *Forssk.* 1775: 186 (CXXIV No. 632; Cent. VII No. 12); Christensen 1922: 31.
LOCALITY: Yemen: Al Hadiyah (Hadîe) ("montium Hadiensium"), 1763.
MATERIAL. *Forsskål* no type specimen found.
NOTE. Unknown; Schweinfurth 1896: 2 considered it might be *Pteris flabellata* i.e. *P. dentata* subsp. *flabellata* (Thunb.) Runemark.

Pteris serrulata *Forssk.* 1775: 187 (CXXIV No. 637; Cent. VII No. 17); Christensen 1922: 31.
LOCALITY. Yemen: Al Hadiyah ("montium Hadiensium"), 1763.
MATERIAL. *Forsskål* no type specimen found.
NOTE. Unknown.

Pteris subciliata *Forssk.* 1775: 185 (CXXIV No. 630; Cent. VII No. 10); Christensen 1922: 31.
LOCALITY: Yemen: Al Hadiyah ("montium Hadiensium"), 1763.
MATERIAL. *Forsskål* no type specimen found.
NOTE. Unknown; Schweinfurth 1896: 3 considered it to be *P. longifolia* L.

Pteris vittata *L.* 1753: 1074; Vahl 1790: 81; F.Z. Pterid.: 115 (1970).
P. obliqua Forssk. 1775: 185 (CXXIV No. 629; Cent. VII No. 9); Christensen 1922: 31.
LOCALITY. Yemen: Al Hadiyah ("montium Hadiensium"), Mar. 1763.
MATERIAL. *Forsskål* 803 (C — 1 sheet, type of *P. obliqua*, microf. 85: II. 3, 4); 802 (C — 1 sheet, type of *P. obliqua*, microf. 85: II. 5, 6).

Selaginella imbricata (*Forssk.*) *Spring ex Decne.* in Arch. Mus. Hist. Nat. 11: 193 (1841–2); Deflers 1889: 222; Schwartz 1939: 22; F.Z. Pterid.: 25 (1970).
Lycopodium imbricatum Forssk. 1775: 187 (CXXV No. 650; Cent. VII No. 20); Christensen 1922: 31.
L. bryopteris Vahl 1790: 81.
VERNACULAR NAME. *Raesen'Schaker rabba* (Arabic).
LOCALITY. Yemen: Al Hadiyah, Wasab ("Hadîe, Uahfad"), 1763.
MATERIAL. *Forsskål* 1443 (C — 1 sheet ex herb. Vahl, type of *L. imbricatum* & *L. bryopteris*, microf. 65(1–2): III. 5, 6).
Forsskål s.n. (LD — 1 sheet ex herb. Retzius, type of *L. imbricatum*).

Selaginella yemensis (*Sw.*) *Spring ex Decne.* in Arch. Mus. Hist. Nat. 11: 191 (1841–2); Deflers 1889: 222.
Lycopodium sanguinolentum sensu Forssk. 1775: CXXV No. 651, non L.
VERNACULAR NAME. *Hocha, Seraidt* (Arabic).
LOCALITY. Yemen: Bolghose ("Blg." = Bolghose), 1763.
MATERIAL. *Forsskål* 1444 (C — 1 sheet, isotype of *S. yemensis*, microf. nil); 1445 (C — 1 sheet ex herb. Vahl, type of *S. yemensis*, microf. nil).
Forsskål s.n. (S — 1 sheet, 85/237 ex herb. Swartz, holotype of *S. yemensis*, written up by Paole Bizzarri 1973).
Forsskål s.n. (LD — 1 sheet ex herb. Retzius, isotype of *S. yemensis*).

Tectaria gemmifera (*Fée*) *Alston* in Journ. Bot. 77: 288 (1939).
LOCALITY. Unspecified.
MATERIAL. *Forsskål* 808 (C — 1 sheet, microf. 133: III. 1, 2).

GYMNOSPERMS

CUPRESSACEAE

Juniperus oxycedrus *L.* 1753: 1038; Forssk. XXXV No. 439; Fl. Turk. 1: 80 (1965).
VERNACULAR NAME. *Agziothida* (Greek).
LOCALITY. Turkey: Gökçeada I. ("Imr" = Imros), July 1761.
MATERIAL. *Forsskål* 796 (C — 1 sheet with field label ? "Bs. 36", microf. 58: II. 5, 6).

EPHEDRACEAE

Ephedra aphylla *Forssk.* 1775: 170 (LXXVII No. 536; Cent. VI No. 64); Christensen 1922: 29; Frietag & Maier-Stolte in Taxon 38(4): 546 (1989).
VERNACULAR NAME. *Sparta* (Greek).
LOCALITY. Egypt: Rashid ("Rosettae in sepibus altissimis"), Oct.–Nov. 1761.
MATERIAL. *Forsskål* no type specimen found (see Note).
NOTE. Freitag & Maier-Stolte (l.c.) have designated a neotype (*Bornmüller* 1749).

Ephedra foeminea *Forssk.* 1775: 219 (XXXV No. 440; Cent. VIII No. 96); Christensen 1922: 35; Freitag & Maier-Stolte in Taxon 38(4): 550 (1989); Fl. Europ. ed. 2, 1: 49 (1993).
Ephedra campylopoda C.A. Meyer, Vers. Mon. Gatt. Eph.: 73 (1846); Fl. Turk. 1: 85 (1965).
LOCALITY. Turkey: Gökçeada ("In Insula Imros"), July 1761.
MATERIAL. *Forsskål* 794 (C — 1 sheet with field label "Imros", lectotype of *foemina*, microf. 41: II. 5, 6); 1245 (C — 1 sheet ex herb. Schumacher, isolectotype of *E. foemina*, microf. 41: II. 7, 8); 1246 (C — 1 sheet ex herb. Schumacher, isolectotype of *E. foemina*, microf. 41: III. 1, 2).
NOTE. The lectotype was selected by Freitag & Maier-Stolte (l.c.).

PINACEAE

Pinus brutia *Ten.*, Prodr. Fl. Nap.: 72 (1811); Fl. Europ. 1: 35 (1964); Fl. Turk. 1: 74 (1965).
LOCALITY. ? Turkey.
MATERIAL. *Forsskål* 795 (C — 1 sheet with field label "Y. 2", microf. 122: I. 5, 6).
NOTE. Although there is a pencil annotation on the sheet "an halepensis", this shoot with young cone is more likely to be *P. brutia* than *P. halepensis* which does not occur naturally in Western Turkey where Forsskål travelled.

Pinus pinea *L.* 1753: 1000; Forssk. 1775: XXXIV No. 419; Fl. Europ. 1: 35 (1964); Fl. Turk. 1: 75 (1965).
LOCALITY. Turkey: Istanbul garden ("Cph"), 1762.
MATERIAL. *Forsskål* s.n. (LD — 1 sheet ex herb. Retzius).

ALGAE

See F. Børgesen, A revision of Forsskål's algae mentioned in Flora Aegyptiaco-Arabica and found in his Herbarium in the Botanical Museum of the University of Copenhagen. Dansk Botanisk Arkiv 8(2): 1–15, plate 1 (1932)

Christensen 1922: 31 (Cent VII Nos 21–60).

LICHEN

See Christensen 1922: 32 (Cent. VII No. 61).
Lichen filamentosus Forssk. 1775: CXXV No. 659.
Lichen foliaceus Forssk. 1775: CXXV No. 658.
Lichen floridus Forssk. 1775: CXXV No. 656.
Lichen leprosus Forssk. 1775: CXXV No. 657.

FUNGI

See Christensen 1922: 36 (No. 480) *Boletus marginalis*.

Numbered sequence of specimens in the "Herb. Forskalii" with references to the IDC microfiche edition of the herbarium and identifications.

(1–890: Old numbers given to the specimens between 1840 and 1880; above 900: Modern numbers).

1.	75:II.5,6	*Phoenix dactylifera* L.
2.	9:III.6,7	*Arisaema flavum* (Forssk.) Schott
3.	10:I.1,2	*Arisaema flavum* (Forssk.) Schott
4.	10:I.3,4	*Arisaema bottae* Schott
5.	117:I.3,4	*Dracunculus vulgaris* Schott & Engl.
6.	4:III.3,4	*Allium sphaerocephalon* L.
7.	4:III.5,6	*Allium sphaerocephalon* L.
8.	74:III.3,4	*Asphodelus fistulosus* L. var. *tenuifolius* (Cav.) Bak.
9.	30:II.8 & III.1	*Sanseviera forskaliana* (Schult.f.) Hepper & Wood
10.	129:III.1,2	*Urginea maritima* (L.) Bak.
11.	129:II.7,8	*Scilla hyacinthina* (Roth) J. F. Macbride
12.	12:II.3,4	*Asparagus africanus* Lam.
13.	12:I.5,6	*Asparagus acutifolius* L.
14.	12:I.7,8	*Asparagus acutifolius* L.
15.	12:II.5,6	*Asparagus stipularis* Forssk.
16.	4:III.7,8	*Aloe pendens* Forssk.
17.	55:II.5,6 & III.7,8	*Dipcadi erythraeum* Webb & Berth.
18.	129:I.7,8	*Allium curtum* Boiss & Gaill.
19.	4:II.7,8	*Allium desertorum* Forssk.
20.	104:III.3,4	*Ottelia alismoides* (L.) Pers.
21.	117:I.1,2	*Crinum album* (Forssk.) Herb.
22.	[no specimen traced]	
23.	75:II.7,8	*Pancratium maximum* Forssk.
24.	57:III.1,2	*Gynandriris sisyrinchium* (L.) Parl.
25.	112:III.7,8	*Ruppia maritima* L.
26.	113:I.7,8 & II.1	*Thalassodendron ciliatum* (Forssk.) den Hartog
27.	28:III.7,8	*Merendera abyssinica* A. Rich.
28.	129:I.3,4	*Juncus bufonius* L.
29.	58:II.1,2	*Juncus acutus* L.
30.	58:II.3,4	*Juncus subulatus* Forssk.
31.	122:I.3,4	*Aneilema forskahlii* Kunth
32.	29:I.7,8	*Aneilema forskahlii* Kunth
33.	29:II.1,2	*Aneilema forskahlii* Kunth
34.	107:II.7,8	*Cyanotis* sp.
35.	122:I.1,2	*Commelina forskahlii* Kunth
36.	121:III.7,8	*Commelina benghalensis* L.
37.	29:I.5,6	*Commelina africana* L.
38.	127:I.1,2	*Eragrostis pilosa* (L.) P. Beauv.
39.	82:II.3,4	*Poa schimperiana* A. Rich.
40.	14:III.1,2	*Centropodia forskalei* (Vahl) Cope
41.	[no specimen traced]	
42.	125:III.3,4	*Aeluropus lagopoides* (L.) Thwait.
43.	14:II.7,8	*Centropodia forskalei* (Vahl) Cope
44.	36:III.1,2	*Eleusine floccifolia* (Forssk.) Spreng.
45.	36:II.7,8	*Eleusine floccifolia* (Forssk.) Spreng.

46.	75:III.1,2	*Ochthochloa compressa* (Forssk.) Hilu
47.	44:II.3,4	*Cutandria dichotoma* (Forssk.) Trabut
48.	90:I.1,2	*Lasiurus scindicus* Henrad
49.	44:III.3,4	*Cutandia maritima* (L.) Barbey
50.	55:I.1,2	*Sorghum halepense* (L.) Pers.
51.	127:II.3,4	*Hordeum marinum* Huds. subsp. *gussonianum* (Parl.) Thell.
52.	78:III.5,6	*Crypsis schoenoides* (L.) Lam.
53.	78:III.3,4	*Crypsis schoenoides* (L.) Lam.
54.	82:I.5,6	*Eragrostis aegyptiaca* (Willd.) Del.
55.	10:III.7,8	*Phragmites australis* (Cav.) Steud.
56.	36:II.3,4	*Cynosurus effusus* Link
57.	36:III.5,6	*Aeluropus lagopoides* (L.) Thwait.
58.	10:II.1,2	*Phragmites australis* (Cav.) Steudel
59.	82:I.7,8	*Eragrostis kiwuensis* Jedwabn.
60.	128:II.3,4	*Catapodium rigidum* (L.) C.E. Hubb.
61.	82:I.3,4	*Eragrostis aegyptiaca* (Willd.) Del.
62.	8:III.7,8	*Stipagrostis plumosa* (L.) T. Anders.
63.	78:II.5,6	*Polypogon viridis* (Gouan) Breistr.
64.	89:III.5,6	*Saccharum spontaneum* L. subsp. *aegyptiacum* (Willd.) Hack.
65.	76:II.3,4	*Panicum coloratum* L.
66.	55:II.1,2	*Hordeum vulgare* L.
67.	125:III.1,2	*Aeluropus lagopoides* (L.) Thwait.
68.	78:I.7,8	*Polypogon monspeliensis* (L.) Desf.
69.	78:II.1,2	*Polypogon monspeliensis* (L. Desf.
70.	128:II.7,8	*Sphenopus divaricatus* (Gouan) Reichb.
71.	14:II.3,4	*Helictotrichon* sp.
72.	41:II.3,4	*Cenchrus biflorus* Roxb.
73.	10:II.5,6	*Imperata cylindrica* (L.) Raeuschel
74.	89:III.7,8	*Saccharum spontaneum* L. subsp. *aegyptiacum* (Willd.)
75.	78:I.5,6	*Polypogon monspeliensis* (L.) Desf.
76.	[no specimen traced]	
77.	61:I.5,6	*Lagurus ovatus* L.
78.	61:I.7,8	*Lagurus ovatus* L.
79.	3:III.7,8	*Sporobolus pyramidalis* P. Beauv.
80.	128:I.5,6	*Polypogon monspeliensis* (L.) Desf.
81.	[no specimen traced]	
82.	[no specimen traced]	
83.	10:I.5,6	*Calamagrostis purpurescens* R. Br.
84.	44:I.5,6	*Schismus arabicus* Nees
85.	76:III.3,4	*Panicum turgidum* Forssk.
86.	76:II.7,8	*Brachiaria mutica* (Forssk.) Stapf
87.	76:II.5,6	*Brachiaria mutica* (Forssk.) Stapf
88.	[no specimens traced]	
89.	16:III.1,2	*Bromus madritensis* L.
90.	16:II.7,8	*Bromus madritensis* L.
91.	16:III.3,4	*Bromus madritensis* L.
92.	16:II.1,2	*Bromus intermedius* Guss.
93.	16:III.7,8	*Bromus madritensis* L.
94.	44:III.5,6	*Odyssea mucronata* (Forssk.) Stapf
95.	16:II.3,4	*Bromus squarrosus* L.
96.	16:II.5,6	*Bromus squarrosus* L.
97.	[no specimen traced]	

98.	127:II.1,2	*Festuca ovina* L.
99.	16:I.3,4	*Brachypodium retusum* (Pers.) P. Beauv.
100.	16:I.1,2	*Brachypodium retusum* (Pers.) P. Beauv.
101.	[no specimen traced]	
102.	[no specimen traced]	
103.	[no specimen traced]	
104.	15:III.7,8	*Trachynia distachya* (L.) Link
105.	[no specimen traced]	
106.	54:III.5,6	*Pennisetum glaucum* (L.) R. Br.
107.	54:III.1,2	*Sorghum bicolor* (L.) Moench
108.	54:III.3,4	*Sorghum bicolor* (L.) Moench
109.	55:I.3,4	*Sorghum halepense* (L.) Pers
110.	55:I.5,6	*Sorghum halepense* (L.) Pers
111.	54:III.7,8	*Sorghum bicolor* (L.) Moench
112.	79:I.1,2	*Digitaria velutina* (Forssk.) P. Beauv.
113.	76:I.3,4	*Paspalidium geminatum* (Forssk.) Stapf
114.	76:III.1,2	*Digitaria sanguinalis* (L.) Scop.
115.	78:III.7,8	*Digitaria velutina* (Forssk.) P. Beauv.
116.	76:I.5,6	*Cenchrus setigerus* Vahl
117.	78:III.1,2	*Pennisetum setaceum* (Forssk.) Chiov.
118.	109:I.7,8	*Trisetaria linearis* Forssk.
119.	9:II.5,6	*Aristida adscensionis* L.
120.	[no specimen traced]	
121.	128:I.7,8	*Rhynchelytrum repens* (Willd.) Hubbard
122.	[no specimen traced]	
123.	8:III.3,4	*Aristida adscensionis* L.
124.	8:III.1,2	*Aristida adscensionis* L.
125.	9:I.1,2	*Stipagrostis plumosa* (L.) T. Anders.
126.	7:II.3,4	*Cymbopogon caesius* (Hook. & Arn.) Stapf
127.	7:I.7,8	*Dichantium annulatum* (Forssk.) Stapf
128.	128:III.1,2	*Stipa capensis* Thunb.
129.	4:I.5,6	*Sporobolus spicatus* (Vahl) Kunth
130.	4:I.1,2	*Sporobolus spicatus* (Vahl) Kunth
131.	[no specimen traced]	
132.	78:II.3,4	*Sporobolus arenarius* (Gouan) Jouve
133.	[no specimen traced]	
134.	[no specimen traced]	
135.	[no specimen traced]	
136.	[no specimen traced]	
137.	[no specimen traced]	
138.	76:III.7,8	*Panicum turgidum* Forssk.
139.	55:II.3,4	*Hordeum murinum* L. subsp. *glaucum* (Steud.) Tzvlev.
140.	125:II.7,8	*Aeluropus lagopoides* (L.) Thwait.
141.	[no specimen traced]	
142.	4:I.3,4	*Sporobolus spicatus* (Vahl) Kunth
143.	91:II.3,4	*Salicornea fruticosa* (L.) L.
144.	77:I.1,2	*Trianthema triquetra* Willd.
145.	105:I.1,2	*Schanginia hortensis* (Forssk. ex J. F. Gmel.) Moq.
146.	91:III.5,6	*Arthrocnemum macrostachyum* (Moric.) K. Koch
147.	91:I.3,4	*Halocnemum strobilaceum* (Pallas) M. Bieb.
148.	94:II.1,2	*Salsola soda* L.
149.	26:I.7,8	*Suaeda vera* Forssk. ex J. F. Gmel.
150.	92:II.3,4	*Anabasis articulata* (Forssk.) Moq.

151.	93:I.3,4	*Salsola inermis* Forssk.
152.	92:III.1,2	*Salsola tetrandra* Forssk.
153.	93:II.3,4	*Salsola kali* L.
154.	94:I.5,6	*Salsola soda* L.
155.	93:III.1,2	*Salsola longifolia* Forssk.
156.	93:I.7,8	*Salsola kali* L.
157.	94:III.5,6	*Salsola longifolia* Forssk.
158.	92:II.7,8	*Salsola inermis* Forssk.
159.	93:III.3,4	*Bassia muricata* (L.) Aschers.
160.	105:II.5,6	*Suaeda vermiculata* Forssk. ex J. F. Gmel.
161.	105:II.3,4	*Suaeda vera* Forssk. ex J. F. Gmel.
162.	104:III.7,8	*Suaeda fruticosa* Forssk. ex J. F. Gmel.
163.	92:III.3,4	*Salsola inermis* Forssk.
164.	104:III.5,6	*Suaeda aegyptiaca* (Hasselq.) Zohary
165.	105:I.3,4	*Schanginia hortensis* (Forssk. ex J.F. Gmel.) Moq.
166.	94:III.1,2	*Salsola longifolia* Forssk.
167.	105:I.7,8	*Suaeda monoica* Forssk. ex J. F. Gmel.
168.	91:I.5,6	*Halochnemum strobilaceum* (Pallas) M. Bieb.
169.	91:I.7,8	*Arthrocnemum macrostachyum* (Moric.) K. Koch
170.	91:II.5,6	*Salicornia fruticosa* (L.) L.
171.	90:III.6,7	*Halocnemum strobilaceum* (Pallas) M. Bieb.
172.	91:I.1,2	*Halocnemum strobilaceum* (Pallas) M. Bieb.
173.	91:III.1,2	*Halopeplis perfoliata* (Forssk.) Bunge
174.	91:III.3,4	*Arthrocnemum macrostachyum* (Moric.) K. Koch
175.	118:III.6,7	*Traganum nudatum* Del.
176.	94:II.3,4	*Salsola tetrandra* Forssk.
177.	94:II.5,6	*Salsola tetrandra* Forssk.
178.	118:III.4,5	*Traganum nudatum* Del.
179.	118:III.2,3	*Suaeda salsa* (L.) Pallas
180.	105:II.1,2	*Suaeda monoica* Forssk. ex J. F. Gmel.
181.	92:III.5,6	*Salsola inermis* Forssk.
182.	94:I.3,4	*Noaea mucronata* (Forssk.) Aschers. & Schweinf.
183.	14:II.1,2	*Halimione portulacoides* (L.) Aellen
184.	26:I.5,6	*Chenopodium schraderianum* Schultes
185.	14:I.3,4	*Atriplex halimus* L. var. *schweinfurthii* Boiss.
186.	13:III.3,4	*Atriplex coriacea* Forssk.
187.	14:I.5,6	*Atriplex leucoclada* Boiss.
188.	92:II.1,2	*Anabasis articulata* (Forssk.) Moq.
189.	105:I.5,6	*Schanginia hortensis* (Forssk. ex J. F. Gmel.) Moq.
190.	14:I.1,2	*Atriplex tartarica* L.
191.	13:III.5,6	*Atriplex coriacea* Forssk.
192.	14:I.7,8	*Atriplex leucoclada* Boiss.
193.	3:II.7,8	*Aerva javanica* (Burm. f.) Juss.
194.	3:III.1,2	*Aerva javanica* (Burm. f.) Juss.
195.	3:I.7,8	*Aerva javanica* (Burm. f.) Juss.
196.	3:II.1,2	*Aerva javanica* (Burm. f.) Juss.
197.	3:I.5,6	*Aerva javanica* (Burm. f.) Juss.
198.	[no specimen traced]	
199.	[no specimen traced]	
200.	5:I.1,2	*Alternanthera sessilis* (L.) DC.
201.	5:I.3,4	*Alternanthera sessilis* (L.) DC.
202.	1:III.5,6	*Celosia trigyna* L.

203.	2:II.1,2	*Aerva lanata* (L.) Juss.
204.	2:I.1,2	*Celosia trigyna* L.
205.	2:I.3,4	*Saltia papposa* (Forssk.) Moq.
206.	2:I.5,6	*Saltia papposa* (Forssk.) Moq.
207.	[no specimen traced]	
208.	1:III.7,8	*Celosia trigyna* L.
209.	107:II.3,4	*Thymus* sp. 2
210.	107:I.7,8	*Micromeria imbricata* (Forssk.) C. Chr.
211.	71:III.3,4,5	*Ocimum forskolei* Benth.
212.	[no specimen traced]	
213.	72:I.3,4	*Becium serpyllifolium* (Forssk.) Wood
214.	[no specimen traced]	
215.	[no specimen traced]	
216.	131:I.7,8	*Nepeta nepetellae* Forssk.
217.	79:II.7,8	*Leucas glabrata* (Vahl) R. Br.
218.	28:I.7,8	*Otostegia fruticosa* (Forssk.) Penzig
219.	66:II.1,2	*Marrubium peregrinum* L.
220.	66:II.3,4	*Marrubium alysson* L.
221.	66:II.5,6	*Marrubium alysson* L.
222.	79:II.5,6	*Leucas alba* (Forssk.) Sebald & *Leucas glabrata* (Vahl) R.Br. [mixed collection]
223.	70:II.1,2	*Ballota acetabulosa* (L.) Benth.
224.	28:I.3,4	*Otostegia fruticosa* (Forssk.) Penzig
225.	74:II.3,4	*Origanum vulgare* L. subsp. *hirtum* (Link) Lets.
226.	74:II.5,6	*Origanum vulgare* L. subsp. *hirtum* (Link) Lets.
227.	74:II.7,8	*Origanum vulgare* L. subsp. *hirtum* (Link) Lets.
228.	79:III.1,2	*Phlomis herba-venti* L.
229.	74:III.1,2	*Origanum majorana* L.
230.	130:III.6,7	*Micromeria imbricata* (Forssk.) C. Chr.
231.	107:I.5,6	*Micromeria imbricata* (Forssk.) C. Chr.
232.	107:I.3,4	*Thymus sp.* 1. (*Th. ciliatus* Forssk.)
233.	107:II.1,2	*Thymus laevigatus* Vahl
234.	8:II.3,4	*Carissa edulis* (Forssk.) Vahl
235.	70:III.7,8	*Adenium obesum* (Forssk.) Roem. & Schult.
236.	70:III.3,4	*Adenium obesum* (Forssk.) Roem. & Schult. (flws) & *Breonadia salicina* (Vahl) Hepper & Wood (leaves) [mixed sheet]
237.	70:III.5,6	*Adenium obesum* (Forssk.) Roem. & Schult. (flws) & *Breonadia salicina* (Vahl) Hepper & Wood (leaves) [mixed sheet]
238.	138:II.6,7	*Apocynaceae* gen. et sp. indet.
239.	125:II.5,6	*Enicostema axillare* (Lam.) A. Raynal
240.	77:II.1,2	*Swertia polynectaria* (Forssk.) Gilg
241.	49:III.7,8	*Centaurium pulchellum* (Sw.) Druce
242.	50:I.1,2	*Centaurium pulchellum* (Sw.) Druce
243.	20:II.5,6	*Campanula edulis* Forssk.
244.	20:II.7,8	*Legousia speculum-veneris* (L.) Chaix
245.	20:II.1,2	*Campanula edulis* Forssk.
246.	20:II.3,4	*Campanula edulis* Forssk.

247.	124:II.1,2	*Cephalaria* sp.
248.	no microf.	*Centranthus calcitrapa* (L.) Dufr.
249.	81:III.1,2	*Plantago coronopus* L. (plant) & *Plantago ovata* Forssk. var. *ovata* (infloresc.) [mixed sheet]
250.	81:II.7,8	*Plantago ovata* Forssk. var. *ovata*
251.	81:III.5,6	*Plantago serraria* L.
252.	80:III.1,2	*Plantago coronopus* L.
253.	81:II.5,6	*Plantago ovata* Forssk.
254.	81:I.7,8	*Plantago ovata* Forssk. var. *decumbens* (Forssk.) Zohary
255.	80:III.7,8	*Plantago cylindrica* Forssk.
256.	81:III.3,4	*Plantago scabra* Moench
257.	81:I.3,4	*Plantago cylindrica* Forssk.
258.	81:III.7,8	*Plantago serraria* L.
259.	81:I.5,6	*Plantago ovata* Forssk. var. *decumbens* (Forssk.) Zohary
260.	81:I.1,2	*Plantago cylindrica* Forssk.
261.	80:III.5,6	*Plantago crassifolia* Forssk.
262.	81:II.3,4	*Plantago major* L.
263.	81:II.1,2	*Plantago lanceolata* L.
264.	11:III.7,8	*Gomphocarpus fruticosus* (L.) Ait. f.
265.	11:II.5,6	*Kanahia laniflora* (Forssk.) R. Br.
266.	[no specimen traced]	
267.	12:I.1,2	*Pentatropis nivalis* (J. F. Gmel.) Field & Wood
268.	11:II.7,8	*Kanahia laniflora* (Forssk.) R. Br.
269.	11:I.3,4	*Pergularia tomentosa* L.
270.	11:I.5,6	*Pergularia tomentosa* L.
271.	35:III.7,8	*Leptadenia arborea* (Forssk.) Schweinf.
272.	11:III.1,2	*Pentatropis nivalis* (J. F. Gmel.) Field & Wood
273.	35:III.5,6	*Leptadenia arborea* (Forssk.) Schweinf.
274.	103:II.1,2	*Stapelia subulata* (Forssk.) Decne.
275.	11:III.3,4	*Odontanthera radians* (Forssk.) Field
276.	11:III.5,6	*Odontanthera radians* (Forssk.) Field
277.	12:I.3,4	*Sarcostemma viminale* (L.) R. Br.
278.	no microf.	*Heliotropium digynum* (Forssk.) C. Chr.
279.	99:II.5,6	*Ehretia cymosa* Thonn.
280.	40:III.3,4	*Echium angustifolium* Miller subsp. *sericeum* (Vahl) Kltz.
281.	117:III.1,2	*Echium* cf. *rauwolfii* Del.
282.	117:I.5,6	*Moltkiopsis ciliata* (Forssk.) I. M. Johnst.
283.	7:I.1,2	*Alkanna lemanii* (Tineo) A. DC.
284.	40:III.7,8	*Echium angustifolium* Miller subsp. *sericeum* (Vahl) Kltz.
285.	117:II.3,4	*Heliotropium digynum* (Forssk.) C. Chr.
286.	117:II.5,6	*Echium angustifolium* Miller subsp. *sericeum* (Vahl) Kltz.
287.	40:II.7,8	*Echium longifolium* Del.
288.	15:II.5,6	*Trichodesma africana* (L.) R. Br.
289.	15:II.3,4	*Borago officinalis* L.
290.	117:II.7,8	*Echium glomeratum* Poir.
291.	40:III.5,6	*Echium angustifolium* Miller subsp. *sericeum* (Vahl) Kltz.
292.	6:II.7,8	*Anchusa aegyptiaca* (L.) DC.

293.	[no specimen traced]	
294.	[no specimen traced]	
295.	7:I.3,4	*Anchusa undulata* L.
296.	6:III.1,2	*Anchusa aegyptiaca* (L.) DC.
297.	41:I.3,4	*Echium rubrum* Forssk.
298.	40:III.1,2	*Echium angustifolium* Miller subsp. *sericeum* (Vahl) Kltz.
299.	no microf.	*Heliotropium ovalifolium* Forssk.
300.	63:II.3,4	*Heliotropium bacciferum* Forssk.
301.	53:II.5,6	*Heliotropium europaeum* L.
302.	no microf.	*Heliotropium supinum* L.
303.	no microf.	*Heliotropium sp.* indet.
304.	6:III.5,6	*Lappula spinocarpos* (Forssk.) Ktze
305.	9:II.7,8	*Arnebia tetrastigma* Forssk.
306.	63:I.5,6	*Moltkiopsis ciliata* (Forssk.) I. M. Johnst.
307.	63:I.1,2	*Moltkiopsis ciliata* (Forssk.) I. M. Johnst.
308.	6:III.7,8	*Alkanna lehmanii* (Tineo) A. DC.
309.	7:I.5,6	*Anchusa undulata* L.
310.	6:III.3,4	*Gastrocotyle hispida* (Forssk.) Bunge
311.	63:I.7,8	*Moltkiopsis ciliata* (Forssk.) I. M. Johnst.
312.	36:I.1,2	*Cynoglossum lanceolatum* Forssk.
313.	[no specimen traced]	
314.	[no specimen traced]	
315.	24:II.7,8	*Lantana viburnoides* (Forssk.) Vahl
316.	[no specimen traced]	
317.	96:II.1,2	*Avicennia marina* (Forssk.) Vierh.
318.	96:II.3,4	*Avicennia marina* (Forssk.) Vierh.
319.	[no specimen traced]	
320.	[no specimen traced]	
321.	[no specimen traced]	
322.	[no specimen traced]	
323.	24:III.1,2	*Lantana viburnoides* (Forssk.) Vahl.
324.	71:II.7,8	*Becium filamentosum* (Forssk.) Chiov.
325.	72:I.7,8	*Becium serpyllifolium* (Forssk.) Wood
326.	72:I.5,6	*Becium serpyllifolium* (Forssk.) Wood
327.	60:III.7,8	*Dorstenia foetida* (Forssk.) Schweinf.
328.	131:I.5,6	*Ocimum tenuiflorum* L.
329.	131:I.3,4	*Labiatae* gen. et sp. indet.
330.	70:II.5,6	*Ajuga iva* (L.) Schreb.
331.	106:III.7,8	*Teucrium scordium* L. subsp. *scordioides* (Schreb.) Maire & Petitmengin
332.	106:III.5,6	*Teucrium polium* L.
333.	67:III.3,4	*Mentha longifolia* (L.) L.
334.	67:III.1,2	*Mentha longifolia* (L.) L.
335.	67:II.7,8	*Mentha* × *piperita* L.
336.	131:II.1,2	*Mentha longifolia* (L.) L. var. *schimperi* (Briq.) Briq.
337.	103:I.7,8	*Stachys recta* L. subsp. *subcrenata* (Vis.) Briq.
338.	130:III.4,5	*Stachys cf. maritima* Gouan
339.	103:I.5,6	*Stachys aegyptiaca* Pers.
340.	103:I.3,4	*Stachys obliqua* Waldst. & Kit.(*S. orientalis* L.)
341.	67:I.7,8	*Salvia aegyptiaca* L.
342.	95:II.1,2	*Salvia lanigera* Poir.

343.	95:II.5,6	*Salvia merjamie* Forssk.
344.	95:I.1,2	*Salvia forskahlei* L.
345.	28:I.5,6	*Otostegia fruticosa* (Forssk.) Penzig
346.	66:II.7,8	*Marrubium alysson* L.
347.	106:III.1,2	*Teucrium polium* L.
348.	72:II.1,2	*Plectranthus hadiensis* (Forssk.) Schweinf. ex Spreng.
349.	131:I.1,2	*Solenostemon latifolius* (Benth.) Morton
350.	71:III.1,2	*Plectranthus ovatus* Benth.
351.	70:II.7,8	*Ajuga iva* (L.) Schreb.
352.	6:II.1,2	*Anagallis arvensis* L.
353.	84:III.3,4	*Primula verticillata* Forssk.
354.	84:III.1,2	*Primula verticillata* Forssk.
355.	111:III.2,3	*Utricularia inflexa* Forssk.
356.	111:II.8 & III.1	*Utricularia inflexa* Forssk.
357.	61:II.5,6	*Orobanche aegyptiaca* Pers.
358.	74:III.7,8	*Orobanche crenata* Forssk.
359.	15:I.7,8	*Mimusops laurifolia* (Forssk.) Friis
360.	15:I.5,6	*Mimusops laurifolia* (Forssk.) Friis
361.	65(2-2):I.5,6	*Maesa lanceolata* Forssk.
362.	65(2-2):I.3,4	*Maesa lanceolata* Forssk.
363.	65(2-2):I.1,2	*Maesa lanceolata* Forssk.
364.	98:II.5,6	*Sesamum indicum* L.
365.	[no specimen traced]	
366.	98:II.3,4	*Sesamum indicum* L.
367.	88:II.7,8	*Phaulopsis imbricata* (Forssk.) Sweet
368.	88:II.5,6	*Phaulopsis imbricata* (Forssk.) Sweet
369.	88:II.3,4	*Asystasia guttata* (Forssk.) Brummitt
370.	38:I.7,8	*Asystasia gangetica* (L.) T. Anders.
371.	88:III.1,2	*Asystasia gangetica* (L.) T. Anders.
372.	116:III.3,4	*Lepidagathis aristata* (Vahl) Nees
373.	59:II.3,4	*Barleria bispinosa* (Forssk.) Vahl
374.	59:I.5,6	*Barleria bispinosa* (Forssk.) Vahl.
375.	59:I.7,8	*Barleria prionitis* L. subsp. *appressa* (Forssk.) Brummitt & Wood
376.	59:II.1,2	*Barleria trispinosa* (Forssk.) Vahl
377.	59:III.5,6	*Barleria lanceata* (Forssk.) C. Chr.
378.	59:I.1,2	*Barleria prionitis* L. subsp. *appressa* (Forssk.) Brummitt & Wood
379.	1:I.5,6	*Acanthus arboreus* Forssk.
380.	60:II.3,4	*Ecbolium gymnostachyum* (Nees) Milne-Redh.
381.	[no specimen traced]	
382.	59:III.3,4	*Dicliptera foetida* (Forssk.) Blatter
383.	138:II.4,5	*Justicia coerulea* Forssk.
384.	38:II.7,8	*Justicia odora* (Forssk.) Vahl
385.	38:III.3,4	*Peristrophe paniculata* (Forssk.) Brummitt
386.	60:II.5,6	*Ecbolium gymnostachyum* (Nees) Milne-Redh.
387.	60:I.1,2	*Hypoestes forsskalii* (Vahl) Roem. & Schult.
388.	116:II.5,6	*Justicia flava* (Vahl) Vahl
389.	38:II.1,2	*Isoglossa punctata* (Vahl) Brummitt & Wood
390.	59:III.1,2	*Megalochlamys violaceum* (Vahl) Vollesen
391.	38:II.3,4	*Monechma debile* (Forssk.) Nees
392.	39:I.1,2	*Dicliptera verticillata* (Forssk.) C. Chr.
393.	39:I.3,4	*Dicliptera verticillata* (Forssk.) C. Chr.
394.	38:I.3,4	*Justicia flava* (Vahl) Vahl

395.	38:I.5,6	*Justicia flava* (Vahl) Vahl
396.	97:II.1,2	*Scutellaria arabica* Jaub. & Spach
397.	8:I.5,6	*Linaria haelava* (Forssk.) F. G. Dietr.
398.	7:III.3,4	*Kickxia aegyptiaca* (L.) Nobl.
399.	97:II.3,4	*Scutellaria arabica* Jaub. & Spach
400.	[no specimen traced]	
401.	17:I.3,4	*Cycniopsis humifusa* (Forssk.) Engl.
402.	112:I.7,8	*Verbascum pinnatifolium* Vahl
403.	112:I.3,4	*Verbascum sinuatum* L.
	(perhaps not a Forsskål specimen; not retrieved)	
404.	17:I.1,2	*Cycniopsis humifusa* (Forssk.) Engl.
405.	101:II.3,4	*Solanum nigrum* L. s.l.
		or *Solanum villosum* Miller
406.	102:II.5,6	*Solanum terminale* Forssk.
407.	101:I.5,6	*Solanum nigrum* L. s.l.
		or *Solanum villosum* Miller
408.	65(1-2):III.1,2	*Lycium europaeum* L.
409.	101:III.5,6	*Solanum coagulans* Forssk.
410.	101:III.3,4	*Solanum coagulans* Forssk.
411.	101:II.5,6	*Solanum glabratum* Dunal
412.	102:I.1,2	*Solanum carense* Dunal
413.	137:I.1,2	*Solanum platacanthum* Dunal
414.	102:II.7,8	*Solanum forskalii* Dunal
415.	102:II.1,2	*Solanum incanum* L.
416.	80:II.3,4	*Withania somnifera* (L.) Dunal
417.	80:I.7,8	*Physalis angulata* L.
418.	102:I.3,4	*Solanum dulcamara* L.
419.	102:II.3,4	*Solanum nigrum* L. s.l.
		or *Solanum villosum* Miller
420.	101:I.3,4	*Solanum nigrum* L. s.l.
		or *Solanum villosum* Miller
421.	101:I.7,8	*Solanum nigrum* L. s.l.
		or *Solanum villosum* Miller
422.	101:II.1,2	*Solanum nigrum* L. s.l.
		or *Solanum villosum* Miller
423.	55:III.3,4	*Hyoscyamus muticus* L.
424.	[no specimen traced]	
425.	136:III.1,2	*Solanum* sp. indet. (? not a Forsskål coll.)
426.	[no specimen traced]	
427.	136:III.5,6	*Solanum* sp. indet. (? not a Forsskål coll.)
428.	[no specimen traced]	
429.	101:II.7,8	*Solanum glabratum* Dunal
430.	[no specimen traced]	
431.	[no specimen traced]	
432.	57:II.1,2	*Ipomoea cairica* (L.) Sweet
433.	[no specimen traced]	
434.	[no specimen traced]	
435.	[no specimen traced]	
436.	[no specimen traced]	
437.	[no specimen traced]	
438.	31:III.1,2	*Convolvulus prostratus* Forssk.
439.	31:III.5,6	*Convolvulus althaeoides* L.
		subsp. *tenuissimus* (Sibth. & Sm.) Stace
440.	32:I.3,4	*Convolvulus hystrix* Vahl
441.	34:I.3,4	*Cressa cretica* L.

442.	34:I.1,2	*Cressa cretica* L.
443.	33:III.7,8	*Cressa cretica* L.
444.	122:II.5,6	*Ipomoea obscura* (L.) Ker-Gawl.
445.	31:II.3,4	*Ipomoea eriocarpa* R. Br.
446.	31:II.7,8	*Convolvulus prostratus* Forssk.
447.	57:I.3,4	*Ipomoea aquatica* Forssk.
448.	57:I.7,8	*Ipomoea cairica* (L.) Sweet
449.	31:III.7,8	*Convolvulus althaeoides* L. subsp. *tenuissimus* (Sibth. & Sm.) Stace
450.	57:II.3,4	*Ipomoea nil* (L.) Roth
451.	57:I.5,6	*Ipomoea pes-caprae* (L.) R. Br.
452.	57:II.5,6	*Ipomoea triflora* Forssk.
453.	57:II.7,8	*Ipomoea verticillata* Forssk.
454.	32:I.1,2	*Convolvulus hystrix* Vahl
455.	30:III.3,4	*Convolvulus lanatus* Vahl
456.	31:I.5,6	*Convolvulus lanatus* Vahl
457.	122:I.7,8	*Convolvulus siculus* L. subsp. *agrestis* (Schweinf.) Verdc.
458.	122:II.1,2	*Convolvulus siculus* L. subsp. *agrestis* (Schweinf.) Verdc.
459.	91:III.7,8	*Salix subserrata* Willd.
460.	92:I.1,2	*Salix cf. tetrasperma* Roxb.
461.	135:III.3,4	*Salix subserrata* Willd.
462.	135:III.5,6	*Salix subserrata* Willd.
463.	77:III.5,6	*Nitraria retusa* (Forssk.) Aschers.
464.	138:I.4,5	*Turraea holstii* Guerke
465.	32:III.1,2	*Cordia sinensis* Lam.
466.	[no specimen traced]	
467.	137:I.5,6	*Grewia villosa* Willd.
468.	138:I.6,7	*Dicotyledones* fam., gen. & sp. indet.
469.	129:III.3,4	*Linum strictum* L.
470.	107:III.1,2	*Dobera glabra* (Forssk.) Poir.
471.	138:I.8 & II.1	*Dicotyledones* fam., gen. & sp. indet.
472.	138:II.2,3	*Dicotyledones* fam., gen. & sp. indet.
473.	82:I.1,2	*Platanus orientalis* L.
474.	113:II.8 & III.1	*Zygophyllum simplex* L.
475.	[no specimen traced]	
476.	[no specimen traced]	
477.	22:III.3,4	*Caucanthus edulis* Forssk.
478.	41:II.1,2	*Trichilia emetica* Vahl
479.	[no specimen traced]	
480.	106:I.1,2	*Tamarix aphylla* (L.) Karsten
481.	106:I.7,8	*Tamarix aphylla* (L.) Karsten
482.	86:II.7,8	*Reamurea hirtella* Jaub. & Spach
483.	109:II.3,4	*Triumfetta rhomboidea* Jacq.
484.	[no specimen traced]	
485.	[no specimen traced]	
486.	[no specimen traced]	
487.	[no specimen traced]	
488.	[no specimen traced]	
489.	24:II.5,6	*Grewia velutina* (Forssk.) Vahl
490.	24:I.7,8	*Grewia arborea* (Forssk.) Lam.
491.	24:II.1,2	*Grewia tenax* (Forssk.) Fiori
492.	56:II.1,2	*Hypericum revolutum* Vahl
493.	56:I.5,6	*Hypericum perforatum* L.

494.	128:III.5,6	*Hypericum bithynicum* Boiss.
495.	56:I.3,4	*Hypericum calycinum* L.
496.	56:I.7,8	*Hypericum revolutum* Vahl
497.	82:III.7,8	*Polygala tinctoria* Vahl
498.	[no specimen traced]	
499.	87:II.5,6	*Ziziphus spina-christi* (L.) Desf.
500.	87:III.1,2	*Ziziphus spina-christi* (L.) Desf.
501.	87:II.7,8	*Ziziphus spina-christi* (L.) Desf.
502.	[no specimen traced]	
503.	22:II.7,8	*Maytenus parviflora* (Vahl) Sebsebe
504.	22:II.1,2	*Catha edulis* (Vahl) Forssk. ex Endl.
505.	22:I.5,6	*Catha edulis* (Vahl) Forssk. ex Endl.
506.	22:I.7,8	*Catha edulis* (Vahl) Forssk. ex Endl.
507.	22:II.5,6	*Maytenus parviflora* (Vahl) Sebsebe
508.	[no specimen traced]	
509.	[no specimen traced]	
510.	67:I.5,6	*Melhania velutina* Forssk.
511.	58:III.1,2	*Ludwigia stolonifera* (Guill. & Perr.) Raven
512.	58:III.3,4	*Corchrus depressus* (L.) C. Chr.
513.	138:II.8 & III.1	*Epilobium angustifolium* L.
514.	58:II.7,8	*Ludwigia stolonifera* (Guill. & Perr.) Raven
515.	58:III.5,6	*Corchrus depressus* (L.) C. Chr.
516.	[no specimen traced]	
517.	[no specimen traced]	
518.	27:II.5,6	*Salvadora persica* L.
519.	103:III.1,2	*Limonium cylindrifolium* (Forssk.) Kuntze
520.	103:III.3,4	*Limonium cylindrifolium* (Forssk.) Kuntze
521.	133:II.3,4	*Limonium sp.*
522.	103:II.7,8	*Limonium axillare* (Forssk.) Kuntze
523.	104:I.7,8	*Goniolimon incanum* (L.) Hepper
524.	103:II.3,4	*Limonium pruinosum* (L.) Kuntze
525.	104:I.3,4	*Limonium gmelinii* (Willd.) Kuntze
526.	104:I.1,2	*Limonium gmelinii* (Willd.) Kuntze
527.	104:II.1,2	*Goniolimon incanum* (L.) Hepper
528.	104:I.5,6	*Goniolimon incanum* (L.) Hepper
529.	41:I.5,6	*Elaeagnus angustifolia* L.
530.	124:II.3,4	*Elaeagnus angustifolia* L.
531.	41:I.7,8	*Elaeagnus angustifolia* L.
532.	77:II.5,6	*Thymelaea hirsuta* (L.) Endl.
533.	111:III.7,8	*Commicarpus plumbaginea* (Cav.) Standl.
534.	111:III.5,6	*Commicarpus plumbaginea* (Cav.) Standl.
534 bis.	51:I.5,6	*Glinus lotoides* L.
535.	112:III.5,6	*Cassytha filiformis* L.
535 bis.	51:II.1,2	*Glinus setiflorus* Forssk.
536.	75:II.1,2	*Osyris alba* L.
536 bis.	77:I.5,6	*Trianthema crystallina* (Forssk.) Vahl
537.	75:I.7,8	*Osyris alba* L.
537 bis.	52:I.5,6	*Gymnocarpos decandrum* Forssk.
538.	107:I.1,2	*Thesium linophyllum* L.
538 bis.	87:III.3,4	*Zaleya pentandra* (L.) C. Jeffrey
539.	77:I.3,4	*Trianthema triquetra* Willd.
539 bis.	84:II.3,4	*Portulaca quadrifida* L.
540.	51:I.7,8	*Glinus setiflorus* Forssk.
540 bis.	84:II.1,2	*Portulaca quadrifida* L.
541.	75:I.5,6	*Corbichonia decumbens* (Forssk.) Exell

542.	51:II.4,5	*Glinus setiflorus* Forssk.
542 bis.	84:II.5,6	*Portulaca quadrifida* L.
543.	51:I.3,4	*Glinus lotoides* L.
543 bis.	51:II.6,7	*Glinus setiflorus* Forssk.
544.	51:I.1,2	*Aizoon canariense* L.
545.	71:II.1,2	*Neurada procumbens* L.
546.	71:I.5,6	Neurada procumbens L.
547.	71:I.7,8	*Neurada procumbens* L.
548.	84:I.7,8	*Portulaca quadrifida* L.
549.	39:II.7,8	*Dianthus uniflorus* Forssk.
550.	100:I.5,6	*Silene villosa* Forssk.
551.	5:I.7,8	*Polycarpon prostratum* (Forssk.) Aschers. & Schweinf.
552.	99:III.3,4	*Silene succulenta* Forssk.
553.	85:I.1,2	*Pteranthus dichotomus* Forssk.
554.	39:III.1,2	*Dianthus uniflorus* Forssk.
555.	39:II.1,2	*Petrorhagia prolifera* (L.) Ball & Heyw.
556.	39:I.7,8	*Petrorhagia prolifera* (L.) Ball & Heyw.
557.	33:I.1,2	*Polycarpaea repens* (Forssk.) Aschers. & Schweinf.
558.	32:III.5,6	*Paronychia arabica* (L.) DC.
559.	5:I.5,6	*Polycarpon prostratum* (Forssk.) Asch. & Schweinf.
560.	100:I.3,4	*Silene villosa* Forssk.
561.	99:III.7,8	*Silene succulenta* Forssk.
562.	84:III.7,8	*Pteranthus dichotomus* Forssk.
563.	53:III.7,8	*Paronychia desertorum* Boiss.
564.	5:II.5,6	*Spergularia marina* (L.) Griseb.
565.	96:I.6,7	*Velezia rigida* L.
566.	39:II.3,4	*Minuartia or ? Arenaria* sp. indet.
567.	99:III.5,6	*Silene succulenta* Forssk.
568.	118:I.5,6	*Silene paradoxa* L.
569.	99:III.1,2	*Silene involuta* Forssk.
570.	118:I.3,4	*Silene biappendiculata* Rohrb.
571.	99:II.7,8	*Silene conica* L.
572.	35:II.5,6	*Silene sp.*
573.	79:I.3,4	*Gisekia pharnaceoides* L.
574.	[no specimen traced]	
575.	104:II.5,6	*Sida alba* L.
576.	[no specimen traced]	
577.	[no specimen traced]	
578.	54:II.5,6	*Hibiscus ovalifolius* (Forssk.) Vahl
579.	54:II.3,4	*Hibiscus ovalifolius* (Forssk.) Vahl
580.	[no specimen traced]	
581.	54:I.5,6	*Pavonia hildebrandtii* (Guerke) Ulb.
582.	[no specimen traced]	
583.	[no specimen traced]	
584.	99:I.5,6	*Sida alba* L.
585.	[no specimen traced]	
586.	104:II.3,4	*Sida alba* L.
587.	65(2-2):II.4,5	*Malva parviflora* L.
588.	4:II.3,4	*Althea rosea* (L.) Cav.
589.	98:III.3,4	*Abutilon bidentatum* A. Rich.
590.	54:II.7,8	*Hibiscus purpureus* Forssk.

591.	65(2-2):I.8 & II.1	*Malva verticillata* L.
592.	66:I.3,4	*Malva sylvestris* L. var. *eriocarpa* Boiss.
593.	66:I.5,6	*Malva sylvestris* L. var. *eriocarpa* Boiss.
594.	65(2-2):II.2,3	*Malva sylvestris* L.
595.	65(2-2):II.7,8 & 66:I.1,2	*Malva sylvestris* L.
596.	98:III.1,2	*Abutilon bidentatum* A. Rich.
597.	99:II.3,4	*Wissadula amplissima* (L.) R. E. Fries var. *rostrata* (Schum. & Thon.) R. E. Fries
598.	130:I.3,4	*Sida urens* L.
599.	52:I.3,4	*Gossypium herbaceum* L.
600.	110:III.5,6	*Urena ovalifolia* Forssk.
601.	54:I.3,4	*Pavonia hildebrandtii* Guerke & Ulbr.
602.	130:I.5,6	*Hibiscus sp.*
603.	54:I.7,8	*Pavonia flavo-ferruginea* (Forssk.) Hepper & Wood
604.	54:II.1,2	*Hibiscus palmatus* Forssk.
605.	35:III.3,4	*Sterculia africana* (Lour.) Fiori
606.	86:III.3,4	*Reseda decursiva* Forssk.
607.	[no specimen traced]	
608.	87:I.3,4	*Reseda luteola* L.
609.	87:II.3,4	*Reseda arabica* Boiss.
610.	86:III.5,6	*Reseda decursiva* Forssk.
611.	83:I.3,4	*Polygala sp. indet.*
612.	87:I.7,8	*Reseda luteola* L.
613.	71:II.5,6	*Nymphaea lotus* L.
614.	27:III.7,8	*Helianthemum cahiricum* Del.
615.	27:III.3,4	*Helianthemum stipulatum* (Forssk.) C. Chr.
616.	27:III.1,2	*Helianthemum stipulatum* (Forssk.) C. Chr.
617.	27:III.5,6	*Helianthemum stipulatum* (Forssk.) C. Chr.
618.	27:II.7,8	*Helianthemum stipulatum* (Forssk.) C. Chr.
619.	88:I.1,2	*Cleome droserifolia* (Forsk.) Del.
620.	88:I.3,4	*Cleome droserifolia* (Forsk.) Del.
621.	112:II.3,4	*Hybanthus enneaspermus* (L.) F. Muell.
622.	112:II.5,6	*Hybanthus enneaspermus* (L.) F. Muell.
623.	45:III.5,6	*Frankenia revoluta* Forssk.
624.	45:III.3,4	*Frankenia revoluta* Forssk.
625.	125:I.7,8	*Frankenia pulverulenta* L.
626.	72:II.3,4	*Oncoba spinosa* Forssk.
627.	72:II.5,6	*Oncoba spinosa* Forssk.
628.	[no specimen traced]	
629.	67:III.7,8	*Mesembryanthemum nodiflorum* L.
630.	68:I.1,2	*Mesembryanthemum nodiflorum* L.
631.	68:I.3,4	*Mesembryanthemum nodiflorum* L.
632.	68:I.5,6	*Mesembryanthemum nodiflorum* L.
633.	19:I.1,2	*Cadaba glandulosa* Forssk.
634.	18:III.7,8	*Cadaba farinosa* Forssk.
635.	19:II.7,8	*Cadaba rotundifolia* Forssk.
636.	19:I.5,6	*Cadaba glandulosa* Forssk.
637.	19:I.3,4	*Cadaba glandulosa* Forssk.
638.	20:III.3,4	*Maerua oblongifolia* (Forssk.) A. Rich.
639.	20:III.1,2	*Maerua oblongifolia* (Forssk.) A. Rich.

640.	100:I.7,8	*Cleome amblyocarpa* Barr. & Murb.
641.	100:III.5,6	*Capparis decidua* (Forssk.) Edgew.
642.	[no specimen traced]	
643.	20:III.5,6	*Maerua oblongifolia* (Forssk.) A. Rich.
644.	19:II.5,6	*Cadaba rotundifolia* Forssk.
645.	18:III.5,6	*Cadaba farinosa* Forssk.
646.	19:I.7,8	*Cadaba glandulosa* Forssk.
647.	101:I.1,2	*Capparis decidua* (Forssk.) Edgew.
648.	94:I.7,8	*Salsola soda* L.
649.	65(1-2):III.7	*Maerua crassifolia* Forssk.
650.	[no specimen traced]	
651.	[no specimen traced]	
652.	123:I.7,8	*Lagenaria siceraria* (Molina) Standley
653.	[no specimen traced]	
654.	[no specimen traced]	
655.	2:III.3,4	*Adenia venenata* Forssk.
656.	123:I.1,2	*Zehneria thwaitesii* (Schweinf.) C. Jeffrey
657.	109:III.1,2	*Cucumis prophetarum* L.
658.	109:II.7,8	*Cucumis prophetarum* L.
659.	109:III.3,4	*Cucumis prophetarum* L.
660.	35:II.7,8	*Coccinia grandis* (L.) Voigt
661.	109:III.5,6	*Cucumis melo* L. & *Kedrostis leloja* (Forssk. ex J. F. Gmel.) C. Jeffrey [mixed sheet]
662.	109:III.7,8	*Coccinia grandis* (L.) Voigt
663.	110:I.1,2	*Coccinia grandis* (L.) Voigt
664.	35:III.1,2	*Citrullus colocynthis* (L.) Schrad.
665.	123:I.5,6	*Kedrostis leloja* (Forssk.) C. Jeffrey
666.	110:I.3,4	*Coccinia grandis* (L.) Voigt
667.	123:I.3,4	*Dicotyledones* fam., gen. & sp. indet.
668.	22:III.1,2	*Torilis arvensis* (Huds.) Link
669.	no microf.	*Echinophora tenuifolia* L.
670.	42:I.3,4	*Eryngium bourgatii* Gouan
671.	137:II.1,2	*Sanicula elata* D. Don
672.	37:III.5,6	*Ammi visnaga* (L.) Lam.
673.	37:III.7,8	*Ammi visnaga* (L.) Lam.
674.	137:II.7,8	*Ammi visnaga* (L.) Lam.
675.	5:III.3,4	*Ammi majus* L.
676.	5:III.5,6	*Ammi majus* L.
677.	98:II.7,8	?*Seseli tortuosum* L.
678.	137:II.5,6	*Pimpinella anisum* L.
679.	137:II.3,4	*Ammi visnaga* Lam.
680.	42:I.5,6	*Eryngium maritimum* L.
681.	17:III.1,2	*Bupleurum flavum* Forssk.
682.	90:II.8 & III.1	*Cissus rotundifolius* (Forssk.) Vahl
683.	27:II.1,2	*Salvadora persica* L.
684.	90:III.2,3	*Cissus rotundifolius* (Forssk.) Vahl
685.	129:III.5,6	*Plicosepalus acaciae* (Zucc.) Weins & Polhill
686.	112:II.7,8	*Viscum schimperi* Engl.
687.	112:III.1,2	*Viscum schimperi* Engl.
688.	32:II.7,8	*Cordia sinensis* Lam.
689.	33:I.6,7	*Kalanchoe lanceolata* (Forssk.) Pers.
690.	33:I.3,4	*Kalanchoe alternans* (Vahl) Pers.
691.	33:I.8 & II.1	*Crassula alba* Forssk.

692.	55:II.7,8	*Dipcadi erythraeum* Webb & Berth.
693.	22:III.5,6	*Cocculus pendulus* (J. R. & G. Forst.) Diels
694.	23:I.1,2	*Cocculus pendulus* (J. R. & G. Forst.) Diels
695.	61:III.7,8	*Cocculus pendulus* (J. R. & G. Forst.) Diels
696.	23:I.3,4	*Cocculus hirsutus* (L.) Theob.
697.	86:II.1,2	*Ranunculus sceleratus* L.
698.	86:I.7,8	*Ranunculus multifidus* Forssk.
699.	134:I.7,8	*Thalictrum foetidum* L.
700.	28:I.1,2	*Clematis vitalba* L.
701.	14:III.7,8	*Berberis forsskaliana* Schneider
702.	70:I.3,4	*Hypecoum aegyptiacum* (Forssk.) Aschers. & Schweinf.
703.	70:I.5,6	*Hypecoum aegyptiacum* (Forssk.) Aschers. & Schweinf.
704.	26:I.3,4	*Roemeria hybrida* (L.) DC.
705.	25:III.5,6	*Roemeria hybrida* (L.) DC. subsp. *dodecandra* (Forssk.) Dur. & Barrat.
706.	26:I.1,2	*Roemeria hybrida* (L.) DC. subsp. *dodecandra* (Forssk.) Dur. & Barrat.
707.	49:I.5,6	*Fumaria densiflora* DC.
708.	65(1-2):I.5,6	*Farsetia aegyptia* Turra
709.	15:III.5,6	*Erucaria crassifolia* (Forssk.) Del.
710.	86:II.5,6	*Enarthrocarpus lyratus* (Forssk.) DC.
711.	15:III.1,2	*Erucaria crassifolia* (Forssk.) Del.
712.	122:III.7,8	*Eremobium aegyptiacum* (Spreng.) Hochr.
713.	15:III.3,4	*Erucaria crassifolia* (Forssk.) Del.
714.	56:II.3,4	*Iberis pinnata* L.
715.	62:II.3,4	*Lepidium armoracia* Fisch. & Mey.
716.	105:III.3,4	*Schouwia purpurea* (Forssk.) Schweinf.
717.	62:II.5,6	*Lepidium graminifolium* L.
718.	6:II.3,4	*Anastatica hierochuntica* L.
719.	105:III.5,6	*Schouwia purpurea* (Forssk.) Schweinf.
720.	25:I.7,8	*Matthiola livida* (Del.) DC.
721.	57:III.5,6	*Cakile maritima* Scop. subsp. *aegyptiaca* (Willd.) Nyman
722.	100:III.1,2	*Diplotaxis harra* (Forssk.) Boiss.
723.	25:I.5,6	*Farsetia linearis* (Forssk.) Boiss.
724.	86:II.3,4	*Enarthrocarpus lyratus* (Forssk.) DC.
725.	113:I.5,6	*Zilla spinosa* (L.) Prantl
726.	41:III.5,6	*Erica arborea* L.
727.	41:III.7,8	*Erica manipuliflora* Salisb.
728.	42:I.1,2	*Erica manipuliflora* Salisb.
729.	17:III.7,8	*Myrsine africana* L.
730.	18:I.1,2	*Myrsine africana* L.
731.	50:II.7,8	*Erodium crassifolium* L'Hérit.
732.	50:II.1,2	*Erodium glaucophyllum* (L.) L'Hérit.
733.	50:III.3,4	*Erodium laciniatum* (Cav.) Willd.
734.	50:III.5,6	*Erodium laciniatum* (Cav.) Willd.
735.	50:I.5,6	*Geranium arabicum* Forssk.
736.	50:I.7,8	*Geranium arabicum* Forssk.
737.	50:III.7,8	*Erodium laciniatum* (Cav.) Willd.
738.	50:III.1,2	*Erodium crassifolium* L'Hérit.
739.	50:II.3,4	*Erodium glaucophyllum* (L.) L'Hérit.

740.	43:III.7,8	*Fagonia cretica* L. [no microf. *Caralluma dentata* (Forssk.) Blatter; also this No? seems lost]
741.	44:I.1,2	*Fagonia cretica* L.
742.	44:I.3,4	*Fagonia cretica* L.
743.	107:III.7,8	*Tribulus pentandrus* Forssk.
744.	107:III.5,6	*Tribulus terrestris* L.
745.	[no specimen traced]	
746.	80:II.5,6	*Pistacia terebinthus* L.
747.	80:II.7,8	*Rhus cf. abyssinica* Oliv.
748.	[no specimen traced]	
749.	[no specimen traced]	
750.	[no specimen traced]	
751.	113:II.4,5	*Zygophyllum simplex* L.
752.	113:II.6,7	*Zygophyllum simplex* L.
753.	113:II.2,3	*Zygophyllum coccineum* L.
754.	[no specimen traced]	
755.	89:II.7,8	*Haplophyllum tuberculatum* (Forssk.) A. Juss.
756.	89:III.1,2	*Haplophyllum tuberculatum* (Forssk.) A. Juss.
757.	[no specimen traced]	
758.	77:III.1,2	*Nitraria retusa* (Forssk.) Aschers.
759.	77:III.3,4	*Nitraria retusa* (Forssk.) Aschers.
760.	43:II.7,8	*Ochna inermis* (Forssk.) Schweinf.
761.	33:II.5,6	*Crataegus monogyna* Jacq.
762.	33:II.3,4	*Crataegus monogyna* Jacq.
763.	102:III.5,6	*Sorbus domestica* L.
764.	85:III.3,4	*Flemingia cf. grahamiana* Wight & Arn.
765.	85:III.5,6	*Pyrus amygdalifolius* Vill.
766.	[no specimen traced]	
767.	61:III.3,4	*Lawsonia inermis* L.
768.	61:III.1,2	*Lawsonia inermis* L.
769.	77:I.7,8	*Parietaria alsinifolia* Del.
770.	111:I.7,8	*Pouzolzia parasitica* (Forssk.) Schweinf.
771.	111:I.5,6	*Girardinia diversifolia* (Link) Friis
772.	111:I.1,2	*Laportea aestuans* (L.) Chew
773.	111:II.1,2	*Urtica urens* L.
774.	111:I.3,4	*Droguetia iners* (Forssk.) Schweinf.
775.	19:III.5,6	*Forsskalea tenacissima* L.
776.	45:III.1,2	*Ficus vasta* Forssk.
777.	45:I.3,4	*Ficus sp.*
778.	44:III.7	*Ficus carica* L.
779.	45:II.7,8	*Ficus sycomorus* L.
780.	no microf.	*Ficus cordata* Thunb. subsp. *salicifolia* (Vahl) C.C. Berg
781.	45:I.5,6	*Ficus palmata* Forssk.
782.	45:II.5,6	*Ficus sur* Forssk.
783.	45:II.1,2	*Ficus populifolia* Vahl
784.	45:I.7,8	*Ficus palmata* Forssk.
785.	45:II.3,4	*Ficus exasperata* Vahl
786.	85:III.7,8	*Quercus ilex* L.
787.	86:I.5,6	*Quercus pubescens* Willd.
788.	125:I.3,4	*Quercus coccifera* L.
789.	124:III.8 & 125:I.1,2	*Quercus coccifera* L.
790.	86:I.1,2	*Quercus coccifera* L.

791.	86:I.3,4	*Quercus ilex* L.
792.	125:I.5,6	*Quercus cerris* L.
793.	15:I.3,4	*Alnus glutinosa* (L.) Gaertn.
794.	41:II.5,6	*Ephedra campylopoda* C. A. Meyer
795.	122:I.5,6	*Pinus brutia* Ten.
796.	58:II.5,6	*Juniperus oxycedrus* L.
797.	41:III.3,4	*Equisetum ramosissimum* Desf.
798.	133:III.5,6	*Pteridium aquilinum* (L.) Kuhn
799.	85:II.7,8	*Pteris cretica* L.
800.	85:III.1,2	*Pteris cretica* L.
801.	13:I.1,2	*Phyllitis scolopendrium* (L.) Newm.
802.	85:II.5,6	*Pteris vittata* L.
803.	85:II.3,4	*Pteris vittata* L.
804.	2:III.7,8	*Adiantum capillus-veneris* L.
805.	2:III.5,6	*Adiantum capillus-veneris* L.
806.	133:III.7,8	*Adiantum poiretii* Wikstr.
807.	84:I.1,2	*Arthropteris orientalis* (J. F. Gmel.) Posthum.
808.	133:III.1,2	*Tectaria gemmifera* (Fée) Alston
809.	83:III.6,7	*Christella dentata* (Forssk.) Brownsey & Jeremy
810.	64:I.1,2	*Asplenium rutifolium* (Berg.) Kunze var. *bipinnatum* (Forssk.) Schelpe
811.	134:I.3,4	*Adiantum flabellatum* L.
812.	134:I.5,6	*Adiantum flabellatum* L.
813.	3:I.1,2	*Adiantum incisum* Forssk.
814.	84:I.5,6	*Polystichum setiferum* (Forssk.) Woynar
815.	2:II.7,8	*Actiniopteris radiata* (Schwartz) Link
816.	12:III.7,8	*Asplenium adiantum-nigrum* L.
817.	12:III.3,4	*Asplenium aethiopicum* (Burm. f.) Becherer
818.	[no specimen traced]	
819.	[no specimen traced]	
820.	2:III.1,2	*Actiniopteris radiata* (Schwartz) Link
821.	133:III.3,4	*Asplenium adiantum-nigrum* L.
822.	84:I.3,4	*Arthropteris orientalis* (J.F. Gmel.) Posthum.
823.	134:I.1,2	*Adiantum poiretii* Wikstr.
824.	103:I.1,2	"*Spongia tubulosa* Forssk." (?Animal).
825.	115:III.3,4	*Algae* fam., gen. & sp. indet.
826.	47:III.1,2	*Actinotrichia fragilis* (Forssk.) Börgesen (Algae)
827.	115:III.1,2	*Algae* fam., gen. & sp. indet.
828.		
829.	102:III.7,8	"*Spongia flabelliformis* Forssk." (?Animal).
830.	62:III.1,2	*Usnea florida* Achar. (Lichen)
831.	115:I.1,2	*Algae* fam., gen. & sp. indet.
832.	47:II.5,6	*Fucus vesiculosus* L. (Algae)
833.	no microf.	*Algae* fam., gen. & sp. indet.
834.	114:III.4,5	*Algae* fam., gen. & sp. indet.
835.	115:I.7,8	*Algae* fam., gen. & sp. indet.
836.	115:II.1,2	*Algae* fam., gen. & sp. indet.
837.	48:I.4,5	*Cytoseira barbata* (Algae)
838.	114:III.2,3	*Algae* fam., gen. & sp. indet.
839.	46:II.3,4	*Phyllophora rubens* (L.) Grev. (Algae)
840.	114:II.8 & III.1	*Algae* fam., gen. & sp. indet.
841.	115:I.5,6	*Algae* fam., gen. & sp. indet.
842.	[no specimen traced]	
843.	114:II.6,7	*Algae* fam., gen. & sp. indet.
844.	47:III.3,4	*Caulerpa serrulata* (Forssk.) Agardh (Algae)

845.	48:II.3,4	*Caulerpa racemosa* (Forssk.) Weber von Bosse (Algae)
846.	[no specimen traced]	
847.	115:II.7,8	*Algae* fam., gen. & sp. indet.
848.	115:I.3,4	*Algae* fam., gen. & sp. indet.
849.	47:I.5,6	*Sargassum denticulatum* (Forssk.) Börgesen (Algae)
850.	48:II.5,6	*Cytoseira myrica* (Gmel.) Agadh (Algae)
851.	48:II.7,8	*Sargassum subrepandum* (Forssk.) Agardh (Algae)
852.	48:III.1,2	*Sargassum subrepandum* (Forssk.) Agardh (Algae)
853.	110:II.7,8	*Ulva reticulata* Forssk. (Algae)
854.	47:I.7,8	*Dictyota dichotoma* (L.) Ktz. (Algae)
855.	[no specimen traced]	
856.	115:II.5,6	*Algae* fam., gen. & sp. indet.
857.	114:I.8 & II.1	*Algae* fam., gen. & sp. indet.
858.	114:I.6,7	*Algae* fam., gen. & sp. indet.
859.	114:I.4,5	*Algae* fam., gen. & sp. indet.
860.	29:III.7,8	*Ceramium rubrum* Ag. (Algae)
861.	114:III.6,7	*Algae* fam., gen. & sp. indet.
862.	47:III.7,8	*Polysiphonia elongata* (Huds.) Harv. (Algae)
863.	110:I.7,8	*Halarachnion ligulatum* (Woodw.) Ktz. (Algae)
864.	110:II.5,6	*Ulva lactuca* L. (Algae)
865.	110:II.3,4	*Chaetomorpha melagonium* (Weber & Mohr) Ktz. (Algae)
866.	30:II.1,2	*Chaetomorpha linum* Ktz. (Algae)
867.	29:III.3,4	*Chaetomorpha linum* Ktz. (??; Algae)
868.	29:III.5,6	*Ceramium diaphanum* Ag. & *Ceramium rubrum* Ag. [mixed sheet] (Algae)
869.	30:II.3,4	*Chaetomorpha linum* Ktz. (Algae)
870.	30:I.6,7	*Hypnea musciformis* (Wulf.) Lamour. (Algae)
871.	46:III.7,8	*Sargassum crispum* (Forssk.) Ag. (Algae)
872.	48:III.3,4	869.
870.	30:I.6,7	*Hypnea musciformis* (Wulf.) Lamour. (Algae)
871.	46:III.7,8	*Sargassum crispum* (Forssk.) Ag. (Algae)
872.	48:III.3,4	*Sargassum subrepandum* (Forssk.) Agardh (Algae)
873.	48:III.5	*Sargassum subrepandum* (Forssk.) Agardh (Algae)
874.	46:I.3,4	*Echinocaulon acerosum* (Forssk.) Brgesen (Algae)
875.	114:I.2,3	*Algae* fam., gen. & sp. indet.
876.	49:I.1,2	*Laurencia obtusa* Ag. (Algae)
877.	29:II.3,4	*Laurencia obtusa* Ag. (Algae)
878.	48:II.1,2	*Caulerpa prolifera* (Forssk.) Lamour. (Algae)
879.	114:II.2,3	*Algae* fam., gen. & sp. indet.
880.	46:III.1,2	*Turbinaria decurrens* Bory (Algae)
881.	46:III.3,4	*Turbinaria decurrens* Bory (Algae)
882.	48:I.8	*Caulerpa setularoides* (Gmel.) Howe (Algae)
883.	46:II.1,2	*Hormosira articulata* (Forssk.) Zanard (Algae)
884.	30:I.1,2	*Bonnemaisonia asparagoides* (Woodw.) Ag. (Algae)
885.	47:II.1,2	*Solieria dura* (Zanard.) Schmitz (Algae)
886.	48:I.6,7	*Chondria papillosa* C. Agardh (Algae)

887.	49:I.3,4	*Laurentia uvifera* (Forssk.) Börgesen (Algae)
888.	47:I.1,2	*Gracilaria debilis* (Forssk.) Börgesen (Algae)
889.	48:III.6,7	*Cystophyllum trinode* (Forssk.) Agardh (Algae)
890.		[not used]
891.		[not used]
892.		[not used]
893.		[not used]
894.		[not used]
895.		[not used]
896.		[not used]
897.		[not used]
898.		[not used]
899.		[not used]
900.	130:II.1,2	*Abutilon pannosum* (Forst. f.) Schlecht.
901.	98:III.7,8	*Abutilon pannosum* (Forst. f.) Schlecht.
902.	1:I.1,2	*Acalypha ciliata* Forssk.
903.	1:I.3,4	*Acalypha fruticosa* Forssk.
904.	1:I.7,8	*Acanthus arboreus* Forssk.
905.	1:II.1,2	*Blepharis ciliaris* (L.) B.L. Burtt
906.	1:II.5,6	*Acanthus hirsutus* Boiss.
907.	1:II.3,4	*Blepharis maderaspatensis* (L.) Roth
908.	1:II.7,8	*Achillea santolina* L.
909.	1:III.1,2	*Achillea santolina* L.
910.	1:III.3,4	*Achyranthes aspera* L. var. *sicula* L.
911.	116:III.5	*Achyranthes aspera* L. var. *sicula* L.
912.	no microf.	*Celosia trigyna* L. var. *fasciculiflora* Fenzl.
913.	2:II.3,4	*Consolida aconiti* (L.) Lindl.
914.	2:II.5,6	*Consolida aconiti* (L.) Lindl.
915.	3:III.5,6	*Aerva javanica* (Burm. f.) Juss.
916.	2:I.7,8	*Aerva javanica* (Burm. f.) Juss.
917.	3:II.3,4	*Aerva javanica* (Burm. f.) Juss.
918.	3:II.5,6	*Aerva javanica* (Burm. f.) Juss.
919.	121:I.3,4	*Aetheorhiza bulbosa* (L.) Cass.
920.	4:I.7,8	*Sporobolus spicatus* (Vahl) Kunth
921.	4:II.1,2	*Sporobolus spicatus* (Vahl) Kunth
922.	4:III.1,2	*Allium desertorum* Forssk.
923.	4:II.5,6	*Alisma plantago-aquatica* L.
924.	5:II.1,2	*Polycarpon prostratum* (Forssk.) Aschers. & Schweinf.
925.	5:II.3,4	*Polycarpon prostratum* (Forssk.) Aschers. & Schweinf.
926.	no microf.	*Polycarpon prostratum* (Forssk.) Aschers. & Schweinf.
927.	130:I.1,2	*Althaea hirsuta* L.
928.	5:II.7,8	*Ambrosia maritima* L.
929.	5:III.1,2	*Ambrosia maritima* L.
930.	no microf.	*Ambrosia maritima* L.
931.	no microf.	*Ambrosia maritima* L.
932.	no microf.	*Ambrosia maritima* L.
933.	126:I.1,2	*Ammophila arenaria* (L.) Link.
934.	5:III.7,8	*Commiphora kataf* (Forssk.) Engl.
935.	6:I.1,2	*Commiphora gileadensis* (L.) C. Chr.

936.	6:I.3,4	*Commiphora gileadensis* (L.) C. Chr.
937.	6:I.5,6	*Commiphora gileadensis* (L.) C. Chr.
938.	6:I.7,8	*Anagallis arvensis* L.
939.	6:II.5,6	*Anastatica hierochuntica* L.
940.	117:III.5,6	*Alkanna lehmannii* (Tineo) A. DC.
941.	7:II.1,2	*Cymbopogon caesius* (Hook. & Arn.) Stapf
942.	7:II.5,6	*Cymbopogon caesius* (Hook. & Arn.) Stapf
943.	7:II.7,8	*Heterpogon contortus* (L.) Roem. & Schult.
944.	7:III.1,2	*Anthemis melampodina* Del.
945.	no microf.	*Urginea maritima* (L.) Bak.
946.	7:III.5,6	*Kickxia aegyptiaca* (L.) Nabelek
947.	7:III.7,8	*Kickxia aegyptiaca* (L.) Nabelek
948.	8:I.1,2	*Linaria haelava* (Forssk.) Del.
949.	8:I.3,4	*Linaria haelava* (Forssk.) Del.
950.	8:I.7,8	*Anticharis linearis* (Benth.) Aschers.
951.	8:II.1,2	*Anticharis linearis* (Benth.) Aschers.
952.	134:III.5,6	*Anthospermum herbaceum* L. f.
953.	134:III.7,8	*Anthospermum herbaceum* L. f.
954.	135:I.7,8	*Anthospermum herbaceum* L. f.
955.	8:II.5,6	*Minuartia filifolia* (Forssk.) Mattf.
956.	8:II.7,8	*Minuartia filifolia* (Forssk.) Mattf.
957.	9:III.1,2	*Arnebia tetrastigma* Forssk.
958.	9:III.3,4,5	*Gerbera piloselloides* (L.) Cass.
959.	120:I.1,2	*Artemisia judaica* L.
960.	no microf.	*Artemisia judaica* L.
961.	119:III.5,6	*Artemisia monosperma* Del.
962.	119:II.7,8	*Artemisia sp.*
963.	10:I.7,8	*Phragmites australis* (Cav.) Trin. & Steud.
964.	10:II.3,4	*Phragmites australis* (Cav.) Trin. & Steud.
965.	126:I.3,4	*Arundo donax* L.
966.	10:II.7,8	*Imperata cylindrica* (L.) Räusch
967.	10:III.1,2	*Phragmites australis* (Cav.) Trin. & Steud.
968.	10:III.3,4	*Phragmites australis* (Cav.) Trin. & Steud.
969.	10:III.5,6	*Phragmites australis* (Cav.) Trin. & Steud.
970.	11:I.1,2	*Phragmites australis* (Cav.) Trin. & Steud.
971.	no microf.	*Pergularia tomentosa* L.
972.	11:I.7,8	*Pergularia tomentosa* L.
973.	11:II.1,2	*Kanahia laniflora* (Forssk.) R. Br.
974.	11:II.3,4	*Kanahia laniflora* (Forssk.) R. Br.
975.	138:III.2,3	*Blyttia spiralis* (Forssk.) Field & Wood
976.	13:III.7,8	*Atriplex farinosa* Forssk.
977.	118:II.7,8	*Atriplex tartarica* L.
978.	119:I.5,6	*Halimione portulacoides* (L.) Aellen
979.	118:III.1	*Atriplex sp.*
980.	14:III.3,4	*Centropodia forskalii* (Vahl) Cope
981.	14:III.5,6	*Centropodia forskalii* (Vahl) Cope
982.	129:II.1,2	*Asparagus officinalis* L.
983.	12:II.1,2	*Asparagus acutifolius* L.
984.	12:II.7,8	*Asparagus stipularis* Forssk.
985.	12:III.1,2	*Asparagus stipularis* Forssk.
986.	12:III.5,6	*Asplenium aethiopicum* (Burm. f.) Becherer
987.	13:I.3,4	*Pulicaria crispa* (Forssk.) Benth. ex Oliver
988.	13:I.5,6	*Pulicaria crispa* (Forssk.) Benth. ex Oliver
989.	13:I.7,8	*Pulicaria crispa* (Forssk.) Benth. ex Oliver
990.	13:II.1,2	*Pulicaria crispa* (Forssk.) Benth. ex Oliver

991.	13:II.3,4	*Pulicaria crispa* (Forssk.) Benth. ex Oliver
992.	13:II.5,6	*Macowania ericifolia* (Forssk.) Burtt & Grau
993.	119:III.7,8	Compositae, tribe Anthemideae, gen. et sp. indet.
994.	120:I.8	Compositae, tribe Anthemideae, gen. et sp. indet.
995.	17:II.3,4	*Asteriscus aquaticus* (L.) Less.
996.	13:II.7,8	*Astragalus annularis* Forssk.
997.	13:III.1,2	*Astragalus fruticosus* Forssk.
998.	no microf.	*Astragalus peregrinus* Vahl
999.	no microf.	*Astragalus peregrinus* Vahl
1000.	132:II.5,6	*Astragalus sp.*
1001.	116:II.7,8	*Barleria acanthoides* Vahl
1002.	116:III.1,2	*Barleria acanthoides* Vahl
1003.	31:I.3,4	Chenopodiaceae, gen. et sp. indet.
1004.	15:I.1,2	*Beta vulgaris* L.
1005.	119:I.1,2	*Beta vulgaris* L.
1006.	15:II.1,2	*Boerhavia repens* L. var. *diffusa* (L.) Boiss.
1007.	120:II.1,2	*Blumea axillaris* (Lam.) DC.
1008.	120:II.3,4	*Blumea axillaris* (Lam.) DC.
1009.	no microf.	*Brachypodium phoenicioides* (L.) Roem. & Schult.
1010.	16:I.5,6	*Brachypodium phoenicioides* (L.) Roem. & Schult.
1011.	15:II.7,8	*Erucaria crassifolia* (Forssk.) Del.
1012.	no microf.	*Bromus sp.*
1013.	126:I.5,6	*Bromus hordeaceus* L.
1014.	126:I.7,8	*Bromus japonicus* Thunb.
1015.	no microf.	*Rostraria cristata* (L.) Tzvelev.
1016.	16:I.8	*Leptochloa fusca* (L.) Kunth
1017.	126:II.1,2	*Bromus rubens* L.
1018.	126:II.3,4	*Bromus rubens* L.
1019.	17:I.5,6	*Nauplius graveolens* (Forssk.) A. Wicklund
1020.	17:I.7,8	*Nauplius graveolens* (Forssk.) A. Wicklund
1021.	17:II.1,2	*Nauplius graveolens* (Forssk.) A. Wicklund
1022.	17:II.7,8	*Nauplius graveolens* (Forssk.) A. Wicklund
1023.	17:III.3,4	*Bupleurum flavum* Forssk.
1024.	17:III.5,6	*Myrsine africana* L.
1025.	no microf.	*Myrsine africana* L.
1026.	18:II.3,4	*Solanecio angulatus* (Vahl) C. Jeffrey
1027.	18:II.5,6	*Solanecio angulatus* (Vahl) C. Jeffrey
1028.	18:II.7,8	*Solanecio angulatus* (Vahl) C. Jeffrey
1029.	18:III.1,2	*Solanecio angulatus* (Vahl) C. Jeffrey
1030.	18:I.3,4	*Kleinia odora* (Forssk.) DC.
1031.	18:I.5,6,7	*Kleinia pendula* (Forssk.) DC.
1032.	18:I.8	*Kleinia pendula* (Forssk.) DC.
1033.	18:II.1,2	*Kleinia semperviva* (Forssk.) DC.
1034.	18:III.3,4	*Cadaba farinosa* Forssk.
1035.	19:II.1,2	*Cadaba glandulosa* Forssk.
1036.	19:II.3,4	*Cadaba rotundifolia* Forssk.
1037.	19:III.1,2	*Cadaba rotundifolia* Forssk.
1038.	19:III.3,4	*Cadia purpurea* (Picciv.) Aiton
1039.	19:III.7,8	*Forsskaolea tenacissima* L.
1040.	20:I.1,2	*Calendula arvensis* L.
1041.	20:I.3,4	*Ruellia grandiflora* (Forssk.) Blatter
1042.	20:I.5,6	*Ruellia grandiflora* (Forssk.) Blatter
1043.	20:I.7,8	*Ruellia grandiflora* (Forssk.) Blatter

1044.	122:III.3,4	*Cardamine africana* L.
1045.	20:III.7,8	*Carthamus glaucus* Bieb.
1046.	21:I.1,2	*Carlina corymbosa* L. subsp. *graeca* (Boiss.) Nyman
1047.	21:I.3,4	*Carlina corymbosa* L. subsp. *graeca* (Boiss.) Nyman
1048.	21:II.5,6	*Senna alexandrina* Miller
1049.	21:II.7,8	*Senna alexandrina* Miller
1050.	21:III.1,2	*Senna alexandrina* Miller
1051.	21:III.3,4	*Senna alexandrina* Miller
1052.	no microf.	*Senna italica* Miller
1053.	21:I.7,8	*Senna sophera* (L.) Roxb.
1054.	21:II.1,2	*Senna sophera* (L.) Roxb.
1055.	21:II.3,4	*Senna sophera* (L.) Roxb.
1056.	no microf.	*Senna sophera* (L.) Roxb.
1057.	21:III.5,6	*Chamaecrista nigricans* (Vahl) Greene
1058.	21:III.7,8	*Chamaecrista nigricans* (Vahl) Greene
1059.	21:I.5,6	*Senna italica* Miller
1060.	132:I.7,8	*Senna italica* Miller
1061.	no microf.	*Senna italica* Miller
1062.	no microf.	*Senna italica* Miller
1063.	22:I.1,2	*Senna obtusifolia* (L.) Irwin & Greene
1064.	22:I.3,4	*Senna obtusifolia* (L.) Irwin & Greene
1065.	128:II.1,2	*Catapodium rigidum* (L.) C.E. Hubb.
1066.	22:II.3,4	*Catha edulis* (Vahl) Forssk. ex Endl.
1067.	no microf.	*Pseudorlaya pumila* (L.) Grande
1068.	no microf.	*Caucanthus edulis* Forssk.
1069.	no microf.	*Cenchrus echinatus* L.
1070.	23:II.3,4	*Atractylis carduus* (Forssk.) C. Chr.
1071.	23:II.5,6	*Atractylis carduus* (Forssk.) C. Chr.
1072.	23:II.1,2	*Centaurea calcitrapa* L.
1073.	23:I.7,8	*Centaurea aspera* L.
1074.	23:II.7,8	*Centaurea aegyptiaca* L.
1075.	23:I.5,6	*Centaurea glomerata* Vahl
1076.	23:III.1,2	*Volutaria lippii* (L.) Cass.
1077.	23:III.3,4	*Volutaria lippii* (L.) Cass.
1078.	23:III.5,6	*Centaurothamnus maximus* (Forssk.) Wagenitz & Dittrich
1079.	23:III.7,8	*Centaurothamnus maximus* (Forssk.) Wagenitz & Dittrich
1080.	24:I.1,2	*Centaurothamnus maximus* (Forssk.) Wagenitz & Dittrich
1081.	50:I.3,4	*Centaurium pulchellum* (Sw.) Druce
1082.	24:I.3,4	*Aegialophila pumilio* (L.) Boiss.
1083.	no microf.	*Centaurium pulchelum* (Sw.) Druce
1084.	24:I.5,6	*Ceruana pratensis* Forssk.
1085.	24:II.3,4	*Grewia tenax* (Forssk.) Fiori
1086.	24:III.3,4	*Lantana viburnoides* (Forssk.) Vahl
1087.	24:III.5,6	*Lantana viburnoides* (Forssk.) Vahl
1088.	no microf.	*Farsetia linearis* (Forssk.) Schweinf.
1089.	24:III.7,8	*Farsetia aegyptia* Turra
1090.	25:I.1,2	*Farsetia aegyptia* Turra
1091.	25:I.3,4	*Farsetia aegyptia* Turra
1092.	25:II.1,2	*Matthiola livida* (Del.) DC.
1093.	25:II.3,4	*Matthiola livida* (Del.) DC.

1094.	25:II.5,6	*Matthiola livida* (Del.) DC.
1095.	25:II.7,8	*Matthiola livida* (Del.) DC.
1096.	25:III.1,2	*Matthiola livida* (Del.) DC.
1097.	25:III.3,4	*Matthiola tricuspidata* (L.) R. Br.
1098.	25:III.7,8	*Roemeria hybrida* (L.) DC.
1099.	127:I.3,4	*Chloris barbata* Sw. mixed with *Eustachys paspaloides* (Vahl) L. & M.
1100.	26:II.1,2	*Iphiona mucronata* (Forssk.) Aschers. & Schweinf.
1101.	26:II.3,4	*Iphiona mucronata* (Forssk.) Aschers. & Schweinf.
1102.	26:II.5,6	*Iphiona mucronata* (Forssk.) Aschers. & Schweinf.
1103.	26:II.7,8	*Iphiona mucronata* (Forssk.) Aschers. & Schweinf.
1104.	26:III.1,2	*Iphiona mucronata* (Forssk.) Aschers. & Schweinf.
1105.	26:III.3,4	*Vernonia spatulata* (Forssk.) Sch.-Bip.
1106.	26:III.5,6	*Vernonia spatulata* (Forssk.) Sch.-Bip.
1107.	26:III.7,8	*Ifloga spicata* (Forssk.) Sch.-Bip.
1108.	27:I.1,2	*Ifloga spicata* (Forssk.) Sch.-Bip.
1109.	120:III.3,4	*Ifloga spicata* (Forssk.) Sch.-Bip.
1110.	121:II.7,8	*Chiliadenus montanus* (Vahl) Brullo
1111.	121:III.1,2	*Chiliadenus montanus* (Vahl) Brullo
1112.	27:I.3,4	*Cichorium spinosum* L.
1113.	27:I.5,6	*Cichorium spinosum* L.
1114.	27:I.7,8	*Cichorium spinosum* L.
1115.	no microf.	*Cichorium spinosum* L.
1116.	119:III.1,2	*Cineraria abyssinica* A. Rich.
1117.	27:II.3,4	*Salvadora persica* L.
1118.	no microf.	*Cladium mariscus* (L.) Pohl
1119.	no microf.	*Otostegia fruticosa* (Forssk.) Penzig
1120.	28:II.1,2	*Clutia lanceolata* Forssk.
1121.	28:II.7,8	*Carthamus dentatus* (Forssk.) DC.
1122.	28:II.3,4	*Carthamus lanatus* L.
1123.	28:II.5,6	*Carthamus lanatus* L.
1124.	28:III.1,2	*Cardopatium corymbosum* (L.) Pers.
1125.	28:III.3,4	*Cardopatium corymbosum* (L.) Pers.
1126.	28:III.5,6	*Coffea arabica* L.
1127.	29:I.1,2	*Astragalus spinosus* (Forssk.) Muschl.
1128.	29:I.3,4	*Astragalus spinosus* (Forssk.) Muschl.
1129.	no microf.	*Commelina forskaolei* Vahl
1130.	30:III.2	*Convolvulus arvensis* L.
1131.	31:II.1,2	*Convolvulus arvensis* L.
1132.	31:II.5,6	*Convolvulus arvensis* L.
1133.	122:II.3,4	*Convolvulus arvensis* L.
1134.	30:III.5,6	*Convolvulus lanatus* Vahl
1135.	31:I.1,2	*Convolvulus lanatus* Vahl
1136.	31:I.7,8	*Convolvulus lanatus* Vahl
1137.	122:II.7,8	*Convolvulus lanatus* Vahl
1138.	31:III.3,4	*Convolvulus prostratus* Forssk.
1139.	120:I.4,5	*Conyza aegyptiaca* (L.) Ait.
1140.	32:I.5,6	*Laggera crispata* (Vahl) Hepper & Wood
1141.	32:I.7,8	*Pluchea dioscoridis* (L.) DC.

1142.	120:II.5,6	*Conyza pyrrhopappa* A. Rich.
1143.	32:II.1,2	*Phagnalon rupestre* (L.) DC.
1144.	32:II.3,4	*Phagnalon rupestre* (L.) DC.
1145.	32:II.5,6	*Phagnalon rupestre* (L.) DC.
1146.	no microf.	*Phagnalon rupestre* (L.) DC.
1147.	no microf.	*Phagnalon harazianum* Defl.
1148.	32:III.7,8	*Paronychia arabica* (L.) DC.
1149.	32:III.3,4	*Cornus sanguinea* L. subsp. *australis* (C.A. Mey.) Jav.
1150.	33:II.2	*Crassula alba* Forssk.
1151.	121:II.5,6	*Crepis micrantha* Czerep.
1152.	33:II.7,8	*Picris asplenioides* L.
1153.	119:II.5,6	*Picris asplenioides* L.
1154.	121:II.3,4	*Picris asplenioides* L.
1155.	34:I.5	*Cressa cretica* L.
1156.	34:I.6,7	*Cressa cretica* L.
1157.	33:III.1,2	*Cressa cretica* L.
1158.	33:III.3,4	*Cressa cretica* L.
1159.	33:III.5,6	*Cressa cretica* L.
1160.	131:II.3,4	*Crotalaria microphylla* Vahl
1161.	131:II.5,6	*Crotalaria microphylla* Vahl
1162.	34:I.8-II.1	*Chrozophora obliqua* (Vahl) Spreng.
1163.	34:III.2,3	*Chrozophora plicata* (Vahl) A. Juss. ex Spreng.
1164.	34:III.4,5	*Chrozophora plicata* (Vahl) A. Juss. ex Spreng.
1165.	34:II.4,5	*Chrozophora oblongifolia* (Del.) Spreng.
1166.	34:II.6,7	*Chrozophora oblongifolia* (Del.) Spreng.
1167.	34:II.8-III.1	*Chrozophora oblongifolia* (Del.) Spreng.
1168.	no microf.	*Crucianella angustifolia* L.
1169.	35:I.3,4	*Crucianella aegyptiaca* L.
1170.	35:I.5,6	*Crucianella aegyptiaca* L.
1171.	no microf.	*Crucianella latifolia* L.
1172.	no microf.	*Crucianella latifolia* L.
1173.	35:I.1,2	*Crucianella latifolia* L.
1174.	35:I.7,8	*Crucianella maritima* L.
1175.	135:I.1,2	*Crucianella maritima* L.
1176.	123:II.3,4	*Ctenolepis cerasiformis* (Stocks) Hook. f. ex Oliv.
1177.	35:II.3,4	*Silene vulgaris* (Moench) Garcke
1178.	35:II.1,2	*Silene sp.*
1179.	36:I.3,4	*Dactyloctenium aristatum* Link
1180.	36:I.5,6	*Desmostachya bipinnata* (L.) Stapf
1181.	36:I.7,8	*Desmostachya bipinnata* (L.) Stapf
1182.	36:II.1,2	*Desmostachya bipinnata* (L.) Stapf
1183.	36:II.5,6	*Cynosurus echinatus* L.
1184.	126:III.6,7	*Eleusine floccifolia* (Forssk.) Spreng.
1185.	36:III.3,4	*Aeluropus lagopoides* (L.) Thwait.
1186.	126:III.5	*Cynosurus ternatus* Forssk.
1187.	no microf.	*Cynosurus ternatus* Forssk.
1188.	no microf.	*Cyperus cruentus* Rottb.
1189.	123:III.7,8	*Cyperus effusus* Rottb.
1190.	124:I.1,2	*Cyperus effusus* Rottb.
1191.	37:I.1,2	*Cyperus fuscus* L.
1192.	37:I.3,4	*Cyperus fuscus* L.
1193.	37:I.5,6	*Cyperus fuscus* L.

1194.	123:III.1,2	*Cyperus fuscus* L. mixed with *Fimbristylis bisumbellata* & *Cyperus laevigatus*
1195.	no microf.	*Cyperus fuscus* L.
1196.	123:III.3,4	*Cyperus conglomeratus* Rottb.
1197.	123:III.5,6	*Cyperus conglomeratus* Rottb.
1198.	36:III.7,8	*Cyperus rotundus* L.
1199.	123:II.5,6	*Cyperus rotundus* L.
1200.	123:II.7,8	*Cyperus rotundus* L.
1201.	124:I.3,4	*Cyperus longus* L.
1202.	124:I.5,6	*Cyperus longus* L.
1203.	no microf.	*Cyperus longus* L.
1204.	37:I.7,8	*Cyperus laevigatus* L.
1205.	37:II.1,2	*Cyperus laevigatus* L.
1206.	37:II.3,4	*Cyperus articulatus* L.
1207.	no microf.	*Cyperus sp. indet.*
1208.	126:III.1,2	*Dactylis glomerata* L.
1209.	37:II.5,6	*Aeluropus lagopoides* (L.) Thwait.
1210.	37:II.7,8	*Aeluropus lagopoides* (L.) Thwait.
1211.	37:III.1,2	*Aeluropus lagopoides* (L.) Thwait.
1212.	37:III.3,4	*Aeluropus lagopoides* (L.) Thwait.
1213.	126:III.3	*Dactyloctenium aegyptiacum* (L.) Willd.
1214.	126:III.4	*Dactyloctenium aegyptiacum* (L.) Willd.
1215.	44:III.1,2	*Leptochloa fusca* (L.) Kunth
1216.	136:I.1,2	*Datura metel* L.
1217.	136:I.3,4	*Datura metel* L.
1218.	38:I.1,2	*Delphinium peregrinum* L.
1219.	132:I.5,6	*Desmodium triflorum* (L.) DC.
1220.	38:III.1,2	*Justicia odora* (Forssk.) Vahl
1221.	38:III.5,6	*Peristrophe paniculata* (Forssk.) Brummitt
1222.	38:III.7,8	*Anisotes trisulcus* (Forssk.) Nees
1223.	71:I.1,2	*Anisotes trisulcus* (Forssk.) Nees
1224.	39:I.5,6	*Dianthus caryophyllus* L.
1225.	118:I.1,2	*Dianthus caryophyllus* L.
1226.	39:II.5,6	*Dianthus uniflorus* Forssk.
1227.	no microf.	*Dicotyledons gen. et sp. indet.*
1228.	39:III.3,4	*Digera muricata* (L.) Mart.
1229.	39:III.5,6	*Digera muricata* (L.) Mart.
1230.	40:I.3,4	*Sesbania grandiflora* (L.) Poir.
1231.	39:III.7,8	*Sesbania sesban* (L.) Merr.
1232.	40:I.1,2	*Sesbania sesban* (L.) Merr.
1233.	40:I.5,6	*Crotalaria retusa* L.
1234.	40:I.7,8	*Vigna luteola* (Jacq.) Benth.
1235.	129:II.3,4	*Dracaena sp. indet.*
1236.	no microf.	*Echinochloa crus-galli* (L.) P. Beauv.
1237.	no microf.	*Echinophora tenuifolia* L.
1238.	40:II.5,6	*Echinops spinosissimus* Turra
1239.	no microf.	*Echinops microcephalus* Sm.
1240.	117:II.1,2	*Echium glomeratum* Poir.
1241.	41:I.1,2	*Echium rubrum* Forssk.
1242.	124:II.5	*Elaeagnus angustifolius* L.
1243.	124:I.7,8	*Eleocharis geniculata* (L.) Roem. & Schult.
1244.	133:II.7,8	*Emex spinosa* (L.) Campd.
1245.	41:II.7,8	*Ephedra campylopoda* C.A. Mey.
1246.	41:III.1,2	*Ephedra campylopoda* C.A. Mey.

1247.	126:III.8	*Eragrostis ciliaris* (L.) R. Br.
1248.	82:II.1,2	*Eragrostis pilosa* (L.) P. Beauv.
1249.	124:II.6,7	*Euphorbia cuneata* Vahl
1250.	3:I.3,4	*Aegilops geniculata* Roth
1251.	43:III.1,2	*Fagonia arabica* L.
1252.	43:III.3,4	*Fagonia arabica* L.
1253.	135:II.5,6	*Fagonia arabica* L. mixed with *Ruta chalepensis* L.
1254.	43:III.5,6	*Fagonia cretica* L.
1255.	127:I.5,6	*Schismus arabicus* Nees mixed with *Schismus barbatus* (L.) Thell.
1256.	44:I.7,8	*Schismus arabicus* Nees
1257.	44:II.5,6	*Cutandia dichotoma* (Forssk.) Trabut
1258.	44:II.7,8	*Cutandia dichotoma* (Forssk.) Trabut mixed with *Cutandia memphitica* (Spreng.) Benth.
1259.	44:II.1,2	*Cutandia memphitica* (Spreng.) Benth.
1260.	no microf.	*Vulpia fasciculata* (Forssk.) Fritch
1261.	127:I.7,8	*Cutandia maritima* (L.) Barbey
1262.	no microf.	*Odyssea mucronata* (Forssk.) Stapf
1263.	44:III.8 45:I.1,2	*Ficus cordata* Thunb. subsp. *salicifolia* (Vahl) C.C. Berg
1264.	45:III.7,8	*Frankenia revoluta* Forssk.
1265.	46:I.1,2	*Frankenia revoluta* Forssk.
1266.	42:III.3,4	*Euphorbia indica* Lam.
1267.	42:III.5,6	*Euphorbia indica* Lam.
1268.	42:I.7,8	*Euphorbia peplis* L.
1269.	42:II.1,2	*Euphorbia peplis* L.
1270.	42:II.3,4	*Euphorbia ? platyphyllos* L.
1271.	no microf.	*Euphorbia granulata* Forssk. var. *glabrata* (Gay) Boiss.
1272.	42:III.7,8	*Euphorbia terracina* L.
1273.	43:I.1,2	*Euphorbia peplus* L.
1274.	no microf.	*Euphorbia peplus* L.
1275.	43:I.3,4	*Euphorbia falcata* L.
1276.	42:II.5,6	*Euphorbia granulata* Forssk. var. *granulata* (Gay) Boiss.
1277.	42:II.7,8	*Euphorbia granulata* Forssk. var. *granulata* (Gay) Boiss.
1278.	42:III.1,2	*Euphorbia granulata* Forssk. var. *granulata* (Gay) Boiss.
1279.	43:I.5,6	*Euphorbia retusa* Forssk.
1280.	43:I.7,8	*Euphorbia retusa* Forssk.
1281.	43:II.1,2	*Euphorbia retusa* Forssk.
1282.	43:II.3,4	*Euphorbia retusa* Forssk.
1283.	no microf.	*Euphorbia schimperi* Presl.
1284.	no microf.	*Euphorbia schimperi* Presl.
1285.	no microf.	*Euphorbia scordifolia* Jacq.
1286.	no microf.	*Euphorbia thymifolia* Burm.
1287.	43:II.5,6	*Euphorbia triaculeata* Forssk.
1288.	49:I.7,8	*Galium album* Miller subsp. *pycnotrichum* (H. Braun) Krendl
1289.	49:II.1,2	*Galium heldreichii* Halacsy
1290.	135:I.5,6	*Galium aparine* L.

1291.	49:II.5,6	*Galium aparinoides* Forssk.
1292.	49:II.3,4	*Galium aparinoides* Forssk.
1293.	no microf.	*Galium glaucum* L.
1294.	49:III.1,2	*Galium paschale* Forssk.
1295.	no microf.	*Galium rubioides* L.
1296.	no microf.	*Galium rubioides* L.
1297.	no microf.	*Galium tricornutum* Dandy
1298.	135:I.3,4	*Galium / Asperula* sp.
1299.	50:II.5,6	*Erodium glaucophyllum* (L.) L'Hérit.
1300.	51:II.8 & 51:III.1	*Glinus setiflorus* Forssk.
1301.	51:II.3	*Glinus setiflorus* Forssk.
1302.	40:II.1,2	*Neonotonia wightii* (Wight & Arn.) Lackey var. *longicauda* (Schweinf.) Lackey
1303.	40:II.3,4	*Neonotonia wightii* (Wight & Arn.) Lackey var. *longicauda* (Schweinf.) Lackey
1304.	51:III.2,3	*Caesalpinia bonduc* (L.) Roxb.
1305.	51:III.6,7	*Gnaphalium crispatulum* Del.
1306.	51:III.4,5	*Helichrysum forskahlii* (Gmel.) Hilliard & Burtt
1307.	51:III.8 & 52:I.1,2	*Pseudognaphalium luteo-album* (L.) Hilliard & Burtt
1308.	120:II.7,8	*Gnaphalium pulvinatum* Del.
1309.	120:III.1,2	*Gnaphalium pulvinatum* Del.
1310.	no microf.	*Phagnalon sordidum* (L.) Reichb.
1311.	138:I.1,2	*Grewia sp. indet.*
1312.	128:III.7,8	*Gymnocarpus decandrum* Forssk.
1313.	52:II.1,2	*Hedera helix* L.
1314.	52:II.3,4	*Kohautia caespitosa* Schnizl.
1315.	52:II.5,6	*Kohautia caespitosa* Schnizl.
1316.	52:II.7,8	*Kohautia caespitosa* Schnizl.
1317.	52:III.1,2	*Alhagi maurorum* Medik.
1318.	52:III.3,4	*Taverniera lappacea* (Forssk.) DC.
1319.	52:III.5,6	*Taverniera lappacea* (Forssk.) DC.
1320.	Vahl 36:II.6,7	*Desmodium repandum* (Vahl) DC.
1321.	52:III.7,8	*Alysicarpus glumaceus* (Vahl) DC.
1322.	53:I.1,2	*Alysicarpus glumaceus* (Vahl) DC.
1323.	53:I.3,4	*Alysicarpus glumaceus* (Vahl) DC.
1324.	53:I.5,6	*Alysicarpus glumaceus* (Vahl) DC.
1325.	53:I.7,8	*Sophora alopecuroides* L.
1326.	no microf.	*Helichrysum schimperi* (A. Rich.) Mocscr
1327.	no microf.	*Helictotrichon bromoides* (Gouan) C.E. Hubb.
1328.	14:II.5,6	*Heictotrichon sp.*
1329.	53:II.1,2	*Heliotropium bacciferum* Forssk.
1330.	53:II.3,4	*Helitropium longiflorum* (A. DC.) DC.
1331.	53:II.7,8	*Herniaria hirsuta* L.
1332.	53:III.1,2	*Herniaria hirsuta* L.
1333.	53:III.3,4	*Herniaria hirsuta* L.
1334.	53:III.5,6	*Paronychia desertorum* Boiss.
1335.	54:I.1,2	*Diplotaxis acris* (Forssk.) Boiss.
1336.	130:I.7,8	*Hibiscus deflersii* Cuf.
1337.	132:II.3,4	*Hildebrandtia africana* Vatke
1338.	no microf.	*Hippocrepis unisiliquosa* L.
1339.	49:III.3,4	*Retama retam* (Forssk.) Webb
1340.	49:III.5,6	*Retama retam* (Forssk.) Webb
1341.	no microf.	*Retama retam* (Forssk.) Webb

1342. 133:I.1,2 *Retama retam* (Forssk.) Webb
1343. 55:I.7,8 *Sorghum halepense* (L.) Pers.
1344. 127:II.5,6 *Hordeum marinum* Huds. subsp. *marinum*
1345. 127:II.7,8 *Hordeum marinum* Huds. subsp. *gussoneanum* (Parl.) Thell.
1346. 127:III.1,2 *Hordeum murinum* L. subsp. *glaucum* (Steud.) Tzveler
1347. 55:III.1,2 *Hyoscyamus muticus* L.
1348. 121:I.1,2 *Hyoseris lucida* L.
1349. 55:III.5,6 *Moringa peregrina* (Forssk.) Fiori
1350. 56:I.1,2 *Moringa peregrina* (Forssk.) Fiori
1351. no microf. *Moringa peregrina* (Forssk.) Fiori
1352. no microf. *Hypericum perforatum* L.
1353. no microf. *Hypericum perforatum* L.
1354. 56:II.5,6 *Indigofera semitrijuga* Forssk.
1355. 56:II.7,8 *Indigofera semitrijuga* Forssk.
1356. 56:III.1,2 *Indigofera spicata* Forssk.
1357. 56:III.3,4 *Indigofera spinosa* Forssk.
1358. 56:III.5,6 *Indigofera spinosa* Forssk.
1359. 56:III.7,8 *Indigofera articulata* Gouan
1360. 57:I.1,2 *Pulicaria arabica* L.
1361. 122:III.1,2 *Quamoclit coccinea* (L.) Moench
1362. 129:I.1,2 *Gynandriris sisyrinchium* (L.) Parl.
1363. 57:III.3,4 *Cakile maritima* Scop. subsp. *aegyptiaca* (Willd.) Nyman
1364. 57:III.7,8 *Cakile maritima* Scop. subsp. *aegyptiaca* (Willd.) Nyman
1365. 58:I.1,2 *Cakile maritima* Scop. subsp. *aegyptiaca* (Willd.) Nyman
1366. 58:I.3,4 *Cakile maritima* Scop. subsp. *aegyptiaca* (Willd.) Nyman
1367. 58:I.5,6 *Cakile maritima* Scop. subsp. *aegyptiaca* (Willd.) Nyman
1368. 133:I.7,8 *Jasminum officinale* L.
1369. 124:III.4,5 *Jatropha curcas* L.
1370. 34:II.2,3 *Jatropha glauca* Vahl
1371. 34:III.6,7 *Jatropha pelargoniifolia* Courbon
1372. no microf. *Juncus bufonius* L.
1373. no microf. *Juncus effusus* L.
1374. 129:I.5,6 *Juncus striatus* E.H.F. Meyer
1375. 58:III.7,8 *Corchorus depressus* (L.) C. Chr.
1376. 59:I.3,4 *Barleria prionitis* L. subsp. *appressa* (Forssk.) Brummitt & Wood
1377. 59:II.7,8 *Barleria prionitis* L. subsp. *appressa* (Forssk.) Brummitt & Wood
1378. 59:II.5,6 *Barleria bispinosa* (Forssk.) Vahl
1379. 60:I.7,8 *Barleria bispinosa* (Forssk.) Vahl
1380. 60:II.1,2 *Barleria bispinosa* (Forssk.) Vahl
1381. 60:I.3,4 *Hypoestes triflora* (Forssk.) Roem. & Schult.
1382. 60:I.5,6 *Hypoestes triflora* (Forssk.) Roem. & Schult.
1383. 38:II.5,6 *Megalochlamys violaceum* (Vahl) Vollesen
1384. 60:II.7,8 *Ethulia conyzoides* L.
1385. 60:III.1,2 *Ethulia conyzoides* L.
1386. 60:III.3,4 *Ethulia conyzoides* L.
1387. 60:III.5,6 *Ethulia conyzoides* L.

1388. 61:I.1,2 *Scariola viminea* (L.) F.W. Schmidt
1389. 61:I.3,4 *Lactuca serriola* L.
1390. no microf. *Lagurus ovatus* L.
1391. 127:III.3 *Lamarckia aurea* (L.) Moench
1392. 127:III.8 *Lamarckia aurea* (L.) Moench
1393. 61:II.1,2 *Hyoseris lucida* L.
1394. 61:II.3,4 *Hyoseris lucida* L.
1395. 61:II.7,8 *Orobanche aegyptiaca* Pers.
1396. 62:I.7,8 *Launaea sp.*
1397. 121:III.5,6 *Zollikoferia pumila* DC.
1398. 62:I.3,4 *Picris asplenioides* L.
1399. 62:II.1,2 *Picris asplenioides* L.
1400. 120:III.7,8 *Leontodon hispidulus* (Del.) Boiss.
1401. no microf. *Launaea mucronata* (Forssk.) Muschl.
1402. no microf. *Launaea mucronata* (Forssk.) Muschl.
1403. 121:I.5,6 *Leontodon tuberosum* L.
1404. 61:III.5,6 *Cocculus pendulus* (J.R. & G. Forst.) Diels
1405. 62:I.1,2 *Cocculus pendulus* (J.R. & G. Forst.) Diels
1406. no microf. *Taraxacum megalorrhizon* (Forssk.) Hand.-Mazz.
1407. 62:II.7,8 *Coronopus squamatus* (Forssk.) Asch.
1408. 62:III.3,4 *Limeum humile* Forssk.
1409. 62:III.5,6 *Limeum humile* Forssk.
1410. 62:III.7,8 *Limeum humile* Forssk.
1411. 133:II.1,2 *Limonium ferulaceum* (L.) O. Kuntze
1412. 63:I.3,4 *Moltkiopsis ciliata* (Forssk.) I.M. Johnst.
1413. 117:III.3,4 *Moltkiopsis ciliata* (Forssk.) I.M. Johnst.
1414. 63:II.1,2 *Moltkiopsis ciliata* (Forssk.) I.M. Johnst.
1415. no microf. *Heliotropium digynum* (Forssk.) C. Chr.
1416. 63:II.5,6 *Heliotropium bacciferum* Forssk.
1417. 63:II.7,8 *Heliotropium bacciferum* Forssk.
1418. 63:III.1,2 *Heliotropium bacciferum* Forssk.
1419. 63:III.3,4 *Heliotropium bacciferum* Forssk.
1420. 63:III.5,6 *Heliotropium bacciferum* Forssk.
1421. 63:III.7,8 *Lolium ? multiflorum* Lam., mixed with *L. perenne* L.
1422. 64:I.3,4 *Dorycnium graecum* (L.) Ser.
1423. 64:I.5,6 *Dorycnium graecum* (L.) Ser.
1424. 64:I.7,8 *Lotus polyphyllus* Clarke
1425. 64:II.1,2 *Lotus polyphyllus* Clarke
1426. 64:II.3,4 *Lotus polyphyllus* Clarke
1427. 132:III.5,6 *Lotus creticus* L.
1428. 64:II.5,6 *Lotus corniculatus* L. var. *ciliatus* Koch
1429. 64:II.7,8 *Lotus arabicus* L.
1430. 64:III.1,2 *Lotus arabicus* L.
1431. no microf. *Lotus corniculatus* L. var. *tenuifolius* L.
1432. 64:III.3,4 65(1-2):I.1,2 *Lotus halophilus* Boiss. & Sprun.
1433. 64:III.5,6 65(1-2):I.3,4 *Lotus halophilus* Boiss. & Sprun.
1434. 132:III.3,4 *Lotus sp.*
1435. 65(1-2):I.7,8 *Farsetia aegyptia* Turra
1436. 64:III.7,8 65(1-2):II.3,4 *Farsetia aegyptia* Turra
1437. 65(1-2):II.1,2 *Farsetia aegyptia* Turra

1438.	65(1-2):II.7,8	*Lupinus digitatus* Forssk.
1439.	65(1-2):II.5,6	*Lupinus albus* L.
1440.	no microf.	*Lupinus albus* L.
1441.	65(1-2):III.3,4	*Lycium europaeum* L.
1442.	129:III.7,8	*Lycopodiella cernua* (L.) Pic. Ser.
1443.	65(1-2):III.5,6	*Selaginella imbricata* (Forssk.) Decne.
1444.	no microf.	*Selaginella yemensis* (Sw.) Decne.
1445.	no microf.	*Selaginella yemensis* (Sw.) Decne.
1446.	135:III.7,8	*Lysimachia nummularia* L.
1447.	65(2-2):I.7	*Maesa lanceolata* Forssk.
1448.	65(2-2):II.6	*Malva parviflora* L.
1449.	66:I.7,8	*Marrubium peregrinum* L.
1450.	no microf.	*Marrubium vulgare* L.
1451.	66:III.1,2	*Tanacetum parthenium* (L.) Sch.
1452.	66:III.3,4	*Tanacetum parthenium* (L.) Sch.
1453.	67:I.3,4	*Medicago littoralis* Loisel.
1454.	66:III.5,6	*Medicago marina* L.
1455.	no microf.	*Medicago marina* L.
1456.	66:III.7,8	*Medicago polymorpha* L.
1457.	133:I.3,4	*Melia azederach* L.
1458.	no microf.	*Melilotus alba* Medik.
1459.	67:II.1	*Salvia aegyptiaca* L.
1460.	67:II.2,3	*Salvia aegyptiaca* L.
1461.	67:II.4,5	*Salvia aegyptiaca* L.
1462.	22:III.7,8	*Cocculus pendulus* (J.R. & G. Forst.) Diels
1463.	130:II.6	*Mentha aquatica* L.
1464.	130:II.5	*Mentha pulegium* L.
1465.	no microf.	*Mentha verticillata* L.
1466.	67:II.6	*Mentha sp.*
1467.	67:III.5,6	*Mesembryanthemum nodiflorum* L.
1468.	68:I.7,8	*Mesembryanthemum nodiflorum* L.
1469.	127:III.6,7	*Mibora minima* (L.) Desv.
1470.	68:II.1,2	*Eclipta prostrata* (L.) L.
1471.	68:II.3,4	*Albizia julibrissin* Durazz.
1472.	68:II.5,6	*Albizia julibrissin* Durazz.
1473.	68:II.7,8	*Acacia asak* (Forssk.) Willd.
1474.	68:III.1,2	*Acacia hockii* De Wild
1475.	68:III.3,4	*Albizia lebbeck* (L.) Benth.
1476.	69:III.7,8	*Acacia mellifera* (Vahl) Benth.
1477.	70:I.1,2	*Acacia mellifera* (Vahl) Benth.
1478.	68:III.5,6	*Acacia nilotica* (L.) Del.
1479.	68:III.7,8	*Acacia nilotica* (L.) Del.
1480.	69:I.3,4	*Acacia farnesiana* (L.) Willd.
1481.	69:I.5,6	*Acacia farnesiana* (L.) Willd.
1482.	69:I.7,8	*Acacia farnesiana* (L.) Willd.
1483.	69:II.1,2	*Acacia farnesiana* (L.) Willd.
1484.	69:II.3,4	*Acacia hamulosa* Benth.
1485.	69:II.5,6	*Acacia hamulosa* Benth.
1486.	69:II.7,8	*Pterolobium stellatum* (Forssk.) Brenan
1487.	69:III.1,2	*Pterolobium stellatum* (Forssk.) Brenan
1488.	69:III.3,4	*Acacia tortilis* (Forssk.) Hayne
1489.	69:III.5,6	*Acacia tortilis* (Forssk.) Hayne
1490.	69:I.1,2	*Acacia sp.*
1491.	70:I.7,8	*Hypecoum aegyptiacum* (Forssk.) Asch. & Schweinf.

1492.	70:II.3,4	*Ballota acetabulosa* (L.) Benth.
1493.	70:III.1,2	*Ajuga iva* (L.) Schreb.
1494.	71:II.3,4	*Neurada procumbens* L.
1495.	134:III.1,2	*Neurada procumbens* L.
1496.	137:I.3,4	*Nicotiana tabacum* L.
1497.	119:II.1,2	*Notobasis syriaca* (L.) Cass.
1498.	125:II.1,2	*Nymphoides cristata* (Griseb.) O. Kuntze
1499.	125:II.3,4	*Nymphoides cristata* (Griseb.) O. Kuntze
1500.	71:III.6,7	*Ocimum forskolei* Benth.
1501.	71:III.8	*Ocimum forskolei* Benth.
1502.	72:I.1,2	*Becium serpyllifolium* (Forssk.) Wood
1503.	no microf.	*Becium serpyllifolium* (Forssk.) Wood
1504.	72:II.7,8	*Ononis spinosa* L.
1505.	72:III.1,2	*Ononis spinosa* L.
1506.	72:III.3,4	*Ononis spinosa* L.
1507.	72:III.5,6	*Ononis spinosa* L.
1508.	72:III.7,8	*Ononis spinosa* L.
1509.	73:I.8-II.1	*Rhynchosia flava* (Forssk.) Thulin
1510.	73:II.2,3	*Rhynchosia flava* (Forssk.) Thulin
1511.	73:II.4,5	*Ononis minutissima* L.
1512.	73:II.6,7	*Ononis minutissima* L.
1513.	73:II.8-III.1	*Ononis natrix* L.
1514.	73:III.2,3	*Ononis natrix* L.
1515.	119:II.3,4	*Onopordum tauricum* Willd.
1516.	73:III.4,5	*Lotus quinatus* (Forssk.) Gillett
1517.	73:III.6,7	*Lotus quinatus* (Forssk.) Gillett
1518.	73:III.8	*Lotus quinatus* (Forssk.) Gillett
1519.	74:I.3,4	*Ononis serrata* Forssk.
1520.	74:I.1,2	*Ononis serrata* Forssk.
1521.	73:I.1,2	*Ononis vaginalis* Vahl
1522.	73:I.3,4	*Ononis vaginalis* Vahl
1523.	73:I.5,6	*Ononis vaginalis* Vahl
1524.	73:I.7	*Ononis vaginalis* Vahl
1525.	74:I.5,6	*Pentas lanceolata* (Forssk.) Defl.
1526.	74:I.7,8	*Pentas lanceolata* (Forssk.) Defl.
1527.	no microf.	*Pentas lanceolata* (Forssk.) Defl.
1528.	74:II.1,2	*Eulophia streptopetala* Lindl.
1529.	no microf.	*Origanum vulgare* L.
1530.	132:II.1,2	*Ormocarpum yemenense* Gillett
1531.	74:III.5,6	*Ornithopus compressus* L.
1532.	132:III.7,8	*Ornithopus sativus* Brot.
1533.	no microf.	*Orobanche crenata* Forssk.
1534.	no microf.	*Orobanche crenata* Forssk.
1535.	no microf.	*Orobanche crenata* Forssk.
1536.	75:I.1,2	*Orobanche ramosa* L.
1537.	75:I.3,4	*Cistanche phelypaea* (L.) Coutinho
1538.	120:I.6,7	*Osteospermum vaillantii* (Decne.) T. Norl.
1539.	no microf.	*Otanthus maritimus* (L.) Hoffm. & Link
1540.	no microf.	*Oxalis corniculata* L.
1541.	75:II.3,4	*Oxalis corniculata* L.
1542.	no microf.	*Pancratium maritimum* L.
1543.	116:III.6,7	*Pancratium maritimum* L.
1544.	75:III.3,4	*Cynodon dactylon* (L.) Pers.
1545.	no microf.	*Cynodon dactylon* (L.) Pers.
1546.	126:II.5,6	*Cynodon dactylon* (L.) Pers.

1547.	126:II.7,8	*Cynodon dactylon* (L.) Pers.
1548.	75:III.5,6	*Pennisetum divisum* (J. F. Gmel.) Henr.
1549.	75:III.7,8	*Pennisetum divisum* (J. F. Gmel.) Henr.
1550.	76:I.1,2	*Pennisetum divisum* (J. F. Gmel.) Henr.
1551.	76:I.7,8	*Panicum repens* L.
1552.	no microf.	*Panicum miliaceum* L.
1553.	76:II.1,2	*Panicum sp.*
1554.	76:III.5,6	*Panicum turgidum* Forssk.
1555.	127:III.4,5	*Parapholis strigosa* (Dumort.) C. E. Hubb.
1556.	77:II.3,4	*Swertia polynectaria* (Forssk.) Gilg
1557.	no microf.	*Swertia polynectaria* (Forssk.) Gilg
1558.	77:II.7,8	*Thymelaea hirsuta* (L.) Endl.
1559.	58:I.7,8	*Pavetta longiflora* Vahl
1560.	135:II.1,2	*Pavetta villosa* Vahl
1561.	135:II.3,4	*Pavetta villosa* Vahl
1562.	no microf.	*Nitraria retusa* (Forssk.) Aschers.
1563.	77:III.7,8	*Lythrum hyssopifolium* L.
1564.	78:I.1,2	*Lythrum hyssopifolium* L.
1565.	78:I.3,4	*Lythrum hyssopifolium* L.
1566.	78:II.7,8	*Polypogon viridis* (Gouan) Breistr.
1567.	79:I.5,6	*Mollugo cerviana* (L.) Ser.
1568.	79:I.7,8	*Vigna aconitifolia* (Jacq.) Marechal
1569.	79:II.1,2	*Vigna aconitifolia* (Jacq.) Marechal
1570.	79:II.3,4	*Vigna radiata* (L.) Wilczek
1571.	no microf.	*Leucas alba* (Forssk.) Sebald
1572.	79:III.3,4	*Phlomis herba-venti* L.
1573.	130:II.7,8	*Leucas urticifolia* (Vahl) R. Br.
1574.	80:I.1,2	*Phyllanthus tenellus* Muell.Arg. var. *arabicus* Muell. Arg.
1575.	80:I.3,4	*Phyllanthus tenellus* Muell.Arg. var. *arabicus* Muell. Arg.
1576.	136:I.7,8	*Physalis alkekengi* L.
1577.	80:II.1,2	*Withania somnifera* (L.) Dunal
1578.	no microf.	*Withania somnifera* (L.) Dunal
1579.	no microf.	*Phytolacca americana* L.
1580.	120:I.3	*Picris sulphurea* Delile
1581.	62:I.5,6	*Picris cf. asplenioides* L.
1582.	no microf.	*Pimpinella menachensis* Wolff
1583.	80:III.3,4	*Plantago coronopus* L.
1584.	no microf.	*Plantago cylindrica* Forssk.
1585.	138:I.3	*Plantago lagopus* L.
1586.	109:II.1,2	*Elymus sp.*
1587.	82:II.5,6	*Delonix elata* (L.) Gamble
1588.	82:II.7,8	*Sphaeranthus suaveolens* (Forssk.) DC.
1589.	82:III.1,2	*Sphaeranthus suaveolens* (Forssk.) DC.
1590.	82:III.3,4	*Sphaeranthus suaveolens* (Forssk.) DC.
1591.	82:III.5,6	*Sphaeranthus suaveolens* (Forssk.) DC.
1592.	83:I.5,6	*Polygala sp.*
1593.	83:I.1,2	*Polygala tinctoria* Vahl
1594.	83:III.4,5	*Polygonum salicifolium* Willd.
1595.	83:II.1,2,3	*Polygonum maritimum* L.
1596.	83:II.4,5	*Polygonum maritimum* L.
1597.	83:II.6,7	*Polygonum equisetiforme* Sibth. & Sm.
1598.	83:III.2,3	*Polygonum equisetiforme* Sibth. & Sm.
1599.	83:II.8-III.1	*Polygonum equisetiforme* Sibth. & Sm.

1600.	83:I.7,8	*Polygonum pulchellum* Lois.
1601.	no microf.	*Potentilla supina* L.
1602.	84:II.7,8	*Potentilla dentata* Forssk.
1603.	84:III.5,6	*Primula verticillata* Forssk.
1604.	134:III.3,4	*Prunus spinosa* L.
1605.	85:I.3,4	*Pteranthus dichotomus* Forssk.
1606.	85:I.5,6	*Pteranthus dichotomus* Forssk.
1607.	85:I.7,8	*Pteranthus dichotomus* Forssk.
1608.	85:II.1,2	*Pteranthus dichotomus* Forssk.
1609.	120:III.5,6	*Pulicaria grandidentata* Jaub. & Spach
1610.	86:III.1,2	*Reamuria hirtella* Jaub. & Spach
1611.	no microf.	*Reichardia tingitana* (L.) Roth
1612.	no microf.	*Reseda alba* L.
1613.	134:II.7,8	*Reseda alba* L.
1614.	134:II.5,6	*Reseda undata* L. (no. 1 & 3), mixed with *R. decursiva* Forssk. (no. 2)
1615.	134:II.3,4	*Reseda decursiva* Forssk. (no. 1), mixed with *R. lutea* L. (no. 2)
1616.	no microf.	*Reseda decursiva* Forssk.
1617.	86:III.7,8	*Caylusea hexagyna* (Forssk.) M.L.Green
1618.	87:I.1,2	*Caylusea hexagyna* (Forssk.) M.L.Green
1619.	134:II.1,2	*Caylusea hexagyna* (Forssk.) M.L.Green
1620.	87:I.5,6	*Reseda luteola* L.
1621.	87:II.1,2	*Reseda arabica* Boiss.
1622.	no microf.	*Rhynchosia malacophylla* (Sprengel) Bojer
1623.	131:III.7,8	*Vigna variegata* Defl.
1624.	no microf.	*Rhynchosia minima* (L.) DC.
1625.	no microf.	*Rhynchosia minima* (L.) DC.
1626.	131:III.3,4	*Rhynchosia minima* (L.) DC.
1627.	131:III.5,6	*Rhynchosia minima* (L.) DC.
1628.	no microf.	*Vigna variegata Defl.*
1629.	87:III.5,6	*Gypsophila capillaris* (Forssk.) C. Chr.
1630.	87:III.7,8	*Cleome droserifolia* (Forssk.) Del.
1631.	88:I.5,6	*Rosa andegavensis* Bast.
1632.	88:I.7,8	*Rosa andegavensis* Bast.
1633.	128:I.3,4	*Rostraria cristata* (L.) Tzvelev
1634.	128:I.1,2	*Rostraria cristata* (L.) Tzvelev
1635.	88:II.1,2	*Rubia peregrina* L.
1636.	no microf.	*Rubus sanguineus* Friv.
1637.	130:III.1,2	*Crossandra johanninae* Fiori
1638.	88:III.3,4	*Ruellia patula* Jacq.
1639.	89:I.1,2	*Rumex nervosus* Vahl
1640.	89:I.3,4	*Rumex nervosus* Vahl
1641.	89:I.5,6	*Rumex nervosus* Vahl
1642.	88:III.5,6	*Rubus aegyptiacus* L.
1643.	88:III.7,8	*Rumex dentatus* L.
1644.	89:I.7,8	*Rumex pictus* Forssk.
1645.	89:II.3,4	*Rumex vesicarius* L.
1646.	89:II.1,2	*Rumex vesicarius* L.
1647.	135:II.7,8	*Ruta chalepensis* L.
1648.	89:II.5,6	*Haplophyllum buxbaumii* (Poir.) G. Don
1649.	89:III.3,4	*Haplophyllum tuberculatum* (Forssk.) A. Juss.
1650.	117:I.7,8	*Haplophyllum tuberculatum* (Forssk.) A. Juss.
1651.	135:III.1,2	*Haplophyllum tuberculatum* (Forssk.) A. Juss.
1652.	90:I.3,4	*Lasiurus scindicus* Henrard

1653.	90:I.5,6	*Lasiurus scindicus* Henrard
1654.	90:I.7	*Lasiurus scindicus* Henrard
1655.	90:I.8-II.1	*Lasiurus scindicus* Henrard
1656.	90:II.2,3	*Lasiurus scindicus* Henrard
1657.	90:II.4,5	*Lasiurus scindicus* Henrard
1658.	90:II.6,7	*Lasiurus scindicus* Henrard
1659.	no microf.	*Lasiurus scindicus* Henrard
1660.	no microf.	*Lasiurus scindicus* Henrard
1661.	90:III.4,5	*Halocnemum strobilaceum* (Pallas) M. Bieb.
1662.	91:II.1,2	*Halocnemum strobilaceum* (Pallas) M. Bieb.
1663.	91:II.7,8	*Halocnemum strobilaceum* (Pallas) M. Bieb.
1664.	92:I.3,4	*Anabasis articulata* (Forssk.) Moq.
1665.	92:I.5,6	*Anabasis articulata* (Forssk.) Moq.
1666.	92:I.7,8	*Anabasis articulata* (Forssk.) Moq.
1667.	92:II.5,6	*Salsola volkensii* Asch. & Schweinf.
1668.	92:III.7,8	*Salsola inermis* Forssk.
1669.	93:I.1,2	*Salsola inermis* Forssk.
1670.	93:I.5,6	*Salsola inermis* Forssk.
1671.	93:II.1,2	*Salsola kali* L.
1672.	93:II.5	*Salsola kali* L.
1673.	93:II.8	*Salsola longifolia* Forssk.
1674.	94:III.3,4	*Salsola longifolia* Forssk.
1675.	93:III.5,6	*Bassia muricata* (L.) Aschers.
1676.	93:III.7,8	*Bassia muricata* (L.) Aschers.
1677.	119:I.3,4	*Bassia muricata* (L.) Aschers.
1678.	94:I.1,2	*Noaea mucronata* (Forssk.) Ascers. & Schweinf.
1679.	94:II.7,8	*Salsola tetrandra* Forssk.
1680.	95:II.3,4	*Salvia aegyptiaca* L.
1681.	94:III.7,8	*Salvia forskahlei* L.
1682.	95:I.3,4	*Salvia forskahlei* L.
1683.	95:I.5,6	*Salvia forskahlei* L.
1684.	95:I.7,8	*Salvia forskahlei* L.
1685.	95:II.7,8	*Anthemis tinctoria* L.
1686.	95:III.1,2	*Achillea fragrantissima* (Forssk.) Sch.Bip.
1687.	95:III.3,4	*Achillea fragrantissima* (Forssk.) Sch.Bip.
1688.	95:III.5,6	*Achillea fragrantissima* (Forssk.) Sch.Bip.
1689.	95:III.7,8	*Anacyclus monanthos* (L.) Thell.
1690.	96:I.1,2	*Anacyclus monanthos* (L.) Thell.
1691.	96:I.3,4	*Anacyclus monanthos* (L.) Thell.
1692.	96:I.5	*Velezia rigida* L.
1693.	96:I.8	*Saponaria officinalis* L.
1694.	no microf.	*Scabiosa argentea* L.
1695.	no microf.	*Scabiosa atropurpurea* L.
1696.	129:II.5,6	*Scilla hyacinthina* (Roth) Macbride
1697.	no microf.	*Chlorophytum tetraphyllum* (L.f.) Bak.
1698.	52:I.7,8	*Chlorophytum tetraphyllum* (L.f.) Bak.
1699.	no microf.	*Scirpoides holoschoenus* (L.) Sojak
1700.	no microf.	*Scirpoides holoschoenus* (L.) Sojak
1701.	96:II.5,6	*Fimbristylis bis-umbellata* (Forssk.) Bub.
1702.	96:II.7,8	*Scirpus maritimus* L.
1703.	96:III.1,2	*Schoenoplectus articulatus* (L.) Palla
1704.	96:III.3,4	*Schoenoplectus articulatus* (L.) Palla
1705.	96:III.5,6	*Cyperus capitatus* Vand.
1706.	96:III.7,8	*Cyperus capitatus* Vand.
1707.	97:I.1,2	*Cyperus capitatus* Vand.

1708.	97:I.3,4	*Scoparia dulcis* L.
1709.	97:I.5,6	*Scoparia dulcis* L.
1710.	no microf.	*Launaea nudicaulis* (L.) Hook. f.
1711.	no microf.	*Launaea nudicaulis* (L.) Hook. f.
1712.	97:I.7,8	*Scorpiurus muricatus* L.
1713.	97:II.5,6	*Dasypyrum villosum* (L.) Cand.
1714.	97:II.7,8	*Senecio hadiensis* Forssk.
1715.	97:III.1,2	*Senecio aegyptius* L.
1716.	97:III.3,4	*Senecio aegyptius* L.
1717.	97:III.5,6	*Senecio aegyptius* L.
1718.	119:III.3,4	*Senecio aegyptius* L.
1719.	97:III.7,8	*Senecio aquaticus* L.
1720.	98:I.1,2	*Senecio lyratus* Forssk.
1721.	98:I.3,4	*Senecio lyratus* Forssk.
1722.	98:I.5,6	*Senecio glaucus* L.
1723.	98:I.7,8	*Inula critmoides* L.
1724.	98:II.1,2	*Centaurea spinosa* L.
1725.	123:II.1,2	*Sicyos angulatus* L.
1726.	98:III.5,6	*Abutilon bidentatum* A.Rich.
1727.	99:I.1,2	*Abutilon bidentatum* A.Rich.
1728.	99:I.3,4	*Sida ovata* Forssk.
1729.	99:I.7,8	*Sida ovata* Forssk.
1730.	99:II.1,2	*Sida ovata* Forssk.
1731.	118:II.3,4	*Silene colorata* Poir.
1732.	118:II.1,2	*Silene gallica* L., *S. nocturna* L. (mixed sheet)
1733.	118:I.7,8	*Agrostemma githago* L.
1734.	118:II.5,6	*Silene cf. italica* (L.) Pers.
1735.	117:III.7,8	*Cleome amblyocarpa* Barr. & Mart.
1736.	100:II.1,2	*Anarrhinum forskaohlii* (J. F. Gmel.) Cuf.
1737.	100:II.3,4	*Anarrhinum forskaohlii* (J. F. Gmel.) Cuf.
1738.	100:II.7	*Diplotaxis harra* (Forssk.) Boiss.
1739.	100:II.8	*Diplotaxis harra* (Forssk.) Boiss.
1740.	100:II.5,6	*Sinapis allionii* Jacq.
1741.	no microf.	*Sisymbrium irio* L.
1742.	100:III.3,4	*Capparis decidua* (Forssk.) Edgew.
1743.	100:III.7,8	*Capparis decidua* (Forssk.) Edgew.
1744.	101:III.1,2	*Solanum coagulans* Forssk.
1745.	102:I.5,6	*Solanum incanum* L.
1746.	136:III.7,8	*Solanum incanum* L.
1747.	102:I.7,8	*Solanum incanum* L.
1748.	136:III.3,4	*Solanum nigrum* L.
1749.	136:II.3,4	*Solanum nigrum* L.
1750.	136:II.1,2	*Solanum palmetorum* Dunal
1751.	101:III.7,8	*Solanum microcarpum* Vahl
1752.	136:II.5,6	*Solanum villosum* Miller
1753.	136:II.7,8	*Solanum villosum* Miller
1754.	102:III.3,4	*Sonchus oleraceus* L.
1755.	102:III.1,2	*Sonchus tenerrimus* L.
1756.	128:II.6	*Sphenopus divaricatus* (Gouan) Rchb.
1757.	no microf.	*Sphenopus divaricatus* (Gouan) Rchb.
1758.	128:II.5	*Sphenopus sp.*
1759.	no microf.	*Stachys obliqua* Waldst. & Kit
1760.	103:II.5,6	*Limonium pruinosum* (L.) Kuntze
1761.	103:III.5,6	*Limonium cylindrifolium* (Forssk.) Kuntze

1762.	103:III.7,8	*Limoniastrum monopetalum* (L.) Boiss.
1763.	104:II.7,8	*Sida alba* L.
1764.	104:III.1,2	*Sida alba* L.
1765.	8:III.5,6	*Stipagrostis plumosa* (L.) T. Anders.
1766.	9:I.3,4	*Stipagrostis plumosa* (L.) T. Anders.
1767.	9:I.5,6	*Stipagrostis plumosa* (L.) T. Anders.
1768.	9:I.7,8	*Stipagrostis plumosa* (L.) T. Anders.
1769.	9:II.1,2	*Stipagrostis plumosa* (L.) T. Anders.
1770.	9:II.3,4	*Stipagrostis plumosa* (L.) T. Anders.
1771.	105:II.7,8	*Schouwia purpurea* (Forssk.) Schweinf.
1772.	105:III.1,2	*Schouwia purpurea* (Forssk.) Schweinf.
1773.	121:III.3,4	*Synedrella nodiflora* (L.) Gaertn.
1774.	105:III.7,8	*Tamarix tetragyna* Ehrenb.
1775.	106:I.3,4	*Tamarix aphylla* (L.) Karsten
1776.	106:I.5,6	*Tamarix aphylla* (L.) Karsten
1777.	106:II.1,2	*Tamarix aphylla* (L.) Karsten
1778.	106:II.3,4	*Cotula anthemoides* L.
1779.	106:II.5,6	*Cotula anthemoides* L.
1780.	106:II.7,8	*Cotula anthemoides* L.
1781.	no microf.	*Teline monspessulanae* (L.) C. Koch
1782.	106:III.3,4	*Teucrium polium L.*
1783.	no microf.	*Teucrium polium* L.
1784.	no microf.	*Thymus vulgaris* L.
1785.	107:II.5,6	*Cyanotis sp.*
1786.	121:I.7,8	*Tolpis virgata* (Desf.) Bertol.
1787.	107:III.3,4	*Urospermum picroides* (L.) F.W. Schmidt
1788.	108:I.1,2	*Tribulus pentandrus* Forssk.
1789.	108:I.3,4	*Trifolium alexandrinum* L.
1790.	108:I.5,6	*Trifolium alexandrinum* L.
1791.	108:I.7,8	*Trifolium alexandrinum* L.
1792.	no microf.	*Trifolium alexandrinum* L.
1793.	108:II.1,2	*Trifolium resupinatum* L.
1794.	67:I.1,2	*Melilotus indica* L.
1795.	108:II.3,4	*Melilotus indica* L.
1796.	132:III.1,2	*Trifolium micranthum* Viv.
1797.	108:II.5,6	*Trigonella hamosa* L.
1798.	108:II.7,8	*Trigonella hamosa* L.
1799.	108:III.1,2	*Trigonella hamosa* L.
1800.	no microf.	*Trigonella procumbens* (Besser) Rchb.
1801.	108:III.3,4	*Trigonella stellata* Forssk.
1802.	108:III.5,6	*Trigonella stellata* Forssk.
1803.	108:III.7,8	*Trigonella stellata* Forssk.
1804.	109:I.1,2	*Trigonella stellata* Forssk.
1805.	109:I.3,4	*Trigonella stellata* Forssk.
1806.	109:I.5,6	*Trigonella stellata* Forssk.
1807.	128:III.3,4	*Triticum durum* Desf.
1808.	109:II.5,6	*Triumfetta rhomboidea* Jacq.
1809.	137:I.7,8	*Triumfetta rhomboidea* Jacq.
1810.	no microf.	*Ulex europaeus* L.
1811.	110:III.7,8	*Hibiscus ovalifolius* (Forssk.) Vahl
1812.	no microf.	*Urochondra setulosa* (Trin.) C.E. Hubb.
1813.	111:II.7	*Utricularia inflexa* Forssk.
1814.	79:III.5,6	*Priva adhaerens* (Forssk.) Chiov.
1815.	79:III.7,8	*Priva adhaerens* (Forssk.) Chiov.
1816.	111:II.3,4	*Valantia hispida* L.

1817.	111:II.5,6	*Valantia muralis* L.
1818.	111:III.4 & 112:I.1,2	*Commicarpus plumbaginea* (Cav.) Standl.
1819.	133:I.5,6	*Commicarpus plumbaginea* (Cav.) Standl.
1820.	112:I.5,6	*Verbascum pinnatifidum* Vahl
1821.	no microf.	*Verbena supina* L.
1822.	112:II.1,2	*Vicia peregrina* L.
1823.	no microf.	*Hybanthus enneaspermus* (L.) F.v.Muell.
1824.	no microf.	*Vitis vinifera* L.
1825.	112:III.3,4	*Cassytha filiformis* L.
1826.	113:I.1,2	*Zilla spinosa* (L.) Prantl
1827.	113:I.3,4	*Zilla spinosa* (L.) Prantl
1828.	113:III.2,3	*Zygophyllum simplex* L.
1829.	113:III.4,5	*Zygophyllum album* L.f.
1830.	113:III.6,7	*Zygophyllum album* L.f.
1831.	16:III.5,6	*Bromus rigidus* Roth
1832.	59:III.7,8	*Barleria lanceata* (Forssk.) C. Chr.
1833.	71:I.3,4	*Adenium obesum* (Forssk.) Roem. & Schult.
1834.	80:I.5,6	*Phyllanthus ovalifolius* Forssk.
1835.	124:III.6,7	*Phyllanthus ovalifolius* Forssk.
1836.	133:II.5,6	*Liminium sp.*
1837.	136:I.5,6	*Hyoscyamus muticus* L.
1838.	[not used]	
1839.	no microf.	*Centaurea calcitrapa* L.
1840.	no microf.	*Traganum nudatum* Del.
1841.	no microf.	*Teucrium scordidum* L. subsp. *scordioides* (Schreber) Maire & Petitmengin
1842.	no microf.	*Asperula arvensis* L.
1843.	no microf.	*Adiantum incisum* Forssk.
1844.	116:II.3,4	*Barleria proxima* Lindau
1845.	no microf.	*Zaleya pentandra* (L.) Jeffrey
1846.	3:II.5,6	*Aerva javanica* (Burm. f.) Juss.

Index to the IDC microfiche edition (IDC 2200) of the 'Herbarium Forsskalii'

1: I.1,2 — 902 *Acalypha ciliata* Forssk.
1: I.3,4 — 903 *Acalypha fruticosa* Forssk.
1: I.5,6 — 379 *Acanthus arboreus* Forssk.
1: I.7,8 — 904 *Acanthus arboreus* Forssk.
1: II.1,2 — 905 *Blepharis ciliaris* (L.) B.L. Burtt
1: II.3,4 — 907 *Blepharis maderaspatensis* (L.) Roth.
1: II.5,6 — 906 *Acanthus hirsutus* Boiss.
1: II.7,8 — 908 *Achillea santolina* L.
1: III.1,2 — 909 *Achillea santolina* L.
1: III.3,4 — 910 *Achyranthes aspera* L. var. *sicula* L.
1: III.5,6 — 202 *Celosia trigyna* L.
1: III.7,8 — 208 *Celosia trigyna* L.
2: I.1,2 — 204 *Celosia trigyna* L.
2: I.3,4 — 205 *Saltia papposa* (Forssk.) Moq.
2: I.5,6 — 206 *Saltia papposa* (Forssk.) Moq.
2: I.7,8 — 916 *Aerva javanica* (Burm. f.) Juss.
2: II.1,2 — 203 *Aerva lanata* (L.) Juss.
2: II.3,4 — 913 *Consolida aconiti* (L.) Lindl.
2: II.5,6 — 914 *Consolida aconiti* (L.) Lindl.
2: II.7,8 — 815 *Actiniopteris radiata* (Schwartz) Link
2: III.1,2 — 820 *Actiniopteris radiata* (Schwartz) Link
2: III.3,4 — 655 *Adenia venenata* Forssk.
2: III.5,6 — 805 *Adiantum capillus-veneris* L.
2: III.7,8 — 804 *Adiantum capillus-veneris* L.
3: I.1,2 — 813 *Adiantum incisum* Forssk.
3: I.3,4 — 1250 *Aegilops geniculata* Roth
3: I.5,6 — 197 *Aerva javanica* (Burm. f.) Juss.
3: I.7,8 — 195 *Aerva javanica* (Burm. f.) Juss.
3: II.1,2 — 196 *Aerva javanica* (Burm. f.) Juss.
3: II.3,4 — 917 *Aerva javanica* (Burm. f.) Juss.
3: II.5,6 — 846 *Aerva javanica* (Burm. f.) Juss.
3: II.7,8 — 193 *Aerva javanica* (Burm. f.) Juss.
3: III.1,2 — 194 *Aerva javanica* (Burm. f.) Juss.
3: III.3,4 — 918 *Aerva javanica* (Burm. f.) Juss.
3: III.5,6 — 915 *Aerva javanica* (Burm. f.) Juss.
3: III.7,8 — 79 *Sporobolus pyramidalis* P. Beauv.
4: I.1,2 — 130 *Sporobolus spicatus* (Vahl) Kunth
4: I.3,4 — 142 *Sporobolus spicatus* (Vahl) Kunth
4: I.5,6 — 129 *Sporobolus spicatus* (Vahl) Kunth
4: I.7,8 — 920 *Sporobolus spicatus* (Vahl) Kunth
4: II.1,2 — 921 *Sporobolus spicatus* (Vahl) Kunth
4: II.3,4 — 588 *Althaea rosea* (L.) Cav.
4: II.5,6 — 923 *Alisma plantago-aquatica* L.
4: II.7,8 — 19 *Allium desertorum* Forssk.
4: III.1,2 — 922 *Allium desertorum* Forssk.
4: III.3,4 — 6 *Allium sphaerocephalon* L.
4: III.5,6 — 7 *Allium sphaerocephalon* L.
4: III.7,8 — 16 *Aloe pendens* Forssk.
5: I.1,2 — 200 *Alternanthera sessilis* (L.) DC.
5: I.3,4 — 201 *Alternanthera sessilis* (L.) DC.

5: I.5,6	559 *Polycarpon prostratum* (Forssk.) Asch. & Schweinf.
5: I.7,8	551 *Polycarpon prostratum* (Forssk.) Aschers. & Schweinf.
5: II.1,2	924 *Polycarpon prostratum* (Forssk.) Aschers. & Schweinf.
5: II.3,4	925 *Polycarpon prostratum* (Forssk.) Aschers. & Schweinf.
5: II.5,6	564 *Spergularia marina* (L.) Griseb.
5: II.7,8	928 *Ambrosia maritima* L.
5:III.1,2	929 *Ambrosia maritima* L.
5:III.3,4	675 *Ammi majus* L.
5:III.5,6	676 *Ammi majus* L.
5:III.7,8	934 *Commiphora kataf* (Forssk.) Engl.
6: I.1,2	935 *Commiphora gileadensis* (L.) C. Chr.
6: I.3,4	936 *Commiphora gileadensis* (L.) C. Chr.
6: I.5,6	937 *Commiphora gileadensis* (L.) C. Chr.
6: I.7,8	938 *Anagallis arvensis* L.
6: II.1,2	352 *Anagallis arvensis* L.
6: II.3,4	718 *Anastatica hierochuntica* L.
6: II.5,6	939 *Anastatica hierochuntica* L.
6: II.7,8	292 *Anchusa aegyptiaca* (L.) DC.
6:III.1,2	296 *Anchusa aegyptiaca* (L.) DC.
6:III.3,4	310 *Gastrocotyle hispida* (Forssk.) Bunge
6:III.5,6	304 *Lappula spinocarpos* (Forssk.) Ktze
6:III.7,8	308 *Alkanna lehmannii* (Tineo) A. DC.
7: I.1,2	283 *Alkanna lehmannii* (Tineo) A. DC.
7: I.3,4	295 *Anchusa undulata* L.
7: I.5,6	309 *Anchusa undulata* L.
7: I.7,8	127 *Dichantium annulatum* (Forssk.) Stapf
7: II.1,2	941 *Cymbopogon caesius* (Hook. & Arn.) Stapf
7: II.3,4	126 *Cymbopogon caesius* (Hook. & Arn.) Stapf
7: II.5,6	942 *Cymbopogon caesius* (Hook. & Arn.) Stapf
7: II.7,8	943 *Heteropogon contortus* (L.) Roem. & Schult.
7:III.1,2	944 *Anthemis melampodina* Del.
7:III.3,4	398 *Kickxia aegyptiaca* (L.) Nobl.
7:III.5,6	946 *Kickxia aegyptiaca* (L.) Nábelek
7:III.7,8	947 *Kickxia aegyptiaca* (L.) Nábelek
8: I.1,2	948 *Linaria haelava* (Forssk.) Del.
8: I.3,4	949 *Linaria haelava* (Forssk.) Del.
8: I.5,6	397 *Linaria haelava* (Forssk.) F. G. Dietr.
8: I.7,8	950 *Anticharis linearis* (Benth.) Aschers.
8: II.1,2	951 *Anticharis linearis* (Benth.) Aschers.
8: II.3,4	234 *Carissa edulis* (Forssk.) Vahl
8: II.5,6	955 *Minuartia filifolia* (Forssk.) Mattf.
8: II.7,8	956 *Minuartia filifolia* (Forssk.) Mattf.
8:III.1,2	124 *Aristida adscensionis* L.
8:III.3,4	123 *Aristida adscensionis* L.
8:III.5,6	1765 *Stipagrostis plumosa* (L.) T. Anders.
8:III.7,8	62 *Stipagrostis plumosa* (L.) T. Anders.
9: I.1,2	125 *Stipagrostis plumosa* (L.) T. Anders.
9: I.3,4	1766 *Stipagrostis plumosa* (L.) T. Anders.
9: I.5,6	1767 *Stipagrostis plumosa* (L.) T. Anders.
9: I.7,8	1768 *Stipagrostis plumosa* (L.) T. Anders.
9: II.1,2	1769 *Stipagrostis plumosa* (L.) T. Anders.
9: II.3,4	1770 *Stipagrostis plumosa* (L.) T. Anders.
9: II.5,6	119 *Aristida adscensionis* L.
9: II.7,8	305 *Arnebia tetrastigma* Forssk.
9:III.1,2	957 *Arnebia tetrastigma* Forssk.

Fiche	No.	Name
9: III.3,4,5	958	*Gerbera piloselloides* (L.) Cass.
9: III.6,7	2	*Arisaema flavum* (Forssk.) Schott
9: III.8		[Blank]
10: I.1,2	3	*Arisaema flavum* (Forssk.) Schott
10: I.3,4	4	*Arisaema bottae* Schott
10: I.5,6	83	*Calamagrostis purpurescens* R. Br.
10: I.7,8	963	*Phragmites australis* (Cav.) Trin. & Steud.
10: II.1,2	58	*Phragmites australis* (Cav.) Trin. & Steud.
10: II.3,4	964	*Phragmites australis* (Cav.) Trin. & Steud.
10: II.5,6	73	*Imperata cylindrica* (L.) Räusch.
10: II.7,8	966	*Imperata cylindrica* (L.) Räusch.
10: III.1,2	967	*Phragmites australis* (Cav.) Trin. & Steud.
10: III.3,4	968	*Phragmites australis* (Cav.) Trin. & Steud.
10: III.5,6	969	*Phragmites australis* (Cav.) Trin. & Steud.
10: III.7,8	55	*Phragmites australis* (Cav.) Trin. & Steud.
11: I.1,2	970	*Phragmites australis* (Cav.) Trin. & Steud.
11: I.3,4	269	*Pergularia tomentosa* L.
11: I.5,6	270	*Pergularia tomentosa* L.
11: I.7,8	972	*Pergularia tomentosa* L.
11: II.1,2	973	*Kanahia laniflora* (Forssk.) R. Br.
11: II.3,4	974	*Kanahia laniflora* (Forssk.) R. Br.
11: II.5,6	265	*Kanahia laniflora* (Forssk.) R. Br.
11: II.7,8	268	*Kanahia laniflora* (Forssk.) R. Br.
11: III.1,2	272	*Pentatropis nivalis* (G. F. Gmel.) Field & Wood
11: III.3,4	275	*Odontanthera radians* (Forssk.) Field
11: III.5,6	276	*Odontanthera radians* (Forssk.) Field
11: III.7,8	264	*Gomphocarpus fruticosus* (L.) Ait. f.
12: I.1,2	267	*Pentatropis nivalis* (G. F. Gmel.) Field & Wood
12: I.3,4	277	*Sarcostemma viminale* (L.) R. Br.
12: I.5,6	13	*Asparagus acutifolius* L.
12: I.7,8	14	*Asparagus acutifolius* L.
12: II.1,2	983	*Asparagus acutifolius* L.
12: II.3,4	12	*Asparagus africanus* Lam.
12: II.5,6	15	*Asparagus stipularis* Forssk.
12: II.7,8	984	*Asparagus stipularis* Forssk.
12: III.1,2	985	*Asparagus stipularis* Forssk.
12: III.3,4	817	*Asplenium aethiopicum* (Burm. f.) Becherer
12: III.5,6	986	*Asplenium aethiopicum* (Burm. f.) Becherer
12: III.7,8	816	*Asplenium adiantum-nigrum* L.
13: I.1,2	801	*Phyllitis scolopendrium* (L.) Newm.
13: I.3,4	987	*Pulicaria crispa* (Forssk.) Benth. ex Oliver
13: I.5,6	988	*Pulicaria crispa* (Forssk.) Benth. ex Oliver
13: I.7,8	989	*Pulicaria crispa* (Forssk.) Benth. ex Oliver
13: II.1,2	990	*Pulicaria crispa* (Forssk.) Benth. ex Oliver
13: II.3,4	991	*Pulicaria crispa* (Forssk.) Benth. ex Oliver
13: II.5,6	992	*Macowania ericifolia* (Forssk.) Burtt & Grau
13: II.7,8	996	*Astragalus annularis* Forssk.
13: III.1,2	997	*Astragalus fruticosus* Forssk.
13: III.3,4	186	*Atriplex coriacea* Forssk.
13: III.5,6	191	*Atriplex coriacea* Forssk.
13: III.7,8	976	*Atriplex farinosa* Forssk.
14: I.1,2	190	*Atriplex tartarica* L.
14: I.3,4	185	*Atriplex halimus* L. var. *schweinfurthii* Boiss.
14: I.5,6	187	*Atriplex leucoclada* Boiss.
14: I.7,8	192	*Atriplex leucoclada* Boiss.

14: II.1,2 183 *Halimione portulacoides* (L.) Aellen
14: II.3,4 71 *Helictotrichon* sp.
14: II.5,6 1328 *Helictotrichon* sp.
14: II.7,8 43 *Centropodia forskalei* (Vahl) Cope
14:III.1,2 40 *Centropodia forskalei* (Vahl) Cope
14:III.3,4 980 *Centropodia forskalei* (Vahl) Cope
14:III.5,6 981 *Centropodia forskalei* (Vahl) Cope
14:III.7,8 701 *Berberis forsskaliana* Schneider
15: I.1,2 1004 *Beta vulgaris* L.
15: I.3,4 793 *Alnus glutinosa* (L.) Gaertn.
15: I.5,6 360 *Mimusops laurifolia* (Forssk.) Friis
15: I.7,8 359 *Mimusops laurifolia* (Forssk.) Friis
15: II.1,2 1006 *Boerhavia repens* L. var. *diffusa* (L.) Boiss.
15: II.3,4 289 *Borago officinalis* L.
15: II.5,6 288 *Trichodesma africana* (L.) R. Br.
15: II.7,8 1011 *Erucaria crassifolia* (Forssk.) Del.
15:III.1,2 711 *Erucaria crassifolia* (Forssk.) Del.
15:III.3,4 713 *Erucaria crassifolia* (Forssk.) Del.
15:III.5,6 709 *Erucaria crassifolia* (Forssk.) Del.
15:III.7,8 104 *Trachynia distachya* (L.) Link
16: I.1,2 100 *Brachypodium retusum* (Pers.) P. Beauv.
16: I.3,4 99 *Brachypodium retusum* (Pers.) P. Beauv.
16: I.5,6 1010 *Brachypodium phoenicioides* (L.) Roem. & Schult.
16: I.7 [Blank]
16: I.8 1016 *Leptochloa fusca* (L.) Kunth
16: II.1,2 92 *Bromus intermedius* Guss.
16: II.3,4 95 *Bromus squarrosus* L.
16: II.5,6 96 *Bromus squarrosus* L.
16: II.7,8 90 *Bromus madritensis* L.
16:III.1,2 89 *Bromus madritensis* L.
16:III.3,4 91 *Bromus madritensis* L.
16:III.5,6 1831 *Bromus rigidus* Roth
16:III.7,8 93 *Bromus madritensis* L.
17: I.1,2 404 *Cycniopsis humifusa* (Forssk.) Engl.
17: I.3,4 401 *Cycniopsis humifusa* (Forssk.) Engl.
17: I.5,6 1019 *Nauplius graveolens* (Forssk.) A. Wicklund
17: I.7,8 1020 *Nauplius graveolens* (Forssk.) A. Wicklund
17: II.1,2 1021 *Nauplius graveolens* (Forssk.) A. Wicklund
17: II.3,4 995 *Asteriscus aquaticus* (L.) Less.
17: II.5,6 *Pterocaulon redolens* (Forst.) Boerl. (not a Forsskål specimen)
17: II.7,8 1022 *Nauplius graveolens* (Forssk.) A. Wicklund
17:III.1,2 681 *Bupleurum flavum* Forssk.
17:III.3,4 1023 *Bupleurum flavum* Forssk.
17:III.5,6 1024 *Myrsine africana* L.
17:III.7,8 729 *Myrsine africana* L.
18: I.1,2 730 *Myrsine africana* L.
18: I.3,4 1030 *Kleinia odora* (Forssk.) DC.
18: I.5,6,7 1031 *Kleinia pendula* (Forssk.) DC.
18: I.8 1032 *Kleinia pendula* (Forssk.) DC.
18: II.1,2 1033 *Kleinia semperviva* (Forssk.) DC.
18: II.3,4 1026 *Solanecio angulatus* (Vahl) C. Jeffrey
18: II.5,6 1027 *Solanecio angulatus* (Vahl) C. Jeffrey
18: II.7,8 1028 *Solanecio angulatus* (Vahl) C. Jeffrey
18:III.1,2 1029 *Solanecio angulatus* (Vahl) C. Jeffrey
18:III.3,4 1034 *Cadaba farinosa* Forssk.

18: III.5,6 645 *Cadaba farinosa* Forssk.
18: III.7,8 634 *Cadaba farinosa* Forssk.
19: I.1,2 633 *Cadaba glandulosa* Forssk.
19: I.3,4 637 *Cadaba glandulosa* Forssk.
19: I.5,6 636 *Cadaba glandulosa* Forssk.
19: I.7,8 646 *Cadaba glandulosa* Forssk.
19: II.1,2 1035 *Cadaba glandulosa* Forssk.
19: II.3,4 1036 *Cadaba rotundifolia* Forssk.
19: II.5,6 644 *Cadaba rotundifolia* Forssk.
19: II.7,8 635 *Cadaba rotundifolia* Forssk.
19: III.1,2 1037 *Cadaba rotundifolia* Forssk.
19: III.3,4 1038 *Cadia purpurea* (Picciv.) Aiton
19: III.5,6 775 *Forsskalea tenacissima* L.
19: III.7,8 1039 *Forsskaolea tenacissima* L.
20: I.1,2 1040 *Calendula arvensis* L.
20: I.3,4 1041 *Ruellia grandiflora* (Forssk.) Blatter
20: I.5,6 1042 *Ruellia grandiflora* (Forssk.) Blatter
20: I.7,8 1043 *Ruellia grandiflora* (Forssk.) Blatter
20: II.1,2 245 *Campanula edulis* Forssk.
20: II.3,4 246 *Campanula edulis* Forssk.
20: II.5,6 243 *Campanula edulis* Forssk.
20: II.7,8 244 *Legousia speculum-veneris* (L.) Chaix
20: III.1,2 639 *Maerua oblongifolia* (Forssk.) A. Rich.
20: III.3,4 638 *Maerua oblongifolia* (Forssk.) A. Rich.
20: III.5,6 643 *Maerua oblongifolia* (Forssk.) A. Rich.
20: III.7,8 1045 *Carthamus glaucus* Bieb.
21: I.1,2 1046 *Carlina corymbosa* L. subsp. *graeca* (Boiss) Nyman
21: I.3,4 1047 *Carlina corymbosa* L. subsp. *graeca* (Boiss) Nyman
21: I.5,6 1059 *Senna italica* Miller
21: I.7,8 1053 *Senna sophera* (L.) Roxb.
21: II.1,2 1054 *Senna sophera* (L.) Roxb.
21: II.3,4 1055 *Senna sophera* (L.) Roxb.
21: II.5,6 1048 *Senna alexandrina* Miller
21: II.7,8 1049 *Senna alexandrina* Miller
21: III.1,2 1050 *Senna alexandrina* Miller
21: III.3,4 1051 *Senna alexandrina* Miller
21: III.5,6 1057 *Chamaecrista nigricans* (Vahl) Greene
21: III.7,8 1058 *Chamaecrista nigricans* (Vahl) Greene
22: I.1,2 1063 *Senna obtusifolia* (L.) Irwin & Greene
22: I.3,4 1064 *Senna obtusifolia* (L.) Irwin & Greene
22: I.5,6 505 *Catha edulis* (Vahl) Forssk. ex Endl.
22: I.7,8 506 *Catha edulis* (Vahl) Forssk. ex Endl.
22: II.1,2 504 *Catha edulis* (Vahl) Forssk. ex Endl.
22: II.3,4 1066 *Catha edulis* (Vahl) Forssk. ex Endl.
22: II.5,6 507 *Maytenus parviflora* (Vahl) Sebsebe
22: II.7,8 503 *Maytenus parviflora* (Vahl) Sebsebe
22: III.1,2 668 *Torilis arvensis* (Huds.) Link
22: III.3,4 477 *Caucanthus edulis* Forssk.
22: III.5,6 693 *Cocculus pendulus* (J. R. & G. Forst.) Diels
22: III.7,8 1462 *Cocculus pendulus* (J.R. & G. Forst.) Diels
23: I.1,2 694 *Cocculus pendulus* (J. R. & G. Forst.) Diels
23: I.3,4 696 *Cocculus hirsutus* (L.) Theob.
23: I.5,6 1075 *Centaurea glomerata* Vahl
23: I.7,8 1073 *Centaurea aspera* L.
23: II.1,2 1072 *Centaurea calcitrapa* L.

23: II.3,4 — 1070 *Atractylis carduus* (Forssk.) C. Chr.
23: II.5,6 — 1071 *Atractylis carduus* (Forssk.) C. Chr.
23: II.7,8 — 1074 *Centaurea aegyptiaca* L.
23:III.1,2 — 1076 *Volutaria lippii* (L.) Cass.
23:III.3,4 — 1077 *Volutaria lippii* (L.) Cass.
23:III.5,6 — 1078 *Centaurothamnus maximus* (Forssk.) Wagenitz & Dittrich
23:III.7,8 — 1079 *Centaurothamnus maximus* (Forssk.) Wagenitz & Dittrich
24: I.1,2 — 1080 *Centaurothamnus maximus* (Forssk.) Wagenitz & Dittrich
24: I.3,4 — 1082 *Aegialophila pumilio* (L.) Boiss.
24: I.5,6 — 1084 *Ceruana pratensis* Forssk.
24: I.7,8 — 490 *Grewia arborea* (Forssk.) Lam.
24: II.1,2 — 491 *Grewia tenax* (Forssk.) Fiori
24: II.3,4 — 1085 *Grewia tenax* (Forssk.) Fiori
24: II.5,6 — 489 *Grewia velutina* (Forssk.) Vahl
24: II.7,8 — 315 *Lantana viburnoides* (Forssk.) Vahl
24:III.1,2 — 323 *Lantana viburnoides* (Forssk.) Vahl
24:III.3,4 — 1086 *Lantana viburnoides* (Forssk.) Vahl
24:III.5,6 — 1087 *Lantana viburnoides* (Forssk.) Vahl
24:III.7,8 — 1089 *Farsetia aegyptia* Turra
25: I.1,2 — 1090 *Farsetia aegyptia* Turra
25: I.3,4 — 1091 *Farsetia aegyptia* Turra
25: I.5,6 — 723 *Farsetia linearis* (Forssk.) Boiss.
25: I.7,8 — 720 *Matthiola livida* (Del.) DC.
25: II.1,2 — 1092 *Matthiola livida* (Del.) DC.
25: II.3,4 — 1093 *Matthiola livida* (Del.) DC.
25: II.5,6 — 1094 *Matthiola livida* (Del.) DC.
25: II.7,8 — 1095 *Matthiola livida* (Del.) DC.
25:III.1,2 — 1096 *Matthiola livida* (Del.) DC.
25:III.3,4 — 1097 *Matthiola tricuspidata* (L.) R. Br.
25:III.5,6 — 705 *Roemeria hybrida* (L.) DC. subsp. *dodecandra* (Forssk.) Dur. & Barrat.
25:III.7,8 — 1098 *Roemeria hybrida* (L.) DC.
26: I.1,2 — 706 *Roemeria hybrida* (L.) DC. subsp. *dodecandra* (Forssk.) Dur. & Barrat.
26: I.3,4 — 704 *Roemeria hybrida* (L.) DC.
26: I.5,6 — 184 *Chenopodium schraderianum* Schultes
26: I.7,8 — 149 *Suaeda vera* Forssk. ex J. F. Gmel.
26: II.1,2 — 1100 *Iphiona mucronata* (Forssk.) Aschers. & Schweinf.
26: II.3,4 — 1101 *Iphiona mucronata* (Forssk.) Aschers. & Schweinf.
26: II.5,6 — 1102 *Iphiona mucronata* (Forssk.) Aschers. & Schweinf.
26: II.7,8 — 1103 *Iphiona mucronata* (Forssk.) Aschers. & Schweinf.
26:III.1,2 — 1104 *Iphiona mucronata* (Forssk.) Aschers. & Schweinf.
26:III.3,4 — 1105 *Vernonia spatulata* (Forssk.) Sch.-Bip.
26:III.5,6 — 1106 *Vernonia spatulata* (Forssk.) Sch.-Bip.
26:III.7,8 — 1107 *Ifloga spicata* (Forssk.) Sch.-Bip.
27: I.1,2 — 1108 *Ifloga spicata* (Forssk.) Sch.-Bip.
27: I.3,4 — 1112 *Cichorium spinosum* L.
27: I.5,6 — 1113 *Cichorium spinosum* L.
27: I.7,8 — 1114 *Cichorium spinosum* L.
27: II.1,2 — 683 *Salvadora persica* L.
27: II.3,4 — 1117 *Salvadora persica* L.
27: II.5,6 — 518 *Salvadora persica* L.
27: II.7,8 — 618 *Helianthemum stipulatum* (Forssk.) C. Chr.
27:III.1,2 — 616 *Helianthemum stipulatum* (Forssk.) C. Chr.
27:III.3,4 — 615 *Helianthemum stipulatum* (Forssk.) C. Chr.

27:III.5,6 617 *Helianthemum stipulatum* (Forssk.) C. Chr.
27:III.7,8 614 *Helianthemum kahiricum* Del.
28: I.1,2 700 *Clematis vitalba* L.
28: I.3,4 224 *Otostegia fruticosa* (Forssk.) Penzig
28: I.5,6 345 *Otostegia fruticosa* (Forssk.) Penzig
28: I.7,8 218 *Otostegia fruticosa* (Forssk.) Penzig
28: II.1,2 1120 *Clutia lanceolata* Forssk.
28: II.3,4 1122 *Carthamus lanatus* L.
28: II.5,6 1123 *Carthamus lanatus* L.
28: II.7,8 1121 *Carthamus dentatus* (Forssk.) DC.
28:III.1,2 1124 *Cardopatium corymbosum* (L.) Pers.
28:III.3,4 1125 *Cardopatium corymbosum* (L.) Pers.
28:III.5,6 1126 *Coffea arabica* L.
28:III.7,8 27 *Merendera abyssinica* A. Rich.
29: I.1,2 1127 *Astragalus spinosus* (Forssk.) Muschl.
29: I.3,4 1128 *Astragalus spinosus* (Forssk.) Muschl.
29: I.5,6 37 *Commelina africana* L.
29: I.7,8 32 *Aneilema forsskahlii* Kunth
29: II.1,2 33 *Aneilema forsskahlii* Kunth
29: II.3,4 877 *Laurencia obtusa* Ag. (Algae)
29: II.5,6 Algae fam., gen. & sp. indet.
29: II.7,8 Algae fam., gen. & sp. indet.
29:III.1,2 Algae fam., gen. & sp. indet.
29:III.3,4 867 *Chaetomorpha linum* Ktz. (Algae)
29:III.5,6 868 *Ceramium diaphanum* Ag. & *Ceramium rubrum* Ag. [mixed sheet] (Algae)
29:III.7,8 860 *Ceramium rubrum* Ag. (Algae)
30: I.1,2 884 *Bonnemaisonia asparagoides* (Woodw.) Ag. (Algae)
30: I.3 Algae fam., gen. & sp. indet.
30: I.4,5 Algae fam., gen. & sp. indet.
30: I.6,7 870 *Hypnea musciformis* (Wulf.) Lamour. (Algae)
30: I.8 Algae fam., gen. & sp. indet.
30: II.1,2 866 *Chaetomorpha linum* Ktz. (Algae)
30: II.3,4 869 *Chaetomorpha linum* Ktz. (Algae)
30: II.5 Algae fam., gen. & sp. indet.
30: II.6,7 Algae fam., gen. & sp. indet.
30: II.8 9 *Sanseviera forskaliana* (Schult.f.) Hepper & Wood
30: III.1 9 *Sanseviera forskaliana* (Schult.f.) Hepper & Wood
30: III.2 1130 *Convolvulus arvensis* L.
30:III.3,4 455 *Convolvulus lanatus* Vahl.
30:III.5,6 1134 *Convolvulus lanatus* Vahl
30:III.7,8 [Blank]
31: I.1,2 1135 *Convolvulus lanatus* Vahl
31: I.3,4 1003 Chenopodiaceae, gen. et sp. indet.
31: I.5,6 456 *Convolvulus lanatus* Vahl
31: I.7,8 1136 *Convolvulus lanatus* Vahl
31: II.1,2 1131 *Convolvulus arvensis* L.
31: II.3,4 445 *Ipomoea eriocarpa* R. Br.
31: II.5,6 1132 *Convolvulus arvensis* L.
31: II.7,8 446 *Convolvulus prostratus* Forssk.
31:III.1,2 438 *Convolvulus prostratus* Forssk.
31:III.3,4 1138 *Convolvulus prostratus* Forssk.
31:III.5,6 439 *Convolvulus althaeoides* L. subsp. *tenuissimus* (Sibth. & Sm.) Stace
31:III.7,8 449 *Convolvulus althaeoides* L. subsp. *tenuissimus* (Sibth. & Sm.) Stace
32: I.1,2 454 *Convolvulus hystrix* Vahl

32: I.3,4 440 *Convolvulus hystrix* Vahl
32: I.5,6 1140 *Laggera crispata* (Vahl) Hepper & Wood
32: I.7,8 1141 *Pluchea discoides* (L.) DC.
32: II.1,2 1143 *Phagnalon rupestre* (L.) DC.
32: II.3,4 1144 *Phagnalon rupestre* (L.) DC.
32: II.5,6 1145 *Phagnalon rupestre* (L.) DC.
32: II.7,8 688 *Cordia sinensis* Lam.
32: III.1,2 465 *Cordia sinensis* Lam.
32: III.3,4 1149 *Cornus sanguinea* L. subsp. *australis* (C.A. Mey.) Jav.
32: III.5,6 558 *Paronychia arabica* (L.) DC.
32: III.7,8 1148 *Paronychia arabica* (L.) DC.
33: I.1,2 557 *Polycarpaea repens* (Forssk.) Aschers & Schweinf.
33: I.3,4 690 *Kalanchoe alternans* (Vahl) Pers.
33: I.5 Letter relating to the missing sheet of Kalanchoe deficiens (Forssk.) Asch. & Schweinf.
33: I.6,7 689 *Kalanchoe lanceolata* (Forssk.) Pers.
33: I.8 691 *Crassula alba* Forssk.
33: II.1 691 *Crassula alba* Forssk.
33: II.2 1150 *Crassula alba* Forssk.
33: II.3,4 762 *Crataegus monogyna* Jacq.
33: II.5,6 761 *Crataegus monogyna* Jacq.
33: II.7,8 1152 *Picris asplenioides* L.
33: III.1,2 1157 *Cressa cretica* L.
33: III.5,6 1156 *Cressa cretica* L.
33: III.7,8 443 *Cressa cretica* L.
34: I.3,4 441 *Cressa cretica* L.
34: I.5 1155 *Cressa cretica* L.
34: I.6,7 1156 *Cressa cretica* L.
34: I.8 1162 *Chrozophora obliqua* (Vahl) Spreng.
34: II.1 1162 *Chrozophora obliqua* (Vahl) Spreng.
34 II.2,3 1370 *Jatropha glauca* Vahl
34: II.4,5 1165 *Chrozophora oblongifolia* (Del.) Spreng.
34: II.6,7 1166 *Chrozophora oblongifolia* (Del.) Spreng.
34: II.8 1167 *Chrozophora oblongifolia* (Del.) Spreng.
34: III.1 1167 *Chrozophora oblongifolia* (Del.) Spreng.
34: III.2,3 1163 *Chrozophora plicata* (Vahl) A. Juss. ex Spreng.
34: III.4,5 1164 *Chrozophora plicata* (Vahl) A. Juss. ex Spreng.
34: III.6,7 1371 *Jatropha pelargoniifolia* Courbon
34: III.8 [Blank]
35: I.1,2 1173 *Crucianella latifolia* L.
35: I.3,4 1169 *Crucianella aegyptiaca* L.
35: I.5,6 1170 *Crucianella aegyptiaca* L.
35: I.7,8 1174 *Crucianella maritima* L.
35: II.1,2 1178 *Silene sp.*
35: II.3,4 1177 *Silene vulgaris* (Moench) Garcke
35: II.5,6 572 *Silene sp.*
35: II.7,8 660 *Coccinia grandis* (L.) Voigt
35: III.1,2 664 *Citrullus colocynthis* (L.) Schrad.
35: III.3,4 605 *Sterculia africana* (Lour.) Fiori
35: III.5,6 273 *Leptadenia arborea* (Forssk.) Schweinf.
35: III.7,8 271 *Leptadenia arborea* (Forssk.) Schweinf.
36: I.1,2 312 *Cynoglossum lanceolatum* Forssk.
36: I.3,4 1179 *Dactyloctenium aristatum* Link
36: I.5,6 1180 *Desmostachya bipinnata* (L.) Stapf
36: I.7,8 1181 *Desmostachya bipinnata* (L.) Stapf

36: II.1,2	1182	*Desmostachya bipinnata* (L.) Stapf
36: II.3,4	56	*Cynosurus effusus* Link
36: II.5,6	1183	*Cynosurus echinatus* L.
36: II.7,8	45	*Eleusine floccifolia* (Forssk.) Spreng.
36: III.1,2	44	*Eleusine floccifolia* (Forssk.) Spreng.
36: III.3,4	1185	*Aeleuropus lagopoides* (L.) Thwait.
36: III.5,6	57	*Aeleuropus lagopoides* (L.) Thwait.
36: III.7,8	1198	*Cyperus rotundus* L.
37: I.1,2	1191	*Cyperus fuscus* L.
37: I.3,4	1192	*Cyperus fuscus* L.
37: I.5,6	1193	*Cyperus fuscus* L.
37: I.7,8	1204	*Cyperus laevigatus* L.
37: II.1,2	1205	*Cyperus laevigatus* L.
37: II.3,4	1206	*Cyperus articulatus* L.
37: II.5,6	1209	*Aeleuropus lagopoides* (L.) Thwait.
37: II.7,8	1210	*Aeleuropus lagopoides* (L.) Thwait.
37: III.1,2	1211	*Aeleuropus lagopoides* (L.) Thwait.
37: III.3,4	1212	*Aeleuropus lagopoides* (L.) Thwait.
37: III.5,6	672	*Ammi visnaga* (L.) Lam.
37: III.7,8	673	*Ammi visnaga* (L.) Lam.
38: I.1,2	1218	*Delphinium peregrinum* L.
38: I.3,4	394	*Justicia flava* (Vahl) Vahl
38: I.5,6	395	*Justicia flava* (Vahl) Vahl
38: I.7,8	370	*Asystasia gangetica* (L.) T. Anders.
38: II.1,2	389	*Isoglossa punctata* (Vahl) Brummitt & Wood
38: II.3,4	391	*Monechma debile* (Forssk.) Nees
38: II.5,6	1383	*Megalochlamys violaceum* (Vahl) Vollesen
38: II.7,8	384	*Justicia odora* (Forssk.) Vahl
38: III.1,2	1220	*Justicia odora* (Forssk.) Vahl
38: III.3,4	385	*Peristrophe paniculata* (Forssk.) Brummitt
38: III.5,6	1221	*Peristrophe paniculata* (Forssk.) Brummitt
38: III.7,8	1222	*Anisotes trisulcus* (Forssk.) Nees
39: I.1,2	392	*Dicliptera verticillata* (Forssk.) C. Chr.
39: I.3,4	393	*Dicliptera verticillata* (Forssk.) C. Chr.
39: I.5,6	1224	*Dianthus caryophyllus* L.
39: I.7,8	556	*Petrorhagia prolifera* (L.) Ball & Heyw.
39: II.1,2	555	*Petrorhagia prolifera* (L.) Ball & Heyw.
39: II.3,4	566	*Minuartia* or ? *Arenaria* sp. indet.
39: II.5,6	1226	*Dianthus uniflorus* Forssk.
39: II.7,8	549	*Dianthus uniflorus* Forssk.
39: III.1,2	554	*Dianthus uniflorus* Forssk.
39: III.3,4	1228	*Digera muricata* (L.) Mart.
39: III.5,6	1229	*Digera muricata* (L.) Mart.
39: III.7,8	1231	*Sesbania sesban* (L.) Merr.
40: I.1,2	1232	*Sesbania sesban* (L.) Merr.
40: I.3,4	1230	*Sesbania grandiflora* (L.) Poir.
40: I.5,6	1233	*Crotalaria retusa* L.
40: I.7,8	1234	*Vignea luteola* (Jacq.) Benth.
40: II.1,2	1302	*Neonotonia wightii* (Wight & Arn.) Lackey var. *longicauda* (Schweinf.) Lackey
40: II.3,4	1303	*Neonotonia wightii* (Wight & Arn.) Lackey var. *longicauda* (Schweinf.) Lackey
40: II.5,6	1238	*Echinops spinosissimus* Turra
40: II.7,8	287	*Echium longifolium* Del.
40: III.1,2	298	*Echium angustifolium* Miller subsp. *sericeum* (Vahl) Kltz.

40:III.3,4	280 *Echium angustifolium* Miller subsp. *sericeum* (Vahl) Kltz.
40:III.5,6	291 *Echium angustifolium* Miller subsp. *sericeum* (Vahl) Kltz.
40:III.7,8	284 *Echium angustifolium* Miller subsp. *sericeum* (Vahl) Kltz.
41: I.1,2	1241 *Echium rubrum* Forssk.
41: I.3,4	297 *Echium rubrum* Forssk.
41: I.5,6	529 *Elaeagnus angustifolia* L.
41: I.7,8	531 *Elaeagnus angustifolia* L.
41: II.1,2	478 *Trichilia emetica* Vahl.
41: II.3,4	72 *Cenchrus biflorus* Roxb.
41: II.5,6	794 *Ephedra campylopoda* C. A. Mey.
41: II.7,8	1245 *Ephedra campylopoda* C.A. Mey.
41:III.1,2	1246 *Ephedra campylopoda* C.A. Mey.
41:III.3,4	797 *Equisetum ramosissimum* Desf.
41:III.5,6	726 *Erica arborea* L.
41:III.7,8	727 *Erica manipuliflora* Salisb.
42: I.1,2	728 *Erica manipuliflora* Salisb.
42: I.3,4	670 *Eryngium bourgatii* Gouan
42: I.5,6	680 *Eryngium maritimum* L.
42: I.7,8	1268 *Euphorbia peplis* L.
42: II.1,2	1269 *Euphorbia peplis* L.
42: II.3,4	1270 *Euphorbia ? platyphyllos* L.
42: II.5,6	1276 *Euphorbia granulata* Forssk. var. *granulata* (Gay) Boiss.
42: II.7,8	1277 *Euphorbia granulata* Forssk. var. *granulata* (Gay) Boiss.
42:III.1,2	1278 *Euphorbia granulata* Forssk. var. *granulata* (Gay) Boiss.
42:III.3,4	1266 *Euphorbia indica* Lam.
42:III.5,6	1267 *Euphorbia indica* Lam.
42:III.7,8	1272 *Euphorbia terracina* L.
43: I.1,2	1273 *Euphorbia peplus* L.
43: I.3,4	1275 *Euphorbia falcata* L.
43: I.5,6	1279 *Euphorbia retusa* Forssk.
43: I.7,8	1280 *Euphorbia retusa* Forssk.
43: II.1,2	1281 *Euphorbia retusa* Forssk.
43: II.3,4	1282 *Euphorbia retusa* Forssk.
43: II.5,6	1287 *Euphorbia triaculeata* Forssk.
43: II.7,8	760 *Ochna inermis* (Forssk.) Schweinf.
43:III.1,2	1251 *Fagonia arabica* L.
43:III.3,4	1252 *Fagonia arabica* L.
43:III.5,6	1254 *Fagonia cretica* L.
43:III.7,8	740 *Fagonia cretica* L.[*Caralluma dentata* (Forssk.) Blatter; also this No? seems lost]
44: I.1,2	741 *Fagonia cretica* L.
44: I.3,4	742 *Fagonia cretica* L.
44: I.5,6	84 *Schismus arabicus* Nees
44: I.7,8	1256 *Schismus arabicus* Nees
44: II.1,2	1259 *Cutandia memphitica* (Spreng.) Benth.
44: II.3,4	47 *Cutandria dichotoma* (Forssk.) Trabut
44: II.5,6	1257 *Cutandia dichotoma* (Forssk.) Trabut
44: II.7,8	1258 *Cutandia dichotoma* (Forssk.) Trabut mixed with *Cutandia memphitica* (Spreng) Benth.
44:III.1,2	1215 *Leptochloa fusca* (L.) Kunth
44:III.3,4	49 *Cutandia maritima* (L.) Barbey
44:III.5,6	94 *Odyssea mucronata* (Forssk.) Stapf
44:III.7	778 *Ficus carica* L.
44:III.8	1263 *Ficus cordata* Thunb. subsp. *salicifolia* (Vahl) C.C. Berg
45: I.1,2	1263 *Ficus cordata* Thunb. subsp. *salicifolia* (Vahl) C.C. Berg

45: I.3,4 777 *Ficus* sp.
45: I.5,6 781 *Ficus palmata* Forssk.
45: I.7,8 784 *Ficus palmata* Forssk.
45: II.1,2 783 *Ficus populifolia* Vahl
45: II.3,4 785 *Ficus exasperata* Vahl
45: II.5,6 782 *Ficus sur* Forssk.
45: II.7,8 779 *Ficus sycomorus* L.
45: III.1,2 776 *Ficus vasta* Forssk.
45: III.3,4 624 *Frankenia revoluta* Forssk.
45: III.5,6 623 *Frankenia revoluta* Forssk.
45: III.7,8 1264 *Frankenia revoluta* Forssk.
46: I.1,2 1265 *Frankenia revoluta* Forssk.
46: I.3,4 874 *Echinocaulon acerosum* (Forssk.) Børgesen (Algae)
46: I.5,6 Algae fam., gen. & sp. indet.
46: I.7,8 Algae fam., gen. & sp. indet.
46: II.1,2 883 *Hormosira articulata* (Forssk.) Zanard (Algae)
46: II.3,4 839 *Phyllophora rubens* (L.) Grev. (Algae)
46: II.5,6 Algae fam., gen. & sp. indet.
46: II.7,8 Algae fam., gen. & sp. indet.
46: III.1,2 880 *Turbinaria decurrens* Bory (Algae)
46: III.3,4 881 *Turbinaria decurrens* Bory (Algae)
46: III.5,6 Algae fam., gen. & sp. indet.
46: III.7,8 871 *Sargassum crispum* (Forssk.) Ag. (Algae)
47: I.1,2 888 *Gracilaria debilis* (Forssk.) Børgesen (Algae)
47: I.3,4 Algae fam., gen. & sp. indet.
47: I.5,6 849 *Sargassum denticulatum* (Forssk.) Børgesen (Algae)
47: I.7,8 854 *Dictyota dichotoma* (L.) Ktz. (Algae)
47: II.1,2 885 *Solieria dura* (Zanard.) Schmitz (Algae)
47: II.3,4 Algae fam., gen. & sp. indet.
47: II.5,6 832 *Fucus vesiculosus* L. (Algae)
47: II.7,8 Algae fam., gen. & sp. indet.
47: III.1,2 826 *Actinotrichia fragilis* (Forssk.) Børgesen (Algae)
47: III.3,4 844 *Caulerpa serrulata* (Forssk.) Agardh (Algae)
47: III.5,6 *Fucus diffusus* Huds. (Algae)
47: III.7,8 862 *Polysiphonia elongata* (Huds.) Harv. (Algae)
48: I.1 Algae fam., gen. & sp. indet.
48: I.2,3 Algae fam., gen. & sp. indet.
48: I.4,5 837 *Cytosira barbata* (Algae)
48: I.6,7 886 *Chondria papillosa* C. Agardh (Algae)
48: I.8 882 *Caulerpa setularoides* (Gmel.) Howe (Algae)
48: II.1,2 878 *Caulerpa prolifera* (Forssk.) Lamour. (Algae)
48: II.3,4 845 *Caulerpa racemosa* (Forssk.) Weber von Bosse (Algae)
48: II.5,6 850 *Cytoseira myrica* (Gmel.) Agadh (Algae)
48: II.7,8 851 *Sargassum subrepandum* (Forssk.) Agardh (Algae)
48: III.1,2 852 *Sargassum subrepandum* (Forssk.) Agardh (Algae)
48: III.3,4 872 *Sargassum subrepandum* (Forssk.) Agardh (Algae)
48: III.5 873 *Sargassum subrepandum* (Forssk.) Agardh (Algae)
48: III.6,7 889 *Cystophyllum trinode* (Forssk.) Agardh (Algae)
48: III.8 [Blank]
49: I.1,2 876 *Laurencia obtusa* Ag. (Algae)
49: I.3,4 887 *Laurencia uvifera* (Forssk.) Børgesen (Algae)
49: I.5,6 707 *Fumaria densiflora* DC.
49: I.7,8 1288 *Galium album* Miller subsp. *pycnotrichum* (H. Braun) Krendl
49: II.1,2 1289 *Galium heldreichii* Halacsy
49: II.3,4 1292 *Galium aparinoides* Forssk.

49: II.5,6	1291 *Galium aparinoides* Forssk.
49: II.7,8	Specimen not retrieved!
49: III.1,2	1294 *Galium paschale* Forssk.
49: III.3,4	1339 *Retama retam* (Forssk.) Webb
49: III.5,6	1340 *Retama retam* (Forssk.) Webb
49: III.7,8	241 *Centaurium pulchellum* (Sw.) Druce
50: I.1,2	242 *Centaurium pulchellum* (Sw.) Druce
50: I.3,4	1081 *Centaurium pulchellum* (Sw.) Druce
50: I.5,6	735 *Geranium arabicum* Forssk.
50: I.7,8	736 *Geranium arabicum* Forssk.
50: II.1,2	732 *Erodium glaucophyllum* (L.) L'Hérit.
50: II.3,4	739 *Erodium glaucophyllum* (L.) L'Hérit.
50: II.5,6	1299 *Erodium glaucophyllum* (L.) L'Hérit.
50: II.7,8	731 *Erodium crassifolium* L'Hérit.
50: III.1,2	738 *Erodium crassifolium* L'Hérit.
50: III.3,4	733 *Erodium laciniatum* (Cav.) Willd.
50: III.5,6	734 *Erodium laciniatum* (Cav.) Willd.
50: III.7,8	737 *Erodium laciniatum* (Cav.) Willd.
51: I.1,2	544 *Aizoon canariense* L.
51: I.3,4	543 *Glinus lotoides* L.
51: I.5,6	534bis *Glinus lotoides* L.
51: I.7,8	540 *Glinus setiflorus* Forssk.
51: II.1,2	535bis *Glinus setiflorus* Forssk.
51: II.3	1301 *Glinus setiflorus* Forssk.
51: II.4,5	542 *Glinus setiflorus* Forssk.
51: II.6,7	543bis *Glinus setiflorus* Forssk.
51: II.8	1300 *Glinus setiflorus* Forssk.
51: III.1	1300 *Glinus setiflorus* Forssk.
51: III.2,3	1304 *Caesalpinia bonduc* (L.) Roxb.
51: III.4,5	1306 *Helichrysum forskahlii* (Gmel.) Hilliard & Burtt
51: III.6,7	1305 *Gnaphalium crispatulum* Del.
51: III.8	1307 *Pseudognaphalium luteo-album* (L.) Hilliard & Burtt
52: I.1,2	1307 *Pseudognaphalium luteo-album* (L.) Hilliard & Burtt
52: I.3,4	599 *Gossypium herbaceum* L.
52: I.5,6	537bis *Gymnocarpos decandrum* Forssk.
52: I.7,8	1698 *Chlorophytum tetraphyllum* (L.f.) Bak.
52: II.1,2	1313 *Hedera helix* L.
52: II.3,4	1314 *Kohautia caespitosa* Schnizl.
52: II.5,6	1315 *Kohautia caespitosa* Schnizl.
52: II.7,8	1316 *Kohautia caespitosa* Schnizl.
52: III.1,2	1317 *Alhagi maurorum* Medik.
52: III.3,4	1318 *Taverniera lappacea* (Forssk.) DC.
52: III.5,6	1319 *Taverniera lappacea* (Forssk.) DC.
52: III.7,8	1321 *Alysicarpus glumaceus* (Vahl) DC.
53: I.1,2	1322 *Alysicarpus glumaceus* (Vahl) DC.
53: I.3,4	1323 *Alysicarpus glumaceus* (Vahl) DC.
53: I.5,6	1324 *Alysicarpus glumaceus* (Vahl) DC.
53: I.7,8	1325 *Sophora alopecuroides* L.
53: II.1,2	1329 *Heliotropium bacciferum* Forssk.
53: II.3,4	1330 *Heliotropium longiflorum* (A. DC.) DC.
53: II.5,6	301 *Heliotropium europaeum* L.
53: II.7,8	1331 *Herniaria hirsuta* L.
53: III.1,2	1332 *Herniaria hirsuta* L.
53: III.3,4	1333 *Herniaria hirsuta* L.
53: III.5,6	1334 *Paronychia desertorum* Boiss.

53:III.7,8 563 *Paronychia desertorum* Boiss.
54: I.1,2 1335 *Diplotaxis acris* (Forssk.) Boiss.
54: I.3,4 601 *Pavonia hildebrandtii* Guerke & Ulbr.
54: I.5,6 581 *Pavonia hildebrandtii* (Guerke) Ulb.
54: I.7,8 603 *Pavonia flavo-ferruginea* (Forssk.) Hepper & Wood
54: II.1,2 604 *Hibiscus palmatus* Forssk.
54: II.3,4 579 *Hibiscus ovalifolius* (Forssk.) Vahl
54: II.5,6 578 *Hibiscus ovalifolius* (Forssk.) Vahl
54: II.7,8 590 *Hibiscus purpureus* Forssk.
54:III.1,2 107 *Sorghum bicolor* (L.) Moench
54:III.3,4 108 *Sorghum bicolor* (L.) Moench
54:III.5,6 106 *Pennisetum glaucum* (L.) R. Br.
54:III.7,8 111 *Sorghum bicolor* (L.) Moench
55: I.1,2 50 *Sorghum halepense* (L.) Pers.
55: I.3,4 109 *Sorghum halepense* (L.) Pers
55: I.5,6 110 *Sorghum halepense* (L.) Pers
55: I.7,8 1343 *Sorghum halepense* (L.) Pers.
55: II.1,2 66 *Hordeum vulgare* L.
55: II.3,4 139 *Hordeum murinum* L. subsp. *glaucum* (Steud.) Tzvlev.
55: II.5,6 17 *Dipcadi erythraeum* Webb & Berth.
55: II.7,8 692 *Dipcadi erythraeum* Webb & Berth.
55:III.1,2 1347 *Hyoscyamus muticus* L.
55:III.3,4 423 *Hyoscyamus muticus* L.
55:III.5,6 1349 *Moringa peregrina* (Forssk.) Fiori
55:III.7,8 17 *Dipcadi erythraeum* Webb & Berth.
56: I.1,2 1350 *Moringa peregrina* (Forssk.) Fiori
56: I.3,4 495 *Hypericum calycinum* L.
56: I.5,6 493 *Hypericum perforatum* L.
56: I.7,8 496 *Hypericum revolutum* Vahl
56: II.1,2 492 *Hypericum revolutum* Vahl
56: II.3,4 714 *Iberis pinnata* L.
56: II.5,6 1354 *Indigofera semitrijuga* Forssk.
56: II.7,8 1355 *Indigofera semitrijuga* Forssk.
56:III.1,2 1356 *Indigofera spicata* Forssk.
56:III.3,4 1357 *Indigofera spinosa* Forssk.
56:III.5,6 1358 *Indigofera spinosa* Forssk.
56:III.7,8 1359 *Indigofera articulata* Gouan
57: I.1,2 1360 *Pulicaria arabica* L.
57: I.3,4 447 *Ipomoea aquatica* Forssk.
57: I.5,6 451 *Ipomoea pes-caprae* (L.) R. Br.
57: I.7,8 448 *Ipomoea cairica* (L.) Sweet
57: II.1,2 432 *Ipomoea cairica* (L.) Sweet
57: II.3,4 450 *Ipomoea nil* (L.) Roth
57: II.5,6 452 *Ipomoea triflora* Forssk.
57: II.7,8 453 *Ipomoea verticillata* Forssk.
57:III.1,2 4 *Gynandriris sisyrinchium* (L.) Parl.
57:III.3,4 1363 *Cakile maritima* Scop. subsp. *aegyptiaca* (Willd.) Nyman
57:III.5,6 721 *Cakile maritima* Scop. subsp. *aegyptiaca* (Willd.) Nyman
57:III.7,8 1364 *Cakile maritima* Scop. subsp. *aegyptiaca* (Willd.) Nyman
58: I.1,2 1365 *Cakile maritima* Scop. subsp. *aegyptiaca* (Willd.) Nyman
58: I.3,4 1366 *Cakile maritima* Scop. subsp. *aegyptiaca* (Willd.) Nyman
58: I.5,6 1367 *Cakile maritima* Scop. subsp. *aegyptiaca* (Willd.) Nyman
58: I.7,8 1559 *Pavetta longiflora* Vahl
58: II.1,2 29 *Juncus acutus* L.
58: II.3,4 30 *Juncus subulatus* Forssk.

58: II.5,6	796 *Juniperus oxycedrus* L.
58: II.7,8	514 *Ludwigia stolonifera* (Guill. & Perr.) Raven
58:III.1,2	511 *Ludwigia stolonifera* (Guill. & Perr.) Raven
58:III.3,4	512 *Corchrus depressus* (L.) C. Chr.
58:III.5,6	515 *Corchrus depressus* (L.) C. Chr.
58:III.7,8	1375 *Corchorus depressus* (L.) C. Chr.
59: I.1,2	378 *Barleria prionitis* L. subsp. *appressa* (Forssk.) Brummitt & Wood
59: I.3,4	376 *Barleria prionitis* L. subsp. *appressa* (Forssk.) Brummitt & Wood
59: I.5,6	374 *Barleria bispinosa* (Forssk.) Vahl
59: I.7,8	375 *Barleria prionitis* L. subsp. *appressa* (Forssk.) Brummitt & Wood
59: II.1,2	376 *Barleria trispinosa* (Forssk.) Vahl
59: II.3,4	373 *Barleria bispinosa* (Forssk.) Vahl
59: II.5,6	1378 *Barleria bispinosa* (Forssk.) Vahl
59: II.7,8	1377 *Barleria prionitis* L. subsp. *appressa* (Forssk.) Brummitt & Wood
59:III.1,2	390 *Megalochlamys violaceum* (Vahl) Vollesen
59:III.3,4	382 *Dicliptera foetida* (Forssk.) Blatter
59:III.5,6	377 *Barleria lanceata* (Forssk.) C. Chr.
59:III.7,8	1832 *Barleria lanceata* (Forssk.) C. Chr.
60: I.1,2	387 *Hypoestes forsskalii* (Vahl) Roem. & Schult.
60: I.3,4	1381 *Hypoestes triflora* (Forssk.) Roem. & Schult.
60: I.5,6	1382 *Hypoestes triflora* (Forssk.) Roem. & Schult.
60: I.7,8	1379 *Barleria bispinosa* (Forssk.) Vahl
60: II.1,2	1380 *Barleria bispinosa* (Forssk.) Vahl
60: II.3,4	380 *Ecbolium gymnostachyum* (Nees) Milne-Redh.
60: II.5,6	386 *Ecbolium gymnostachyum* (Nees) Milne-Redh.
60: II.7,8	1384 *Ethulia conyzoides* L.
60:III.1,2	1385 *Ethulia conyzoides* L.
60:III.3,4	1386 *Ethulia conyzoides* L.
60:III.5,6	1387 *Ethulia conyzoides* L.
60:III.7,8	327 *Dorstenia foetida* (Forssk.) Schweinf.
61: I.1,2	1388 *Scariola viminea* (L.) F.W. Schmidt
61: I.3,4	1389 *Lactuca serriola* L.
61: I.5,6	77 *Lagurus ovatus* L.
61: I.7,8	78 *Lagurus ovatus* L.
61: II.1,2	1393 *Hyoseris lucida* L.
61: II.3,4	1394 *Hyoseris lucida* L.
61: II.5,6	357 *Orobanche aegyptiaca* Pers.
61: II.7,8	1395 *Orobanche aegyptiaca* Pers.
61:III.1,2	768 *Lawsonia inermis* L.
61:III.3,4	767 *Lawsonia inermis* L.
61:III.5,6	1404 *Cocculus pendulus* (J.R. & G. Forst.) Diels
61:III.7,8	695 *Cocculus pendulus* (J. R. & G. Forst.) Diels
62: I.1,2	1405 *Cocculus pendulus* (J.R. & G. Forst.) Diels
62: I.3,4	1398 *Picris asplenioides* L.
62: I.5,6	1581 *Picris cf. asplenioides* L.
62: I.7,8	1396 *Launaea* sp.
62: II.1,2	1399 *Picris asplenioides* L.
62: II.3,4	715 *Lepidium armoracia* Fisch. & Mey.
62: II.5,6	717 *Lepidium graminifolium* L.
62: II.7,8	1407 *Coronopus squamatus* (Forssk.) Asch.
62:III.1,2	830 *Usnea florida* Achar. (Lichen)
62:III.3,4	1408 *Limeum humile* Forssk.
62:III.5,6	1409 *Limeum humile* Forssk.
62:III.7,8	1410 *Limeum humile* Forssk.
63: I.1,2	307 *Moltkiopsis ciliata* (Fortssk.) I. M. Johnst.

63: I.3,4 1412 *Moltkiopsis ciliata* (Forssk.) I.M. Johnst.
63: I.5,6 306 *Moltkiopsis ciliata* (Forssk.) I. M. Johnst.
63: I.7,8 311 *Moltkiopsis ciliata* (Forssk.) I. M. Johnst.
63: II.1,2 1414 *Moltkiopsis ciliata* (Forssk.) I.M. Johnst.
63: II.3,4 300 *Heliotropium bacciferum* Forssk.
63: II.5,6 1416 *Heliotropium bacciferum* Forssk.
63: II.7,8 1417 *Heliotropium bacciferum* Forssk.
63: III.1,2 1418 *Heliotropium bacciferum* Forssk.
63: III.3,4 1419 *Heliotropium bacciferum* Forssk.
63: III.5,6 1420 *Heliotropium bacciferum* Forssk.
63: III.7,8 1421 *Lolium ? multiflorum* Lam., mixed with *L. perenne* L.
64: I.1,2 810 *Asplenium rutifolium* (Berg.) Kunze var. *bipinnatum* (Forrsk.) Schelpe
64: I.3,4 1422 *Dorycnium graecum* (L.) Ser.
64: I.5,6 1423 *Dorycnium graecum* (L.) Ser.
64: I.7,8 1424 *Lotus polyphyllus* Clarke
64: II.1,2 1425 *Lotus polyphyllus* Clarke
64: II.3.4 1426 *Lotus polyphyllus* Clarke
64: II.5,6 1428 *Lotus corniculatus* L. var. *ciliatus* Koch
64: II.7,8 1429 *Lotus arabicus* L.
64: III.1,2 1430 *Lotus arabicus* L.
64: III.3,4 1432 *Lotus halophilus* Boiss. & Sprun.
64: III.5,6 1433 *Lotus halophilus* Boiss. & Sprun.
64: III.7,8 1436 *Farsetia aegyptia* Turra
65(1-2): I.1,2 1432 *Lotus halophilus* Boiss. & Sprun.
65(1-2): I.3,4 1433 *Lotus halophilus* Boiss. & Sprun.
65(1-2): I.5,6 708 *Farsetia aegyptia* Turra
65(1-2): I.7,8 1435 *Farsetia aegyptia* Turra
65(1-2): II.1,2 1437 *Farsetia aegyptia* Turra
65(1-2): II.3,4 1436 *Farsetia aegyptia* Turra
65(1-2): II.5,6 1439 *Lupinus albus* L.
65(1-2): II.7,8 1438 *Lupinus digitatus* Forssk.
65(1-2): III.1,2 408 *Lycium europaeum* L.
65(1-2): III.3,4 1441 *Lycium europaeum* L.
65(1-2): III.5,6 1443 *Selginella imbricata* (Forssk.) Decne.
65(1-2): III.7 649 *Maerua crassifolia* Forssk.
65(1-2): III.8 [Blank]
65(2-2): I.1,2 363 *Maesa lanceolata* Forssk.
65(2-2): I.3,4 362 *Maesa lanceolata* Forssk.

65(2-2): I.5,6 — 361 *Maesa lanceolata* Forssk.
65(2-2): I.7 — 1447 *Maesa lanceolata* Forssk.
65(2-2): I.8
591 *Malva verticillata* L.
65(2-2): II.1 — 591 *Malva verticillata* L.
65(2-2): II.2,3 — 594 *Malva sylvestris* L.
65(2-2): II.4,5 — 587 *Malva parviflora* L.
65(2-2): II.6 — 1448 *Malva parviflora* L.
65(2-2): II.7,8 — 595 *Malva sylvestris* L.
65(2-2): III.1-8 — [Blank]
66: I.1,2 — 595 *Malva sylvestris* L.
66: I.3,4 — 592 *Malva sylvestris* L. var. *eriocarpa* Boiss.
66: I.5,6 — 593 Malva sylvestris L. var. *eriocarpa* Boiss.
66: I.7,8 — 1449 *Marrubium peregrinum* L.
66: II.1,2 — 219 *Marrubium peregrinum* L.
66: II.3,4 — 220 *Marrubium alysson* L.
66: II.5,6 — 221 *Marrubium alysson* L.
66: II.7,8 — 346 *Marrubium alysson* L.
66:III.1,2 — 1451 *Tanacetum parthenium* (L.) Sch.
66:III.3,4 — 1452 *Tanacetum parthenium* (L.) Sch.
66:III.5,6 — 1454 *Medicago marina* L.
66:III.7,8 — 1456 *Medicago polymorpha* L.
67: I.1,2 — 1794 *Melilotus indica* L.
67: I.3,4 — 1453 *Medicago littoralis* Loisel.
67: I.5,6 — 510 *Melhania velutina* Forssk.
67: I.7,8 — 341 *Salvia aegyptiaca* L.
67: II.1 — 1459 *Salvia aegyptiaca* L.
67: II.2,3 — 1460 *Salvia aegyptiaca* L.
67: II.4,5 — 1461 *Salvia aegyptiaca* L.
67: II.6 — 1466 *Mentha* sp.
67: II.7,8 — 335 *Mentha x piperita* L.
67:III.1.2 — 334 *Mentha longifolia* (L.) L.
67:III.3,4 — 333 *Mentha longifolia* (L.) L.
67:III.5,6 — 1467 *Mesembryanthemum nodiflorum* L.
67:III.7,8 — 629 *Mesembryanthemum nodiflorum* L.
68: I.1,2 — 630 *Mesembryanthemum nodiflorum* L.
68: I.3,4 — 631 *Mesembryanthemum nodiflorum* L.
68: I.5,6 — 632 *Mesembryanthemum nodiflorum* L.
68: I.7,8 — 1468 *Mesembryanthemum nodiflorum* L.
68: II.1,2 — 1470 *Eclipta prostrata* (L.) L.
68: II.3,4 — 1471 *Albizia julibrissin* Durazz.
68: II.5,6 — 1472 *Albizia julibrissin* Durazz.
68: II.7,8 — 1473 *Acacia asak* (Forssk.) Willd.
68:III.1,2 — 1474 *Acacia hockii* De Wild.
68:III.3,4 — 1475 *Albizia lebbeck* (L.) Benth.
68:III.5,6 — 1478 *Acacia nilotica* (L.) Del.

68: III.7,8	1479 *Acacia nilotica* (L.) Del.
69: I.1,2	1490 *Acacia* sp.
69: I.3,4	1480 *Acacia farnesiana* (L.) Willd.
69: I.5,6	1481 *Acacia farnesiana* (L.) Willd.
69: I.7,8	1482 *Acacia farnesiana* (L.) Willd.
69: II.1,2	1483 *Acacia farnesiana* (L.) Willd.
69: II.3,4	1484 *Acacia hamulosa* Benth.
69: II.5,6	1485 *Acacia hamulosa* Benth.
69: II.7,8	1486 *Pterolobium stellatum* (Forssk.) Brenan
69: III.1,2	1487 *Pterolobium stellatum* (Forssk.) Brenan
69: III.3,4	1488 *Acacia tortilis* (Forssk.) Hayne
69: III.5,6	1489 *Acacia tortilis* (Forssk.) Hayne
69: III.7,8	1476 *Acacia mellifera* (Vahl) Benth.
70: I.1,2	1477 *Acacia mellifera* (Vahl) Benth.
70: I.3,4	702 *Hypecoum aegyptiacum* (Forssk.) Aschers. & Schweinf.
70: I.5,6	703 *Hypecoum aegyptiacum* (Forssk.) Aschers. & Schweinf.
70: I.7,8	1491 *Hypecoum aegyptiacum* (Forssk.) Asch. & Schweinf.
70: II.1,2	223 *Ballota acetabulosa* (L.) Benth.
70: II.3,4	1492 *Ballota acetabulosa* (L.) Benth.
70: II.5,6	330 *Ajuga iva* (L.) Schreb.
70: II.7,8	351 *Ajuga iva* (L.) Schreb.
70: III.1,2	1493 *Ajuga iva* (L.) Schreb.
70: III.3,4	236 *Adenium obesum* (Forssk.) Roem. & Schult. (flws) & *Breonadia salicina* (Vahl) Hepper & Wood (leaves) [mixed sheet]
70: III.5,6	237 *Adenium obesum* (Forssk.) Roem. & Schult. (flws) & *Breonadia salicina* (Vahl) Hepper & Wood (leaves) [mixed sheet]
70: III.7,8	235 *Adenium obesum* (Forssk.) Roem. & Schult.
71: I.1,2	1223 *Anisotes trisulcus* (Forssk.) Nees
71: I.3,4	1833 *Adenium obesum* (Forssk.) Roem. & Schult.
71: I.5,6	546 *Neurada procumbens* L.
71: I.7,8	547 *Neurada procumbens* L.
71: II.1,2	545 *Neurada procumbens* L.
71: II.3,4	1494 *Neurada procumbens* L.
71: II.5,6	613 *Nymphaea lotus* L.
71: II.7,8	324 *Becium filamentosum* (Forssk.) Chiov.
71: III.1,2	350 *Plectranthus ovatus* Benth.
71: III.3,4,5	211 *Ocimum forskolei* Benth.
71: III.6,7	1500 *Ocimum forskolei* Benth.
71: III.8	1501 *Ocimum forskolei* Benth.
72: I.1,2	1502 *Becium serpyllifolium* (Forssk.) Wood
72: I.3,4	213 *Becium serpyllifolium* (Forssk.) Wood
72: I.5,6	326 *Becium serpyllifolium* (Forssk.) Wood
72: I.7,8	325 *Becium serpyllifolium* (Forssk.) Wood
72: II.1,2	348 *Plectranthus hadiensis* (Forssk.) Schweinf. ex Spreng.
72: II.3,4	626 *Oncoba spinosa* Forssk.
72: II.5,6	627 *Oncoba spinosa* Forssk.
72: II.7,8	1504 *Ononis spinosa* L.
72: III.1,2	1505 *Ononis spinosa* L.
72: III.3,4	1506 *Ononis spinosa* L.
72: III.5,6	1507 *Ononis spinosa* L.
72: III.7,8	1508 *Ononis spinosa* L.
73: I.1,2	1521 *Ononis vaginalis* Vahl
73: I.3,4	1522 *Ononis vaginalis* Vahl
73: I.5,6	1523 *Ononis vaginalis* Vahl

73: I.7	1524 *Ononis vaginalis* Vahl
73: I.8	1509 *Rhynchosia flava* (Forssk.) Thulin
73: II.1	1509 *Rhynchosia flava* (Forssk.) Thulin
73: II.2,3	1510 *Rhynchosia flava* (Forssk.) Thulin
73: II.4,5	1511 *Ononis minutissima* L.
73: II.6,7	1512 *Ononis minutissima* L.
73: II.8	1513 *Ononis natrix* L.
73: III.1	1513 *Ononis natrix* L.
73: III.2,3	1514 *Ononis natrix* L.
73: III.4,5	1516 *Lotus quinatus* (Forssk.) Gillett
73: III.6,7	1517 *Lotus quinatus* (Forssk.) Gillett
73: III.8	1518 *Lotus quinatus* (Forssk.) Gillett
74: I.1,2	1520 *Ononis serrata* Forssk.
74: I.3,4	1519 *Ononis serrata* Forssk.
74: I.5,6	1525 *Pentas lanceolata* (Forssk.) Defl.
74: I.7,8	1526 *Pentas lanceolata* (Forssk.) Defl.
74: II.1,2	1528 *Eulophia streptopetala* Lindl.
74: II.3,4	225 *Origanum vulgare* L. subsp. *hirtum* (Link) Lets.
74: II.5,6	226 *Origanum vulgare* L. subsp. *hirtum* (Link) Lets.
74: II.7,8	227 *Origanum vulgare* L. subsp. *hirtum* (Link) Lets.
74: III.1,2	229 *Origanum majorana* L.
74: III.3,4	8 *Asphodelus fistulosus* L. var *tenuifolius* (Cav.) Bak.
74: III.5,6	1531 *Ornithopus compressus* L.
74: III.7,8	358 *Orobanche crenata* Forssk.
75: I.1,2	1536 *Orobanche ramosa* L.
75: I.3,4	1537 *Cistanche phelypaea* (L.) Coutinho
75: I.5,6	541 *Corbichonia decumbens* (Forssk.) Exell
75: I.7,8	537 *Osyris alba* L.
75: II.1,2	536 *Osyris alba* L.
75: II.3,4	1541 *Oxalis corniculata* L.
75: II.5,6	1 *Phoenix dactylifera* L.
75: II.7,8	23 *Pancratium maximum* Forssk.
75: III.1,2	46 *Ochthochloa compressa* (Forssk.) Hilu
75: III.3,4	1544 *Cynodon dactylon* (L.) Pers.
75: III.5,6	1548 *Pennisetum divisum* (J. F. Gmel.) Henr.
75: III.7,8	1549 *Pennisetum divisum* (J. F. Gmel.) Henr.
76: I.1,2	1550 *Pennisetum divisum* (J. F. Gmel.) Henr.
76: I.3,4	113 *Paspalidium geminatum* (Forssk.) Stapf
76: I.5,6	116 *Cenchrus setigerus* Vahl
76: I.7,8	1551 *Panicum repens* L.
76: II.1,2	1553 *Panicum* sp.
76: II.3,4	65 *Panicum coloratum* L.
76: II.5,6	87 *Brachiaria mutica* (Forssk.) Stapf
76: II.7,8	86 *Brachiaria mutica* (Forssk.) Stapf
76: III.1,2	114 *Digitaria sanguinalis* (L.) Scop.
76: III.3,4	85 *Panicum turgidum* Forssk.
76: III.5,6	1554 *Panicum turgidum* Forssk.
76: III.7,8	138 *Panicum turgidum* Forssk.
77: I.1,2	144 *Trianthema triquetra* Willd.
77: I.3,4	539 *Trianthema triquetra* Willd.
77: I.5,6	536bis *Trianthema crystallina* (Forssk.) Vahl
77: I.7,8	769 *Parietaria alsinifolia* Del.
77: II.1,2	240 *Swertia polynectaria* (Forssk.) Gilg
77: II.3,4	1556 *Swertia polynectaria* (Forssk.) Gilg
77: II.5,6	532 *Thymelaea hirsuta* (L.) Endl.

77: II.7,8	1558 *Thymelaea hirsuta* (L.) Endl.
77:III.1,2	758 *Nitraria retusa* (Forssk.) Aschers.
77:III.3,4	759 *Nitraria retusa* (Forssk.) Aschers.
77:III.5,6	463 *Nitraria retusa* (Forssk.) Aschers.
77:III.7,8	1563 *Lythrum hyssopifolium* L.
78: I.1,2	1564 *Lythrum hyssopifolium* L.
78: I.3,4	1565 *Lythrum hyssopifolium* L.
78: I.5,6	75 *Polypogon monspeliensis* (L.) Desf.
78: I.7,8	68 *Polypogon monspeliensis* (L.) Desf.
78: II.1,2	69 *Polypogon monspeliensis* (L.) Desf.
78: II.3,4	132 *Sporobolus arenarius* (Gouan) Jouve
78: II.5,6	63 *Polypogon viridis* (Gouan) Breistr.
78: II.7,8	1566 *Polypogon viridis* (Gouan) Breistr.
78:III.1,2	117 *Pennisetum setaceum* (Forssk.) Chiov.
78:III.3,4	53 *Crypsis schoenoides* (L.) Lam.
78:III.5,6	52 *Crypsis schoenoides* (L.) Lam.
78:III.7,8	115 *Digitaria velutina* (Forssk.) P. Beauv.
79: I.1,2	112 *Digitaria velutina* (Forssk.) P. Beauv.
79: I.3,4	573 *Gisekia pharnaceoides* L.
79: I.5,6	1567 *Mollugo cerviana* (L.) Ser.
79: I.7,8	1568 *Vigna aconitifolia* (Jacq.) Marechal
79: II.1,2	1569 *Vigna aconitifolia* (Jacq.) Marechal
79: II.3,4	1570 *Vigna radiata* (L.) Wilczek
79: II.5,6	222 *Leucas alba* (Forssk.) Sebald & *Leucas glabrata* (Vahl) R.Br. [mixed collection]
79: II.7,8	217 *Leucas glabrata* (Vahl) R. Br.
79:III.1,2	228 *Phlomis herba-venti* L.
79:III.3,4	1572 *Phlomis herba-venti* L.
79:III.5,6	1814 *Priva adhaerens* (Forssk.) Chiov.
79:III.7,8	1815 *Priva adhaerens* (Forssk.) Chiov.
80: I.1,2	1574 *Phyllanthus tenellus* Muell.Arg. var. *arabicus* Muell.Arg.
80: I.3,4	1575 *Phyllanthus tenellus* Muell.Arg. var. *arabicus* Muell.Arg.
80: I.5,6	1834 *Phyllanthus ovalifolius* Forssk.
80: I.7,8	417 *Physalis angulata* L.
80: II.1,2	1577 *Withania somnifera* (L.) Dunal
80: II.3,4	416 *Withania somnifera* (L.) Dunal
80: II.5,6	746 *Pistacia terebinthus* L.
80: II.7,8	747 *Rhus cf. abyssinica* Oliv.
80:III.1,2	252 *Plantago coronopus* L.
80:III.3,4	1583 *Plantago coronopus* L.
80:III.5,6	261 *Plantago crassifolia* Forssk.
80:III.7,8	255 *Plantago cylindrica* Forssk.
81: I.1,2	260 *Plantago cylindrica* Forssk.
81: I.3,4	257 *Plantago cylindrica* Forssk.
81: I.5,6	259 *Plantago ovata* Forssk. var. *decumbens* (Forssk.) Zohary
81: I.7,8	254 *Plantago ovata* Forssk. var. *decumbens* (Forssk.) Zohary
81: II.1,2	263 *Plantago lanceolata* L.
81: II.3,4	262 *Plantago major* L.
81: II.5,6	253 *Plantago ovata* Forssk.
81: II.7,8	250 *Plantago ovata* Forssk. var. *ovata*
81:III.1,2	249 *Plantago coronopus* L. (plant) & *Plantago ovata* Forssk. var. *ovata* (infloresc.) [mixed sheet]
81:III.3,4	256 *Plantago scabra* Moench
81:III.5,6	251 *Plantago serraria* L.
81:III.7,8	258 *Plantago serraria* L.

82: I.1,2	473 *Platanus orientalis* L.
82: I.3,4	61 *Eragrostis aegyptiaca* (Willd.) Del.
82: I.5,6	54 *Eragrostis aegyptiaca* (Willd.) Del.
82: I.7,8	59 *Eragrostis kiwuensis* Jedwabn.
82: II.1,2	1248 *Eragrostis pilosa* (L.) P. Beauv.
82: II.3,4	39 *Poa schimperiana* A. Rich.
82: II.5,6	1587 *Delonix elata* (L.) Gamble
82: II.7,8	1588 *Sphaeranthus suaveolens* (Forssk.) DC.
82:III.1,2	1589 *Sphaeranthus suaveolens* (Forssk.) DC.
82:III.3,4	1590 *Sphaeranthus suaveolens* (Forssk.) DC.
82:III.5,6	1591 *Sphaeranthus suaveolens* (Forssk.) DC.
82:III.7,8	497 *Polygala tinctoria* Vahl
83: I.1,2	1593 *Polygala tinctoria* Vahl
83: I.3,4	611 *Polygala* sp. indet.
83: I.5,6	1592 *Polygala* sp.
83: I.7,8	1600 *Polygonum pulchellum* Lois.
83: II.1,2,3	1595 *Polygonum maritimum* L.
83: II.4,5	1596 *Polygonum maritimum* L.
83: II.6,7	1597 *Polygonum equisetiforme* Sibth. & Sm.
83: II.8	1599 *Polygonum equisetiforme* Sibth. & Sm.
83: III.1	1599 *Polygonum equisetiforme* Sibth. & Sm.
83:III.2,3	1598 *Polygonum equisetiforme* Sibth. & Sm.
83:III.4,5	1594 *Polygonum salicifolium* Willd.
83:III.6,7	809 *Christella dentata* (Forssk.) Brownsey & Jermy
84: III.8	[Blank]
84: I.1,2	807 *Arthropteris orientalis* (J. F. Gmel.) Posthum.
84: I.3,4	822 *Arthropteris orientalis* (J. F. Gmel.) Posthum.
84: I.5,6	814 *Polystichum setiferum* (Forssk.) Woynar
84: I.7,8	548 *Portulaca quadrifida* L.
84: II.1,2	540bis *Portulaca quadrifida* L.
84: II.3,4	539bis *Portulaca quadrifida* L.
84: II.5,6	542bis *Portulaca quadrifida* L.
84: II.7,8	1602 *Potentilla dentata* Forssk.
84:III.1,2	354 *Primula verticillata* Forssk.
84:III.3,4	353 *Primula verticillata* Forssk.
84:III.5,6	1603 *Primula verticillata* Forssk.
84:III.7,8	562 *Pteranthus dichotomus* Forssk.
85: I.1,2	553 *Pteranthus dichotomus* Forssk.
85: I.3,4	1605 *Pteranthus dichotomus* Forssk.
85: I.5,6	1606 *Pteranthus dichotomus* Forssk.
85: I.7,8	1607 *Pteranthus dichotomus* Forssk.
85: II.1,2	1608 *Pteranthus dichotomus* Forssk.
85: II.3,4	803 *Pteris vittata* L.
85: II.5,6	802 *Pteris vittata* L.
85: II.7,8	799 *Pteris cretica* L.
85:III.1,2	800 *Pteris cretica* L.
85:III.3,4	764 *Flemingia* cf. *grahamiana* Wight & Arn.
85:III.5,6	765 *Pyrus amygdalifolius* Vill.
85:III.7,8	786 *Quercus ilex* L.
86: I.1,2	790 *Quercus coccifera* L.
86: I.3,4	791 *Quercus ilex* L.
86: I.5,6	787 *Quercus pubescens* Willd.
86: I.7,8	698 *Ranunculus multifidus* Forssk.
86: II.1,2	697 *Ranunculus sceleratus* L.
86: II.3,4	724 *Enarthrocarpus lyratus* (Forssk.) DC.

86: II.5,6 710 *Enarthrocarpus lyratus* (Forssk.) DC.
86: II.7,8 482 *Reamurea hirtella* Jaub. & Spach
86:III.1,2 1610 *Reamuria hirtella* Jaub. & Spach
86:III.3,4 606 *Reseda decursiva* Forssk.
86:III.5,6 610 *Reseda decursiva* Forssk.
86:III.7,8 1617 *Caylusea hexagyna* (Forssk.) M.L.Green
87: I.1,2 1618 *Caylusea hexagyna* (Forssk.) M.L.Green
87: I.3,4 608 *Reseda luteola* L.
87: I.5,6 1620 *Reseda luteola* L.
87: I.7,8 612 *Reseda luteola* L.
87: II.1,2 1621 *Reseda arabica* Boiss.
87: II.3,4 609 *Reseda arabica* Boiss.
87: II.5,6 499 *Ziziphus spina-christi* (L.) Desf.
87: II.7,8 501 *Ziziphus spina-christi* (L.) Desf.
87:III.1,2 500 *Ziziphus spina-christi* (L.) Desf.
87:III.3,4 538bis *Zalaya pentandra* (L.) C. Jeffrey
87:III.5,6 1629 *Gypsophila capillaris* (Forssk.) C. Chr.
87:III.7,8 1630 *Cleome droserifolia* (Forssk.) Del.
88: I.1,2 619 *Cleome droserifolia* (Forsk.) Del.
88: I.3,4 620 *Cleome droserifolia* (Forsk.) Del.
88: I.5,6 1631 *Rosa andegavensis* Bast.
88: I.7,8 1632 *Rosa andegavensis* Bast.
88: II.1,2 1635 *Rubia peregrina* L.
88: II.3,4 369 *Asystasia guttata* (Forssk.) Brummitt
88: II.5,6 368 *Phaulopsis imbricata* (Forssk.) Sweet
88: II.7,8 367 *Phaulopsis imbricata* (Forssk.) Sweet
88:III.1,2 371 *Asystasia gangetica* (L.) T. Anders.
88:III.3,4 1638 *Ruellia patula* Jacq.
88:III.5,6 1642 *Rubus aegyptiacus* L.
88:III.7,8 1643 *Rumex dentatus* L.
89: I.1,2 1639 *Rumex nervosus* Vahl
89: I.3,4 1640 *Rumex nervosus* Vahl
89: I.5,6 1641 *Rumex nervosus* Vahl
89: I.7,8 1644 *Rumex pictus* Forssk.
89: II.1,2 1646 *Rumex vesicarius* L.
89: II.3,4 1645 *Rumex vesicarius* L.
89: II.5,6 1648 *Haplophyllum buxbaumii* (Poir.) G. Don
89: II.7,8 755 *Haplophyllum tuberculatum* (Forssk.) A. Juss.
89:III.1,2 756 *Haplophyllum tuberculatum* (Forssk.) A. Juss.
89:III.3,4 1649 *Haplophyllum tuberculatum* (Forssk.) A. Juss.
89:III.5,6 64 *Saccharum spontaneum* L. subsp. *aegyptiacum* (Willd.) Hack.
89:III.7,8 74 *Saccharum spontaneum* L. subsp. *aegyptiacum* (Willd.) Hack.
90: I.1,2 48 *Lasiurus scindicus* Henrard
90: I.3,4 1652 *Lasiurus scindicus* Henrard
90: I.5,6 1653 *Lasiurus scindicus* Henrard
90: I.7 1654 *Lasiurus scindicus* Henrard
90: I.8 1655 *Lasiurus scindicus* Henrard
90: II.1 1655 *Lasiurus scindicus* Henrard
90: II.2,3 1656 *Lasiurus scindicus* Henrard
90: II.4,5 1657 *Lasiurus scindicus* Henrard
90: II.6,7 1658 *Lasiurus scindicus* Henrard
90: II.8 682 *Cissus rotundifolius* (Forssk.) Vahl
90:III.1 682 *Cissus rotundifolius* (Forssk.) Vahl
90:III.2,3 684 *Cissus rotundifolius* (Forssk.) Vahl
90:III.4,5 1661 *Halocnemum strobilaceum* (Pallas) M. Bieb.

90: III.6,7 171 *Halocnemum strobilaceum* (Pallas) M. Bieb.
90: III.8 [Blank]
91: I.1,2 172 *Halocnemum strobilaceum* (Pallas) M. Bieb.
91: I.3,4 147 *Halocnemum strobilaceum* (Pallas) M. Bieb.
91: I.5,6 168 *Halocnemum strobilaceum* (Pallas) M. Bieb.
91: I.7,8 169 *Arthrocnemum macrostachyum* (Moric.) K. Koch
91: II.1,2 1662 *Halocnemum strobilaceum* (Pallas) M. Bieb.
91: II.3,4 143 *Salicornia fruticosa* (L.) L.
91: II.5,6 170 *Salicornia fruticosa* (L.) L.
91: II.7,8 1663 *Halocnemum strobilaceum* (Pallas) M. Bieb.
91: III.1,2 173 *Halopeplis perfoliata* (Forssk.) Bunge
91: III.3,4 174 *Arthrocnemum macrostachyum* (Moric.) K. Koch
91: III.5,6 146 *Arthrocnemum macrostachyum* (Moric.) K. Koch
91: III.7,8 459 *Salix subserrata* Willd.
92: I.1,2 460 *Salix cf. tetrasperma* Roxb.
92: I.3,4 1664 *Anabasis articulata* (Forssk.) Moq.
92: I.5,6 1665 *Anabasis articulata* (Forssk.) Moq.
92: I.7,8 1666 *Anabasis articulata* (Forssk.) Moq.
92: II.1,2 188 *Anabasis articulata* (Forssk.) Moq.
92: II.3,4 150 *Anabasis articulata* (Forssk.) Moq.
92: II.5,6 1667 *Salsola volkensii* Asch. & Schweinf.
92: II.7,8 158 *Salsola inermis* Forssk.
92: III.1,2 152 *Salsola tetrandra* Forssk.
92: III.3,4 163 *Salsola inermis* Forssk.
92: III.5,6 181 *Salsola inermis* Forssk.
92: III.7,8 1668 *Salsola inermis* Forssk.
93: I.1,2 1669 *Salsola inermis* Forssk.
93: I.3,4 151 *Salsola inermis* Forssk.
93: I.5,6 1670 *Salsola inermis* Forssk.
93: I.7,8 156 *Salsola kali* L.
93: II.1,2 1671 *Salsola kali* L.
93: II.3,4 153 *Salsola kali* L.
93: II.5 1672 *Salsola kali* L.
93: II.6,7 *Salsola longifolia* Forssk. (not a Forsskål specimen).
93: II.8 1673 *Salsola longifolia* Forssk.
93: III.1,2 155 *Salsola longifolia* Forssk.
93: III.3,4 159 *Bassia muricata* (L.) Aschers.
93: III.5,6 1675 *Bassia muricata* (L.) Aschers.
93: III.7,8 1676 *Bassia muricata* (L.) Aschers.
94: I.1,2 1678 *Noaea mucronata* (Forssk.) Ascers. & Schweinf.
94: I.3,4 182 *Noaea mucronata* (Forssk.) Aschers. & Schweinf.
94: I.5,6 154 *Salsola soda* L.
94: I.7,8 648 *Salsola soda* L.
94: II.1,2 148 *Salsola soda* L.
94: II.3,4 176 *Salsola tetrandra* Forssk.
94: II.5,6 177 *Salsola tetrandra* Forssk.
94: II.7,8 1679 *Salsola tetrandra* Forssk.
94: III.1,2 166 *Salsola longifolia* Forssk.
94: III.3,4 1674 *Salsola longifolia* Forssk.
94: III.5,6 157 *Salsola longifolia* Forssk.
94: III.7,8 1681 *Salvia forskahlei* L.
95: I.1,2 344 *Salvia forskahlei* L.
95: I.3,4 1682 *Salvia forskahlei* L.
95: I.5,6 1683 *Salvia forskahlei* L.
95: I.7,8 1684 *Salvia forskahlei* L.

95: II.1,2 342 *Salvia lanigera* Poir.
95: II.3,4 1680 *Salvia aegyptiaca* L.
95: II.5,6 343 *Salvia merjamie* Forssk.
95: II.7,8 1685 *Anthemis tinctoria* L.
95:III.1,2 1686 *Achillea fragrantissima* (Forssk.) Sch.Bip.
95:III.3,4 1687 *Achillea fragrantissima* (Forssk.) Sch.Bip.
95:III.5,6 1688 *Achillea fragrantissima* (Forssk.) Sch.Bip.
95:III.7,8 1689 *Anacyclus monanthos* (L.) Thell.
96: I.1,2 1690 *Anacyclus monanthos* (L.) Thell.
96: I.3,4 1691 *Anacyclus monanthos* (L.) Thell.
96: I.5 1692 *Velezia rigida* L.
96: I.6,7 565 *Velezia rigida* L.
96: I.8 1693 *Saponaria officinalis* L.
96: II.1,2 317 *Avicennia marina* (Forssk.) Vierh.
96: II.3,4 318 *Avicennia marina* (Forssk.) Vierh.
96: II.5,6 1701 *Fimbristylis bis-umbellata* (Forssk.) Bub.
96: II.7,8 1702 *Bolboschoenus maritimus* (L.) Palla
96:III.1,2 1703 *Schoenoplectus articulatus* (L.) Palla
96:III.3,4 1704 *Schoenoplectus articulatus* (L.) Palla
96:III.5,6 1705 *Cyperus capitatus* Vand.
96:III.7,8 1706 *Cyperus capitatus* Vand.
97: I.1,2 1707 *Cyperus capitatus* Vand.
97: I.3,4 1708 *Scoparia dulcis* L.
97: I.5,6 1709 *Scoparia dulcis* L.
97: I.7,8 1712 *Scorpiurus muricatus* L.
97: II.1,2 396 *Scutellaria arabica* Jaub. & Spach
97: II.3,4 399 *Scutellaria arabica* Jaub. & Spach
97: II.5,6 1713 *Dasypyrum villosum* (L.) Cand.
97: II.7,8 1714 *Senecio hadiensis* Forssk.
97:III.1,2 1715 *Senecio aegyptius* L.
97:III.3,4 1716 *Senecio aegyptius* L.
97:III.5,6 1717 *Senecio aegyptius* L.
97:III.7,8 1719 *Senecio aquaticus* L.
98: I.1,2 1720 *Senecio lyratus* Forssk.
98: I.3,4 1721 *Senecio lyratus* Forssk.
98: I.5,6 1722 *Senecio glaucus* L.
98: I.7,8 1723 *Inula critmoides* L.
98: II.1,2 1724 *Centaurea spinosa* L.
98: II.3,4 366 *Sesamum indicum* L.
98: II.5,6 364 *Sesamum indicum* L.
98: II.7,8 677 ?*Seseli tortuosum* L.
98:III.1,2 596 *Abutilon bidentatum* A. Rich.
98:III.3,4 589 *Abutilon bidentatum* A. Rich.
98:III.5,6 1726 *Abutilon bidentatum* A.Rich.
98:III.7,8 901 *Abutilon pannosum* (Forst. f.) Schlecht.
99: I.1,2 1727 *Abutilon bidentatum* A.Rich.
99: I.3,4 1728 *Sida ovata* Forssk.
99: I.5,6 584 *Sida alba* L.
99: I.7,8 1729 *Sida ovata* Forssk.
99: II.1,2 1730 *Sida ovata* Forssk.
99: II.3,4 597 *Wissadula amplissima* (L.) R. E. Fries var. *rostrata* (Schum. & Thon.) R. E. Fries
99: II.5,6 279 *Ehretia cymosa* Thonn.
99: II.7,8 571 *Silene conica* L.
99:III.1,2 569 *Silene involuta* Forssk.

99:III.3,4	552 *Silene succulenta* Forssk.
99:III.5,6	567 *Silene succulenta* Forssk.
99:III.7,8	561 *Silene succulenta* Forssk.
100: I.1,2	*Silene succulenta* Forssk. (not a Forsskål specimen).
100: I.3,4	560 *Silene villosa* Forssk.
100: I.5,6	550 *Silene villosa* Forssk.
100: I.7,8	640 *Cleome amblyocarpa* Barr. & Urb.
100: II.1,2	1736 *Anarrhinum forskaohlii* (J. F. Gmel.) Cuf.
100: II.3,4	1737 *Anarrhinum forskaohlii* (J. F. Gmel.) Cuf.
100: II.5,6	1740 *Sinapis allionii* Jacq.
100: II.7	1738 *Diplotaxis harra* (Forssk.) Boiss.
100: II.8	1739 *Diplotaxis harra* (Forssk.) Boiss.
100:III.1,2	722 *Diplotaxis harra* (Forssk.) Boiss.
100:III.3,4	1742 *Capparis decidua* (Forssk.) Edgew.
100:III.5,6	641 *Capparis decidua* (Forssk.) Edgew.
100:III.7,8	1743 *Capparis decidua* (Forssk.) Edgew.
101: I.1,2	647 *Capparis decidua* (Forssk.) Edgew.
101: I.3,4	420 *Solanum nigrum* L. s.l. or *Solanum villosum* Miller
101: I.5,6	407 *Solanum nigrum* L. s.l. or *Solanum villosum* Miller
101: I.7,8	421 *Solanum nigrum* L. s.l. or *Solanum villosum* Miller
101: II.1,2	422 *Solanum nigrum* L. s.l. or *Solanum villosum* Miller
101: II.3,4	405 *Solanum nigrum* L. s.l. or *Solanum villosum* Miller
101: II.5,6	411 *Solanum glabratum* Dunal
101: II.7,8	429 *Solanum glabratum* Dunal
101:III.1,2	1744 *Solanum coagulans* Forssk.
101:III.3,4	410 *Solanum coagulans* Forssk.
101:III.5,6	409 *Solanum coagulans* Forssk.
101:III.7,8	1751 *Solanum microcarpum* Vahl
102: I.1,2	412 *Solanum carense* Dunal
102: I.3,4	418 *Solanum dulcamara* L.
102: I.5,6	1745 *Solanum incanum* L.
102: I.7,8	1747 *Solanum incanum* L.
102: II.1,2	415 *Solanum incanum* L.
102: II.3,4	419 *Solanum nigrum* L. s.l. or *Solanum villosum* Miller
102: II.5,6	406 *Solanum terminale* Forssk.
102: II.7,8	414 *Solanum forskalii* Dunal
102:III.1,2	1755 *Sonchus tenerrimus* L.
102:III.3,4	1754 *Sonchus oleraceus* L.
102:III.5,6	763 *Sorbus domestica* L.
102:III.7.8	829 "*Spongia flabelliformis* Forssk." (?Animal).
103: I.1,2	824 "*Spongia tubulosa* Forssk." (?Animal).
103: I.3,4	340 *Stachys obliqua* Waldst. & Kit.(*S. orientalis* L.)
103: I.5,6	339 *Stachys aegyptiaca* Pers.
103: I.7,8	337 *Stachys recta* L. subsp. *subcrenata* (Vis.) Briq.
103: II.1,2	274 *Caralluma subulata* (Forssk.) Decn.
103: II.3,4	524 *Limonium pruinosum* (L.) Kuntze
103: II.5,6	1760 *Limonium pruinosum* (L.) Kuntze
103: II.7,8	522 *Limonium axillare* (Forssk.) Kuntze
103:III.1,2	519 *Limonium cylindrifolium* (Forssk.) Kuntze
103:III.3,4	520 *Limonium cylindrifolium* (Forssk.) Kuntze
103:III.5,6	1761 *Limonium cylindrifolium* (Forssk.) Kuntze
103:III.7,8	1762 *Limoniastrum monopetalum* (L.) Boiss.
104: I.1,2	526 *Limonium gmelinii* (Willd.) Kuntze
104: I.3,4	525 *Limonium gmelinii* (Willd.) Kuntze
104: I.5,6	528 *Goniolimon incanum* (L.) Hepper

104: I.7,8 523 *Goniolimon incanum* (L.) Hepper

104: II.1,2 527 *Goniolimon incanum* (L.) Hepper

104: II.3,4 586 *Sida alba* L.

104: II.5,6 575 *Sida alba* L.

104: II.7,8 1763 *Sida alba* L.

104: III.1,2 1764 *Sida alba* L.

104: III.3,4 20 *Ottelia alismoides* (L.) Pers.

104: III.5,6 164 *Suaeda aegyptiaca* (Hasselq.) Zohary

104: III.7,8 162 *Suaeda fruticosa* Forssk. ex J. F. Gmel.

105: I.1,2 145 *Schanginia hortensis* (Forssk. ex J. F. Gmel.) Moq.

105: I.3,4 165 *Schanginia hortensis* (Forssk. ex J. F. Gmel.) Moq.

105: I.5,6 189 *Schanginia hortensis* (Forssk. ex J. F. Gmel.) Moq.

105: I.7,8 167 *Suaeda monoica* Forssk. ex J. F. Gmel.

105: II.1,2 180 *Suaeda monoica* Forssk. ex J. F. Gmel.

105: II.3,4 161 *Suaeda vera* Forssk. ex J. F. Gmel.

105: II.5,6 160 *Suaeda vermiculata* Forssk. ex J. F. Gmel.

105: II.7,8 1771 *Schouwia purpurea* (Forssk.) Schweinf.

105: III.1,2 1772 *Schouwia purpurea* (Forssk.) Schweinf.

105: III.3,4 716 *Schouwia purpurea* (Forssk.) Schweinf.

105: III.5,6 719 *Schouwia purpurea* (Forssk.) Schweinf.

105: III.7,8 1774 *Tamarix tetragyna* Ehrenb.

106: I.1,2 480 *Tamarix aphylla* (L.) Karsten

106: I.3,4 1775 *Tamarix aphylla* (L.) Karsten

106: I.5,6 1776 *Tamarix aphylla* (L.) Karsten

106: I.7,8 481 *Tamarix aphylla* (L.) Karsten

106: II.1,2 1777 *Tamarix aphylla* (L.) Karsten

106: II.3,4 1778 *Cotula anthemoides* L.

106: II.5,6 1779 *Cotula anthemoides* L.

106: II.7,8 1780 *Cotula anthemoides* L.

106: III.1,2 347 *Teucrium polium* L.

106: III.3,4 1782 *Teucrium polium* L.

106: III.5,6 332 *Teucrium polium* L.

106: III.7,8 331 *Teucrium scordium* L. subsp. *scordioides* (Schreb.) Maire & Petitmengin

107: I.1,2 538 *Thesium linophyllum* L.

107: I.3,4 232 *Thymus* sp. 1. (Th. *ciliatus* Forssk.)

107: I.5,6 231 *Micromeria imbricata* (Forssk.) C. Chr.

107: I.7,8 210 *Micromeria imbricata* (Forssk.) C. Chr.

107: II.1,2 233 *Thymus laevigatus* Vahl

107: II.3,4 209 *Thymus* sp. 2

107: II.5,6 1785 *Cyanotis* sp.

107: II.7,8 34 *Cyanotis* sp.

107: III.1,2 470 *Dobera glabra* (Forssk.) Poir.

107: III.3,4 1787 *Urospermum picroides* (L.) F.W. Schmidt

107: III.5,6 744 *Tribulus terrestris* L.

107: III.7,8 743 *Tribulus pentandrus* Forssk.

108: I.1,2 1788 *Tribulus pentandrus* Forssk.

108: I.3,4 1789 *Trifolium alexandrinum* L.

108: I.5,6 1790 *Trifolium alexandrinum* L.

108: I.7,8 1791 *Trifolium alexandrinum* L.

108: II.1,2 1793 *Trifolium resupinatum* L.

108: II.3,4 1795 *Melilotus indica* L.

108: II.5,6 1797 *Trigonella hamosa* L.

108: II.7,8 1798 *Trigonella hamosa* L.

108: III.1,2 1799 *Trigonella hamosa* L.

108: III.3,4 — 1801 *Trigonella stellata* Forssk.
108: III.5,6 — 1802 *Trigonella stellata* Forssk.
108: III.7,8 — 1803 *Trigonella stellata* Forssk.
109: I.1,2 — 1804 *Trigonella stellata* Forssk.
109: I.3,4 — 1805 *Trigonella stellata* Forssk.
109: I.5,6 — 1806 *Trigonella stellata* Forssk.
109: I.7,8 — 118 *Trisetaria linearis* Forssk.
109: II.1,2 — 1586 *Elymus* sp.
109: II.3,4 — 483 *Triumfetta rhomboidea* Jacq.
109: II.5,6 — 1808 *Triumfetta rhomboidea* Jacq.
109: II.7,8 — 658 *Cucumis prophetarum* L.
109: III.1,2 — 657 *Cucumis prophetarum* L.
109: III.3,4 — 659 *Cucumis prophetarum* L.
109: III.5,6 — 661 *Cucumis melo* L. & *Kedrostis leloja* (Forssk. ex J. F. Gmel.) C. Jeffrey [mixed sheet]
109: III.7,8 — 662 *Coccinia grandis* (L.) Voigt
110: I.1,2 — 663 *Coccinia grandis* (L.) Voigt
110: I.3,4 — 666 *Coccinia grandis* (L.) Voigt
110: I.5,6 — Algae fam., gen. & sp. indet.
110: I.7,8 — 863 *Halarachnion ligulatum* (Woodw.) Ktz. (Algae)
110: II.1,2 — Algae fam., gen. & sp. indet.
110: II.3,4 — 865 *Chaetomorpha melagonium* (Weber & Mohr) Ktz. (Algae)
110: II.5,6 — 864 *Ulva lactuca* L. (Algae)
110: II.7,8 — 853 *Ulva reticulata* Forssk. (Algae)
110: III.1,2 — Algae fam., gen. & sp. indet.
110: III.3 — Algae fam., gen. & sp. indet.
110: III.4 — Algae fam., gen. & sp. indet.
110: III.5,6 — 600 *Urena ovalifolia* Forssk.
110: III.7,8 — 1811 *Hibiscus ovalifolius* (Forssk.) Vahl
111: I.1,2 — 772 *Laportea aestuans* (L.) Chew
111: I.3,4 — 774 *Droguetia iners* (Forssk.) Schweinf.
111: I.5,6 — 771 *Girardinia diversifolia* (Link) Friis
111: I.7,8 — 770 *Pouzolzia parasitica* (Forssk.) Schweinf.
111: II.1,2 — 773 *Urtica urens* L.
111: II.3,4 — 1816 *Valantia hispida* L.
111: II.5,6 — 1817 *Valantia muralis* L.
111: II.7 — 1813 *Utricularia inflexa* Forssk.
111: II.8 — 356 *Utricularia inflexa* Forssk.
111: III.1 — 356 *Utricularia inflexa* Forssk.
111: III.2,3 — 355 *Utricularia inflexa* Forssk.
111: III.4 — 1818 *Commicarpus plumbaginea* (Cav.) Standl.
111: III.5,6 — 534 *Commicarpus plumbaginea* (Cav.) Standl.
111: III.7,8 — 533 *Commicarpus plumbaginea* (Cav.) Standl.
112: I.1,2 — 1818 *Commicarpus plumbaginea* (Cav.) Standl.
112: I.3,4 — 403 Specimen named *Verbascum sinuatum* L. (probably not a Forsskål specimen; no longer in Herb. Forsskålii, and not retrieved)
112: I.5,6 — 1820 *Verbascum pinnatifidum* Vahl
112: I.7,8 — 402 *Verbascum pinnatifolium* Vahl
112: II.1,2 — 1822 *Vicia peregrina* L.
112: II.3,4 — 621 *Hybanthus enneaspermus* (L.) F. Muell.
112: II.5,6 — 622 *Hybanthus enneaspermus* (L.) F. Muell.
112: II.7,8 — 686 *Viscum schimperi* Engl.
112: III.1,2 — 687 *Viscum schimperi* Engl.
112: III.3,4 — 1825 *Cassytha filiformis* L.

112: III.5,6	535	*Cassytha filiformis* L.
112: III.7,8	25	*Ruppia maritima* L.
113: I.1,2	1826	*Zilla spinosa* (L.) Prantl
113: I.3,4	1827	*Zilla spinosa* (L.) Prantl
113: I.5,6	725	*Zilla spinosa* (L.) Prantl
113: I.7,8	26	*Thalassodendron ciliatum* (Forssk.) den Hartog
113: II.1	26	*Thalassodendron ciliatum* (Forssk.) den Hartog
113: II.2,3	753	*Zygophyllum coccineum* L.
113: II.4,5	751	*Zygophyllum simplex* L.
113: II.6,7	752	*Zygophyllum simplex* L.
113: II.8	474	*Zygophyllum simplex* L.
113: III.1	474	*Zygophyllum simplex* L.
113: III.2,3	1828	*Zygophyllum simplex* L.
113: III.4,5	1829	*Zygophyllum album* L.f.
113: III.6,7	1830	*Zygophyllum album* L.f.
113: III.8		[Blank]
114: I.1		Algae fam., gen. & sp. indet.
114: I.2,3	875	Algae fam., gen. & sp. indet.
114: I.4,5	859	Algae fam., gen. & sp. indet.
114: I.6,7	858	Algae fam., gen. & sp. indet.
114: I.8	857	Algae fam., gen. & sp. indet.
114: II.1	857	Algae fam., gen. & sp. indet.
114: II.2,3	879	Algae fam., gen. & sp. indet.
114: II.4,5		Algae fam., gen. & sp. indet.
114: II.6,7	843	Algae fam., gen. & sp. indet.
114: II.8	840	Algae fam., gen. & sp. indet.
114: III.1	840	Algae fam., gen. & sp. indet.
114: III.2,3	838	Algae fam., gen. & sp. indet.
114: III.4,5	834	Algae fam., gen. & sp. indet.
114: III.6,7	861	Algae fam., gen. & sp. indet.
114: III.8		[Blank]
115: I.1,2	831	Algae fam., gen. & sp. indet.
115: I.3,4	848	Algae fam., gen. & sp. indet.
115: I.5,6	841	Algae fam., gen. & sp. indet.
115: I.7,8	835	Algae fam., gen. & sp. indet.
115: II.1,2	836	Algae fam., gen. & sp. indet.
115: II.3,4		Algae fam., gen. & sp. indet.
115: II.5,6	856	Algae fam., gen. & sp. indet.
115: II.7,8	847	Algae fam., gen. & sp. indet.
115: III.1,2	827	Algae fam., gen. & sp. indet.
115: III.3,4	825	Algae fam., gen. & sp. indet.
115: III.5,6		*Fucus soboliferus* (Algae)
115: III.7,8		*Chondria ovalis* Agarth (Algae)
116: I.1,2		*Chondria tenuissima* Agarth (Algae)
116: I.3,4		Algae fam., gen. & sp. indet.
116: I.5,6		*Ceramium inflexum* Roth. (Algae)
116: I.7,8		Algae fam., gen. & sp. indet.
116: II.1,2		Algae fam., gen. & sp. indet.
116: II.3,4	1844	*Barleria proxima* Lindau
116: II.5,6	388	*Justicia flava* (Vahl) Vahl
116: II.7,8	1001	*Barleria acanthoides* Vahl
116: III.1,2	1002	*Barleria acanthoides* Vahl
116: III.3,4	372	*Lepidagathis aristata* (Vahl) Nees
116: III.5	911	*Achyranthes aspera* L. var. *sicula* L.
116: III.6,7	1543	*Pancratium maritimum* L.

116: III.8 [Blank]

117: I.1,2 21 *Crinum album* (Forssk.) Herb.

117: I.3,4 5 *Dracunculus vulgaris* Schott & Engl.

117: I.5,6 282 *Moltkiopsis ciliata* (Forssk.) I. M. Johnst.

117: I.7,8 1650 *Haplophyllum tuberculatum* (Forssk.) A. Juss.

117: II.1,2 1240 *Echium glomeratum* Poir.

117: II.3,4 285 *Heliotropium digynum* (Forssk.) C. Chr.

117: II.5,6 286 *Echium angustifolium* Miller subsp. *sericeum* (Vahl) Kltz.

117: II.7,8 290 *Echium glomeratum* Poir.

117: III.1,2 281 *Echium cf. rauwolfii* Del.

117: III.3,4 1413 *Moltkiopsis ciliata* (Forssk.) I.M. Johnst.

117: III.5,6 940 *Alkanna lehmannii* (Tineo) A. DC.

117: III.7,8 1735 *Cleome amblyocarpa* Barr. & Mart.

118: I.1,2 1225 *Dianthus caryophyllus* L.

118: I.3,4 570 *Silene biappendiculata* Rohrb.

118: I.5,6 568 *Silene paradoxa* L.

118: I.7,8 1733 *Agrostemma githago* L.

118: II.1,2 1732 *Silene gallica* L. & *S. nocturna* L. (mixed sheet)

118: II.3,4 1731 *Silene colorata* Poir.

118: II.5,6 1734 *Silene cf. italica* (L.) Pers.

118: II.7,8 977 *Atriplex tartarica* L.

118: III.1 979 *Atriplex* sp.

118: III.2,3 179 *Suaeda salsa* (L.) Pallas

118: III.4,5 178 *Traganum nudatum* Del.

118: III.6,7 175 *Traganum nudatum* Del.

118: III.8 [Blank]

119: I.1,2 1005 *Beta vulgaris* L.

119: I.3,4 1677 *Bassia muricata* (L.) Aschers.

119: I.5,6 978 *Halimione portulacoides* (L.) Aellen

119: I.7,8 *Conyza ivifolia* (L.) Less. (not a Forsskål specimen).

119: II.1,2 1497 *Notobasis syriaca* (L.) Cass.

119: II.3,4 1515 *Onopordum tauricum* Willd.

119: II.5,6 1153 *Picris asplenioides* L.

119: II.7,8 962 *Artemisia* sp.

119: III.1,2 1116 *Cineraria abyssinica* A. Rich.

119: III.3,4 1718 *Senecio aegyptius* L.

119: III.5,6 961 *Artemisia monosperma* Del.

119: III.7,8 993 *Compositae*, tribe *Anthemideae*, gen. et sp. indet.

120: I.1,2 959 *Artemisia judaica* L.

120: I.3 1580 *Picris sulphurea* Delile

120: I.4,5 1139 *Conyza aegyptiaca* (L.) Ait.

120: I.6,7 1538 *Osteospermum vaillantii* (Decne.) T. Norl.

120: I.8 994 *Compositae*, tribe *Anthemideae*, gen. et sp. indet.

120: II.1,2 1007 *Blumea axillaris* (Lam.) DC.

120: II.3,4 1008 *Blumea axillaris* (Lam.) DC.

120: II.5,6 1142 *Conyza pyrrhopappa* A. Rich.

120: II.7,8 1308 *Gnaphalium pulvinatum* Del.

120: III.1,2 1309 *Gnaphalium pulvinatum* Del.

120: III.3,4 1109 *Ifloga spicata* (Forssk.) Sch.-Bip.

120: III.5,6 1609 *Pulicaria grandidentata* Jaub. & Spach

120: III.7,8 1400 *Leontodon hispidulus* (Del.) Boiss.

121: I.1,2 1348 *Hyoseris lucida* L.

121: I.3,4 919 *Aetheorhiza bulbosa* (L.) Cass.

121: I.5,6 1403 *Leontodon tuberosum* L.

121: I.7,8 1786 *Tolpis virgata* (Desf.) Bertol.

121: II.1,2		*Senecio* sp. (not a Forsskål specimen)
121: II.3,4	1154	*Picris asplenioides* L.
121: II.5,6	1151	*Crepis micrantha* Czerep.
121: II.7,8	1110	*Chiliadenus montanus* (Vahl) Brullo
121: III.1,2	1111	*Chiliadenus montanus* (Vahl) Brullo
121: III.3,4	1773	*Synedrella nodiflora* (L.) Gaertn.
121: III.5,6	1397	*Launaea* sp.
121: III.7,8	36	*Commelina benghalensis* L.
122: I.1,2	35	*Commelina forskahlii* Kunth
122: I.3,4	31	*Aneilema forsskahlii* Kunth
122: I.5,6	795	*Pinus brutia* Ten.
122: I.7,8	457	*Convolvulus siculus* L. subsp. *agrestis* (Schweinf.) Verdc.
122: II.1,2	458	*Convolvulus siculus* L. subsp. *agrestis* (Schweinf.) Verdc.
122: II.3,4	1133	*Convolvulus arvensis* L.
122: II.5,6	444	*Ipomoea obscura* (L.) Ker-Gawl.
122: II.7,8	1137	*Convolvulus lanatus* Vahl
122: III.1,2	1361	*Quamoclit coccinea* (L.) Moench
122: III.3,4	1044	*Cardamine africana* L.
122: III.5,6		*Zilla spinosa* (L.) Prantl (probably not a Forsskål specimen).
122: III.7,8	712	*Eremobium aegyptiacum* (Spreng.) Hochr.
123: I.1,2	656	*Zehneria thwaitesii* (Schweinf.) C. Jeffrey
123: I.3,4	667	*Dicotyledones* fam., gen. & sp. indet.
123: I.5,6	665	*Kedrostis leloja* (Forssk.) C. Jeffrey
123: I.7,8	652	*Lagenaria siceraria* (Molina) Standley
123: II.1,2	1725	*Sicyos angulatus* L.
123: II.3,4	1176	*Ctenolepis cerasiformis* (Stocks) Hook. f. ex Oliv.
123: II.5,6	1199	*Cyperus rotundus* L.
123: II.7,8	1200	*Cyperus rotundus* L.
123: III.1,2	1194	*Cyperus fuscus* L. mixed with *Fimbristylis bisumbellata* & *Cyperus laevigatus*
123: III.3,4	1196	*Cyperus conglomeratus* Rottb.
123: III.5,6	1197	*Cyperus conglomeratus* Rottb.
123: III.7,8	1189	*Cyperus effusus* Rottb.
124: I.1,2	1190	*Cyperus effusus* Rottb.
124: I.3,4	1201	*Cyperus longus* L.
124: I.5,6	1202	*Cyperus longus* L.
124: I.7,8	1243	*Eleocharis geniculata* (L.) Roem. & Schult.
124: II.1,2	247	*Cephalaria* sp.
124: II.3,4	530	*Elaeagnus angustifolia* L.
124: II.5	1242	*Elaeagnus angustifolius* L.
124: II.6,7	1249	*Euphorbia cuneata* Vahl
124: II.8		*Euphorbia glomerifera* (Milsp.) Wheeler (not a Forsskål specimen)
124: III.1		*Euphorbia glomerifera* (Milsp.) Wheeler (not a Forsskål specimen)
124: III.2,3		*Euphorbia glaucophylla* Poir. (not a Forsskål specimen).
124: III.4,5	1369	*Jatropha curcas* L.
124: III.6,7	1835	*Phyllanthus ovalifolius* Forssk.
124: III.8	789	*Quercus coccifera* L.
125: I.1,2	789	*Quercus coccifera* L.
125: I.3,4	788	*Quercus coccifera* L.
125: I.5,6	792	*Quercus cerris* L.
125: I.7,8	625	*Frankenia pulverulenta* L.
125: II.1,2	1498	*Nymphoides cristata* (Griseb.) O. Kuntze
125: II.3,4	1499	*Nymphoides cristata* (Griseb.) O. Kuntze

125: II.5,6	239 *Enicostema axillare* (Lam.) A. Raynal
125: II.7,8	140 *Aeleuropus lagopoides* (L.) Thwait.
125: III.1,2	67 *Aeleuropus lagopoides* (L.) Thwait.
125: III.3,4	42 *Aeleuropus lagopoides* (L.) Thwait.
125: III.5,6	*Andropogon bicornis* L. (not a Forsskål specimen).
125: III.7,8	*Iseilema prostrata* (L.) Anderss. (not a Forsskål specimen).
126: I.1,2	933 *Ammophila arenaria* (L.) Link.
126: I.3,4	965 *Arundo donax* L.
126: I.5,6	1013 *Bromus hordeaceus* L.
126: I.7,8	1014 *Bromus japonicus* Thunb.
126: II.1,2	1017 *Bromus rubens* L.
126: II.3,4	1018 *Bromus rubens* L.
126: II.5,6	1546 *Cynodon dactylon* (L.) Pers.
126: II.7,8	1547 *Cynodon dactylon* (L.) Pers.
126: III.1,2	1208 *Dactylis glomerata* L.
126: III.3	1213 *Dactyloctenium aegyptiacum* (L.) Willd.
126: III.4	1214 *Dactyloctenium aegyptiacum* (L.) Willd.
126: III.5	1186 *Cynosurus ternatus* Forssk.
126: III.6,7	1184 *Eleusine floccifolia* (Forssk.) Spreng.
126: III.8	1247 *Eragrostis ciliaris* (L.) R. Br.
127: I.1,2	38 *Eragrostis pilosa* (L.) P. Beauv.
127: I.3,4	1099 *Chloris barbata* Sw. mixed with *Eustachys paspaloides* (Vahl) L. & M.
127: I.5,6	1255 *Schismus arabicus* Nees mixed with *Schismus barbatus* (L.) Thell.
127: I.7,8	1261 *Cutandia maritima* (L.) Barbey
127: II.1,2	98 *Festuca ovina* L.
127: II.3,4	51 *Hordeum marinum* Huds. subsp. *gussonianum* (Parl.) Thell.
127: II.5,6	1344 *Hordeum marinum* Huds. subsp. *marinum*
127: II.7,8	1345 *Hordeum marinum* Huds. subsp. *gussoneanum* (Parl.) Thell.
127: III.1,2	1346 *Hordeum murinum* L. subsp. *glaucum* (Steud.) Tzvelev
127: III.3	1391 *Lamarckia aurea* (L.) Moench
127: III.4,5	1555 *Parapholis strigosa* (Dumort.) C. E. Hubb.
127: III.6,7	1469 *Mibora minima* (L.) Desv.
127: III.8	1392 *Lamarckia aurea* (L.) Moench
128: I.1,2	1634 *Rostraria cristata* (L.) Tzvelev
128: I.3,4	1633 *Rostraria cristata* (L.) Tzvelev
128: I.5,6	80 *Polypogon monspeliensis* (L.) Desf.
128: I.7,8	121 *Rhynchelytrum repens* (Willd.) Hubbard
128: II.1,2	1065 *Catapodium rigidum* (L.) C.E. Hubb.
128: II.3,4	60 *Catapodium rigidum* (L.) C.E. Hubb.
128: II.5	1758 *Sphenopus* sp.
128: II.6	1756 *Sphenopus divaricatus* (Gouan) Reichb.
128: II.7,8	70 *Sphenopus divaricatus* (Gouan) Reichb.
128: III.1,2	128 *Stipa capensis* Thunb.
128: III.3,4	1807 *Triticum durum* Desf.
128: III.5,6	494 *Hypericum bithynicum* Boiss.
128: III.7,8	1312 *Gymnocarpus decandrum* Forssk.
129: I.1,2	1362 *Gynandriris sisyrinchium* (L.) Parl.
129: I.3,4	28 *Juncus bufonius* L.
129: I.5,6	1374 *Juncus striatus* E.H.F. Meyer
129: I.7,8	18 *Allium curtum* Boiss & Gaill.
129: II.1,2	982 *Asparagus officinalis* L.
129: II.3,4	1235 *Dracaena* sp. indet.
129: II.5,6	1696 *Scilla hyacinthina* (Roth) Macbride
129: II.7,8	11 *Scilla hyacinthina* (Roth) J. F. Macbride

129: III.1,2 10 *Urginea maritima* (L.) Bak.
129: III.3,4 469 *Linum strictum* L.
129: III.5,6 685 *Plicosepalus acaciae* (Zucc.) Weins & Polhill
129: III.7,8 1442 *Lycopodiella cernua* (L.) Pic.-Ser.
130: I.1,2 927 *Althaea hirsuta* L.
130: I.3,4 598 *Sida urens* L.
130: I.5,6 602 *Hibiscus* sp.
130: I.7,8 1336 *Hibiscus deflersii* Cuf.
130: II.1,2 900 *Abutilon pannosum* (Forst. f.) Schlecht.
130: II.3,4 *Anisomeles indica* (L.) O. Kuntze (not a Forsskål specimen).
130: II.5 1464 *Mentha pulegium* L.
130: II.6 1463 *Mentha aquatica* L.
130: II.7,8 1573 *Leucas urticifolia* (Vahl) R. Br.
130: III.1,2 1637 *Crossandra johanninae* Fiori
130: III.3 *Salvia phlomoides* Asso (not a Forsskål specimen).
130: III.4,5 338 *Stachys cf. maritima* Gouan
130: III.6,7 230 *Micromeria imbricata* (Forssk.) C. Chr.
130: III.8 [Blank]
131: I.1,2 349 *Solenostemon latifolius* (Benth.) Morton
131: I.3,4 329 *Labiatae* gen. et sp. indet.
131: I.5,6 328 *Ocimum tenuiflorum* L.
131: I.7,8 216 *Nepeta nepetellae* Forssk.
131: II.1,2 336 *Mentha longifolia* (L.) L. var. *schimperi* (Briq.) Briq.
131: II.3,4 1160 *Crotalaria microphylla* Vahl
131: II.5,6 1161 *Crotalaria microphylla* Vahl
131: II.7,8 *Crotalaria albida* Heyne (not a Forsskål specimen).
131: III.1,2 *Crotalaria albida* Heyne (not a Forsskål specimen).
131: III.3,4 1626 *Rhynchosia minima* (L.) DC.
131: III.5,6 1627 *Rhynchosia minima* (L.) DC.
131: III.7,8 1623 *Vigna variegata* Defl.
132: I.1,2 *Kummerowia striata* (Thunb.) Schindl. (not a Forsskål specimen).
132: I.3,4 *Kummerowia striata* (Thunb.) Schindl. (not a Forsskål specimen).
132: I.5,6 1219 *Desmodium triflorum* (L.) DC.
132: I.7,8 1060 *Senna italica* Miller
132: II.1,2 1530 *Ormocarpum yemenense* Gillett
132: II.3,4 1337 *Hildebrandtia africana* Vatke
132: II.5,6 1000 *Astragalus* sp.
132: II.7,8 *Indigofera tetrasperma* Pers. (not a Forsskål specimen).
132: III.1,2 1796 *Trifolium micranthum* Viv.
132: III.3,4 1434 *Lotus* sp.
132: III.5,6 1427 *Lotus creticus* L.
132: III.7,8 1532 *Ornithopus sativus* Brot.
133: I.1,2 1342 *Retama retam* (Forssk.) Webb
133: I.3,4 1457 *Melia azederach* L.
133: I.5,6 1819 *Commicarpus plumbaginea* (Cav.) Standl.
133: I.7,8 1368 *Jasminum officinale* L.
133: II.1,2 1411 *Limonium ferulaceum* (L.) O. Kuntze
133: II.3,4 521 *Limonium* sp.
133: II.5,6 1836 *Limonium* sp.
133: II.7,8 1244 *Emex spinosa* (L.) Campd.
133: III.1,2 808 *Tectaria gemmifera* (Fée) Alston
133: III.3,4 821 *Asplenium adiantum-nigrum* L.
133: III.5,6 798 *Pteridium aquilinum* (L.) Kuhn

133:III.7,8 806 *Adiantum poiretii* Wikstr.
134: I.1,2 823 *Adiantum poiretii* Wikstr.
134: I.3,4 811 *Adiantum flabellatum* L.
134: I.5,6 812 *Adiantum flabellatum* L.
134: I.7,8 699 *Thalictrum foetidum* L.
134: II.1,2 1619 *Caylusea hexagyna* (Forssk.) M.L.Green
134: II.3,4 1615 *Reseda decursiva* Forssk. (no. 1), mixed with *R. lutea* L. (no. 2)
134: II.5,6 1614 *Reseda undata* L. (no. 1 & 3), mixed with *R. decursiva* Forssk. (no. 2)
134: II.7,8 1613 *Reseda alba* L.
134:III.1,2 1495 *Neurada procumbens* L.
134:III.3,4 1604 *Prunus spinosa* L.
134:III.5,6 952 *Anthospermum herbaceum* L. f.
134:III.7,8 953 *Anthospermum herbaceum* L. f.
135: I.1,2 1175 *Crucianella maritima* L.
135: I.3,4 1298 *Galium / Asperula* sp.
135: I.5,6 1290 *Galium aparine* L.
135: I.7,8 954 *Anthospermum herbaceum* L. f.
135: II.1,2 1560 *Pavetta villosa* Vahl
135: II.3,4 1561 *Pavetta villosa* Vahl
135: II.5,6 1253 *Fagonia arabica* L. mixed with *Ruta chalepensis* L.
135: II.7,8 1647 *Ruta chalepensis* L.
135:III.1,2 1651 *Haplophyllum tuberculatum* (Forssk.) A. Juss.
135:III.3,4 461 *Salix subserrata* Willd.
135:III.5,6 462 *Salix subserrata* Willd.
135:III.7,8 1446 *Lysimachia nummularia* L.
136: I.1,2 1216 *Datura metel* L.
136: I.3,4 1217 *Datura metel* L.
136: I.5,6 1837 *Hyoscyamus muticus* L.
136: I.7,8 1576 *Physalis alkekengi* L.
136: II.1,2 1750 *Solanum palmetorum* Dunal
136: II.3,4 1749 *Solanum nigrum* L.
136: II.5,6 1752 *Solanum villosum* Miller
136: II.7,8 1753 *Solanum villosum* Miller
136:III.1,2 425 *Solanum* sp. indet. (? not a Forsskål coll.)
136:III.3,4 1748 *Solanum nigrum* L.
136:III.5,6 427 *Solanum* sp. indet. (? not a Forsskål coll.)
136:III.7,8 1746 *Solanum incanum* L.
137: I.1,2 413 *Solanum platacanthum* Dunal
137: I.3,4 1496 *Nicotiana tabacum* L.
137: I.5,6 467 *Grewia villosa* Willd.
137: I.7,8 1809 *Triumfetta rhomboidea* Jacq.
137: II.1,2 671 *Sanicula elata* D. Don
137: II.3,4 679 *Ammi visnaga* (L.) Lam.
137: II.5,6 678 *Pimpinella anisum* L.
137: II.7,8 674 *Ammi visnaga* (L.) Lam.
138: I.1,2 1311 *Grewia* sp. indet.
138: I.3 1585 *Plantago lagopus* L.
138: I.4,5 464 *Turraea holstii* Guerke
138: I.6,7 468 *Dicotyledones* fam., gen. & sp. indet.
138: I.8 471 *Dicotyledones* fam., gen. & sp. indet.
138: II.1 471 *Dicotyledones* fam., gen. & sp. indet.
138: II.2,3 472 *Dicotyledones* fam., gen. & sp. indet.
138: II.4,5 383 *Justicia coerulea* Forssk.
138: II.6,7 238 *Apocynaceae* gen. et sp. indet.

138: II.8 513 *Epilobium angustifolium* L.
138: III.1 513 *Epilobium angustifolium* L.
138:III.2,3 975 *Blyttia spiralis* (Forssk.) Field & Wood
138:III.4-8 [Blank]

A Forsskål-specimen in 'Herb. Vahlii' (IDC 2201).
Vahl

36: II.6,7 1320 *Desmodium repandum* (Vahl) DC

References

Al-Hubaishi, A. & K. Müller-Hohenstein (1984). *An introduction to the vegetation of Yemen.* Eschborn.

[Anonymous] (1789). Nogle Efterretninger om Etatsraad og Finantzdeputeret J. Zoëgas Levned. *Minerva* [Copenhagen], May 1789: 68–204.

Ascherson, P. (1884). Forskal über die Metamorphose der Pflanzen. *Berichte der Deutschen Botanischen Gesellschaft* 2: 293–297.

Ascherson, P.F.A. & Schweinfurth, G. (1887–89). Illustration de la Flore d'Egypte. *Mémoires de l'Institut Egyptien.* 2, Suppl.: 25–260 (1887) & 745–785 (1889).

Blatter, E. (1914–16). Flora of Aden. *Records of the botanical Survey of India* 7,1: 1–79 (1914), 7,2: "77"–336 (1916) & 7,3: 337–418 (1916).

— (1919–36). Flora Arabica. *Records of the botanical Survey of India* 8,1: 1–123 (1919), 8,2: 124–282 (1921), 8,3: 283–365 (1922), 8,4: 366–450 (1923) & 8,5: 451–519 (1936).

Börgesen, F. (1932). A revision of Forsskål's algae mentioned in Flora Aegyptiaco-Arabica and found in his Herbarium in the Botanical Museum of the University of Copenhagen. *Dansk Botanisk Arkiv* 8(2): 1–15.

Brummitt, R. K. (1978). Forsskal's *Ruellia guttata* identified with *Asystasia somalensis* (Acanthaceae). *Kew Bulletin* 32: 453.

— & Powell, C.E. (1992). *Authors of Plant Names* - A list of authors of scientific names of plants, with recommended standard forms of their names, including abbreviations. Royal Botanic Gardens, Kew.

Burdet, H. M. & P. Perret. (1983). Grimm, Asso, Aublet et Forsskål, sont-ils "linnéns"?, ou la portée de l'article 23.6(c) du Code international de la nomenclature botanique. *Candollea* 38: 699–707.

Chiovenda, E. (1923). Note sulle Flora Aegyptiaco-Arabica. *Bollettino della società botanica italiana* 1923: 112–117.

Christensen, C. (1918). *Naturforskeren Pehr Forsskål.* Hans Rejse til Aegypten og Arabien 1761–63 og hans botaniske Arbejder og Samlinger. Copenhagen.

— (1922). Index to Pehr Forsskål: Flora Aegyptiaco-Arabica 1775, with a revision of Herbarium Forsskålii. *Dansk Botanisk Arkiv* 4,3: 1–54.

— (1924–26). *Den danske Botaniks Historie med tilhørende Bibliografi.* Vol. 1 (1924–26) & 2 (1924–26). Copenhagen.

— (1935). Forsskål, Petrus. In: P. Engelstoft and S. Dahl, eds., *Dansk Biografisk Leksikon.* Ed. 2, vol. 7. Copenhagen.

Davis, P. ed. (1965–88). *Flora of Turkey.* Vol. 1–10. Edinburgh University Press.

Decaisne, J. (1841). Plantes de l'Arabie Heureuse receuillee par M. P. E. Botta. *Achives du Museum d'histoire naturelles, Paris* 2: 89–119.

Deflers, A. (1889). *Voyage aux Yemen.* Journal d'une expedition botanique faite en 1887 dans les montagnes de l'Arabie hereuse. Paris.

— (1895). Esquisse de geographie botanique. La végétation de l'Arabie tropicale au dela du Yemen. *Revue de l'Egypte* 1: 349–370, 400–430.

— (1896). Plantes de l'Arabie méridionale receuillis pendant les annés 1889, 1890, 1893 et 1894. *Bulletin de la société botanique de France* 43: 321–331.

Delile, A. R. (1813). *Description de l'Egypte,* Fasc. 2: 145–462.

Farr, E. R., J. A. Leussink and F. A. Stafleu, eds. (1979). Index Nominum Genericorum (Plantarum). Vol. 1–3. *Regnum Vegetabile* 100–102.

Feinbrun-Dothan, N. (1978–1986). *Flora Palaestina.* Vol. 3 (1978), Vol. 4 (1986).

Forsskåhl, G. (1927). *Några blad ur släkten Forsskåhls historia.* Borgå.

Forsskål, P. (1775a). *Decriptiones animalium,* avium, amphibiorum, piscium, insectorum, vermium, quae in itinere orientali observavit Pt. Forskål. Post mortem auctoris edidit Carsten Niebuhr. Adjuncta est materia medica Kahirina atque tabula Mare Rubris geographica. Copenhagen.

— 1775b). *Flora Aegyptiaco-Arabica,* sive descriptiones plantarum quas per Aegyptum inferiorem et Arabiam Felicem detexit, illustravit Pt. Forskål. Post mortem auctoris edidit Carsten Niebuhr. Acc. tabula Arabiae Felicis geographico-botanica. Copenhagen.

— (1776). *Icones rerum naturalium,* quas in itinere orientali depingi curavit Petrus Forsskål. Post mortem auctoris ad regis mandatum aeri incisas edidit Carsten Niebuhr. Copenhagen.
— (1950). *Resa til Lycklige Arabien.* Petrus Forsskål's dagbok 1761-1763. Med anmärkningar utgiven av Svenska Linné-Sällskapet. Uppsala.
Fox Maule, A. (1974). Danish botanical expeditions and collections in foreign countries. *Botanisk Tidsskrift* 69: 167–205.
— (1979). Botanik. In: S. Ellehøj, ed., Københavns Universitet 1479–1979. Vol. 13: 163–259.
Fries, T. M., ed. (1912). *Bref och Skrifvelser af och till Carl von Linńe.* Vol. 1,6: 110–169. Stockholm.
Friis, I. (1981). The taxonomy and distribution of *Mimusops laurifolia* (Sapotaceae). *Kew Bulletin* 35: 785–792.
— (1983a). Notes on the botanical collections and publications of Pehr Forsskal. *Kew Bulletin* 38: 457–467.
— (1983b). The acaulescent and succulent species of *Dorstenia* sect. *Kosaria* (Moraceae) from NE tropical Africa and Arabia. *Nordic Journal of Botany* 3: 533–538.
— (1984). An analysis of the protologues of new genera with new species in Forsskal's *Flora Aegyptiaco-Arabica. Taxon* 33: 655–667.
— (1992). Proposals to amend the ICBN (63–73). Ten proposals to amend the Code, and report of the Special Committee on Binary Combinations. *Taxon* 41: 343–350.
— , R. K. Brummitt, F. N. Hepper, C. Jeffrey, N. P. Taylor, B. Vercourt, R. A. Howard & D. H. Nicolson (1984). Validity of names published by Forsskål and Aublet. *Taxon* 33: 495–496.
— & C. Jeffrey (1986). Proposal to revise ICBN (93). The problem of non-binary species names in works published after 1 May 1753, with a proposal to amend ICBN. *Taxon* 35: 397–400.
— & M. Thulin (1984). The spelling of Pehr Forsskål's family name. *Taxon* 33: 668–672.
Gertz, O. (1945). Växter ur Peter Forsskåls herbarium å Lunds Botaniska Institution. *Svenska Linné-Sällskapets årsskrift* 28: 77–89.
Gilbert, M. G. (1989). A review of Pehr Forsskål's stapeliads and the typification of *Caralluma quadrangula* and *C. subulata. Bradleya* 7: 79–83.
Gmelin, J. F. (1788–1793). *Systema naturae.* Leipzig.
Greuter, W. (1984). Warning against misinterpreting the rule on "Non-Linnean" works (Art. 23.6 (c)). *Taxon* 33: 493–495.
Hansen, T. (1962). *Det lykkelige Arabien. En dansk ekspedition 1761*–67. Copenhagen. (Translated into English 1964 by J. and K. McFarlane as 'Arabia felix, The Danish Expedition of 1761–1767', London).
Hedberg, I. & Edwards, S. (1989). *Flora of Ethiopia.* Vol. 3. Addis Ababa & Uppsala.
Heine, H. (1968). A propos de la nomenclature d'un sébestier de l'ancien monde. Adansonia, sér. 2,8: 181–187.
Hepper, F. N. and J. R. I. Wood (1983). New combinations and notes based on Forsskal's Arabian collections. *Kew Bulletin* 38: 83–86.
Jeffrey, C. (1985). Forsskål and the interpretation of Article 23. *Taxon* 34: 144–147.
Kaulfuss, G. F. (1824). *Enumeratio filicum quas in itinere circa terram legit Cl. Adalbertus de Chamisso* ... Leipzig.
Lagus, W. (1877). *Strödda blad, nytt och gammalt i historiska och humanistiska ämnen,* I. Petter Forskåls lefnad. Helsingfors.
Lange, J. (1875). Erindringer fra Universitetets Botaniske Have ved Charlottenborg. *Botanisk Tidsskrift* 9: 1–68.
Legré, L. (1900). *La botanique en Provence au XVIIIe siècle.* Pierre Forskal et le Florula Estaciensis. Marseille.
Linnaeus, C. (1753). *Species Plantarum.* Stockholm.
— (1758–59). *Systema naturae.* Ed. 10, vol. 1 (1758), 2 (1759). Stockholm.
— (1762–63). *Species Plantarum.* Ed. 2, vol. 1 (1762), 2 (1763). Stockholm.
— (1764). *Opobalsamum declaratum.* Uppsala.
— (1767a). *Systema naturae.* Ed. 12, vol. 1 'Regnum Vegetabile.' Stockholm.
— (1767b). *Mantissa plantarum.* Stockholm.
— (1771). *Mantissa plantarim altera.* Stockholm.
Linné, C. filius (1781). *Supplementum plantarum.* Braunschweig.

Matinolli, E. (1960). Petter Forsskål. Luova ihminen 1700-luvun Pohjolasta. *Annales Universitates Turkuensis*, Ser. B, 79: 1-167.
Meikle, R.D. (1977, 1985). *Flora of Cyprus*. Vol. 1 (1977), Vol. 2 (1985). Bentham-Moxon Trust, Royal Botanic Gardens, Kew.
Michaelis, J. D. (1762). *Fragen an eine Gesellschaft Gelehrter Männer, die auf Befehl Ihro Majestät des Königes von Dänemark nach Arabien Reisen*. Frankfurt a.M.
— (1794-96). *Literarischer Briefwechsel. Herausgegeben von J. G. Buhle*. Vol. 1: 297-492 & Vol. 2: 117-192.
Miller, A. G., I. C. Hedge & R. A. King (1982). Studies in the flora of Arabia I. A botanical bibliography of the Arabian Peninsula. *Notes from the Royal Botanic Garden, Edinburgh* 40: 43-61.
Muschler, R. (1912). *A Manual Flora of Egypt*. Vol. 1-2. Berlin.
Niebuhr, B. G. (1817). Carsten Niebuhrs Levnet. *Athene, et Maaneds skrift* [*Copenhagen*] 8: 1-36 & 117-166.
Niebuhr, C. (1772). *Beschreibung von Arabien*. Aus eigenen Beobachtungen und im Lande selbst gesammelten Nachrichten abgefasset. With 24 plates and 1 map. Copenhagen.
— (1774-78). *Reisebeschreibung nach Arabien und anderen umliegenden Ländern*. Vol. 1-2, with 124 plates and 1 map. Copenhagen.
Norden, F. L. (1750-55). *Voyage d'Egypte et de Nubie, par M. Frédéric Louis Norden, Capitaine des Vasseaux du Roi. Ouvrage enrichi de Cares et de Figures dessinées sur les lieux, par l'Auteur même*. Vol. 1-2. Copenhagen.
Pedersen, O. (1992). *Lovers of Learning. A History of the Royal Danish Academy of Sciences and Letters*. Copenhagen.
Pedrol, J. (1992). Proposal to amend 2261 Suaeda, nom. cons. (Chenopodiaceae). *Taxon* 41: 337-338.
Rasmussen, S. T. ed. (1990). *Den Arabiske Rejse 1761*-1767. En dansk ekspedition set i videnskabshistorisk perspektiv. Munksgaard, Copenhagen.
Rottböll, C. F. (1772). *Descriptiones plantarum rariorum iconibus illustrandas, cum earum, quae primo proximeque prodituro fasciculo continebuntur, elencho, programmate, quo lectiones in horto botanico anno 1772 auspicatur indicit*. Copenhagen.
— (1773). *Descriptionum et iconum rariores et pro maxima partes novas plantas illustrantium. Liber primus*. Copenhagen.
Schioedte, J. C. (1870-71). Af Linnés Brevvexling. Aktstykker til Naturstudiets Historie i Danmark. *Naturhistorisk Tidende* 3,7: 333-522.
Schück, H. (1923). *Från Linnés tid. Petter Forsskål*. Stockholm. [also in: *Svenska Vetenskapsakademiens Handlingar, Nye Serie, Del XX*, 1923].
Schwartz, O. (1939). Flora des tropischen Arabien. *Mitteilungen aus dem Institut für allgemeine Botanik in Hamburg* 10. Hamburg.
Schweinfurth, G. (1889). Ueber seine Reise nach dem glücklichen Arabien. *Verhandlingen der Gesselschaft für Erdkunde zu Berlin* 1889: 299-308.
— (1891). Ueber die Florengemeinschaft von Südarabien und Nordabessinien. *Verhandlungen der Gesselschaft für Erdkunde zu Berlin* 1891: 531-550.
— (1894-99). Sammlung arabisch-Äthiopischer Pflanzen. Ergebnisse von Reisen in den Jahren 1881, 88, 89, 91 und 92. *Bulletin de l'herbier Boissier* 2, Appendix 2: 1-10 (1894), 4, Appendix 2: 21-266 (1896) & 7, Appendix 2: 267-340 (1899).
— (1912). *Arabische Pflanzennamen aus Aegypten, Algerien und Yemen*. Berlin.
— & P. Ascherson (1867). *Beitrag zur Flora Aethiopiens*. Erster Abtheilung, pp. 253-303. Berlin.
Spärck, R. (1963). Peter Forsskåls arabiske rejse og zoologiske samlinger. *Nordenskiöld-samfundets Tidsskrift* 23: 110-136.
Stafleu, F. A. and R. S. Cowan. (1976). Taxonomic Literature. Ed. 2., vol. 1, A-G. *Regnum Vegetabile* 94: 1-1136.
Täckholm, V. (1974). *Students' Flora of Egypt*. Ed. 2. Beirut.
Täckholm, V. & M. Drar (1941-69). *Flora of Egypt*. Vol. 1 (1941), vol. 2 (1950), vol. 3 (1954), vol. 4 (1969). Cairo.
Tewolde, B. G. E. & O. Hedberg (1981). The Ethiopian Flora Project. *A.E.T.F.A.T. Bulletin* 30: 23-25.
Thulin, M. (1993). *Flora of Somalia*. Vol. 1. Royal Botanic Gardens, Kew.
Vahl, M. (1790-1794). *Symbolae Botanicae, sive plantarum tam earum quas in itinere imprimis orientali collegit Petrus Forskål*. Vol. 1 (1790), 2 (1791) & 3 (1794). Copenhagen.

— (1804–05). Enumeratio Plantarum. Vol. 1 (1804) & 2 (1805). Copenhagen.

Warming, E. (1880). Den danske botaniske Litteratur. *Botanisk Tidsskrift* 12: 42–131, 158–247.

Wickens, G. E. (1982). Studies in the flora of Arabia: III A biographical index of plant collectors in the Arabian peninsula (including Socotra). *Notes from the Royal Botanic Garden, Edinburgh* 40: 301–330.

Wolff, T. (1980). Forsskål, Pehr. In: S. Cedergreen Beck, ed., *Dansk Biografisk Leksikon*. Ed. 3., vol. 4. Copenhagen.

Wood, J. R. I. (1982). The identity of *Acacia oerfota* (Forssk.) Schweinf. (Leguminosae-Mimosoideae). *Kew Bulletin* 37: 451–454.

Wood, J. R. I., D. Hillcoat and R. K. Brummitt. (1983). Notes on the types of some names of Arabian Acanthaceae in the Forsskal Herbarium. *Kew Bulletin* 38: 429–456.

Zohary, M. (1966–1972). *Flora Palaestina*. Vol. 1 (1966), Vol. 2 (1972). Jerusalem.

Taxonomic Index

This is an index to species and infraspecific taxa in the Catalogue. All names which have been given a full entry in the Catalogue are indicated with bold type, irrespective of whether they are names in current use, or unidentified or preliminary names proposed by Forsskål. All other names mentioned in the Catalogue are given in italics, both those now considered synonyms and the currently accepted names which in the Catalogue are only mentioned in discussions in the notes.

Abelmoschus esculentus (L.) Moench, 195, 197
Abrus precatorius L., 174
Abutilon bidentatum A. Rich., 196
Abutilon indicum sensu C. Chr., non (L.) Decne., 196
Abutilon pannosum (Forst. f.) Schlecht., 196
Acacia arabica (Lam.) Willd., 176
Acacia asak (Forssk.) Willd., 174
Acacia ehrenbergiana Hayne, 174
Acacia farnesiana (L.) Willd., 175
Acacia flava (Forssk.) Schweinf., 174
Acacia hadiensis DC., 176
Acacia hamulosa Benth., 175
Acacia hockii De Wild., 175
Acacia mellifera (Vahl) Benth., 175
Acacia nilotica (L.) Del., 175
Acacia nubica Benth., 175
Acacia oerfota (Forssk.) Schweinf., 176
Acacia oerfota sensu Brenan, 175
Acacia senegal sensu Schwartz partly, non (L.) Willd., 175
Acacia seyal Del., 176
Acacia spirocarpa Hochst., 176
Acacia stellata (Forssk.) Willd., 187
Acacia tortilis (Forssk.) Hayne subsp. **tortilis**, 176
Acalypha betulina Retz., 153
Acalypha brachystachya Hornem., 153
Acalypha ciliata Forssk., 152
Acalypha decidua Forssk., 153
Acalypha fruticosa Forssk., 153
Acalypha indica L., 153
Acalypha spicata Forssk., 153
Acalypha supera Forssk., 153
ACANTHACEAE, 63
Acanthus arboreus Forssk., 63
Acanthus edulis Forssk., 64
Acanthus hirsutus Boiss., 63
Acanthus maderasp[atensis] sensu Forssk., non L., 65
Acanthus (sp. without epithet) Forssk., 63
Achillea falcata sensu Forssk., 112
Achillea fragrantissima (Forssk.) Sch. Bip., 112
Achillea lobata Forssk., 112
Achillea santolina L., 112
Achyranthes aspera L. var. **sicula** L., 73
Achyranthes aspera sensu Forssk., 73
Achyranthes capitata Forssk., 73
Achyranthes decumbens Forssk., 74
Achyranthes paniculata Forssk., 74
Achyranthes papposa Forssk., 75
Achyranthes polystachia Forssk., 74
Achyranthes villosa Forssk., 73
Acinos arvensis (Lam.) Dandy, 165
Aconitum monogyn[um] Forssk., 216
Acrostichum australe L.f., 289
Acrostichum dichotomum sensu Forssk., non L., 289
Acrostichum filiare Forssk., 289, 290
Actiniopteris australis (L.f.) Link, 289
Actiniopteris radiata (Schwartz) Link, 289
Actinotrichia fragilis (Forssk.) Børgesen (Algae), 315
Adenia venenata Forssk., 209
Adenium obesum (Forssk.) Roem. & Schult., 76
Adhatoda flava (Vahl) Nees, 67
Adhatoda odora (Forssk.) Nees, 67
Adhatoda sulcata (Vahl) Nees, 67
Adiantum capillus-veneris L., 289
Adiantum caudatum sensu Schwartz, 289
Adiantum flabellatum L., 289
Adiantum incisum Forssk., 289
Adiantum poiretii Wikstr., 289
Adiantum viride(Forssk.) Vahl, 292
Adini microcephala (Del.) Hiern, 221
Aegialophila pumilio (L.) Boiss., 113
Aegilops bicornis (Forssk.) Jaub. & Spach, 260
Aegilops geniculata Roth, 260
Aegilops kotschyi Boiss., 260
Aegilops ovata L., 260
Aeluropus lagopoides (L.) Trin. ex Thwait., 260
Aeluropus mucronatus (Forssk.) Asch. & Schweinf., 272
Aerva [sp. without epithet] Forssk., 73
Aerva javanica (Burm. f.) Juss., 73
Aerva lanata (L.) Juss., 73
Aerva tomentosa Forssk., 73
Aeschynomene grandiflora L., 190
Aeschynomene sesban L., 190
Aetheorhiza bulbosa (L.) Cass., 113
AGAVACEAE, 253
Agrostemma githago L., 94
Agrostis disticha (Forssk.) Schweig. ex Steud., 278
Agrostis indica sensu Forssk., non L., 278

Agrostis semiverticillata (Forssk.) C. Chr., 276
Agrostis spicata Vahl, 278
Agrostis virginica sensu Forssk., non L., 278
Agrostis viridis Gouan, 275
AIZOACEAE, 69
Aizoon canariense L., 69, 70 (Fig. 10)
Ajuga iva (L.) Schreb., 165
Albizia julibrissin Durazz., 176
Albizia lebbeck (L.) Benth., 176
Alcea ficifolia L., 196
Alcea rosea L., 196
Alchemilla sp. indet., 219
Alectra parasitica A. Rich., 227
ALGAE, 297, 315
Alhagi maurorum Medik., 176
Alisma plantago-aquatica L., 253
ALISMATACEAE, 253
Alkanna lehmanii (Tineo) A. DC., 84
Alkanna strigosa Boiss. & Hohrn., 85
Alkanna tinctoria (L.) Tausch, 41
Alkanna tuberculata (Forssk.) Meikle, 84
Allium curtum Boiss & Gaill., 253
Allium desertorum Forssk., 253
Allium marit[imum] Forssk., 253
Allium sphaerocephalon L., 253
Alnus glutinosa (L.) Gaertn., 84
Aloe arborea Forssk., 283
Aloe audhalica Lavranos & Hardy, 284
Aloe dhalensis Lavranos, 284
Aloe inermis Forssk., 283
Aloe maculata Forssk., 284
Aloe officinalis Forssk., 284
Aloe pendens Forssk., 283, 284
Aloe rivierei Lavranos & Newton, 283
Aloe vacillans Forssk., 284
Aloe variegata sensu Forssk., 283, 284
Aloe vera (L.) Burm.f. var. **officinalis** (Forssk.) Bak., 284
Alsine filifolia (Forssk.) Schweinf., 96
Alsine procumbens (Vahl) Fenzl, 96
Alsine prostrata Forssk., 97
Alsine segetalis sensu Forssk., non L., 100
Alternanthera [sp. without epithet] Forssk., 74
Alternanthera achyranth. Forssk., 74
Alternanthera repens J.F. Gmel., 74
Alternanthera sessilis (L.) DC., 74
Althaea hirsuta L., 196
Althaea rosea (L.) Cav., 196
Alysicarpus glumaceus (Vahl) DC., 177
Alysicarpus rugosus sensu Schwartz, non DC., 177
AMARANTHACEAE, 73
Amaranthus aristatus Forssk., 74
AMARYLLIDACEAE, 253
Amaryllis alba Forssk., 253
Amaryllis coccineus sensu Forssk., non L., 254
Amberboa lippii (L.) DC., 135
Ambrosia maritima L., 113
Ambrosia villosissima Forssk., 113
Ammi majus L., 237
Ammi visnaga (L.) Lam., 237
Ammophila arenaria (L.) Link, 260
Amyris gileadensis L., 90
Amyris kafal Forssk., 90
Amyris kataf Forssk., 90
Amyris opobalsamum L., 90
Anabasis articulata (Forssk.) Moq., 101, 102 (Fig. 15A)
Anabasis spinosissima L.f., 105
ANACARDIACEAE, 75
Anacyclus alexandrinus Willd., 113
Anacyclus monanthos (L.) Thell., 113
Anagallis arvensis L., 215
Anagallis latifolia L., 215
Anarrhinum arabicum (Poir.) Jaub. & Spach, 227
Anarrhinum forskaohlii (J.F. Gmel.) Cufod., 227
Anarrhinum orientale Benth., 227
Anastatica hierochuntica L., 140
Anchusa aegyptiaca (L.) DC., 85
Anchusa bugloss Forssk., 85
Anchusa flava Forssk., 85
Anchusa hispida Forssk., 87
Anchusa hybrida Ten. var. *pubescens* Gusul., 85
Anchusa spinocarpos Forssk., 89
Anchusa tuberculata Forssk., 84
Anchusa undulata L., 85
Andrachne telephioides sensu Vahl, non L., 71
Andrachne telephioides L., 146 (Fig. 21B), 153
Andropogon annulatum Forssk., 267
Andropogon bicorne sensu Forssk., non L., 265
Andropogon bicornis L., 260
Andropogon contortum L., 269
Andropogon ischaemum L., 261
Andropogon ramosum Forssk., 261
Andropogonoides [sp. without epithet] Forssk., 281
Aneilema forskalei Kunth, 255
Anisomeles indica (L.) O. Kuntze, 165
Anisotes trisulcus (Forssk.) Nees, 63
Annona glabra Forssk., 76, 77 (Fig. 11)
Annona squamosa L., 76, 77 (Fig. 11)
ANNONACEAE, 76
Anthemis Frach om ali Forssk., 114
Anthemis melampodina Del., 114
Anthemis tinctoria L., 114
Anthericum aphyllum Forssk., 286
Anthericum asphodelum Forssk., 284
Anthistiria forskalii Kunth, 280
Anthospermum herbaceum L. f., 220
Anthospermum muricatum A. Rich., 220
Anticharis linearis (Benth.) Hochst. ex Asch., 227
Antichorus depressus L., 236
Antirrhinum aegyptiacum L., 228
Antirrhinum haelava Forssk., 228
Antirrhinum linaria sensu Forssk., non L., 227
Antirrhinum supinum sensu Forssk., non L., 227

Antura [sp. without epithet] Forssk., 76
Antura edulis Forssk., 76
APOCYNACEAE, 76
Apocynaceae gen. et sp. indet., 78
ARACEAE, 254
ARALIACEAE, 78
Arenaria fasciculata sensu Forssk., non L., 96
Arenaria filifolia Forssk., 96
Arenaria geniculata Poir., 96
Arenaria procumbens Vahl, 96
Arisaema bottae Schott, 254
Arisaema flavum (Forssk.) Schott, 254
Arisaema schimperianum Schott, 254
Aristida adscensionis L., 261
Aristida forskolii Henrard, 279
Aristida lanata Forssk., 279, 280
Aristida lanata sensu C. Chr., non Forssk., 279
Aristida paniculata Forssk., 261, 279
Aristida plumosa L., 279
Aristolochia bracteata Retz., 78
Aristolochia bracteolata Lam., 78
Aristolochia sempervirens sensu Forssk., 78
ARISTOLOCHIACEAE, 78
Arnebia tetrastigma Forssk., 85
Arnebia tinctoria Forssk., 85
Arnica hirsuta Forssk., 121
Artedia muricata sensu Forssk., non L., 238
Artemisia abrotanum L., 114
Artemisia judaica L., 114
Artemisia monosperma Del., 114
Artemisia semsek Forssk., 114
Artemisia sp. indet., 114
Arthrixia ericifolia (Forssk.) DC., 127
Arthrocnemum glaucum (Del.) Ung.-Sternb., 101
Arthrocnemum macrostachyum (Moric.) K. Koch, 101
Arthrocnemum strobilaceum (Moric.) Koch, 105
Arthropteris orientalis (J.F. Gmel.) Posthumus, 289
Arum flavum Forssk., 254
Arum pentaphyllum Forssk., 254
Arundo [donax] maxima Forssk., 275
Arundo calamagrostis sensu Forssk., non L., 263
Arundo donax L., 261
Arundo donax sensu Forssk., partly, non L., 275
Arundo epigejos sensu Forssk., non L., 270
Arundo isiaca Del., 275
Arundo phragmit[es] L., 275
ASCLEPIADACEAE, 78
Asclepias aphylla Forssk., 83
Asclepias contorta Forssk., 78
Asclepias cordata Forssk., 83
Asclepias daemia Forssk., 83
Asclepias forskalei Schult., 82
Asclepias glabra Forssk., 78
Asclepias laniflora Forssk., 82
Asclepias nivalis J.F. Gmel., 82
Asclepias nivea Forssk., non L., 82
Asclepias radians Forssk., 82
Asclepias setosa Forssk., 79
Asclepias spiralis Forssk., 78
Asclepias stipitacea Forssk., 83
Asparagus acutifolius L., 284
Asparagus africanus Lam., 284
Asparagus agul Forssk., 285
Asparagus asiaticus L., 285
Asparagus falcatus L., 285
Asparagus officinalis L., 285
Asparagus retrofractus sensu Forssk., non L., 284
Asparagus scaberulus sensu Defl., non A. Rich., 284
Asparagus stipularis Forssk., 285
Asperula arvensis L., 220
Asphodelus fistulosus L. var. **tenuifolius** (Cav.) Bak., 285
Asphodelus tenuifolius Cav., 285
Aspidium crenatum (Forssk.) Kuhn, 291
Aspidopteris yemensis Defl., 195
Asplenium achilleifolium (Lam.) C. Chr. var. *bipinnatum* (Forssk.) C. Chr., 290
Asplenium adiantum-nigrum L., 290
Asplenium aethiopicum (Burm. f.) Becherer, 290
Asplenium bipinnatum (Forssk.) C. Chr. ex Hieron., 290
Asplenium lanceolatum Forssk., 290
Asplenium praemorsum Sw., 290
Asplenium ruta muraria sensu Forssk.., non L., 290
Asplenium rutifolium (Berg.) Kunze var. **bipinnatum** (Forssk.) Schelpe, 290
Asplenium scolopendrium L., 292
Aster crispus Forssk., 130
Aster ericifolius Forssk., 127
Asteriscus aquaticus (L.) Less., 114
Asteriscus graveolens (Forssk.) Less., 127
Asteriscus spinosus (L.) Sch. Bip., 115
Asthenatherum forskalei (Vahl) Nevski, 264
Astragalus annularis Forssk., 177
Astragalus christianus sensu Vahl, non L., 177
Astragalus forskalei Boiss., 177
Astragalus fruticosus Forssk., 177
Astragalus peregrinus Vahl, 177
Astragalus rauwolfii Vahl, 177
Astragalus sp. indet., 178
Astragalus spinosus (Forssk.) Muschl., 177
Astragalus tomentosus Lam., 177
Astragalus trimestris sensu Forssk., non L., 177
Asystasia gangetica (L.) T. Anders., 63
Asystasia guttata (Forssk.) Brummitt, 63
Asystasia intrusa (Forssk.) Blume, 63
Atractylis carduus (Forssk.) C. Chr., 115
Atractylis flava Desf., 115
Atractylis humilis sensu Vahl, non L., 115
Atriplex coriacea Forssk., 101
Atriplex farinosa Forssk., 101
Atriplex glauca sensu Forssk., partly, non L., 103
Atriplex halimus L. var. **schweinfurthii** Boiss., 103
Atriplex hastata sensu Forssk., non L., 101

Atriplex leucoclada Boiss., 103
Atriplex portulacoides L., 105
Atriplex sp. indet., 103
Atriplex tartarica L., 103
Avena elata Forssk., 261
Avena forskalei Vahl, 264
Avena penssylvannica sensu Forssk., non L., 264
Avenula bromoides (Gouan) H. Scholz, 269
Avicennia marina (Forssk.) Vierh., 84
AVICENNIACEAE, 84
Axyris ceratoides sensu Vahl, non L., 75

Bacopa monnieri (L.) Pennell, 227
Baeobothrys lanceolata (Forssk.) Vahl, 205
Ballota acetabulosa (L.) Benth., 165
Ballota forskahlei Benth., 166
Balsamodendron opobalsamum (L.) Kunth, 90
Barleria acanthoides Vahl, 64
Barleria appressa (Forssk.) Defl., 64
Barleria bispinosa (Forssk.) Vahl, 64
Barleria lanceata (Forssk.) C. Chr., 64
Barleria noctiflora sensu Vahl, non L.f., 64
Barleria prionitis L. subsp. **appressa** (Forssk.) Brummitt & J.R.I. Wood, 64
Barleria proxima Lindau, 64
Barleria spinicyma Nees in DC., 64
Barleria trispinosa (Forssk.) Vahl, 64
Bassia eriophora (Schrad.) Asch. & Schweinf., 112
Bassia muricata (L.) Asch., 103
Bauhinia inermis Forssk., 178
Bauhinia tomentosa L., 178
Becium filamentosum (Forssk.) Chiov., 165
Becium serpyllifolium (Forssk.) J.R.I. Wood, 166
Bellis flava Forssk., 115
BERBERIDACEAE, 84
Berberis [sp. without epithet] Forssk., 84
Berberis forsskaliana Schneider, 84
Beta vulgaris L., 104
Betula alnus L., 84
BETULACEAE, 84
Bidens apiifolia Forssk., 115
Binectaria [sp. without epithet] Forssk., 226
Binectaria laurifolia Forssk., 226
Blepharis ciliaris (L.) B.L. Burtt, 64
Blepharis edulis (Forssk.) Pers., 64
Blepharis maderaspatensis (L.) Hayne ex Roth, 65
Blumea axillaris (Lam.) DC., 115
Blumea solidaginoides (Poir.) DC., 115
Blyttia spiralis (Forssk.) D. V. Field & J.R.I. Wood, 78
Boerhavia diandra sensu Forssk., non L., 206
Boerhavia dichotoma Vahl, 207
Boerhavia diffusa L., 207
Boerhavia plumbaginea Cav. var. *dichotoma* (Vahl) Asch. & Schweinf., 207
Boerhavia plumbagineus Cav. var. **forskalei** Schweinf., 206
Boerhavia repens L. var. **diffusa** (L.) Boiss., 206
Boerhavia repens L. var. *viscosa* Choisy, 207
Boerhavia scandens Forssk., 206
Bolboschoenus maritimus (L.) Palla, 256
Boletus marginalis Forssk. (Fungi), 297
Bonnemaisonia asparagoides (Woodw.) Ag. (Algae), 316
BORAGINACEAE, 84
Borago officinalis L., 85
Borago verrucosa Forssk., 89
Bothriochloa ischaemum (L.) Keng, 261
Boucerosia dentata (Forssk.) Defl., 79
Boucerosia forskalii (Decne.) Decne., 79
Boucerosia quadrangula (Forssk.) Decne., 79
Brachiaria mutica (Forssk.) Stapf, 261
Brachypodium distachyum (L.) P. Beauv., 280
Brachypodium phoenicioides (L.) Roem. & Schult., 262
Brachypodium retusum (Pers.) P. Beauv., 262
Brassica crassifolia Forssk., 143
Breonadia salicina (Vahl.) Hepper & J.R.I. Wood, 221
Bromus arvens[is] sensu Forssk., non L., 263
Bromus distachyos L., 280
Bromus hordeaceus L., 262
Bromus intermedius Guss., 262
Bromus japonicus Thunb., 262
Bromus madritensis L., 262
Bromus pinnatus sensu Forssk., non L., 262
Bromus poi-formis Forssk., 276
Bromus polystachios Forssk., 271
Bromus rigidus Roth, 262
Bromus rubens L., 263
Bromus rubens sensu Forssk., non L., 262
Bromus sp. indet., 263
Bromus squarrosus L., 263
Bromus villosus Forssk., 262
Browallia humifusa Forssk., 228
Buchnera humifusa (Forssk.) Vahl, 228
Bunias cakile L., 140
Bunias orientalis L., 140
Buphthalmum graveolens Forssk., 127
Buphthalmum melitense Forssk., 115
Buphthalmum pratense (Forssk.) Vahl, 118
Buphthalmum ramosum Forssk., 115
Bupleurum flavum Forssk., 238
BURSERACEAE, 90
Buxus dioica Forssk., 206

Cacalia angulata Vahl, 134
Cacalia odora Forssk., 124
Cacalia pendula Forssk., 124
Cacalia sempervirens Vahl, 124
Cacalia semperviva Forssk., 124
Cacalia sonchifolia sensu Forssk., non L., 134
Cadaba farinosa Forssk., 91
Cadaba glandulosa Forssk., 91
Cadaba rotundifolia Forssk., 91
Cadia [sp. without epithet] Forssk., 178
Cadia arabica Raeusch., 178

Cadia purpurea (Picciv.) Aiton, 178
Cadia varia L'Hérit., 178
Caesalpinia bonduc (L.) Roxb., 178
Caesalpinia bonducella (L.) Fleming, 178
Caesalpinia crista sensu Schwartz, non L., 178
CAESALPINIACEAE see Leguminosae, 174
Caidbeja adhaerens Forssk., 240
Cakile maritima Scop. subsp. **aegyptiaca** (Willd.) Nyman, 140
Calamagrostis purpurascens R. Br., 263
Calendula arvensis L., 116
Calendula officinalis sensu Forssk., non L., 116
Calystegia soldanella (L.) R. Br., 136
Camellia grandiflora Forssk., 68
Campanula edulis Forssk., 91
Campanula speculum[*-veneris*] L., 91
CAMPANULACEAE, 91
Camphorosma pteranthus L., 98
Canavalia africana Dunn, 178
Canavalia gladiata DC., 179
Canavalia polystachyos (Forssk.) Schweinf., 178
Canavalia virosa (Roxb.) Wight & Arn., 178
CAPPARIDACEAE, 91
Capparis cartilaginea Decne., 92
Capparis dahi Forssk., 92
Capparis decidua (Forssk.) Edgew., 92
Capparis galeata Fres., 92
Capparis galeata Fres. var. *montana* Schweinf., 92
Capparis inermis Forssk., 92
Capparis mithridatica Forssk., 93
Capparis oblongifolia Forssk., 93
Capparis sodada R. Br., 92
Capparis spinosa sensu Forssk., non L., 92
Caralluma dentata (Forssk.) Blatter, 79
Caralluma quadrangula (Forssk.) N. E. Br., 79, 80 (Fig. 12)
Caralluma sprengeri (Damm.) N.E. Br., 79
Caralluma subulata (Forssk.) Decne., 79, 80 (Fig. 13)
Cardamine africana L., 141
Cardopatium corymbosum (L.) Pers., 116
Carduus lanatus Forssk., 116
Carissa edulis (Forssk.) Vahl, 76
Carlina ch. lanatae Forssk., 116
Carlina corymbosa L. subsp. **graeca** (Boiss.) Nyman, 116
Carlina rubra Forssk., 116
Caroxylon imbricatum (Forssk.) Moq., 107
Carthamus corymbosus L., 116
Carthamus dentatus (Forssk.) DC., 116
Carthamus glaucus Bieb., 116
Carthamus lanatus L. subsp. **baeticus** (Boiss. & Reuter) Nyman, 116
CARYOPHYLLACEAE, 94
Caryophyllaceae gen. et sp. indet., 100
Cassia angustifolia Vahl, 188
Cassia aschrek Forssk., 189
Cassia italica (Miller) Spreng., 189
Cassia lanceolata Forssk., 189
Cassia ligustrina sensu Forssk., non L., 188
Cassia medica Forssk., 188
Cassia nigricans Vahl, 179
Cassia obtusifolia L., 189
Cassia occidentalis L., 189
Cassia procumbens sensu Forssk., 179
Cassia senna L., 188
Cassia senna sensu Forssk., non L., 189
Cassia sophera L., 189
Cassia [sp. without epithet], 189
Cassia sunsub Forssk., 190
Cassia tora L., 190
Cassia tora sensu Forssk., non L., 189
Cassytha filiformis L., 174
Catapodium rigidum (L.) C.E. Hubb., 263
Catha [sp. without epithet] Forssk., 100
Catha edulis (Vahl) Forssk. ex Endl., 100
Catha forskalii A. Rich., 100
Catha spinosa Forssk., 101
Caucalis angustifolia Forssk., 239
Caucalis glabra Forssk., 238
Caucalis pumila L., 239
Caucanthus [sp. without epithet] Forssk., 195
Caucanthus arabicus Lam., 195
Caucanthus edulis Forssk., 195
Caucanthus forskahlei Raeusch., 195
Caulerpa prolifera (Forssk.) Lamour. (Algae), 316
Caulerpa racemosa (Forssk.) Weber (Algae), 316
Caulerpa serrulata (Forssk.) Agardh (Algae), 315
Caulerpa setularoides (Gmel.) Howe (Algae), 316
Caylusea hexagyna (Forssk.) M. L. Green, 217
Cebatha a) *foliis glabris* Forssk., 201
Cebatha b) *foliis pubescentibus* Forssk., 200
Cebatha villosa (Lam.) C. Chr., 200
CELASTRACEAE, 100
Celastrus edulis Vahl, 100
Celastrus parviflorus Vahl, 100
Celosia caudata Vahl, 74
Celosia polystachia (Forssk.) C.C. Townsend, 74
Celosia populifolia Moq., 74
Celosia trigyna L., 74
Celosia trigyna L. var. **fasciculiflora** Fenzl, 74
Celsia ramosa Forssk., 227
Celtis integrifolia Lam., 240
Celtis toka (Forssk.) Hepper & J.R.I. Wood, 239
Cenchrus biflorus Roxb., 263
Cenchrus ciliaris L., 263
Cenchrus echinatus L., 264
Cenchrus setigerus Vahl, 264
Centaurea acaulis sensu Forssk., non L., 117
Centaurea aegyptiaca L., 117
Centaurea alexandrina Del., 117
Centaurea aspera L., 117
Centaurea calcitrapa L., 117

Centaurea carduus Forssk., 115
Centaurea eriophora Forssk., 117
Centaurea glomerata Vahl, 117
Centaurea lippii L., 135
Centaurea maxima Forssk., 118
Centaurea mucronata Forssk., 113
Centaurea sp. indet. 1, 117
Centaurea sp. indet. 2, 118
Centaurea spinosa L., 117
Centaurea verbascifolia Vahl, 118
Centaurium pulchellum (Sw.) Druce, 162
Centaurium spicatum (L.) Fritsch, 162
Centaurothamnus maximus (Forssk.) Wagenitz & Dittrich, 118
Centranthus calcitrapa (L.) Dufr., 241
Centropodia forskalei (Vahl) Cope, 264
Cephalaria sp., 151
Cephalaria transsylvanica (L.) Schrader, 151
Ceramium diaphanum Ag. (Algae), 316
Ceramium inflexum Roth. (Algae), 363
Ceramium rubrum Ag. (Algae), 316
Cerastium procumbens Forssk., 94
Ceropegia variegata Decne., 79
Ceruana [sp. without epithet] Forssk., 118
Ceruana pratensis Forssk., 118
Chadara [sp. without epithet] Forssk., 236
Chadara arborea Forssk., 236
Chadara tenax Forssk., 236
Chadara velutina Forssk., 237
Chaetomorpha linum Ktz. (Algae), 316
Chaetomorpha melagonium (Weber & Mohr) Ktz. (Algae), 316
Chamaecrista nigricans (Vahl) Greene, 179
Chamaerops humilis L., 287
Charachera [sp. without epithet] Forssk., 242
Charachera tetragona Forssk., 242
Charachera viburnoides Forssk., 241
Cheilanthes farinosa (Forssk.) Kaulf., 290
Cheiranthus farsetia L., 143
Cheiranthus linearis Forssk., 143
Cheiranthus tristis sensu Forssk., non L., 144
Cheiranthus villosus Forssk., 144
Chelidonium dodecandrum Forssk., 209
Chelidonium hybridum L., 208
CHENOPODIACEAE, 101
Chenopodiaceae gen. et sp. indet., 111
Chenopodium album L., 104
Chenopodium botrys sensu Forssk., non L., 104
Chenopodium foetidum Schrad., 104
Chenopodium fruticosum L., 111
Chenopodium murale L., 104
Chenopodium schraderianum Schultes, 104
Chenopodium serotinum Forssk., 104
Chenopodium triangulare Forssk., 104
Chenopodium viride sensu Forssk., non L., 104
Cherleria sedoides sensu Forssk., non L., 96
Chiliadenus montanus (Vahl) Brullo, 118
Chloris barbata Sw., 264
Chloris virgata Sw., 264
Chlorophytum tetraphyllum (L.f.) Bak., 254, 285
Chondria ovalis Agardh (Algae), 363
Chondria papillosa C. Agardh (Algae), 316
Chondria tenuissima Agardh (Algae), 363
Christella dentata (Forssk.) Brownsey & Jermy, 290
Chrozophora obliqua (Vahl) A. Juss. ex Spreng., 153
Chrozophora obliqua sensu Tackholm, non (Vahl) A. Juss. ex Spreng., 153
Chrozophora oblongifolia (Del.) A. Juss. ex Spreng., 153
Chrozophora plicata (Vahl) A. Juss. ex Spreng., 154
Chrozophora verbascifolia (Willd.) A. Juss., 153
Chrysanthellum sp., 115
Chrysocoma corymbosa Forssk., 118
Chrysocoma montana Vahl, 118
Chrysocoma mucronata Forssk., 124
Chrysocoma ovata Forssk., 118
Chrysocoma spatulata Forssk., 135
Chrysocoma spicata Forssk., 123
Chrysocoma spinosa (Vahl) Del., 124
Cichorium spinosum L., 118
Cineraria abyssinica Sch. Bip. ex A. Rich., 119
Cinna arundinacea sensu Forssk., non L., 252
Cissus arborea Forssk., 226
Cissus digitatus Lam., 247
Cissus glandulosa (Forssk.) J.F. Gmel., 243
Cissus quadrangularis L., 243, 244 (Fig. 25)
Cissus rotundifolia (Forssk.) Vahl, 243, 245 (Fig. 26)
Cissus sp. indet., 243
Cissus ternata (Forssk.) J.F. Gmel., 247
CISTACEAE, 112
Cistanche lutea Hoffm. & Link, 228
Cistanche phelypaea (L.) Coutinho, 228
Cistanche tubulosa (Schenk) Wight, 228
Cistus stipulatus Forssk., 112
Cistus stipulatus β Forssk., 112
Cistus thymifol. Forssk., 112
Citrullus [sp. without epithet] Forssk., 147
Citrullus battich Forssk., 147
Citrullus colocynthis (L.) Schrad., 147
Citrullus lanatus (Thunb.) Matsum. & Nakai, 147, 149
Citrus aurantium L., 224
Citrus medica L., 224
Cladium jamaicense Crantz, 256
Cladium mariscus (L.) Pohl, 256
Clematis simensis Fres., 216
Clematis vitalba L., 216
Clematis vitalba sensu Forssk., non L., 216
Cleome africana Botsch., 92
Cleome amblyocarpa Barr. & Murb., 92, 142 (Fig. 19B)
Cleome angustifolia Forssk., 93
Cleome arabica L., 92
Cleome digitata Forssk., 93

Cleome droserifolia (Forssk.) Del., 93
Cleome filifolia Vahl, 93
Cleome roridula R. Br., 93
Clinopodium fruticosum Forssk., 169
Clitoria ternatea L., 179
Clutia lanceolata Forssk., 154
Cnicus arbortivus Forssk., 119
Cnicus arcana (L.) L., 128
Cnicus dentatus Forssk., 116
Cnicus horridus Forssk., 116
Cnicus spinosissimus sensu Forssk., non L., 119
Coccinia grandis (L.) Voigt, 148
Coccinia moghadd (Forssk.) Schweinf., 148
Cocculus cebatha DC., 201
Cocculus hirsutus (L.) Theob., 200
Cocculus leaeba (Del.) DC., 200
Cocculus pendulus (J.R. & G. Forst.) Diels, 200
Cocculus villosus (Lam.) DC., 200
Coffea arabica L., 221
Colchicum montanum Forssk., 286
Colutea spinosa Forssk., 177
Commelina africana L., 255
Commelina benghalensis sensu Forssk., non L., 255
Commelina benghalensis L., 255
Commelina canescens Vahl, 255
Commelina commelinoides Forssk., 255
Commelina divaricata Vahl, 255
Commelina forskalii Kunth, 255
Commelina paniculata Vahl, non Hill, 255
Commelina tuberosa sensu Forssk., non L., 255
COMMELINACEAE, 255
Commicarpus plumbagineus (Cav.) Standl., 207
Commiphora abyssinica Engl., 90
Commiphora erythraea (Ehrby.) Engl., 90
Commiphora gileadensis (L.) C. Chr., 90
Commiphora kataf (Forssk.) Engl., 90
Commiphora opobalsamum (L.) Engl., 90
COMPOSITAE, 112
Compositae (trib. Anthemideae) gen. et sp. indet. 1, 135
Compositae (trib. Anthemideae) gen. et sp. indet. 2, 136
Consolida aconiti (L.) Lindl., 216
Convallaria racemosa Forssk., 286
CONVOLVULACEAE, 136
Convolvulus althaeoides L. subsp. **tenuissimus**, (Sibth. & Sm.) Stace, 136
Convolvulus arvensis L., 136
Convolvulus biflorus sensu Forssk., non L., 136
Convolvulus cneorum sensu Forssk., non L., 136
Convolvulus cneorum sensu Hornem., 112
Convolvulus forskahlii Sprengel, 139
Convolvulus hastatus Forssk., 136
Convolvulus hystrix Vahl, 136
Convolvulus lanatus Vahl, 136
Convolvulus pentapetaloides sensu Vahl, non L., 137
Convolvulus prostratus Forssk., 137
Convolvulus sericeus sensu Forssk., non L., 136
Convolvulus siculus L. subsp. **agrestis** (Schweinf.) Verdc., 137
Convolvulus spinosus Forssk., non Burm., 136
Conyza aegyptiaca (L.) Ait., 119
Conyza arabica Willd., 125
Conyza caule alato Forssk., 125
Conyza crispata Vahl, 125
Conyza dioscoridis (L.) Desf., 129
Conyza incana (Vahl) Willd., 119
Conyza ivifolia (L.) Less., 119
Conyza odora (*odorata* L.) Forssk., 129
Conyza pyrrhopappa Sch. Bip. ex A. Rich., 119
Conyza rupestris L., 128
Conyza tomentosa Forssk., 127, 128
Corallocarpus gijef (Forssk.) Hook. f., 149
Corbichonia decumbens (Forssk.) Exell, 69
Corchorus aestuans Forssk., 236
Corchorus antichorus Raeusch., 236
Corchorus depressus (L.) C. Chr., 236
Corchorus olitorius L., 236
Corchorus trilocularis L., 236
Cordia myxa sensu Vahl, non L., 85
Cordia sinensis Lam., 85
CORNACEAE, 139
Cornus gharaf Ehr. ex Asch., 86
Cornus sanguinea sensu Forssk., non L., 86
Cornus sanguinea L. subsp. **australis** (C.A. Meyer) Jáv., 139
Coronopus squamatus (Forssk.) Asch., 141
Corrigiola albella Forssk., 97
Corrigiola repens Forssk., 97
Cotula anthemoides L., 119
Cotyledon aegyptiaca Lam., 140
Cotyledon alternans Vahl, 140
Cotyledon deficiens Forssk., 140
Cotyledon lanceolata Forssk., 140
Cotyledon orbiculata sensu Forssk., non L., 140
Crassula alba Forssk., 139
CRASSULACEAE, 139
Crataegus monogyna Jacq., 219
Crataegus oxyac[*antha*] sensu Forssk., non L., 219
Crepis micrantha Czerep., 120
Crepis radicata Forssk., 129
Cressa arabica Forssk., 139
Cressa cretica L., 137
Crinum album (Forssk.) Herb., 253
Crinum yemense Defl., 253
Crossandra infundibuliformis sensu Defl., 65
Crossandra johanninae Fiori, 65
Crotalaria albida Heyne ex A.W. Roth, 179
Crotalaria incana L. subsp. **purpurascens** (Lam.) Milne-Redh., 179
Crotalaria microphylla Vahl, 179
Crotalaria retusa L., 180
Crotalaria thebaica (Del.) DC., 180
Croton argenteum sensu Forssk., non L., 153
Croton lobatum sensu Forssk., non L., 159

Croton lobatus L., 154
Croton obliquum Vahl, 153
Croton plicatum Vahl, 154
Croton sp. indet., 154
Croton spinosum sensu Forssk., non L., 159
Croton tinctorium sensu Forssk., non L. 1, 154
Croton tinctorium sensu Forssk., non L. 2, 154
Croton trilobatus Forssk., 154
Croton variegatum sensu Forssk., non L., 159
Croton villosum Forssk., 159
Crucianella aegyptiaca L., 221
Crucianella angustifolia L., 221
Crucianella herbacea Forssk., 221
Crucianella latifolia L., 221
Crucianella maritima L., 221
CRUCIFERAE, 140
Crypsis aculeata (L.) Ait., 264
Crypsis schoenoides (L.) Lam., 264
Crypsis vaginiflora (Forssk.) Opiz, 264
Ctenolepis cerasiformis (Stocks) Hook. f. ex Oliv., 148
Cucubalus behen L., 99
Cucumis [sp. without epithet] Forssk., 149
Cucumis angulatus Forssk., 148
Cucumis anguria sensu Forssk., non L., 149
Cucumis chaete L., 148
Cucumis colocynthis L., 147
Cucumis inedulis Forssk., 148
Cucumis melo L., 148, 149
Cucumis melo L. var. **chaete** (L.) Sageret, 148
Cucumis prophetarum L., 149
Cucumis sativus L., 148
Cucumis trilobatus sensu Forssk., non L., 149
Cucurbita citrullus Forssk., 147
Cucurbita lagenaria L., 150
Cucurbita maxima Forssk., 149
Cucurbita pepo L., 149
Cucurbita pepo longa Forssk., 149
CUCURBITACEAE, 147
Culhamia [sp. without epithet] Forssk., 234
Culhamia hadiensis J.F. Gmel., 234
Cullen corylifolia (L.) Medikus, 180
CUPRESSACEAE, 295
Cutandia dichotoma (Forssk.) Trabut, 265
Cutandia maritima (L.) Barbey, 265
Cutandia memphitica (Spreng.) Benth., 265
Cyanotis sp. indet., 255
Cyanotis tuberosa (Roxb.) Schult.f., 255
Cyclosorus dentatus (Forssk.) Ching, 291
Cycniopsis humifusa (Forssk.) Engl., 228
Cycnium humifusum (Forssk.) Benth. & Hook. f., 228
Cymbopogon caesius (Nees ex Hook. & Arn.) Stapf, 265
Cymodocea ciliata (Forssk.) Ehrh. ex Asch., 288
Cynanchum arboreum Forssk., 82
Cynanchum pyrotechnicum Forssk.82
Cynodon dactylon (L.) Pers., 265
Cynoglossum dubium Forssk., 86
Cynoglossum lanceolatum Forssk., 86
Cynoglossum linifolium sensu Forssk., non L., 86
Cynosurus aegyptiacus Forssk., 266
Cynosurus aegyptius L., 266
Cynosurus durus sensu Forssk., non L., 267
Cynosurus echinatus L., 266
Cynosurus echinatus sensu Forssk., non L., 266
Cynosurus effusus Link, 266
Cynosurus floccifolius Forssk., 268
Cynosurus lima sensu Forssk. non L., 260
Cynosurus paspaloides Vahl, 269
Cynosurus ternatus Forssk., 266
CYPERACEAE, 256
Cyperus alopecuroides Rottb., 256
Cyperus alternifolius L. subsp. *flabelliformis* (Rottb.) Kük., 258
Cyperus articulatus L., 256
Cyperus capitatus Vand., 256
Cyperus complanatus Forssk., 257
Cyperus conglomeratus Rottb., 257
Cyperus conglomeratus Rottb. var. *effusus* (Rottb.) Kük., 257
Cyperus conglomeratus Rottb. var. *jeminicus* (Rottb.) Kük., 257
Cyperus conglomeratus Rottb. var. *multi-glumis* (Boeck.) Kük., 257
Cyperus cruentus Rottb., 257
Cyperus difformis L., 257
Cyperus dives Del., 256
Cyperus effusus Rottb., 257
Cyperus fastigiatus Forssk., 256
Cyperus ferrugineus Forssk., 257, 259
Cyperus flabelliformis Rottb., 258
Cyperus forskolei Dietr., 257
Cyperus fuscus L., 257
Cyperus fuscus L. forma **virescens** (Hoffm.) Vahl, 258
Cyperus globosus Forssk., 257
Cyperus gradatus Forssk., 258
Cyperus hexastachyos Rottb., 259
Cyperus involucratus Rottb., 258
Cyperus jeminicus Rottb., 257
Cyperus laevigatus L., 258
Cyperus lateralis Forssk., 258
Cyperus longus L., 258
Cyperus minimus sensu Forssk., non L., 258
Cyperus mucronatus Rottb., 258
Cyperus niloticus Forssk., 256
Cyperus rotundus L., 259
Cyphostemma digitata (Forssk.) Desc., 246 (Fig. 27), 247
Cyphostemma ternatum (Forssk.) Desc., 247
Cystophyllum trinode (Forssk.) Agardh (Algae), 317
Cytoseira myrica (Gmel.) Agardh (Algae), 316
Cytosira barbata (Algae), 315

Dactylis glomerata L., 266
Dactyloctenium aegyptium (L.) Willd., 266
Dactyloctenium aristatum Link, 266

Dactylus trapezuntius Forssk., 151
Daemia ? glabra (Forssk.) Schult., 78
Daemia cordata (Forssk.) R. Br., 83
Daemia forskali Schult., 83
Danthonia forskalei(Vahl) R. Br., 264
Dasypyrum villosum (L.) Cand., 266
Datura metel L., 229
Daucus b) [sp. without epithet] Forssk., 237
Daucus broteri Ten., 238
Daucus gingidium sensu Forssk., non L., 237
Daucus glaber (Forssk.) Thell., 238
Daucus littoralis Sibth. & Sm. var. *forskahlei* Boiss., 238
Daucus muricata (L.) L., 238
Debregeasia bicolor (Roxb.) Wedd., 240
Debregeasia saeneb (Forssk.) Hepper & J.R.I. Wood, 240
Delonix elata (L.) Gamble, 180
Delphinium aconiti L., 216
Delphinium forskolii Reichenb., 216
Delphinium grandiflorum sensu Forssk., non L., 216
Delphinium nanum DC., 217
Delphinium peregrinum L., 216
Desmazeria rigida (L.) Tutin, 263
Desmidorchis forskalii Decne., 79
Desmodium gangeticum (L.) DC., 180
Desmodium ospriostreblum Chiov., 180
Desmodium repandum (Vahl) DC., 180
Desmodium triflorum (L.) DC., 181
Desmostachya bipinnata (L.) Stapf, 267
Dianthera americana fl. albo sensu Forssk., 66
Dianthera americana flava sensu Forssk., 67
Dianthera bicalyculata Retz., 68
Dianthera debilis Forssk., 68
Dianthera flava Vahl, 67
Dianthera micranthes Nees, 65
Dianthera odora Forssk., 67
Dianthera paniculata Forssk., 68
Dianthera punctata Vahl, 66
Dianthera sulcata Vahl, 67
Dianthera trisulca Forssk., 63
Dianthera verticillata Forssk., 65
Dianthera violacea Vahl, 68
Dianthus caryophyllus L., 94
Dianthus prolifera L. 97
Dianthus pumilus Vahl, 94
Dianthus uniflorus Forssk., 94
Dichantium annulatum (Forssk.) Stapf, 267
Dichrostachys cinerea (L.) Wight & Arn. var. **cinerea**, 181
Dichrostachys nutans (Pers.) Benth., 181
Dicliptera bivalvis sensu Nees, 65
Dicliptera chinensis sensu C. Chr., non (L.) Juss., 65
Dicliptera foetida (Forssk.) Blatter, 65
Dicliptera sexangularis sensu Blatter, non (L.) Juss., 65
Dicliptera verticillata (Forssk.) C. Chr., 65
DICOTYLEDONS, 63
Dicotyledon fam., gen. et sp. indet. 1, 251
Dicotyledon fam., gen. et sp. indet. 2, 251
Dicotyledon fam., gen. et sp. indet. 3, 251
Dicotyledon fam., gen. et sp. indet. 4, 251
Dicotyledon fam., gen. et sp. indet. 5, 251
Dicotyledon fam., gen. et sp. indet. 6, 251
Dicotyledon fam., gen. et sp. indet. 7, 251
Dicotyledon fam., gen. et sp. indet. 8, 252
Dictyota dichotoma (L.) Ktz. (Algae), 316
Digera alternifolia (L.) Asch., 75
Digera arvensis Forssk., 75
Digera muricata (L.) Mart., 75
Digitaria sanguinalis (L.) Scop., 267
Digitaria velutina (Forssk.) P. Beauv., 267
Diospyros lotus L., 152
Diospyros mespiliformis Hochst. ex A. DC., 152
Dipcadi erythraeum Webb & Berth., 286
Diplachne fusca (L.) P. Beauv., 271
Diplanthera uninervis (Forssk.) Asch., 288
Diplotaxis acris (Forssk.) Boiss., 141
Diplotaxis harra (Forssk.) Boiss., 141
DIPSACACEAE, 151
Dobera glabra (Forssk.) Juss. ex Poir., 226
Dolichos aeschynomene sesban Forssk., 190
Dolichos arboreus Forssk., 190
Dolichos cultratus Forssk., 183
Dolichos cuneifolius Forssk., 180
Dolichos didjre Forssk., 193
Dolichos faba indica Forssk., 179
Dolichos faba nigrita Forssk., 181
Dolichos lablab L., 183
Dolichos lubia Forssk., 193
Dolichos polystachyos Forssk., non L., 178
Dolichos sinensis sensu Forssk., non L., 193
Dorstenia foetida (Forssk.) Schweinf., 202, 203 (Fig. 23)
Dorycnium argenteum Del., 185
Dorycnium graecum (L.) Ser., 181
Dorycnium latifolium Willd., 185
Dorycnium quinatum (Forssk.) C. Chr., 185
Doryopteris concolor (Langsd. & Fisch.) Kuhn, 291
Dracaena sp. indet., 253
Dracunculus canariensis Kunth, 254
Dracunculus vulgaris Schott & Engl., 254
Droguetia iners (Forssk.) Schweinf., 240
Dryopteris crenata (Forssk.) Kuntze, 291
Dryopteris dentata (Forssk.) C. Chr., 291
Dryopteris orientalis (J.F. Gmel.) C. Chr., 290

EBENACEAE, 152
Ecbolium gymnostachyum (Nees), Milne-Redh., 65
Ecbolium violaceum (Vahl) Hillcoat & J.R.I. Wood, 68
Ecbolium viride (Forssk.) Alston, 66
Echidnopsis multangula (Forssk.) Chiov., 79
Echinocaulon acerosum (Forssk.) Børgesen (Algae), 316

Echinochloa colona (L.) Link, 267
Echinochloa crus-galli (L.) P. Beauv., 267
Echinochloa geminata (Forssk.) Roberty, 273
Echinophora tenuifolia L. subsp. **sipthorpiana** (Guss.) Tutin, 238
Echinops microcephalus Sm., 120
Echinops ritro L., 120
Echinops sphaerocephalus sensu Forssk., non L., 120
Echinops spinosissimus Turra, 120
Echinops strigosus L., 120
Echinospermum spinocarpos (Forssk.) Boiss., 89
Echinospermum vahlianum Lehm., 89
Echium angustifolium Miller subsp. **sericeum** (Vahl) Klotz, 86
Echium creticum L., 86
Echium creticum sensu Forssk., non L., 87
Echium fruticosum sensu Forssk., non L., 86
Echium glomeratum Poir., 86
Echium longifolium Del., 87
Echium cf. **rauwolfii** Del., 87
Echium rubrum Forssk., 87
Echium sericeum Vahl, 86
Echium setosum sensu Del., non Vahl, 87
Eclipta alba (L.) Hassk., 120
Eclipta erecta (L.) L., 120
Eclipta prostrata (L.) L., 120
Ehretia cymosa Thonn., 87
ELAEAGNACEAE, 152
Elaeagnus angustifolia L., 152
Elbunis alba (Forssk.) Raf., 166
Elcaja [sp. without epithet] Forssk., 200
Eleocharis geniculata (L.) Roem. & Schult., 259
Eleusine compressa (Forssk.) Asch. & Schweinf., 272
Eleusine flagellifera Nees, 272
Eleusine floccifolia (Forssk.) Spreng., 268
Elionurus hirsutus (Forssk.) Munro, 280
Elymus caput-medusae sensu Forssk., non L., 263
Elymus delileana Schult., 268
Elymus sp. indet., 268
Elymus subulatus Forssk., 268
Emex spinosa (L.) Campd., 213
Emilia angulata (Vahl) DC., 134
Enarthrocarpus lyratus (Forssk.) DC., 141
Enicostema axillare (Lam.) A. Raynal, 162
Ephedra aphylla Forssk., 295
Ephedra campylopoda C. A. Mey., 295
Ephedra foemina Forssk., 295
EPHEDRACEAE, 295
Epilobium angustifolium L., 208
Epilobium tetragonum sensu Forssk., non L., 208
Equisetum arvense sensu Forssk., non L., 291
Equisetum ramosissimum Desf., 291
Eraclissa hexagyna Forssk., 146 (Fig. 21B), 153
Eragrostis aegyptiaca (Willd.) Del., 268
Eragrostis cilianensis (All.) Vignolo & Janchen, 268
Eragrostis ciliaris (L.) R. Br., 268
Eragrostis kiwuensis Jedwabn., 269
Eragrostis mucronata (L.) Roem. & Schult., 272
Eragrostis multiflora (Forssk.) Asch., 269
Eragrostis pilosa (L.) P. Beauv., 269
Eragrostis pungens (Vahl) Benth., 272
Eremobium aegyptiacum (Spreng.) Hochr., 143
Erica arborea L.,152
Erica manipuliflora Salisb., 152
Erica scoparia sensu Forssk., non L., 152
Erica verticillata Forssk., 152
ERICACEAE, 152
Erigeron aegyptiacum L., 119
Erigeron decurrens Vahl, 125
Erigeron incanum Vahl, 119
Erigeron serratum Forssk., 119
Erodium ciconium (L.) L'Hérit., 163
Erodium crassifolium L'Hérit., 163
Erodium glaucophyllum (L.) L'Hérit., 163
Erodium hirtum (Forssk.) Willd., 163
Erodium laciniatum (Cav.) Willd., 163
Erodium malacoides (L.) L'Hérit., 164
Erodium triangulare (Forssk.) Muschl., 163
Erucaria crassifolia (Forssk.) Del., 143
Eryngium amethystinum, L., 238
Eryngium bourgatii Gouan, 238
Eryngium foetidum sensu Forssk., non L., 238
Eryngium maritimum L., 238
Ethulia conyzoides L., 121
Eulophia streptopetala Lindl. var. **rueppelii** (Rchb. f.) Cribb, 287
Euonymus inermis Forssk., 207
Euphorbia aculeata Forssk., 154
Euphorbia antiquorum β minor inarticulata Forssk., 156
Euphorbia belgradica Forssk., 155
Euphorbia cuneata Vahl, 155
Euphorbia decumbens Forssk., 156
Euphorbia dichotoma Forssk., 156
Euphorbia esula sensu Forssk., non L., 156
Euphorbia prob. **exigua** L., 155
Euphorbia falcata L., 155
Euphorbia fruticosa Forssk., 156
Euphorbia glaucophylla Poir., 155
Euphorbia glomerifera (Millsp.) Wheeler, 155
Euphorbia granulata Forssk. var. **glabrata** (Gay) Boiss., 155
Euphorbia granulata Forssk. var. **granulata**, 155
Euphorbia hirta L., 155
Euphorbia inarticulata Schweinf., 156
Euphorbia indica Lam., 156
Euphorbia monticola sensu Schweinf., non Hochst., 156
Euphorbia obliquata L., 157
Euphorbia officinalis β caespitosa Forssk., 156
Euphorbia peplis L., 156
Euphorbia peplus L., 156
Euphorbia platyphyllos L., 155, 156
Euphorbia polygonifolia sensu Forssk., non L., 155

Euphorbia pubescens Vahl, 155
Euphorbia retusa Forssk., 157. 158 (Fig. 22)
Euphorbia schimperi Presl, 157
Euphorbia schimperiana Scheele, 157
Euphorbia scordifolia Jacq., 157
Euphorbia serrata sensu Vahl, non L., 157
Euphorbia simplex sensu Forssk., non L., 157
Euphorbia sp. indet., 157
Euphorbia suffruticosa Forssk., 157
Euphorbia terracina L., 157
Euphorbia thymifolia Forssk., non Burm., 157
Euphorbia thymifolia L., 157
Euphorbia tirucallii sensu Forssk., non L., 157
Euphorbia triaculeata Forssk., 157
EUPHORBIACEAE, 152
Eustachys paspaloides (Vahl) Lanza & Mattei, 269

FAGACEAE, 160
Fagonia arabica L., 247
Fagonia cretica L., 247
Fagonia cretica sensu Forssk., non L., 247
Fagonia scabra Forssk., 247
Farsetia aegyptia Turra, 142 (Fig. 19A), 143
Farsetia linearis (Forssk.) Decne. ex Boiss., 143
Farsetia longisiliqua Decne., 143
FERNS AND FERN-ALLIES, 289
Festuca calycina Forssk., 276
Festuca dichotoma Forssk., 265
Festuca fasciculata Forssk., 281
Festuca fusca L., 271
Festuca lanceolata Forssk., 265
Festuca mucronata Forssk., 272
Festuca ovina L., 269
Festuca pungens Vahl, 272
Ficus benghalensis sensu Vahl, non L., 204
Ficus capensis Thunb., 204
Ficus carica L., 202
Ficus chanas Forssk., 204
Ficus cordata Thunb. subsp. **salicifolia** (Vahl) C.C. Berg, 202
Ficus exasperata Vahl, 202
Ficus forskalei Vahl, 205
Ficus indica sensu Forssk., non L., 202
Ficus morifolia Forssk., non Lam., 205
Ficus palmata Forssk., 202, 205
Ficus cf. **palmata** Forssk., 205
Ficus populifolia Vahl, 204
Ficus religiosa sensu Forssk., non L., 204
Ficus salicifolia Vahl, 202
Ficus serrata sensu Forssk., non L., 202
Ficus sur Forssk., 204
Ficus sycomorus L., 204
Ficus taab Forssk., 204
Ficus toka Forssk., 239
Ficus vasta Forssk., 204
Filago pyramidata L., 121
Filago spathulata C. Presl, 121
Fimbristylis bis-umbellata (Forssk.) Bubani, 259
Fimbristylis dichotoma Vahl, 259
FLACOURTIACEAE, 161
Flemingia cf. **grahamiana** Wight & Arn., 181
Flemingia rhodocarpa Bak., 181
Fleurya aestuans (L.) Gaud., 241
Flueggea obovata (Willd.) Wall. ex F. Vill., 159
Flueggea virosa (Willd.) Voigt, 159
Forsskaolea tenacissima L., 240
Francoeuria crispa (Forssk.) Cass., 130
Frankenia hirsuta L. var. *revoluta* (Forssk.) Boiss., 161
Frankenia laevis L. var. *revoluta* (Forssk.) Dur. & Bar., 162
Frankenia pulverulenta L., 161
Frankenia revoluta Forssk., 161
FRANKENIACEAE, 161
Fucus diffusus Huds. (Algae), 347
Fucus soboliferus (Algae), 363
Fucus vesiculosus L. (Algae), 315
Fumaria densiflora DC., 162
Fumaria officinalis sensu Forssk., 162
FUMARIACEAE, 162
FUNGI, 297

Galega tomentosa (Forssk.) Vahl, 191
Galium album Miller subsp. **pycnotrichum** (H. Braun) Krendl, 221
Galium album sensu Forssk., non Miller, 221, 222
Galium aparinoides Forssk., 222
Galium cf. **aparine** L., 222
Galium glaucum L., 222
Galium hamatum Hochst. ex A. Rich., 222
Galium heldreichii Halácsy, 222
Galium or **Asperula** sp. indet., 222
Galium paschale Forssk., 222
Galium rubioides L., 222
Galium tricornutum Dandy, 222
Gastrocotyle hispida (Forssk.) Bunge, 87
Gendarussa debilis (Forssk.) Nees, 68
Genista raetam Forssk., 188
Genista scorpius (L.) DC., 181
Genista spartium sensu Forssk., 188
Gentiana centaur[ium] sensu Forssk., non L., 162
GENTIANACEAE, 162
GERANIACEAE, 163
Geranium arabicum Forssk., 164
Geranium ciconium L., 163
Geranium crassifolium Forssk., 163
Geranium hirtum Forssk., 163
Geranium malacoides L., 164
Geranium simense Hochst., 164
Geranium triangulare Forssk., 163
Gerbera hirsuta (Forssk.) Less., 121
Gerbera piloselloides (L.) Cass., 121
Geruma alba Forssk., 69
Girardinia condensata (Hochst. ex Steud.) Wedd., 240
Girardinia diversifolia (Link) Friis, 240

Girardinia heterophylla (Vahl) Decne., 240
Girardinia palmata Blume, 240
Gisekia pharnaceoides L., 71
Glinus chrystallinus Forssk., 69
Glinus lotoides L., 71
Glinus lotoides L. var. *setiflorus* (Forssk.) Fenzl, 71
Glinus setiflorus Forssk., 71
Glycine abrus L., 174
Glycine javanica auctt., non L., 186
Glycyrrhiza aculeata Forssk., 178
Glycyrrhiza astragaloides Vahl, 190
Gnaphalium arabicum J.F. Gmel., 121
Gnaphalium cinereum J.F. Gmel., 121
Gnaphalium crispatulum Del., 121
Gnaphalium cuneifolium J.F. Gmel., 122
Gnaphalium forskalii J.F. Gmel., 122
Gnaphalium fruticosum Forssk., 122
Gnaphalium kurmense Mart., 122
Gnaphalium luteo-album L., 129
Gnaphalium margaritaceum sensu Forssk., non L., 121
Gnaphalium obtusifolium sensu Forssk., non L., 129
Gnaphalium orientale sensu Forssk., non L., 122
Gnaphalium pulvinatum Del., 122
Gnaphalium sp. 1 [sp. without epithet] Forssk., 121
Gnaphalium sp. 2 [sp. without epithet] Forssk., 121
Gnaphalium sp. 3 [sp. without epithet] Forssk., 122
Gnaphalium spicatum (Forssk.) Vahl, 123
Gomphocarpus fruticosus (L.) Ait. f. var. **setosus** (Forssk.) Schwartz, 78
Gomphocarpus setosus (Forssk.) R. Br., 79
Goniolimon collinum (Griseb.) Boiss., 211
Goniolimon incanum (L.) Hepper, 211
Gossypium arboreum L., 196
Gossypium herbaceum L., 196
Gossypium rubrum Forssk., 196
Gracilaria debilis (Forssk.) Børgesen (Algae), 317
Gramen gen. et sp. indet. 1, 281
Gramen gen. et sp. indet. 2, 281
GRAMINEAE, 260
Grewia arborea (Forssk.) Lam., 236
Grewia bicolor Juss., 237
Grewia canescens Hochst. ex A. Rich., 237
Grewia excelsa Vahl, 236
Grewia populifolia Vahl, 236
Grewia sp. indet., 237
Grewia tenax (Forssk.) Fiori, 236
Grewia velutina (Forssk.) Vahl, 237
Grewia villosa Willd., 237
Guilandina bonducella L., 178
GUTTIFERAE, 164
Gymnocarpos decandrum Forssk., 94, 95 (Fig. 14)
Gymnocarpos deserti Forssk., 94
Gymnocarpos fruticosus (Vahl) Pers., 94
Gymnocladus arabica Lam., 205
GYMNOSPERMS, 295
Gymnosporia senegalensis (Lam.) Loes. var. *spinosa* (Forssk.) Engl., 101
Gymnosporia spinosa (Forssk.) Fiori, 101
Gynandriris sisyrinchium (L.) Parl., 282
Gypsophila capillaris (Forssk.) C. Chr., 96
Gypsophila rokejeka Del., 96

Habenaria aphylla (Forssk.) R. Br., 287
Haemanthus coccineus sensu Forssk., non L., 254, 285
Halarachnion ligulatum (Woodw.) Ktz. (Algae), 316
Halimione portulacoides (L.) Aellen, 105
Halocnemum strobilaceum (Pallas) M. Bieb., 105
Halodule uninervis (Forssk.) Asch., 288
Halopeplis perfoliata (Forssk.) Bunge, 105, 106 (Fig. 16)
Halophila stipulacea (Forssk.) Asch., 288
Haplophyllum buxbaumii (Poir.) G. Don, 224
Haplophyllum tuberculatum (Forssk.) A. Juss., 224
Hedera helix L., 78
Hedyotis herbacea sensu Forssk., non L., 222
Hedypnois cretica (L.) Dum.-Cours., 122
Hedypnois rhagadioloides (L.) F.W. Schmidt, 122
Hedysarum alhagi L., partly, 176
Hedysarum glumaceum Vahl, 177
Hedysarum immaculatum Forssk., 182
Hedysarum lappaceum Forssk., 190
Hedysarum repandum Vahl, 180
Hedysarum violaceum sensu Forssk., non L., 177
Hedysarum virginicum sensu Forssk., 190
Helianthemum kahiricum Del., 112
Helianthemum stipulatum (Forssk.) C. Chr., 112
Helichrysum cymosum (L.) D. Don subsp. *fruticosum* (Forssk.) Hedberg, 122
Helichrysum forskahlii (J.F. Gmel.) Hilliard & Burtt, 122
Helichrysum fruticosum (Forssk.) Vatke, 122
Helichrysum orientale (L.) Gaertn., 122
Helichrysum schimperi (Sch. Bip. ex A. Rich.) Moeser, 122
Helictotrichon bromoides (Gouan) C.E. Hubb., 261, 269
Helictotrichon sp. indet., 269
Heliotropium coromandelianum Retz., 88
Heliotropium bacciferum Forssk., 87
Heliotropium curassavicum sensu Forssk., non L., 88
Heliotropium digynum (Forssk.) Asch. ex C. Chr., 88
Heliotropium europaeum L., 88
Heliotropium fruticosum sensu Forssk., non L., 88

Heliotropium lineatum Vahl, 88
Heliotropium longiflorum (A. DC.) Steud. & Hochst. ex DC., 88
Heliotropium ovalifolium Forssk., 88
Heliotropium sp., 88, 89
Heliotropium supinum L., 88
Heliotropium undulatum Vahl, 87
Herniaria hirsuta L., 96
Herniaria lenticulata Forssk., 97
Herpestis monniera (L.) H.B.K., 227
Hesperis acris Forssk., 141
Heteropogon contortus (L.) P. Beauv. ex Roem. & Schult., 269
Hibiscus cannabinus L., 196
Hibiscus deflersii Cufod., 197
Hibiscus esculentus L., 195
Hibiscus ficulneus L., 197
Hibiscus flavo-ferrugineus Forssk., 199
Hibiscus flavus sensu Forssk., non L., 199
Hibiscus flavus β Forssk., 199
Hibiscus microphyllus Vahl, 199
Hibiscus ovalifolius (Forssk.) Vahl, 197
Hibiscus palmatus Forssk., 197
Hibiscus praecox Forssk., 197
Hibiscus purpureus Forssk., 197
Hibiscus sp. indet., 198
Hibiscus syriacus L., 197
Hibiscus tripartitus Forssk., 198
Hibiscus vitifolius L., 198
Hieracium forskohlei Froel., 122
Hieracium hispidum Forssk., 122
Hieracium multiflorum Forssk., 123
Hieracium sp. indet., 123
Hieracium uniflorum Forssk., 123
Hildebrandtia africana Vatke, 137
Hippocrepis bisiliqua Forssk., 182
Hippocrepis unisiliquosa L. var. **bisiliqua** (Forssk.) Bornm., 182
Holcus dochna Forssk., 277
Holcus durra Forssk., 277
Holcus exiguus Forssk., 277
Holcus halepensis L., 277
Holcus racemosus Forssk., 274
Holcus sorghum L., 277
Holothrix aphylla (Forssk.) Rchb. f., 287
Hordeum geniculatum All., 269
Hordeum glaucum Steud., 270
Hordeum hexastichon L., 270
Hordeum imrinum Forssk., 270
Hordeum marinum Huds. subsp. **gussoneanum** (Parl.) Thell., 269
Hordeum marinum Huds. var. **marinum**, 269
Hordeum maritimum Vahl, 270
Hordeum murinum L. subsp. **glaucum** (Steud.) Tzveler, 270
Hordeum peruersum Forssk., 270
Hordeum vulgare L., 270
Hormosira articulata (Forssk.) Zanard (Algae), 316
Hyacinthus serotinus Forssk., 286
Hybanthus enneaspermus (L.) F. Muell., 243
HYDROCHARITACEAE, 282
Hyoscyamus datora Forssk., 230, 231 (Fig. 24A)
Hyoscyamus muticus L., 230, 231 (Fig. 24A)
Hyoseris lucida L., 123
Hypecoum aegyptiacum (Forssk.) Asch. & Schweinf., 208
Hyperanthera [sp. without epithet] Forssk., 205
Hyperanthera peregrina Forssk., 205
Hyperanthera semidecandra Vahl, 205
Hypericum asc[yron] sensu Forssk., non L., 164
Hypericum bithynicum Boiss., 164
Hypericum c) *foliis linearibus* sensu Forssk., 164
Hypericum calycinum L., 164
Hypericum kalmii sensu Forssk., non L., 165
Hypericum perforatum L., 164
Hypericum perforatum L. forma *veronense* Schr., 164
Hypericum pubescens Boiss., 164
Hypericum revolutum Vahl, 165
Hypnea musciformis (Wulf.) Lamour. (Algae), 316
Hypodematium crenatum (Forssk.) Kuhn, 291
Hypoestes forsskalii (Vahl) R. Br., 66
Hypoestes paniculata (Forssk.) Schweinf., 66
Hypoestes triflora (Forssk.) Roem. & Schult., 66
HYPOXIDACEAE, 282
Hypoxis [sp. without epithet] Forssk., 282

Iberis pinnata L., 143
Ifloga spicata (Forssk.) Sch. Bip., 123
Illecebrum arabicum L., 97
Imperata cylindrica (L.) Raeusch., 270
Indigofera argentea sensu Vahl, non Burm. f., 182
Indigofera articulata Gouan, 182
Indigofera houer Forssk., 183
Indigofera oblongifolia Forssk., 182
Indigofera semitrijuga Forssk., 182
Indigofera spicata Forssk., 182
Indigofera spinosa Forssk., 183
Indigofera tetrasperma Pers., 183
Indigofera tinctoria L., 183
Indigofera tinctoria sensu Forssk., non L., 182
Inula arabica L., 130
Inula crispa (Forssk.) Pers., 130
Inula crithmifolia Vahl, 123
Inula crithmoides L., 123
Inula dysenterica sensu Forssk., non L., 130
Inula odora sensu Forssk., non L., 130
Iphiona mucronata (Forssk.) Asch. & Schweinf., 124
Ipomoea aquatica Forssk., 137
Ipomoea biloba Forssk., 138
Ipomoea cairica (L.) Sweet, 138
Ipomoea eriocarpa R. Br., 138
Ipomoea nil (L.) Roth, 138
Ipomoea obscura (L.) Ker-Gawl., 138
Ipomoea palmata Forssk., 138

Ipomoea pes-caprae (L.) R. Br., 138
Ipomoea quamoclit L., 138
Ipomoea reptans sensu Roem. & Schult., non L., 137
Ipomoea scabra Forssk., 138
Ipomoea triflora Forssk., 139
Ipomoea verticillata Forssk., 139
IRIDACEAE, 282
Iris sisyrinchium L., 282
Isatis aegyptia[ca] L., 140
Isatis pinnata Forssk., 141
Iseilema prostrata (L.) Anderss., 271
Isoglossa punctata (Vahl) Brummitt & J.R.I. Wood, 66
Isolepis uninodis Del., 260
Ixora occidentalis sensu Forssk., non L., 223

Jasminum officinale L., 207
Jasonia montana (Vahl) Botsch., 118
Jatropha curcas L., 159
Jatropha glandulosa Vahl, 159
Jatropha glauca Vahl, 159
Jatropha lobata Muell. Arg., 159
Jatropha pelargoniifolia Courbon, 159
Jatropha pungens Forssk., 160
Jatropha spinosa Vahl, 159
Jatropha variegata Vahl, 159
Jatropha villosa (Forssk.) Muell. Arg., non Wight, 159
JUNCACEAE, 282
Juncus acutus L., 282
Juncus bufonius L., 282
Juncus effusus L., 283
Juncus inflexus L., 283
Juncus spinosus Forssk., 282
Juncus striatus Schousb. ex E.H.F. Meyer, 283
Juncus subulatus Forssk., 283
Juniperus oxycedrus L., 295
Jussiaea diffusa Forssk., 208
Jussiaea edulis Forssk., 236
Jussiaea repens L. var. *diffusa* (Forssk.) Brenan, 208
Justicia appressa Forssk., 64
Justicia bicalyculata (Retz.) Vahl, 68
Justicia bispinosa Forssk., 64
Justicia bivalvis sensu Vahl, non L., 65
Justicia caerulea Forssk., 66
Justicia chinensis sensu Vahl, non L., 65
Justicia cuspidata Vahl, 65
Justicia debilis (Forssk.) Vahl, 68
Justicia dubia Forssk., 67, 68
Justicia flava (Vahl) Vahl, 67
Justicia foetida Forssk., 65
Justicia forskalei Vahl, 66
Justicia lanceata Forssk., 64
Justicia odora (Forssk.) Vahl, 67
Justicia paniculata Forssk., 66
Justicia punctata (Vahl) Vahl, 66
Justicia resupinata Forssk., 67
Justicia sexangularis sensu Forssk., non L., 65
Justicia sulcata (Vahl) Vahl, 67
Justicia triflora Forssk., 66
Justicia trispinosa Forssk., 64
Justicia trisulca (Forssk.) Vahl, 63
Justicia violacea (Vahl) Vahl, 68
Justicia viridis Forssk., 66

Kahiria [sp. without epithet] Forssk., 121
Kalanchoe aegyptiaca (Lam.) DC., 140
Kalanchoe alternans (Vahl) Pers., 139
Kalanchoe deficiens (Forssk.) Asch. & Schweinf., 140
Kalanchoe glaucescens Britten var. *deficiens* (Forssk.) Senni, 140
Kalanchoe lanceolata (Forssk.) Pers., 140
Kalanchoe rosulata Raadts, 140
Kanahia forskalii Decne., 82
Kanahia kannah Roem. & Schult., 82
Kanahia laniflora (Forssk.) R. Br., 82
Kedrostis gijef (Forssk.) C. Jeffrey, 149
Kedrostis hirtella (Naud.) Cogn., 150
Kedrostis leloja (Forssk.) C. Jeffrey, 150
Kentrophyllum dentatum (Forssk.) DC., 116
Keura odorifera Forssk., 287
Kickxia aegyptiaca (L.) Nábelek, 228
Kleinia angulata (Vahl) Willd., 134
Kleinia odora (Forssk.) DC., 124
Kleinia pendula (Forssk.) DC., 124
Kleinia semperviva (Forssk.) DC., 124
Kochia muricata (L.) Schrad., 103
Kohautia caespitosa Schnizl., 222
Kohautia caespitosa Schnizl. var. *schimperi* (Presl) Bremek., 223
Kosaria [sp. without epithet] Forssk., 202, 203 (Fig. 23)
Kosaria foetida Forssk., 202
Kummerowia striata (Thunb.) Schindl., 183
Kyllinga monocephala Rottb., 259

LABIATAE, 165
Labiatae gen. et sp. indet., 174
Lablab purpureus (L.) Sweet, 183
Lablab vulgaris Savi, 183
Lactuca capensis Thunb., 125
Lactuca decorticata Forssk., 131
Lactuca flava Forssk., 126
Lactuca inermis Forssk., 125
Lactuca serriola L., 125
Lactuca sinuata Forssk., 125
Lactuca virosa sensu Vahl, 125
Lagenaria siceraria (Molina) Standley, 150
Laggera arabica (Willd.) Defl., 125
Laggera crispata (Vahl) Hepper & J.R.I. Wood, 125
Laggera decurrens (Vahl) Hepper & J.R.I. Wood, 125
Lagurus ovatus L., 271
Lamarckia aurea (L.) Moench, 271
Lantana tetragona (Forssk.) Schweinf., 242
Lantana viburnoides (Forssk.) Vahl, 241

Laportea aestuans (L.) Chew, 240
Lappula spinocarpos (Forssk.) Asch. ex Kuntze, 89
Lapsana taraxacoides Forssk., 123
Lasiurus hirsutus (Forssk.) Boiss., 271
Lasiurus scindicus Henrard, 271
Lastrea crenata (Forssk.) Bedd., 291
Lathraea phelypaea sensu Forssk., non L., 229
Lathraea quinquefida Forssk., 228
Lathyrus spectabilis Forssk., 179
Lathyrus tomentosus Forssk., 191
Launaea cervicornis (Boiss.) Font Quer & Rothm., 125
Launaea mucronata (Forssk.) Muschl., 125
Launaea nudicaulis (L.) Hook. f., 126
Launaea resedifolia (L.) O. Kuntze var. *mucronata* (Forssk.) Beauverd, 125
Launaea sp. indet. 1, 126
Launaea sp. indet. 2, 126
Launaea spinosa (Forssk.) Sch. Bip. ex O. Kuntze, 125, 126
LAURACEAE, 174
Laurencia obtusa Ag. (Algae), 316
Laurencia uvifera (Forssk.) Børgesen (Algae), 317
Lawsonia inermis L., 195
Lawsonia spinosa L., 195
Leaeba [sp. without epithet] Forssk., 201
Legousia speculum-veneris (L.) Chaix, 91
LEGUMINOSAE, 174
LENTIBULARIACEAE, 194
Leontodon asperum Forssk., 126, 128
Leontodon hispidulus (Del.) Boiss., 126
Leontodon maritim[um] Forssk., 126
Leontodon megalorhizon Forssk., 135
Leontodon mucronatum Forssk., 125
Leontodon tuberosus L., 126
Lepidagathis aristata (Vahl) Nees, 67
Lepidium alpinum ? sensu Forssk., 144
Lepidium armoracia Fisch. & Mey., 144
Lepidium graminifolium L., 144
Lepidium sativum L., 252
Lepidium squamatum Forssk., 141
Lepidotis cernua (L.) Beauv., 291
Leptadenia arborea (Forssk.) Schweinf., 82
Leptadenia pyrotechnica (Forssk.) Decne., 82
Leptochloa fusca (L.) Kunth, 271
Leucas alba (Forssk.) Sebald, 166
Leucas glabrata (Vahl) R. Br., 166
Leucas urticifolia (Vahl) R. Br., 166
Lichen filamentosus Forssk., 297
Lichen floridus Forssk., 297
Lichen foliaceus Forssk., 297
Lichen leprosus Forssk., 297
LICHENS, 297
LILIACEAE, 283
Limeum humile Forssk., 71
Limoniastrum monopetalum (L.) Boiss., 211
Limonium axillare (Forssk.) O. Kuntze, 212
Limonium cylindrifolium (Forssk.) O Kuntze, 212
Limonium ferulaceum (L.) O. Kuntze, 212
Limonium gmelinii (Willd.) O. Kuntze, 212
Limonium pruinosum (L.) O. Kuntze, 212
Limonium psilocladum (Boiss.) O. Kuntze, 212
Limonium sp. indet. 1, 212
Limonium sp. indet. 2, 212
Limosella calycina Forssk., 227
LINACEAE, 194
Linaria aegyptiaca (L.) Dum.-Cours., 228
Linaria haelava (Forssk.) F. G. Dietr., 228
Lindernia crustacea (L.) Muell., 135
Linosyris hastata (Vahl) DC., 135
Linosyris montana (Vahl) DC., 118
Linum corymbulosum Reichenb., 194
Linum sp. indet., 194
Linum strictum L., 194
Lippia nodiflora (L.) A. Rich., 242
Lissochilus arabicus Lindl., 287
Lithospermum angustifolium Forssk., 89
Lithospermum callosum Vahl, 89
Lithospermum ciliatum Forssk., 89
Lithospermum digynum Forssk., 89
Lithospermum heliotropoides Forssk., 88
Lithospermum hispidum Forssk., 87
Lithospermum tetrastigmum (Forssk.) Lam., 85
Lithospermum tinctorium (Forssk.) Vahl, 85
Lolium multiflorum Lam.?, 272
Lolium perenne L., 272
Lolium perenne sensu Forssk., non L., 272
Lonchitis bipinnata Forssk., 290
LORANTHACEAE, 194
Loranthus acaciae Zucc., 194
Lotus arabicus L., 183
Lotus argenteus (Del.) Webb & Berth., 184
Lotus belgradica Forssk., 181
Lotus corniculatus L. var. **ciliatus** Koch, 184
Lotus corniculatus L. var. **tenuifolius** L., 184
Lotus corniculatus sensu Forssk., 184
Lotus cretica sensu Forssk., non L., 184
Lotus creticus L., 184
Lotus dorycnium Vahl, non L., 185
Lotus graecus L., 181
Lotus halophilus Boiss. & Sprun., 184
Lotus hirsutus sensu Forssk., non L., 184
Lotus peregrinus sensu Vahl, non L., 184
Lotus polyphyllus Clarke, 184
Lotus quinatus (Forssk.) Gillett, 185
Lotus rosea Forssk., 183
Lotus sp. indet., 185
Lotus tenuis Waldst. & Kit. ex Willd., 184
Lotus villosus Forssk., non Burm. f., 184
Ludwigia diffusa (Forssk.) Greene, non Buch-Ham., 208
Ludwigia stolonifera (Guill. & Perr.) Raven, 208
Luffa cordata (J.F. Gmel.) Meissn., 150
Luffa cylindrica (L.) M. J. Roem., 150
Lunaria scabra Forssk., 142 (Fig. 19A), 143
Lupinus albus L., 185
Lupinus albus L. var. *termis* (Forssk.) Alef., 185
Lupinus digitatus Forssk., 185

Lupinus forskahlei Boiss., 185
Lupinus pilosus Murr. subsp. *digitatus* (Forssk.) Maire, 185
Lupinus termis Forssk., 185
Lycium europaeum L., 230
Lycopodiella cernua (L.) Pic. Ser., 291
Lycopodium bryopteris Vahl, 293
Lycopodium cernuum L., 291
Lycopodium imbricatum Forssk., 293
Lycopodium sanguinolentum sensu Forssk., non L., 293
Lygos raetam (Forssk.) Heywood, 188
Lysimachia nummularia L., 216
LYTHRACEAE, 195
Lythrum hyssopifolium L., 195

Macowania ericifolia (Forssk.) Burtt & Grau, 127
Maerua [sp. without epithet] Forssk., 93
Maerua arabica J.F. Gmel., 93
Maerua crassifolia Forssk., 93
Maerua oblongifolia (Forssk.) A. Rich., 93
Maerua racemosa Vahl, 94
Maerua uniflora Vahl, 93
Maesa [sp. without epithet] Forssk., 93, 205
Maesa lanceolata Forssk., 205
MALPIGHIACEAE, 195
Malva montana Forssk., 198
Malva moschata L., 198
Malva nicaeensis sensu Vahl, non L., 198
Malva parviflora L., 198
Malva rotundifolia sensu Forssk., non L., 198
Malva sylvestris L., 198
Malva sylvestris L. var. **eriocarpa** Boiss., 198
Malva tournef[ourtiana] sensu Forssk., non L., 198
Malva verticillata L., 198
MALVACEAE, 195
Manettia lanceolata (Forssk.) Vahl, 223
Mangifera amba Forssk., 75
Mangifera indica L., 75
Marrubium alysson L., 166
Marrubium candid[issimum] sensu Forssk., non L., 166
Marrubium peregrinum L., 166
Marrubium plicatum Forssk., 166
Marrubium vulgare L., 167
Matricaria parth[enium] L., 134
Matthiola livida (Del.) DC., 144
Matthiola tricuspidata (L.) R. Br., 144
Maytenus parviflora (Vahl) Sebsebe, 100
Medicago littoralis Rohde ex Loisel., 185
Medicago marina L., 185
Medicago polymorpha L., 186
Megalochlamys violacea (Vahl) Vollesen, 68
Melhania abyssinica sensu Defl., non A. Rich., 234
Melhania ferruginea A. Rich., 234
Melhania velutina Forssk., 234
Melia azedarach L., 200
MELIACEAE, 200
Melilotus alba Medik., 185
Melilotus indica L., 186
Melissa perennis Forssk., 170
MENISPERMACEAE, 200
Menispermum edule Vahl, 201
Menispermum leaeba (Del.) DC., 201
Mentha aquatica L., 167
Mentha aquatica L. var. *capitata* Briq. forma, 167
Mentha gentilis L. var. *tenuiceps* Briq., 168
Mentha gentilis sensu Forssk., 168
Mentha glabrata Vahl, 167
Mentha kahirina Forssk., 167
Mentha longifolia (L.) L., 167
Mentha longifolia (L.) L. var. **schimperi** (Briq.) Briq., 167
Mentha pulegium L., 167
Mentha sp. indet., 168
Mentha spicata sensu Forssk., non L., 167
Mentha × piperita L., 167
Mentha × verticillata L., 167
MENYANTHACEAE, 201
Merendera abyssinica A. Rich., 286
Mesembryanthemum forskalei Hochst., 71
Mesembryanthemum forsskalii Hochst. ex Boiss., 71
Mesembryanthemum geniculiflorum sensu Forssk., 71
Mesembryanthemum nodiflorum L., 71
Mesembryanthemum sp. indet., 72
Mesua glabra Forssk., 252
Mibora minima (L.) Desv., 272
Micrelium [sp. without epithet] Forssk., 120
Micrelium asteroides Forssk., 120
Micrelium tolak Forssk., 120
Micromeria forskahlei Benth., 168
Micromeria graeca (L.) Benth. ex Reichenb., 168
Micromeria imbricata (Forssk.) C. Chr., 168
Mimosa arborea sensu Forssk., non L., 176
Mimosa asak Forssk., 174
Mimosa flava Forssk., 174
Mimosa glomerata Forssk., 181
Mimosa gummifera Forssk., 175
Mimosa lebbeck L., 176
Mimosa mellifera Vahl, 175
Mimosa nilotica L., 175
Mimosa oerfota Forssk., 176
Mimosa scorpioides sensu Forssk., non L., 175
Mimosa sejal Forssk., 176
Mimosa senegal L., 175
Mimosa senegalensis Forssk., 175
Mimosa stellata Forssk., non Lour., 187
Mimosa tortilis Forssk., 176
Mimosa unguis cati sensu Forssk., non L., 175
MIMOSACEAE see Leguminosae, 174
Mimusops kauki sensu Vahl, non L., 226
Mimusops laurifolia (Forssk.) Friis, 226
Mimosops schimperi Hochst., 226
Minuartia filifolia (Forssk.) Mattf., 96

Minuartia geniculata (Poir.) Thell., 96
Minuartia procumbens (Vahl) Zohary, 96
Minuartia sp. indet., 96
Mnemosilla aegyptiaca Forssk., 208
MOLLUGINACEAE, 201
Mollugo cerviana (L.) Ser. var. **spathulifolia** Fenzl, 201
Mollugo glinus Rich., 71
Mollugo setiflora (Forssk.) Chiov., 71
Moltkiopsis ciliata (Forssk.) I. M. Johnston, 89
Moluccella fruticosa Forssk., 165
Monechma debile (Forssk.) Nees, 68
Monechma violaceum (Vahl) Nees, 68
MONOCOTYLEDONS, 253
MORACEAE, 202
Moringa arabica (Lam.) Pers., 205
Moringa oleifera Lam., 205
Moringa peregrina (Forssk.) Fiori, 205
MORINGACEAE, 205
Morus sylvestris Forssk., 205
Moscharia [sp. without epithet] Forssk., 165
Moscharia asperifolia Forssk., 165
Myosotis sp. sensu Vahl & C. Chr., 89
Myosotis spinocarpos (Forssk.) Vahl, 89
Myrica montana Vahl, 206
MYRSINACEAE, 205
Myrsine africana L., 206

Nauplius graveolens (Forssk.) A. Wicklund, 127
Neonotonia wightii (Wight & Arn.) Lackey var. **longicauda** (Schweinf.) Lackey, 186
Nepeta nepetellae Forssk., 168
Nephrodium crenatum (Forssk.) Bak., 291
Nephrolepis biserrata (Sw.) Schott, 291
Nerium foliis integris Forssk., nom. inval., 76
Nerium foliis ternatis Forssk., 221
Nerium obesum Forssk., 76
Nerium salicinum Vahl, 221
Neurada procumbens L., 206
NEURADACEAE, 206
Nicotiana tabacum L., 230
Nitraria retusa (Forssk.) Asch., 248
Noaea mucronata (Forssk.) Asch. & Schweinf., 105
Notobasis syriaca (L.) Cass., 127
Notonia pendula (Forssk.) Chiov., 124
Notonia semperviva (Forssk.) Asch., 124
NYCTAGINACEAE, 206
Nymphaea lotus L., 207
NYMPHAEACEAE, 207
Nymphoides cristata (Griseb.) O. Kuntze, 201

Ochna inermis (Forssk.) Schweinf., 207
Ochna parvifolia Vahl, 207
OCHNACEAE, 207
Ochthocloa compressa (Forssk.) Hilu, 272
Ocimum aegyptiacum Forssk., 169
Ocimum forskolei Benth., 168
Ocimum gratissimum Forssk., non L., 168
Ocimum hadiense Forssk., 170
Ocimum hadiense sensu E.A. Bruce, non Forssk., 168
Ocimum tenuiflorum L., 168
Ocimum vaalae Forssk., 169
Ocimum villosum Forssk., 170
Ocimum α zatarhendi Forssk., 169
Ocimum β zatarhendi sensu auct., non Forssk., 170
Ocymum filamentosum Forssk., 165
Ocymum serpyllifolium Forssk., 166
Odontanthera radians (Forssk.) D.V. Field, 82
Odontospermum graveolens (Forssk.) Sch. Bip., 127
Odyssea mucronata (Forssk.) Stapf, 272
Oldenlandia capensis L.f., 223
Olea europaea L., 207
OLEACEAE, 207
ONAGRACEAE, 208
Oncoba [sp. without epithet] Forssk., 161
Oncoba spinosa Forssk., 161
Ononis alternifolia Forssk., 187
Ononis antiquorum L., 187
Ononis cherleri sensu Forssk., non L., 187
Ononis flava Forssk., 188
Ononis minutissima L., 186
Ononis natrix L., 186
Ononis quinata Forssk., 185
Ononis serrata Forssk., 186
Ononis spinosa L., 187
Ononis spinosa L. subsp. *antiquorum* (L.) Briq., 187
Ononis spinosa L. subsp. *leiosperma* (Boiss.) Sirj., 187
Ononis vaginalis Vahl, 187
Onopordum tauricum Willd., 127
Ophiorrhiza lanceolata Forssk., 223
Opophytum forskahlii (Hochst. ex Boiss.) N. E. Br., 71
ORCHIDACEAE, 287
Orchis aphylla Forssk., 287
Orchis flava Forssk., 287
Orchis viridis Forssk., 287
Origanum heracleot[*icum*] L., 169
Origanum majorana L., 169
Origanum vulgare L. subsp. **hirtum** (Link) Ietswaart, 169
Orygia decumbens Forssk., 69
Orlaya maritima (Gou.) Koch, 239
Ormocarpum yemenense Gillett, 187
Ornithogalum flavum Forssk., 285
Ornithogalum narbonense L., 286
Ornithopus compressus L., 187
Ornithopus sativus Brot., 187
Orobanche aegyptiaca Pers., 229
Orobanche crenata Forssk., 229
Orobanche ramosa L., 229
Orobanche ramosa sensu Vahl, non L., 229
Orobanche tinctoria Forssk., 228
Orobus volubilis Forssk., 179

Orygia decumbens Forssk., 69, 215
Orygia portulacifolia Forssk., 215
Orygia villosa Forssk., 201
Osteospermum vaillantii (Decne.) T. Norl., 127
Osyris alba L., 226
Otanthus maritimus (L.) Hoffm. & Link, 127
Otostegia fruticosa (Forssk.) Schweinf. ex Penzig, 169
Ottelia alismoides (L.) Pers., 282
OXALIDACEAE, 208
Oxalis corniculata L., 208

Pallenis spinosa (L.) Cass., 115
PALMAE, 287
Pancratium illyricum sensu Forssk., 254
Pancratium maritimum L., 254
Pancratium maximum Forssk., 254
PANDANACEAE, 287
Pandanus odoratissimus L.f., 287
Pandanus odoriferus (Forssk.) Chiov., 287
Pandanus tinctorius Sol., 287
Panicum adhaerens Forssk., 277
Panicum appressum Forssk., 261
Panicum coloratum L., 272
Panicum compressum Forssk., 272
Panicum crus-galli L., 267
Panicum crus-galli L. subsp. *microstachys*, 268
Panicum dactylon L., 265
Panicum dichotomum Forssk., 274
Panicum divisum Forssk. ex J.F. Gmel., 274
Panicum fluitans Retz., 274
Panicum forskalii C. Chr., 267
Panicum geminatum Forssk., 273
Panicum glaucum α Forssk., 264
Panicum glaucum β Forssk., 263
Panicum grossarium sensu Forssk., non L., 272, 273
Panicum miliaceum L., 273
Panicum miliaceum sensu Forssk., non L., 272
Panicum muricatum Forssk., 273
Panicum muticum Forssk., 261
Panicum polygamum Forssk., 273
Panicum repens L., 273
Panicum sangu[inale] L., 267
Panicum sanguinale sensu Vahl, non L., 267
Panicum setigerum Forssk., 264
Panicum tetrastichon Forssk., 273
Panicum turgidum Forssk., 273
PAPAVERACEAE, 208
Papularia crystallina Forssk., 72
Parapholis strigosa (Dumort.) C. E. Hubb., 273
Parietaria alsinifolia Del., 241
Parietaria judaica sensu Forssk., non L., 241
Parnassia polynectaria Forssk., 162, 231 (Fig. 24B)
Paronychia arabica (L.) DC., 97
Paronychia arabica (L.) DC. var. *longiseta* (Bertol.) Asch. & Schweinf., 97
Paronychia desertorum Boiss., 97
Paronychia lenticulata (Forssk.) Asch. & Schweinf., 97
Paspalidium desertorum sensu Schwartz, non (A. Rich.) Stapf, 273
Paspalidium geminatum (Forssk.) Stapf, 273
Passerina metnan Forssk., 235
PASSIFLORACEAE, 209
Pavetta longiflora Vahl, 223
Pavetta villosa Vahl, 223
Pavonia flava Spring ex Mart., 199
Pavonia flavo-ferruginea (Forssk.) Hepper & J.R.I. Wood, 199
Pavonia hildebrandtii Guerke & Ulbr., 199
PEDALIACEAE, 209
Peganum retusum Forssk., 248
Pellaea quadripinnata (Forssk.) Prantl, 292
Pellaea viridis (Forssk.) Prantl, 263
Pennisetum ciliare (L.) Link, 274
Pennisetum dichotomum (Forssk.) Del., 274
Pennisetum divisum (J.F. Gmel.) Henr., 274
Pennisetum glaucum (L.) R. Br., 274
Pennisetum setaceum (Forssk.) Chiov., 274
Pennisetum typhoides (Burm.) Stapf & C.E. Hubb., 274
Pentaglossum linifolium Forssk., 195
Pentapetes velutina (Forssk.) Vahl, 234
Pentas lanceolata (Forssk.) Defl., 223
Pentatropis nivalis (J.F. Gmel.) D.V. Field & J.R.I. Wood, 82
Pentatropis spiralis (Forssk.) Decne., 78, 83
Pepo [sp. without epithet] Forssk., 149
Pepo longa Forssk., 149
Pergularia daemia (Forssk.) Chiov., 83
Pergularia glabra (Forssk.) Schwartz, 78
Pergularia tomentosa L., 83
Peristrophe bicalyculata (Retz.) Nees, 68
Peristrophe paniculata (Forssk.) Brummitt, 68
Petrorhagia prolifera (L.) Ball & Heywood, 97
Phagnalon harazianum Defl., 127
Phagnalon rupestre (L.) DC., 128
Phagnalon sordidum (L.) Reichenb., 128
Phalaris crinita Forssk., 275
Phalaris cristata [?*aristata*] Forssk., 275
Phalaris disticha Forssk., 278
Phalaris muricata Forssk., 280
Phalaris semiverticillata Forssk., 275
Phalaris setacea Forssk., 274
Phalaris vaginiflora Forssk., 264
Phalaris velutina Forssk., 267
Pharnaceum occultum Forssk., 71
Pharnaceum umbellatum Forssk., 201
Phaseolus aconitifolius Jacq., 193
Phaseolus palmatus Forssk., 193
Phaseolus radiatus L., 193
Phaulopsis imbricata (Forssk.) Sweet, 68
Phelypaea tinctoria (Forssk.) Walp., 228
Phleum paniculatum Hudson, 274
Phlomis alba Forssk., 166
Phlomis glabrata Vahl, 166
Phlomis herba-venti L., 170

Phlomis molluccoides Vahl, 169
Phlomis urticifolia Vahl, 166
Phoenix dactylifera L., 287
Phragmites australis (Cav.) Steud., 274
Phragmites australis (Cav.) Steud. subsp. **altissimus** (Benth.) W.D. Clayton, 275
Phragmites communis (L.) Trin., 275
Phragmites mauritianus Kunth, 275
Phryma [sp. without epithet] Forssk., 242
Phyla nodiflora (L.) Greene, 242
Phyllanthus maderaspatens[is] sensu Forssk., non L., 160
Phyllanthus hamrur Forssk., 159
Phyllanthus niruri L., 160
Phyllanthus ovalifolius Forssk., 160
Phyllanthus tenellus Muell. Arg. var. **arabicus** Muell. Arg., 160
Phyllitis scolopendrium (L.) Newman, 292
Phyllophora rubens (L.) Grev. (Algae), 315
Physalis alkekengi L., 230
Physalis angulata L., 230
Physalis curassiv[ica] sensu Forssk., non L., 234
Physalis ramosa Forssk., 230
Physalis somnifera L., 234
Phytolacca americana L., 209
PHYTOLACCACEAE, 209
Picnomon acarna (L.) Cass., 128
Picris abyssinica Sch. Bip., 129
Picris altissima Del., 128
Picris asplenioides L., 128
Picris laevis Forssk., 128
Picris radicata (Forssk.) Less., 129
Picris scabra Forssk., 129
Picris sprengeriana (L.) Chaix, 128
Picris sulphurea Del., 129
Piloselloides hirsuta (Forssk.) C. Jeffrey ex Cufod., 121
Pimpinella anisum L., 239
Pimpinella menachensis Schweinf. ex Wolff, 238
PINACEAE, 295
Pinus brutia Ten., 295
Pinus pinea L., 295
Pistacia terebinthus L., 75
PLANTAGINACEAE, 209
Plantago coronopus L., 209
Plantago crassifolia Forssk., 210
Plantago cylindrica Forssk., 210
Plantago decumbens Forssk., 210
Plantago lagopus L., 210
Plantago lanceolata L., 210
Plantago major L., 210
Plantago ovata Forssk., 210
Plantago ovata Forssk. var. **decumbens** (Forssk.) Zohary, 210
Plantago ovata Forssk. var. **ovata**, 210
Plantago psyllium L., 211
Plantago scabra Moench, 211
Plantago serraria L., 211
PLATANACEAE, 211
Platanus orientalis L., 211
Plectranthus amboinicus (Lour.) Spreng., 169
Plectranthus arabicus E. A. Bruce, 170
Plectranthus forskalaei Vahl, 170
Plectranthus hadiensis (Forssk.) Schweinf. ex Spreng., 170
Plectranthus ovatus Benth., 170
Plectranthus quadridentatus sensu O. Schwartz, non Schweinf. ex Baker, 170
Plectranthus tenuiflorus (Vatke) Agnew, 169
Plectranthus villosus Sieb. ex Benth., 170
Plectranthus zatarhendi sensu E.A. Bruce, 170
Plectranthus zatarhendi (Forssk.) E.A. Bruce, 194
Plicosepalus acaciae (Zucc.) Weins & Polhill, 194
Pluchea dioscorides (L.) DC., 129
PLUMBAGINACEAE, 211
Poa amabilis Forssk., non L., 268
Poa multiflora Forssk., 269
Poa schimperiana A. Rich., 275
Poa spiculis 3-floris Forssk., 275
Poinciana elata L., 180
Polycarpaea repens (Forssk.) Asch. & Schweinf., 97
Polycarpia prostrata (Forssk.) C. Chr., 97
Polycarpon prostratum (Forssk.) Asch. & Schweinf., 97
Polycephalos [sp. without epithet] Forssk., 134
Polycephalos suaveolens Forssk., 134
Polygala abyssinica R. Br. ex Fres., 213
Polygala bracteolata Forssk., non L., 213
Polygala paniculata sensu Forssk., non L., 213
Polygala tinctoria Vahl, 213
POLYGALACEAE, 213
POLYGONACEAE, 213
Polygonum aviculare sensu Forssk., non L., 213
Polygonum equisetiforme Sibth. & Sm., 213
Polygonum marit[imum] sensu Forssk., non L., 213
Polygonum maritimum L., 213
Polygonum persicaria sensu Forssk., non L., 214
Polygonum cf. **pulchellum** Lois., 213
Polygonum salicifolium Bruss. ex Willd., 214
Polypodium aculeatum Vahl, non (L.) Roth, 292
Polypodium crenatum Forssk., 291
Polypodium dentatum Forssk., 291
Polypodium orientale J.F. Gmel., 289
Polypodium pectinatum sensu Forssk., non L., 290
Polypodium setiferum Forssk., 292
Polypodium speluncae sensu Vahl, non L., 291
Polypogon monspeliensis (L.) Desf., 275
Polypogon semiverticillatus (Forssk.) Hyl., 276
Polypogon viridis (Gouan) Breistr., 275
Polysiphonia elongata (Huds.) Harv. (Algae), 316
Polystichum setiferum (Forssk.) Woynar, 292
Portulaca cuneifolia Vahl, 215
Portulaca hareschta Forssk., 215
Portulaca imbricata Forssk., 215

Portulaca linifolia Forssk., 215
Portulaca quadrifida L., 215
PORTULACACEAE, 215
Potentilla dentata Forssk., 219
Potentilla hispanica Zimm., 219
Potentilla supina L., 219
Potentilla supina L. var. *aegyptiaca* Vis., 219
Pouzolzia parasitica (Forssk.) Schweinf., 241
Prenanthes spinosa Forssk., 125
Primula verticillata Forssk., 216
PRIMULACEAE, 215
Priva adhaerens (Forssk.) Chiov., 242
Priva dentata Juss., 242
Priva forskalii Jaub. & Spach, 242
Priva forskaolii (Vahl) E. Mey., 242
Prunus spinosa L., 219
Pseudognaphalium luteo-album (L.) Hilliard & Burtt, 129
Pseudorlaya pumila (L.) Grande, 239
Psiadia arabica Jaub. & Spach, 129
Psiadia punctulata (DC.) Vatke, 129
Psoralea corylifolia L., 180
Pteranthus [sp. without epithet] Forssk., 98
Pteranthus dichotomus Forssk., 98
Pteranthus echinatus Desf., 98
Pteranthus forskaolii Mirb., 98
Pteridium aquilinum (L.) Kuhn, 292
Pteris cretica L., 292
Pteris decursiva Forssk., 290
Pteris dentata Forssk., 292
Pteris dentata Forssk. subsp. *flabellata* (Thunb.) Runemark, 292
Pteris farinosa Forssk., 290
Pteris longifolia L., 293
Pteris obliqua Forssk., 293
Pteris quadripinnata Forssk., 291
Pteris regularis Forssk., 292
Pteris semiserrata Forssk., 292
Pteris serrulata Forssk., 292
Pteris subciliata Forssk., 293
Pteris viridis Forssk., 292
Pteris vittata L., 292
Pterocaulon redolens (Forst.) Boerl., 129
Pterolobium stellatum (Forssk.) Brenan, 187
Pulicaria arabica (L.) Cass., 130
Pulicaria crispa (Forssk.) Benth. ex Oliv., 130
Pulicaria grandidentata Jaub. & Spach, 130
Pulicaria incisa (Lam.) DC., 130
Pulicaria undulata sensu auct., non (L.) DC., 219
Pyrus amygdalifolius Vill., 219
Pyrus hadiensis Forssk., 181
Pyrus spinosa sensu Forssk., non L., 219

Quamoclit coccinea (L.) Moench, 139
Quamoclit pennata Bojer, 138
Quercus cerris L., 160
Quercus coccifera L., 161
Quercus ilex a) sensu Forssk. *Quercus ilex* b) sensu Forssk. *Quercus ilex* c) sensu Forssk., 161
Quercus ilex L., 161
Quercus pubescens Willd.
Quercus robur, an cervis sensu Forssk., 161

RANUNCULACEAE, 216
Ranunculus chaerophyllos sensu Forssk., non L., 217
Ranunculus forskoehlii DC., 217
Ranunculus multifidus Forssk., 217
Ranunculus sceleratus L., 217
Raphanus lyratus Forssk., 141
Reamuria hirtella Jaub. & Spach, 235
Reamuria vermiculata sensu Forssk., non L., 235
Reichardia tingitana (L.) Roth, 130
Reseda alba L., 217
Reseda arabica Boiss., 218
Reseda canescens L. (1767), non L. (1753), 217
Reseda decursiva Forssk., 218
Reseda hexagyna Forssk., 217
Reseda lutea L., 218
Reseda luteola L., 218
Reseda phyteuma sensu Forssk., non L., 218
Reseda tetragyna Forssk., 218
Reseda undata L., 218
RESEDACEAE, 217
Retama retam (Forssk.) Webb, 188
RHAMNACEAE, 218
Rhamnus divaricatus Forssk., 218
Rhamnus nabeca sensu Forssk., non L., 218
Rhamnus rectus Forssk., 219
Rhodalsine procumbens (Vahl) J. Gay, 96
Rhus cf. **abyssinica** Hochst. ex Oliv., 75
Rhus natalensis Krauss, 76
Rhus saeneb Forssk., 240
Rhynchelytrum repens (Willd.) C. E. Hubb., 276
Rhynchosia cf. *mensensis* Schweinf., 194
Rhynchosia flava (Forssk.) Thulin, 188
Rhynchosia malacophylla (Sprengel) Bojer, 188
Rhynchosia minima (L.) DC., 188
Ricinus communis L., 160
Ricinus medicus Forssk., 160
Rocama [sp. without epithet] Forssk., 72
Rocama prostrata Forssk., 72
Roemeria dodecandra (Forssk.) Stapf, 209
Roemeria hybrida (L.) DC., 208
Roemeria hybrida (L.) DC. subsp. **dodecandra** (Forssk.) Durand & Baratte, 209
Rokejeka [sp. without epithet] Forssk., 96
Rokejeka capillaris Forssk., 96
Rokejeka deserti J.F. Gmel., 96
Roridula [sp. without epithet] Forssk., 93
Roridula droserifolia Forssk., 93
Rosa andegavensis Bast., 219
Rosa canina sensu Forssk., non L., 219
Rosa ephemera Forssk., 220
Rosa moschata J. Herrm., 220
ROSACEAE, 219
Rostraria cristata (L.) Tzvelev, 276
Rottboellia hirsuta Vahl, 271, 280

Rubia peregrina L., 223
Rubia tinctor[ia] sensu Forssk., non L., 223
RUBIACEAE, 220
Rubus arabicus (Defl.) Schweinf., 220
Rubus fruticosus sensu Forssk., non L., 220
Rubus sanctus Schreber, 220
Ruellia adhaerens Forssk., 242
Ruellia alba Forssk., 68
Ruellia aristata Vahl, 67
Ruellia grandiflora (Forssk.) Blatter, 68
Ruellia guttata Forssk., 63
Ruellia hispida Forssk., 69
Ruellia imbricata Forssk., 68
Ruellia intrusa Forssk., 63
Ruellia longiflora Vahl, 68
Ruellia pallida Vahl, 69
Ruellia patula Jacq., 69
Ruellia strepens sensu Forssk., non L., 69
Rumex aegypt[iacus] vel *comosus* sensu Forssk., 214
Rumex aegyptiacus L., 214
Rumex dentatus L., 214
Rumex glaber Forssk., 213
Rumex lachanus Forssk., 214
Rumex nervosus Vahl, 214
Rumex obtusifolius sensu Forssk., non L., 214
Rumex persicarioides sensu Forssk., non L., 214
Rumex pictus Forssk., 214
Rumex spinosus L., 213
Rumex vesicarius L., 215
Ruppia maritima L., 287
RUPPIACEAE, 287
Ruta chalepensis L., 225
Ruta graveolens sensu Forssk., non L., 225
Ruta linifolia sensu Forssk., non L., 224
Ruta tuberculata Forssk., 224
RUTACEAE, 224

Saccharum biflorum Forssk., 276
Saccharum hirsutum Forssk., 271
Saccharum officinarum L., 276
Saccharum spontaneum L. subsp. **aegyptiacum** (Willd.) Hack., 276
Saelanthus digitatus Forssk., 246 (Fig. 27), 247
Saelanthus glandulosus Forssk., 243
Saelanthus quadrangonus Forssk., 243, 244 (Fig. 25)
Saelanthus rotundifolius Forssk., 243, 245 (Fig. 26)
Saelanthus ternatus Forssk., 247
SALICACEAE, 225
Salicornia cruciata Forssk., 105
Salicornia europaea L., 105
Salicornia europaea sensu Forssk., p.p., 105
Salicornia fruticosa (L.) L., 105
Salicornia herbacea L., 105
Salicornia perfoliata Forssk., 105, 106 (Fig. 16)
Salicornia virginica sensu Forssk., non L., 101
Salix acmophylla Boiss., 225
Salix aegyptiaca L., 225
Salix cf. **tetrasperma** Roxb., 225
Salix glabra Forssk., 225
Salix safsaf baelledi Forssk., 225
Salix subserrata Willd., 225
Salsola articulata Forssk., 101
Salsola forskalii Schweinf., 107
Salsola imbricata Forssk., 107
Salsola inermis Forssk., 107
Salsola kali L., 107
Salsola kali L. var. *hispida* Forssk., 107
Salsola kali L. var. *hispida polygama* Forssk., 108
Salsola kali L. var. *glabra* Forssk., 107
Salsola longifolia Forssk., 108
Salsola monobracteata Forssk., 103
Salsola mucronata Forssk., 105
Salsola muricata L., 103
Salsola soda L., 108
Salsola spinescens Moq., 107
Salsola tetrandra Forssk., 108
Salsola vermiculata sensu Forssk., non L., 108
Salsola volkensii Asch. & Schweinf., 107
Saltia papposa (Forssk.) Moq., 75
Salsola volkensii Asch. & Schweinf., 108
Salvadora persica L., 226
SALVADORACEAE, 226
Salvia aegyptiaca L., 170
Salvia aegyptiaca sensu L. (1767), non L. (1753), 171
Salvia bifida Forssk., 171
Salvia ceratophyt. Forssk., 171
Salvia forskahlei L., 171
Salvia lanigera Poir., 171
Salvia merjamie Forssk., 171
Salvia merjamie Forssk., partly, 171
Salvia nudicaulis Vahl, 171
Salvia nudicaulis Vahl var. *nubia* (Ait.) Baker, 171
Salvia phlomoides Asso, 171
Salvia spinosa L., 171
Salvia virgata Jacq., 171
Sanicula elata D. Don, 239
Sanicula europaea sensu Schwartz, non L., 239
Sansevieria forskaliana (Schult.f.) Hepper & J.R.I. Wood, 286
Sansevieria guineensis sensu Schwartz, 286
SANTALACEAE, 226
Santolina ambigua Forssk., 130
Santolina flava Forssk., 114
Santolina fragrantissima Forssk., 112
Santolina terrestris Forssk., 113
Saponaria ocymoid. sensu Forssk., 100
Saponaria officinalis L., 98
SAPOTACEAE, 226
Sarcostemma forskolianum Schult., 83
Sarcostemma sp. indet., 83
Sarcostemma stipitaceum (Forssk.) R. Br., 83
Sarcostemma viminale (L.) R. Br., 83
Sargassum crispum (Forssk.) Ag. (Algae), 316
Sargassum denticulatum (Forssk.) Børgesen (Algae), 316

Sargassum subrepandum (Forssk.) Ag. (Algae), 316
Satureja abyssinica (A. Rich.) Briq., 172
Satureja graeca L., 168
Satureja imbricata (Forssk.) Briq., 168
Sauromatum venosum (Ait.) Kunth, 254
Scabiosa arenaria Forssk., 151
Scabiosa argentea L., 151
Scabiosa atropurpurea L., 151
Scabiosa columbaria L., 151
Scabiosa lyrata Forssk., 151
Scabiosa sicula L., 151
Scabiosa sp. indet., 151
Scadoxus multiflorus (Martyn) Raf., 254
Scandix infesta L., 239
Scariola viminea (L.) F.W. Schmidt, 131
Sceura marina Forssk., 84
Schanginia aegyptiaca (Hasselq.) Aellen, 109
Schanginia hortensis (Forssk. ex J.F. Gmel.) Moq., 109
Schismus arabicus Nees, 276
Schismus barbatus (L.) Thell., 276
Schismus barbatus sensu Tackholm & Drar, non (L.) Thell., 276
Schoenoplectus articulatus (L.) Palla, 259
Schoenus incanus Forssk., 259
Schoenus mucronatus L., 256
Schouwia purpurea (Forssk.) Schweinf., 144
Scilla hyacinthina (Roth) J.F. Macbride, 286
Scilla tetraphylla L.f., 285
Scirpoides holoschoenus (L.) Soják, 259
Scirpus articulatus L., 259
Scirpus bis-umbellatus Forssk., 259
Scirpus corymbosus Forssk., 256
Scirpus fistulosus Forssk., 259
Scirpus holoschoenus L., 259
Scirpus kalli [3 Alpini] Forssk., 256
Scirpus lateralis Forssk., 260
Scirpus maritimus L., 256
Scirpus supinus L. var. uninodis (Del.) Asch. & Schweinf., 260
Scirpus tuberosus Desf., 256
Scirpus uninodis (Del.) Boiss., 260
Scleropoa dichotoma (Forssk.) Parl., 265
Scleropoa maritima (L.) Parl., 265
Scolymus hispanicus L., 131
Scolymus maculatus L., 131
Scoparia dubia? Forssk., 229
Scoparia dulcis L., 229
Scoparia ternata Forssk., 229
Scorpiurus muricatus L., 188
Scorpiurus subvillosa L., 188
Scorzonera ciliata Forssk., 126
Scorzonera hispida Forssk., 131
Scorzonera resedifolia sensu Vahl, non L., 125
Scorzonera tingitana L.?, 131
SCROPHULARIACEAE, 227
Scutellaria arabica Jaub. & Spach, 172
Scutellaria galeric[ulata] sensu Forssk., non L., 172
Scutellaria peregrina sensu Schwartz, non L., 172
Secale cretica sensu Forssk., non L., 266
Seddera arabica (Forssk.) Choisy, 139
Sehima ischaemoides Forssk., 277
Selaginella imbricata (Forssk.) Spring ex Decne., 293
Selaginella yemensis (Sw.) Spring ex Decne., 293
Senecio aegyptius L. var. **discoideus** Boiss., 131
Senecio anteuphorbium (L.) Sch. Bip. var. *odorus* (Forssk.) Rowley, 124
Senecio aquaticus L. var. **erraticus** (Bertol.) Matthews, 131
Senecio arabicus L., 131
Senecio auriculatus Vahl, 132
Senecio biflorus Vahl, 132
Senecio coronopifolius Desf., 132
Senecio desfontanei Bruce, 132
Senecio glaucus L., 132
Senecio hadiensis Forssk., 132, 133 (Fig. 18)
Senecio hieracifolius sensu Forssk., non L., 131
Senecio jacobaea sensu Forssk., non L., 131
Senecio linifolius Forssk., 132
Senecio linifolius radio nullo Forssk., 132
Senecio lyratus Forssk., 132
Senecio lyratus L.f., 132
Senecio odorus (Forssk.) Sch. Bip., 124
Senecio pendulus (Forssk.) Sch. Bip., 124
Senecio sempervivus (Forssk.) Sch. Bip., 124
Senecio sp. indet. 1, 132
Senecio sp. indet. 2, 132
Senecio squalidus sensu Forssk., non L., 132
Senecio succulentus Forssk., 123
Senna alexandrina Miller, 188
Senna italica Miller, 189
Senna obtusifolia (L.) Irwin & Barneby, 189
Senna occidentalis (L.) Link, 189
Senna sophera (L.) Roxb., 189
Senna tora (L.) Roxb., 190
Serratula ciliata Vahl, 132
Serratula spinosa Forssk., 117
Sesamum indicum L., 209
Sesamum orientale L., 209
Sesbania aegyptiaca Pers., 190
Sesbania grandiflora (L.) Poir., 190
Sesbania sesban (L.) Merr., 190
Seseli tortuosum L., 239
Setaria adhaerens (Forssk.) Chiov., 277
Setaria verticillata (L.) P. Beauv., 277
Sicyos angulatus L., 150
Sida alba L., 199
Sida ciliata Forssk., 199
Sida cordifolia Forssk., non L., 196
Sida ovata Forssk., 199
Sida paniculata sensu Forssk., non L., 200
Sida spinosa sensu Vahl, non L., 199
Sida urens L., 199
Sideritis montana L., 172

Sideritis romana L., 172
Sideroxylon inerme sensu Forssk., non L., 87
Sigesbeckia orientalis L., 134
Silene biappendiculata Rohrb., 98
Silene canopica Del., 98
Silene colorata Poir., 98
Silene conica L.98
Silene gallica L., 98
Silene involuta Forssk., 98
Silene cf. **italica** (L.) Pers., 98
Silene nocturna L., 99
Silene cf. **paradoxa** L., 99
Silene succulenta Forssk., 99
Silene villosa Forssk., 99
Silene vulgaris (Moench) Garcke, 99
Silene sp. indet. 1, 99
Silene sp. indet. 2, 100
Siliquaria glandulosa Forssk., 92, 142 (Fig. 19B)
Simbuleta [sp. without epithet] Forssk., 227
Simbuleta arabica Poir., 227
Simbuleta forskahlii J.F. Gmel., 227
Sinapis allionii Jacq., 144
Sinapis arvensis sensu Forssk., non L., 144
Sinapis harra Forssk., 141
Sinapis turgida Del., 147
Sisymbrium hispidum Vahl, 141
Sisymbrium irio L., 147
Sisymbrium pinnatifidum Forssk., 147
Smilacina forskaliana Schult.f., 286
Sodada decidua Forssk., 92
SOLANACEAE, 229
Solanecio angulatus (Vahl) C. Jeffrey, 134
Solanum aegyptiacum a) Forssk., 233
Solanum aegyptiacum b) Forssk., 233, 234
Solanum albicaule Kotschy ex Dunal, 232
Solanum arabicum Dunal, 230
Solanum armatum Forssk., 230
Solanum bahamense sensu Forssk., non L., 232
Solanum carense Dunal, 230
Solanum coagulans Forssk., 232
Solanum cordatum Forssk., 232
Solanum diphyllum sensu Forssk., non L., 233
Solanum dubium Fres., 232
Solanum dulcamara L., 232
Solanum forskalii Dunal, 232
Solanum glabratum Dunal, 232
Solanum gracilipes Decne., 232
Solanum hadaq Defl., 232
Solanum incanum L., 232
Solanum microcarpum Vahl, 233
Solanum nigrum L., 233
Solanum nigrum L. var. *villosum* Miller, 233
Solanum palmetorum Dunal, 233
Solanum platacanthum Dunal, 233
Solanum sepicula Dunal, 232
Solanum sp. indet. 1, 234
Solanum sp. indet. 2, 234
Solanum terminale Forssk., 233
Solanum villosum Miller, 233
Solanum villosum sensu Forssk., non Miller, 232
Solanum virginianum L., 230
Solanum xanthocarpum Schrad. & Wendl., 230
Solenostemon latifolius (Hochst. ex Benth.) Morton, 172
Solieria dura (Zanard.) Schmitz (Algae), 316
Sonchus oleraceus L., 134
Sonchus tenerrimus L., 134
Sophora alopecuroides L., 190
Sorbus domestica L., 220
Sorghum bicolor (L.) Moench, 277
Sorghum dochna (Forssk.) J.D. Snowden var. *obovatum* (Hack.) J.D. Snowden, 277
Sorghum durra (Forssk.) Stapf var. *eois* (Burkill) J.D. Snowden, 277
Sorghum halepense (L.) Pers., 277
Sorghum membranaceum Chiov. var. *ehrenbergianum* (Koern.) J.D. Snowden, 277
Spergularia marina (L.) Griseb., 100
Sphaeranthus suaveolens (Forssk.) DC., 134
Sphenopus divaricatus (Gouan) Reichenb., 278
Sphenopus sp. aff. **divaricatus** (Gouan) Reichenb., 278
Spongia flabelliformis Forssk. (animal), 315
Spongia tubulosa Forssk. (animal), 315
Sporobolus arenarius (Gouan) Duval-Jouve, 278
Sporobolus pyramidalis P. Beauv., 278
Sporobolus spicatus (Vahl) Kunth, 278
Sporobolus virginicus sensu Tackholm, non (L.) Kunth, 278
Stachys aegyptiaca Pers., 172
Stachys annua (L.) L., 172
Stachys cf. **cretica** L., 172
Stachys cf. **maritima** Gouan, 172
Stachys cretica sensu Forssk., non L., 173
Stachys obliqua Waldst. & Kit., 173
Stachys oriental[is] sensu Forssk., non L., 172
Stachys orientalis sensu Vahl, non L., 173
Stachys recta L. subsp. **subcrenata** (Vis.) Briq., 173
Stachys sylvestris Forssk., 173
Staehelina hastata Vahl, 135
Staehelina spinosa Vahl, 124
Stapelia [sp. without epithet] Forssk., 79
Stapelia dentata Forssk., 79
Stapelia multangula Forssk., 79
Stapelia quadrangula Forssk., 79, 80 (Fig. 12)
Stapelia subulata Forssk., 79, 81 (Fig. 13)
Stapelia variegata Forssk., 79
Statice aphylla Forssk., 212
Statice axillaris Forssk., 212
Statice cylindrifolia Forssk., 212
Statice incana L., 211
Statice monopetala L., 211
Statice pruinosa L., 212
Statice speciosa sensu Forssk., non L., 212
Steinheilia radians (Forssk.) Decne., 82
Stenotaphrum dimidiatum (L.) Brogn., 279
Stenotaphrum secundatum (Walt.) Kuntze, 279
Sterculia africana (Lour.) Fiori, 234

Sterculia arabica T. Anders., 234
Sterculia platanifolia Vahl, 234
STERCULIACEAE, 234
Stewartia corchoroides Forssk., 199
Stipa capensis Thunb., 279
Stipagrostis lanata (Forssk.) De Winter, 279, 280
Stipagrostis plumosa (L.) Munro ex T. Anders., 279
Stratiotes alismoides L., 282
Striga humifusa (Forssk.) Benth., 228
Stroemia farinosa (Forssk.) Vahl, 91
Stroemia glandulosa (Forssk.) Vahl, 91
Stroemia rotundifolia (Forssk.) Vahl, 91
Suaeda aegyptiaca (Hasselq.) Zohary, 109
Suaeda baccata Forssk., 109
Suaeda fruticosa Forssk. ex J.F. Gmel., 109, 110 (Fig. 17)
Suaeda hortensis Forssk. ex J.F. Gmel., 109
Suaeda monoica Forssk. ex J.F. Gmel., 109
Suaeda salsa (L.) Pallas, 111
Suaeda sp. sensu Forssk., 111
Suaeda vera Forssk. ex J.F. Gmel., 111
Suaeda vermiculata Forssk. ex J.F. Gmel., 111, 145 (Fig. 20B)
Subularia purpurea Forssk., 144
Swertia decumbens Vahl, 162
Swertia polynectaria (Forssk.) Gilg, 162, 231 (Fig. 24B)
Synedrella nodiflora (L.) Gaertn., 134

Talinum cuneifolium (Vahl) Willd., 215
Talinum decumbens (Forssk.) Willd., 69
Talinum portulacifolium (Forssk.) Asch. ex Schweinf., 215
TAMARICACEAE, 235
Tamarix aphylla (L.) Karsten, 235
Tamarix articulata Vahl, 235
Tamarix gallica sensu Forssk., non L., 235
Tamarix orientalis Forssk., 235
Tamarix tetragyna Ehrenb., 235
Tanacetum humile Forssk., 119
Tanacetum monanthos L., 113
Tanacetum parthenium (L.) Sch. Bip., 134
Taraxacum megalorrhizon (Forssk.) Hand.-Mazz., 135
Taverniera lappacea (Forssk.) DC., 190
Tectaria coadunata (J. Sm.) C. Chr., 293
Teline monspessulana (L.) C. Koch, 190
Tephrosia tomentosa (Forssk.) Pers., 191
Teramnus repens (Taub.) Bak. subsp. **gracilis** (Chiov.) Verdc., 191
Tetragonia sp. indet., 111
Teucrium iva L., 165
Teucrium polium L., 173
Teucrium scordioides Schreber, 173
Teucrium scordium L. subsp. **scordioides** (Schreber) Maire & Petitmengin, 173
Thalassia ciliata (Forssk.) Koenig, 288
Thalassodendron ciliatum (Forssk.) den Hartog, 288
Thalictrum foetidum L., 217
Thelypteris dentata (Forssk.) E. St. John, 290
Themeda forsskalii Hack., 280
Themeda polygama J. F. Gmel., 280
Themeda triandra Forssk., 280
Thesium linophyllon L., 226
Thespesia populnea (L.) Soland. ex Corr., 199
Thymelaea hirsuta (L.) Endl., 235
THYMELAEACEAE, 235
Thymus bovei Benth., 173
Thymus ciliatus Forssk., 173
Thymus imbricatus Forssk., 168
Thymus laevigatus Vahl, 173
Thymus piperella Vahl, non L., 168
Thymus serpyllum sensu Forssk., non L., 173
Thymus vulgaris L., 173
Thymus sp. 1., 173
Thymus sp. 2., 174
TILIACEAE, 236
Tillandsia decumbens Forssk., 255
Tolpis virgata (Desf.) Bertol., 135
Tomex glabra Forssk., 226
Torilis arvensis (Huds.) Link, 239
Torilis infesta (L.) Hoffm., 239
Trachynia distachya (L.) Link, 280
Traganum nudatum Del., 111
Tragia cordifolia Vahl, 160
Tragia pungens (Forssk.) Muell. Arg., 160
Tragopogon picroid[es] L., 135
Tragus muricatus (Forssk.) Moench, 280
Tragus racemosus (L.) All., 280
Trianthema crystallina (Forssk.) Vahl, 72
Trianthema fruticosa Vahl, 94
Trianthema pentandrum L., 72
Trianthema triquetra Rottl. ex Willd., 72
Tribulus lanuginosus L., 248
Tribulus longipetalus Viv., 248
Tribulus pentandrus Forssk., 248
Tribulus terrestris L., 248
Trichilia emetica Vahl, 200
Trichilia roka Forssk. ex Chiov., 200
Trichilia roka Forssk., non rite publ., 200
Trichodesma africana (L.) R. Br., 89
Trifolium alexandrinum L., 191
Trifolium bicorne Forssk., 191
Trifolium fragiferum L., 191
Trifolium melilothus Forssk., 186
Trifolium melilotus indicus L., 186
Trifolium micranthum Viv., 191
Trifolium repens sensu Forssk., non L., 191
Trifolium resupinatum L., 191
Trifolium semipilosum Fres., 191
Trifolium uniflorum L., 191
Trifolium unifolium Forssk., 180
Trigonella hamosa L., 192
Trigonella monspeliaca sensu Vahl, non L., 192
Trigonella procumbens (Besser) Reichenb., 192
Trigonella stellata Forssk., 192
Trisetaria [sp. without epithet] Forssk., 280
Trisetaria forskahlei J.F. Gmel., 280

Trisetaria linearis Forssk., 280
Trisetum lineare (Forssk.) Boiss., 280
Triticum aegilopoides Forssk., 280
Triticum bicorne Forssk., 260
Triticum dicoccum (Schrank) Schübl., 281
Triticum durum Desf., 281
Triticum junceum sensu Forssk., non (L.) L., 268
Triticum maritimum L., 265
Triticum maritimum sensu Vahl, non L., 265
Triticum spelta L. α **villosum** Forssk., 281
Triticum spelta L. β **glabrum** Forssk., 281
Triumfetta glandulosa Forssk., 237
Triumfetta rhomboidea Jacq., 237
Turbinaria decurrens Bory (Algae), 316
Turia [sp. without epithet] Forssk., 150
Turia cordata J. F. Gmel., 150
Turia foliis cordatis Forssk., 150
Turia gijef Forssk., 149
Turia leloja Forssk., 150
Turia moghadd Forssk., 148
Turia sativa Forssk., 150
Turraea holstii Guerke, 200

Ulex europaeus L., 192
ULMACEAE, 239
Ulva lactuca L. (Algae), 316
Ulva reticulata Forssk. (Algae), 316
UMBELLIFERAE, 237
Umbelliferae gen. et sp. indet., 239
Urena ovalifolia Forssk., 197
Urginea maritima (L.) Bak., 286
Urochloa mutica (Forssk.) Nguyen, 261
Urochondra setulosa (Trin.) C.E. Hubb., 281
Urospermum picroides (L.) Scop. ex F.W. Schmidt, 135
Urtica caudata Vahl, 241
Urtica divaricata sensu Forssk., non L., 241
Urtica dubia ?Forssk., 241
Urtica heterophylla Vahl, non Decne., 240
Urtica hirsuta Vahl, 241
Urtica iners Forssk., 240
Urtica muralis Vahl, 241
Urtica palmata Forssk., 240
Urtica parasitica Forssk., 241
Urtica urens L., 241
Urtica urens L. γ *iners* (Forssk.) Wedd., 240
Urtica verticillata Vahl, 240
URTICACEAE, 240
Usnea florida Achar. (Lichen), 315
Utricularia inflexa Forssk., 194

Valantia [sp. without epithet] Forssk., 223
Valantia aspera Forssk., 223
Valantia hispida L., 223
Valantia muralis L., 223
Valeria scandens sensu Forssk., non Loefl., 207
VALERIANACEAE, 241
Varthemia montana (Vahl) Boiss., 118
Velezia rigida L., 100
Verbascum pinnatifidum Vahl, 229
Verbascum sinuatum L., 229
Verbascum sinuatum sensu Forssk., non L., 229
Verbena capitata Forssk., 242
Verbena forskalaei Vahl, 242
Verbena procumbens Forssk., 242
Verbena supina L., 242
VERBENACEAE, 241
Vereia alternans (Vahl) Spreng., 140
Vernonia cinerascens sensu auct., non Sch. Bip., 135
Vernonia spatulata (Forssk.) Hochst. ex Sch. Bip., 135
Vicia monantha Retz., 192
Vicia peregrina L., 192
Vicia sativa L. var. **angustifolia** L., 192
Vigna aconitifolia (Jacq.) Marechal, 193
Vigna luteola (Jacq.) Benth., 193
Vigna radiata (L.) Wilczek, 193
Vigna radiata (L.) Wilczek var. *setulosa* (Dalz.) Ohwi & Chashi, 193
Vigna unguiculata (L.) Walp., 193
Vigna unguiculata (L.) Walp. subsp. **sesquipedalis** (L.) Verdc., 193
Vigna variegata Defl., 194
Viola arborea sensu Forssk., non L., 243
Viola repens Forssk., 243
VIOLACEAE, 243
Viscum [sp. without epithet] Forssk., 195
Viscum schimperi Engl., 195
VITACEAE, 243
Vitex agnus-castus L., 242
Vitis digitata Defl., 247
Vitis quadrangularis (L.) Wall., 243
Vitis rotundifolius (Forssk.) Defl., 243
Vitis vinifera L., 247
Volutaria lippii (L.) Cass., 135
Volutella aphylla Forssk., 174
Vulpia fasciculata (Forssk.) Fritsch, 281
Vulpia membranacea (L.) Link, 281

Wissadula amplissima (L.) R.E. Fries var. **rostrata** (Schum. & Thonn.) R.E. Fries, 200
Withania somnifera (L.) Dunal, 234

Zaleya pentandra (L.) C. Jeffrey, 72
Zannichellia [sp. without epithet] Forssk., 287
Zapania arabica Poir., 242
Zehneria thwaitesii (Schweinf.) C. Jeffrey, 151
Zilla [sp. without epithet] Forssk., 145 (Fig. 20A), 147
Zilla myagrioides Forssk., 146 (Fig. 21A), 147
Zilla spinosa (L.) Prantl, 145 (Fig. 20A), 146 (Fig. 21A), 147
Ziziphus spina-christi (L.) Desf., 218
Zollikoferia mucronata (Forssk.) Boiss., 125
Zollikoferia spinosa (Forssk.) Boiss., 126
Zostera ciliata Forssk., 288
Zostera dubia Forssk., 288
Zostera stipulacea Forssk., 288
Zostera uninervis Forssk., 288

ZOSTERACEAE, 288
ZYGOPHYLLACEAE, 247
Zygophyllum album L.f., 248, 249 (Fig. 28A)
Zygophyllum coccineum L., 248, 250 (Fig. 29)
Zygophyllum desertorum Forssk., 248, 250 (Fig. 29)
Zygophyllum portulacoides Forssk., 248, 249 (Fig. 28B)
Zygophyllum proliferum Forssk., 248, 249 (Fig. 28A)
Zygophyllum simplex L., 248, 249 (Fig. 28B)